8th Argentinean Bioengineering Society Conference (SABI 2011) and the 7th Clinical Engineering Meeting

Journal of Physics: Conference Series Volume 332

Mar del Plata, Argentina
28-30 September 2011

Editors:

Gustavo Javier Meschino
Virginia L Ballarin

ISBN: 978-1-62276-679-6
ISSN: 1742-6588

Printed from e-media with permission by:

Curran Associates, Inc.
57 Morehouse Lane
Red Hook, NY 12571

Some format issues inherent in the e-media version may also appear in this print version.

Printed by Curran Associates, Inc. (2013)

Institute of Physics
Dirac House, Temple Back
Bristol BS1 6BE UK

Phone: 44 1 17 929 7481
Fax: 44 1 17 920 0979

techtracking@iop.org

Additional copies of this publication are available from:

Curran Associates, Inc.
57 Morehouse Lane
Red Hook, NY 12571 USA
Phone: 845-758-0400
Fax: 845-758-2634
Email: curran@proceedings.com
Web: www.proceedings.com

TABLE OF CONTENTS

CLINICAL ENGINEERING, PROFESSIONAL EXPERIENCES

EQUIPMENT, INSTRUMENTATION, SENSORS AND MEASUREMENTS

BIOMECHANICS

BIOMEDICAL SIGNAL PROCESSING

REHABILITATION ENGINEERING

MODELING AND SIMULATION OF BIOMEDICAL SYSTEMS

MEDICAL INFORMATICS

8th Argentinean Bioengineering Society Conference (SABI 2011) and 7th Clinical Engineering Meeting

Gustavo Javier Meschino and Virginia L Ballarin

Abstract Metrics

INTRODUCTION

In September 2011, the Eighteenth Edition of the Argentinean Bioengineering Society Conference (SABI 2011) and Seventh Clinical Engineering Meeting were held in Mar del Plata, Argetina. The Mar del Plata SABI Regional and the School of Engineering of the Universidad Nacional de Mar del Plata invited All bioengineers, engineers, physicists, mathematicians, biologists, physicians and health professionals working in the field of Bioengineering to participate in this event.

The overall objectives of the Conference were:

- To provide discussion of scientific research results in Bioengineering and Clinical Engineering.
- To promote technological development experiences.
- To strengthen the institutional and scientific communication links in the area of Bioengineering, mainly between Universities of Latin America.
- To encourage students, teachers, researchers and professionals to establish exchanges of experiences and knowledge.
- To provide biomedical engineering technology solutions to the society and contributing ideas for low cost care.

XVIII Congreso Argentino de Bioingeniería
2° Taller de Órganos Artificiales, Biomateriales e Ingeniería de Tejidos
SABI 2011
OBI 2011
de Septiembre 2011
Mar del Plata

sabi 2011
XVIII Congreso Argentino de Bioingeniería
2° Taller de Órganos Artificiales, Biomateriales e Ingeniería de Tejidos
OBI 2011
de Septiembre 2011
Mar del Plata, Argentina

EXECUTIVE COMMITTEE SABI 2011

Chair
Dra Virginia L Ballarin
Universidad Nacional de Mar del Plata

Co-Chair
Dra Teresita R Cuadrado
Universidad Nacional de Mar del Plata – CONICET

Local Comittee
Dr Gustavo Abraham
Universidad Nacional de Mar del Plata – CONICET
Dra Josefina Ballarre
Universidad Nacional de Mar del Plata – CONICET
Dr Eduardo Blotta
Universidad Nacional de Mar del Plata
Dra Agustina Bouchet
Universidad Nacional de Mar del Plata
Dr Marcel Brun
Universidad Nacional de Mar del Plata
Dra Silvia Ceré
Universidad Nacional de Mar del Plata – CONICET
Dra Mariela Azul Gonzalez
Universidad Nacional de Mar del Plata – CONICET
Dra Lucia Isabel Passoni
Universidad Nacional de Mar del Plata
Dr Juan Ignacio Pastore
Universidad Nacional de Mar del Plata – CONICET
Dra Adriana Scandurra
Universidad Nacional de Mar del Plata

SCIENTIFIC ADVISORY COMMITTEE

President
Dr Gustavo Meschino
Universidad Nacional de Mar del Plata
Comittee
Dr Gustavo Abraham
Universidad Nacional de Mar del Plata – CONICET
Mg Rubén Acevedo
Universidad Nacional de Entre Ríos
Ing Pablo Agüero
Universidad Nacional de Mar del Plata
Ing Mariela Ambrustolo
Universidad Nacional de Mar del Plata
Dr Ricardo Armentano
Universidad Favaloro
Dra Virginia L Ballarin
Universidad Nacional de Mar del Plata
Dra Josefina Ballarre
Universidad Nacional de Mar del Plata – CONICET
Dr Eduardo Blotta
Universidad Nacional de Mar del Plata

Ing Marco Benalcázar
Universidad Nacional de Mar del Plata
Mg Freddy Geovanny Benalcázar Palacios
Escuela Superior Politécnica de Chimborazo, Ecuador
Dr Roberto Boeri
Universidad Nacional de Mar del Plata – CONICET – INTEMA
Dra Agustina Bouchet
Universidad Nacional de Mar del Plata
Dr Ariel Braidot
Universidad Nacional de Entre Ríos
Dr Marcel Brun
Universidad Nacional de Mar del Plata
Dra Silvia Ceré
Universidad Nacional de Mar del Plata – CONICET
Ing Fernando Clara
Universidad Nacional de Mar del Plata
Dr Raúl Correa Prado
Universidad Nacional de San Juan
Bioing Pablo Cortez
Universidad Nacional de Mar del Plata
Dra Teresita R Cuadrado
Universidad Nacional de Mar del Plata – CONICET
Ing Eduardo De Forteza
Universidad Favaloro
Dra Mariana Del Fresno
Universidad Nacional del Centro de la Provincia de Buenos Aires
Dr Martín Diaz
Informática Médica Hospital Aleman de Buenos Aires – GIBBA
Ing Julio César Doumecq
Universidad Nacional de Mar del Plata
Mg Ana María Echenique
Universidad Nacional de San Juan
Bioing Pedro Escobar
Universidad Nacional del Centro, Olavarría, Pcia de Buenos Aires
Dr Fernando Daniel Farfán
Universidad Nacional de Tucumán
Dr Carmelo Felice
Universidad Nacional de Tucumán – CONICET
Dr Elmer Fernández
Universidad Católica de Córdoba – CONICET
Ing José Flores
Universidad Nacional de Entre Ríos
Dr Arturo Gayoso
Universidad Nacional de Mar del Plata
Dra Bioing Agustina Garcés
Universidad Nacional de San Juan ¬– CONICET
Bioing Luciano Gentile
Universidad Favaloro
Mg María Eugenia Gómez
Universidad Nacional de San Juan
Dr Claudio González

Universidad Nacional de Mar del Plata
Mg Esteban González
Universidad Nacional de Mar del Plata
Dra Mariela A Gonzalez
Universidad Nacional de Mar del Plata – CONICET
Dr Juan Pablo Graffigna
Universidad Nacional de San Juan
Dra Myriam Herrera
Universidad Nacional de Tucumán – CONICET
Dr Roberto Hidalgo
Universidad Nacional de Mar del Plata
Dr Roberto Isoardi
Fundación Escuela de Medicina Nuclear de Mendoza – CNEA
Dra Susana Jerez
Universidad Nacional de Tucumán
Dr Eric Laciar
Universidad Nacional de San Juan – CONICET
Bioing Roberto Leonarduzzi
Universidad Nacional de Entre Ríos
Mg Norberto Lerendegui
Instituto Tecnológico de Buenos Aires
Dra Natalia López
Universidad Nacional de San Juan – CONICET
Dra Rossana Madrid
Universidad Nacional de Tucuman – CONICET
Ing Florencia Montini Ballarin
Universidad Nacional de Mar del Plata – CONICET
Dra Emilce Moler
Universidad Nacional de Mar del Plata
Dr Jorge Castiñieira Moreira
Universidad Nacional de Mar del Plata
Dr Silvia Murialdo
Universidad Nacional de Mar del Plata – CIC
Dr Juan Manuel Olivera
Universidad Nacional de Tucumán
Dra Lucia Isabel Passoni
Universidad Nacional de Mar del Plata
Dr Juan Ignacio Pastore
Universidad Nacional de Mar del Plata – CONICET
Dra María Elisa Pérez
Universidad Nacional de San Juan
Mg Franco M Pessana
Universidad Favaloro
Dr Julio Politti
Universidad Nacional de Tucumán
Dr Marcelo Risk
Universidad Nacional de Buenos Aires – CONICET
Ing Raúl Rivera
Universidad Nacional de Mar del Plata
Mg Luis Rocha
Universidad Nacional de Tucumán – SIPROSA

Dra Silvia Rodrigo
Universidad Nacional de San Juan
Dra Viviana Rotger
Universidad Nacional de Tucumán
Dr Leonardo Rufiner
Universidad Nacional de Entre Rios – CONICET
Dra Estela Ruiz
Universidad Nacional de Tucumán
Dr Martín Santiago
Universidad Nacional del Centro de la Provincia de Buenos Aires
Dra Adriana Scandurra
Universidad Nacional de Mar del Plata
Ing Graciela Secreto
Universidad Favaloro
Mg Pablo Solarz
Universidad Nacional de Tucumán
Mg Carolina Tabernig
Universidad Nacional de Entre Ríos
Ing Ricardo Taborda
Universidad Nacional de Córdoba
Dra María Eugenia Torres
Universidad Nacional de Entre Ríos – CONICET
Ing Juan Carlos Tulli
Universidad Nacional de Mar del Plata
Dr Gerardo Tusman
Hospital Privado de Comunidad, Mar del Plata
Dr Santiago Urquiza
Universidad Nacional de Mar del Plata
Mg Andrés Valdez
Universidad Nacional de San Juan
Dr Máximo Valentinuzzi
INSIBIO – CONICET – Universidad Nacional de Tucumán

Peer review statement

All papers published in this volume of *Journal of Physics: Conference Series* have been peer reviewed through processes administered by the proceedings Editors. Reviews were conducted by expert referees to the professional and scientific standards expected of a proceedings journal published by IOP Publishing.

Computational analysis of anterior communicating artery aneurysm shear stress before and after aneurysm formation

Marcelo A. Castro[1]
CONICET. Universidad Tecnológica Nacional, Facultad Regional Buenos Aires, Grupo de Investigación y Desarrollo en Bioingeniería (GIBIO)
Medrano 951, Ciudad Autónoma de Buenos Aires, CP 1179, Argentina

E-mail: marcelo.a.castro@gmail.com

Christopher M. Putman
Inova Fairfax Hospital, Fairfax Radiological Consultants
3300 Gallows Road, Falls Church, VA, 22042, USA

Juan R. Cebral
Department of Computational and Data Sciences, College of Sciences, George Mason University
4400 University Drive, Fairfax, VA, 22030, USA

Abstract. It is widely accepted that complexity in the flow pattern at the anterior communicating artery (AComA) is associated with the high rate of aneurysm formation at that location observed in large studies. The purpose of this work is to study associations between hemodynamic patterns, and AComA aneurysm initiation by comparing hemodynamics in the aneurysm and the normal model where the aneurysm was computationally removed. Vascular models of both right and left circulation were independently reconstructed from three-dimensional rotational angiography images using deformable models after image registration of both images, and fused using a surface merging algorithm. The geometric models were then used to generate high-quality volumetric finite element grids of tetrahedra with an advancing front technique. For each patient, the second anatomical model was created by digitally removing the aneurysm. It was iteratively achieved by applying a Laplacian smoothing filter and remeshing the surface. Finite element blood flow numerical simulations were performed for both the pathological and normal models under the same personalized pulsatile flow conditions imposed at the inlets of both models. The Navier-Stokes equations were numerically integrated by using a finite-element formulation. It was observed that aneurysms initiated in regions of high and moderate WSS in the counterpart normal models. Adjacent or close to those regions, low WSS portions of the arterial wall were not affected by the disease. These results are in line with previous observations at other vascular locations.

[1] To whom any correspondence should be addressed.

1. Introduction

The anterior communicating artery (AComA) is a unique vascular location that ideally receives blood from two sources of inflow and redistributes it toward the anterior part of the brain through two efferent arteries [1]. However, aneurysms of the AcomA complex are more likely to have hypoplastic A1 segments [2] or to have exclusive filling angiographically from one A1 segment in up to 78% cases [3]. Those AComA aneurysms receiving blood from both A1 segments have flow patterns dependent of individual flow rates and wave forms, which may result in regions of elevated wall shear stress (WSS) that change their locations during the cardiac cycle [4]. It is accepted that complexity in the flow pattern is associated with the high rate of aneurysm formation observed in large studies [5-7].

Wall shear stress (WSS) play an important role in the formation, growth and ultimately rupture of cerebral aneurysms [8,9]. A previous computational hemodynamic study showed a possible association between high maximum WSS at the systolic peak with rupture in a cohort of AComA aneurysms [10]. In that study ruptured aneurysms had in average as much as twice maximum WSS as in the unruptured group. This result is in line with previous experimental studies [11] and computational studies at other locations [12]. Additionally, in a large study including 210 cerebral aneurysms at different locations it was found a statistical significant association between rupture and WSS distributions with elevated maximum WSS, high flow concentration and small impingement size [13]. Cebral et al. also showed a connection between location of aneurysm blebs and regions of high WSS in models where blebs were virtually removed [14]. In another study, Kulcsar et al. presented a relationship between initiation, and coexistence of high WSS and high positive spatial WSS gradient, observed in three-patient scanned before and after aneurysm formation [15]. The normal vasculature had been incidentally imaged when investigating other pathologies. Other investigators reported possible associations between low shear stress and either rupture [16,17] or blister formation [18].

The purpose of this work is to study associations between hemodynamic patterns and aneurysm initiation by comparing AComA hemodynamics between the original aneurysm model and the vascular model where the aneurysm was computationally removed. Patients included in this study were not scanned before aneurysm formation. Therefore, there is no evidence to determine whether or not the original vascular configurations were affected by the disease. In this first pilot study, in order to minimize the effects of the reconstruction method, only a selected group of patients whose aneurysms affected a limited region of the AComA and the parent artery was considered.

2. Materials and methods

2.1 Patients, images and vascular reconstruction

Three patients with cerebral aneurysms in the AComA were selected from our data base. In order to minimize the dependency of the results on the reconstruction of the normal vascular models, all images presenting an extension of the disease towards further regions beyond the aneurysm neck were rejected for this study. Bilateral rotational scans were obtained to visualize both avenues of flow into the anterior communicating artery (AcomA) aneurysm using a Philips Integris System (Philips Medical Systems, Best, The Netherlands). These images were obtained during a 180° rotation and imaging at 15 frames per second for a total of 8 seconds. The corresponding 120 projection images were reconstructed into a 3D dataset of 128 128 128 voxels covering a field of view of 54.02 mm on a dedicated Philips Integris workstation. The voxel resolution was therefore 0.422 mm. These data were exported into a PC for mathematic vascular modelling using a previously presented methodology [19,20,10]. Each pair of images was aligned using landmark-based rigid registration, and each model was reconstructed using isosurface extraction and deformable models. Models were merged and edited, and aneurysms were virtually removed. In order to achieve that, neck points were identified. Aneurysm was smoothed using a Gaussian filter starting from the smaller structures, without moving points beyond the neck. Remeshing was required in order to preserve element aspect ratio after point relocation. The procedure continued until the aneurysm was completely removed.

2.2 Hemodynamic modeling

Finite element blood flow numerical simulations were performed for both the original and the pre-aneurysm models under the same flow conditions. Blood was modelled as an incompressible Newtonian fluid with attenuation 1.0 g/cm^3 and viscosity 0.04 Poise. The governing equations were the unsteady Navier-Stokes equations in 3D [26]. Vessel walls were assumed rigid, and no slip boundary conditions were applied at the walls. Pulsatile flow conditions derived from phase-contrast magnetic resonance measurement in a healthy subject were imposed at the inlet of the models. Flow waveforms were scaled with the inlet area to achieve a mean WSS of 15 dyne/cm^2 at the inflow boundary of each model. This choice is consistent with studies relating vessel area and flow rates in internal carotid and vertebral arteries [27], as well as with the principle of minimal work expressed by Murray's law [28]. Fully developed pulsatile velocity profiles were prescribed with use of the Womersley solution [29,30]. Assuming that all distal vascular beds have similar total resistance to flow, traction-free boundary conditions with the same pressure level were applied at outlet boundaries. The Navier-Stokes equations were numerically integrated by using a fully implicit finite element formulation [19]. We computed 2 cardiac cycles using 100 time-steps per cycle, and all of the results reported correspond to the second cardiac cycle. All elements in the normal model whose distance to the aneurysmal model was greater than the typical size of two elements were marked. In that way, those elements are located in a region of the original arterial wall where the aneurysm started to form. WSS distributions were computed from the blood velocity field and studied over those regions. Average WSS values were compared to those typical values in the surrounding regions.

The following list summarizes the steps presented in Sections 1.1 and 1.2 [10,14]:

- Image acquisition for both left and right circulation
- Image rigid registration using landmarks
- Image segmentation
- Isosurface extraction
- Isosurface deformable models to adjust surface point location to vascular boundary in image
- Surface merging to generate a single model with containing the whole anterior vasculature
- Identification of neck points
- Aneurysm smoothing and remeshing iteratively within neck until aneurysm is removed
- 3D tetrahedral grid generation using an advancing front technic for both pre and post models
- Flow rate scaling to impose at both internal carotid arteries (ICA)
- Computation of Womersley profile assumed as the solution at the model inflows
- Model assumptions: traction free boundary conditions at outlets, no slip boundary condition at arterial walls, Newtonian rheology and rigid walls
- Numerical integration over two cardiac cycles using a fully implicit finite element formulation
- Computation of wall shear stress from velocity field
- Wall shear stress visualization

3. Results

For all three cases, model A (without aneurysm) and model B (with aneurysm) were created. The geometric information is shown in Table 1. All the patients considered in this study had a unilateral configuration, having a dominant ICA that provides blood to the whole anterior arterial tree. Consequently, there is almost no flow through the A1 segment of the anterior cerebral artery contralateral to the dominant internal carotid artery (ICA), resulting in a reduced arterial diameter (see Figure 1). Original models for all patients are shown in Figure 1 (first and third columns), while models after digital intervention to remove the aneurysm from the original models are shown in the second and fourth columns of the same figure.

Table 1. Geometric information of vascular models

(a) Model #	(b) # elem (A)	(c) # elem(B)	(d) Area ICA left (cm^2)	(e) Area ICA right (cm^2)	(f) Max dist ⊥ (cm)	(g) Max dist ∥ (cm)	(h) AR	(i) Loc.	(j) Unilateral
1	1,431,826	1,680,751	0.116	0.140	0.90	0.50	1.80	right	yes
2	2,427,619	2,532,393	0.251	0.177	0.57	0.44	1.29	left	yes
3	2,056,090	2,151,738	0.238	0.185	0.48	0.38	1.26	left	yes

a) number of model; b) number of tetrahedra elements in the model A (without aneurysm); c) number of tetrahedra elements in the model B (with aneurysm); d) cross-sectional area of the left ICA at the inflow; e) cross-sectional area of the right ICA at the inflow; f) maximum length of the aneurysm perpendicular to the vessel; g) maximum length of the aneurysm parallel to the vessel; h) aspect ratio; i) location of the aneurysm at the juncture of the A1 and A2 segments of the anterior cerebral artery, and the anterior communicating artery based on the acquired images; j) the system is considered unilateral if there is basically no flow through the A1 segment contralateral to the dominant ICA. In all the three cases, both A2 segments are filled by the dominant ICA.

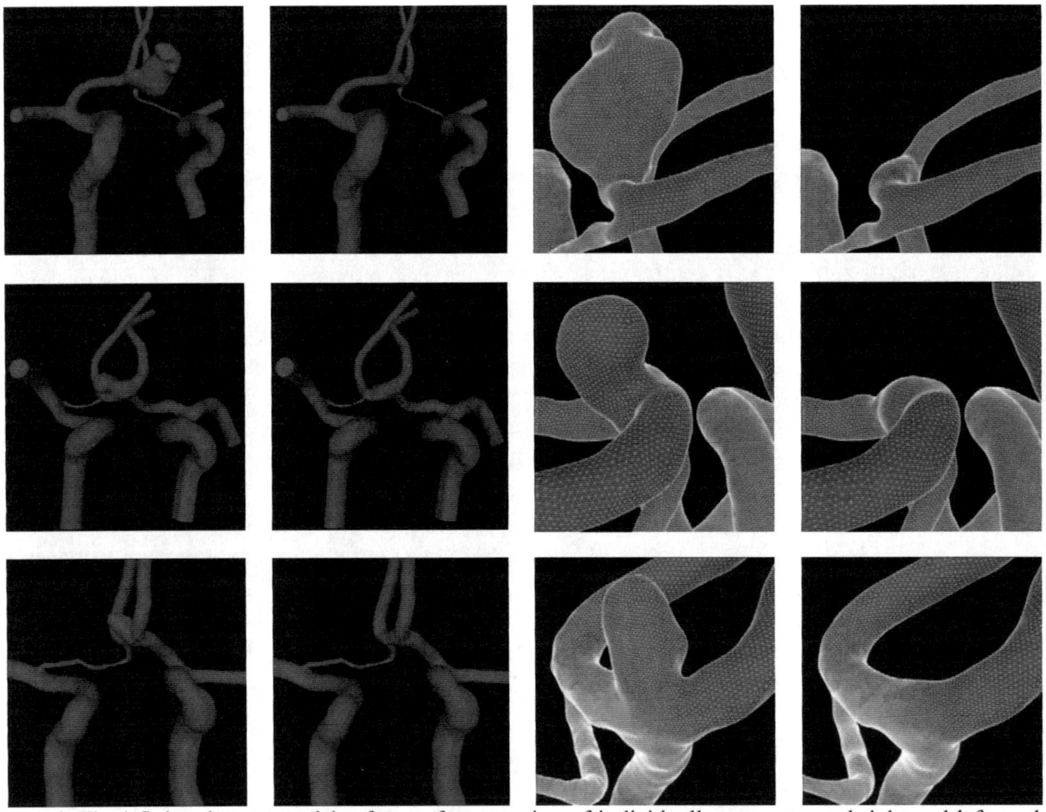

Figure 1. Final finite element models after surface merging of individually reconstructed right and left models: original (first and third columns) and after virtual intervention (second and fourth columns) for patients 1, 2 and 3 (first, second and third row, respectively).

WSS distributions at the systolic peak are displayed in Figure 2 after for model A (first column) and model B (second column) for all patients.

For Patient #1 a high WSS region with a peak of 120 dyne/cm^2 is observed in the AcomA, surrounded by a larger region with an average WSS of 50 dyne/cm^2. Comparison between both models shows that the aneurysm started to form in the high WSS region. A low WSS region of 25 dyne/cm^2 in the feeding artery opposite to the aneurysm is also observed.

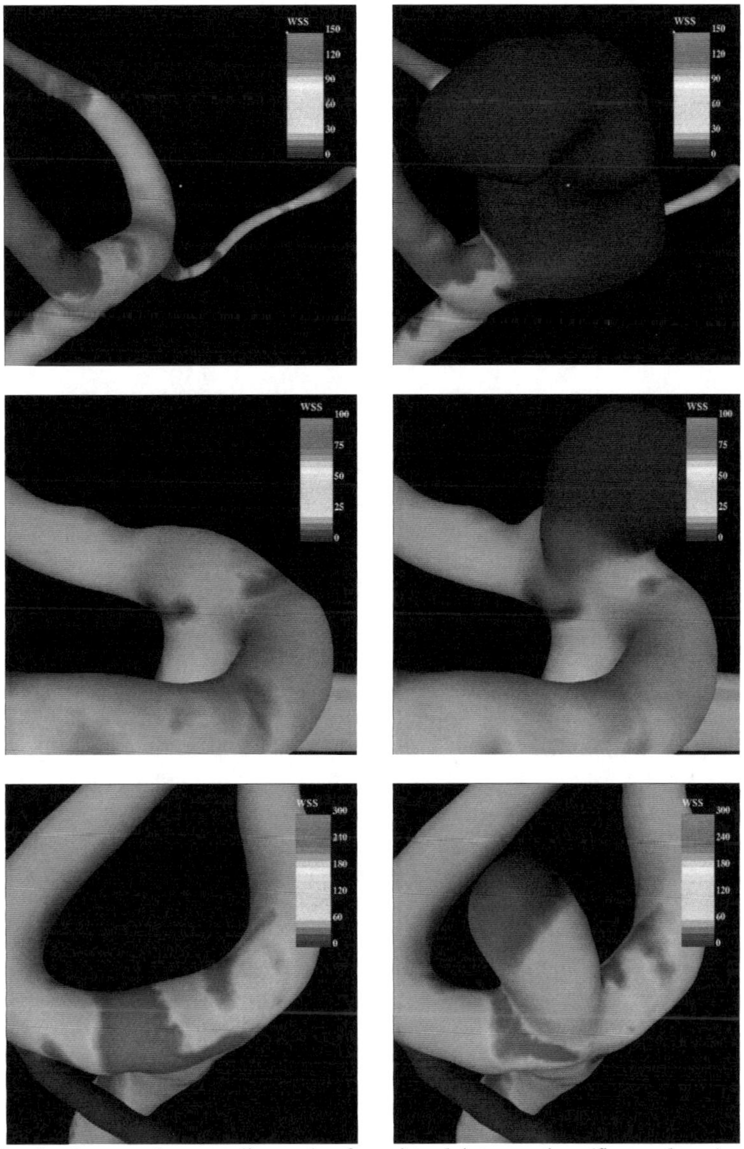

Figure 2: WSS distributions at the systolic peak after virtual intervention (first column) and in the original aneurysm model (second column), for patients 1, 2 and 3 (first, second and third rows, respectively), using the same colour scale for each patient.

For patient #2, two disconnected high WSS regions with peak values of 75 dyne/cm^2 and 50 dyne/cm^2 were observed. The extended area affected by the aneurysm includes both regions. Adjacent to them there is a low WSS portion of the feeding vessel with values less than 10 dyne/cm^2. For patient #3, a extended region with WSS values ranging between 200 dyn/cm^2 and 340 dyn/cm^2 was observed, which surrounds a smaller area of moderate WSS values (120 dyne/cm^2). The aneurysm formed over a region with both high a moderate WSS values, which also had high WSS gradient. A low WSS region in arterial wall opposite to the aneurysm has WSS values less than 50 dynm/cm^2.

4. Discussion

The methodology has some potential limitations. The computational removal of AComA aneurysms may be not applied to large aneurysms with wide necks where the original vasculature may not be easily reconstructed. Ford et al. presented an approach to digital removal of saccular aneurysms [31]. Although the methodology presented there is automatic and may be helpful to process large amount of data, it may generate simplified bifurcation geometries due to the centreline based reconstruction. That may affect WSS computation in models where aneurysms where removed from those locations. Images before aneurysm formation are usually acquired by chance when the patients are screened because of another pathology or aneurysms at other locations [15], therefore they can not be used to carry out a large study. Another potential limitation is the use of non-patient-specific flow rate waveforms imposed at the inlet of our models. However, this limitation is suffered by most computational studies since flow rate measurements are not routinely acquired in clinical evaluations. This limitation is partially overcome by WSS normalization at inflow segments. Other model assumptions had limited impact on aneurismal hemodynamics, as shown in previous works [19].

5. Conclusion

The present work shows a possible association between regions of elevated WSS before aneurysm formation, and aneurysm location. This observation is in line with previously reported results from experimental studies and patient-specific blood flow finite element simulations at other vascular locations. This trend will have to be corroborated by larger studies, which also include other hemodynamic quantities and differences between unilateral and bilateral aneurysms.

References

[1] Perlmutter D, Rhoton ALJ. Microsurgical anatomy of the anterior-anterior communicating-recurrent artery complex. J Neurosurg 1976;45:259 –72

[2] Kasuya H, Shimizu T, Nakaya K, et al. Angles between A1 and A2 segments of the anterior cerebral artery visualized by three-dimensional computed tomographic angiography and association of anterior communicating artery aneurysms. Neurosurgery 1999;45:89 –93

[3] Charbel FT, Seyfried D, Metha B, et al. Dominant A1: angiographic and clinical correlations with anterior communicating artery aneurysms. Neurol Res 1991;13:253–56

[4] Castro MA, Putman CM, Cebral JR. Patient-Specific Computational Fluid Dynamics Modeling of Anterior Communicating Artery Aneurysms: A Study of the Sensitivity of Intra-Aneurysmal Flow Patterns to Flow Conditions in the Carotid Arteries. Am J Neurorad 2006:27:2061-2068

[5] Horiuchi T, Tanaka T, Hongo K. Surgical treatment for aneurysmal subarachnoid hemorrhage in the 8th and 9th decade of life. Neurosurgery 2005;56:469 –75

[6] Leipzip TJ, Morgan J, Horner TG. Analysis of intraoperative rupture in the surgical treatment of 1674 saccular aneurysms. Neurosurgery 2005;56:455– 68

[7] Brilstra EH, Rinkel GJ, van der Graff Y. Treatment of intracranial aneurysms by embolization with coils: a systematic review. Stroke 1999;30:470 –76

[8] Crompton M. Mechanisms of growth and rupture in cerebral berry aneurysms. Br J Med 1966:1:1138-42

[9] Nakatani H, Hashimoto N, Kang H, Yamazoe N, Kikuchi H et al. Cerebral blood flow patterns at major vessel bifurcations and aneurysms in rats. J Neurosurg 1991:74:258-262

[10] Castro MA, Putman CM, Cebral JR. Hemodynamic Patterns of Anterior Communicating Artery Aneurysms: A Possible Association with Rupture. Am J Neurorad 2009:30(2): 297-302

[11] Hashimoto N, Handa H, Nagata I, et al. Experimentally induced cerebral aneurysms in rats. Part V. Relation of hemodynamics in the circle of Willis to formation of aneurysms. Surg Neurol 1980;13:41– 45

[12] Castro MA, Putman CM, Radaelli A, Frangi A, Cebral JR. Hemodynamics and rupture of terminal cerebral aneurysms. Acad Radiol 2009:16 (19):1201-1207

[13] Cebral JR, Mut F, Weir J, Putman CM. Quantitative characterization of the hemodynamic environment in ruptured and unruptured brain aneurysms. Am J Neuroradiol 2011:32:145-151

[14] Cebral JR, Sheridan M, Putman CM. Hemodynamics and Bleb Formation in Intracranial Aneurysms. Am J Neuroradiol 2010:31:304-310

[15] Kulcsar Z, Ugron A, Marosfo M, Berentei Z, Paal G, Szikora I. Hemodynamics of Cerebral Aneurysm Initiation: The Role of Wall Shear Stress and Spatial Wall Shear Stress Gradient. Am J Neuroradiol 2011:32(3):587-594

[16] Jou LD, Lee DH, Morsi H, Mawad ME. Wall Shear Stress on Ruptured and Unruptured Intracranial Aneurysms at the Internal Carotid Artery. Am J Neuroradiol 2008:29:1761-1767

[17] Shojima M et al. Magnitude and role of wall shear stress on cerebral aneurysm. Computational fluid dynamic study of 20 middle cerebral aneurysms. Stroke 2004:35:2500-2505

[18] Shojima M, Nemoto S, Morita A, Oshima M, Watanabe E, Saito N. Role of Shear Stress in the Blister Formation of Cerebral Aneurysms. Neurosurgery 2010: 67(5):1268-1275

[19] Cebral JR, Castro MA, Appanaboyina S, Putman C, Millán D, Frangi A. Efficient Pipeline for Image-Based Patient-Specific Analysis of Cerebral Aneurysms Hemodynamics: Technique and Sensitivity. IEEE - Transactions on Medical Imaging - Special Issue on Vascular Imaging 2005: 24(4):457-467

[20] Yim PJ, Vasbinder B, Ho VH, et al. A deformable isosurface and vascular applications. In: Sonka M and Fitzpatrick JM, ed. Medical Imaging 2002: Image Processing, SPIE Vol. 4684. San Diego, Calif: SPIE; 2002:1390 –97

[21] Castro MA, Putman CM, Cebral JR. Patient-specific computational modeling of cerebral aneurysms with multiple avenues of flow from 3D rotational angiography images. Acad Radiol 2006;13:811–21 ,

[22] Cebral JR, Löhner R, Choyke PL, et al. Merging of intersecting triangulations for finite element modeling. J Biomech 2001;34:815–19

[23] Löhner R. Automatic unstructured grid generators. Finite Elements Analysis Design 1997;25: 111–34

[24] Löhner R. Extensions and improvements of the advancing front grid generation technique. Comp Methods Appl Mech Eng 1996;5:119 –32

[25] Löhner R. Regridding surface triangulations. J Comp Phys 1996;126:1–10

[26] Mazumdar JN. Biofluid Mechanics. World Scientific, Singapore 1992

[27] Cebral JR, Castro MA, Putman CM, Alperin N. Flow-area relationship in internal carotid and vertebral arteries. Physiol Meas 2008:29(10):585-594

[28] Sherman T F 1981 On connecting large vessels to small. The meaning of Murray's law. J Gen Physiol 1981:78:431–53

[29] Womersley JR. Method for the calculation of velocity, rate of flow and viscous drag in arteries when the pressure gradient is known. J Physiol 1955;127:553–563

[30] Taylor CA, Hughes TJR, Zarins CK. Finite element modeling of blood flow in arteries. Comp Methods Appl Mech Engin 1998:158:155–196

[31] Ford MD, Hoi Y, Piccinelli M, Antiga L, Steinman AD, An objective approach to digital removal of saccular aneurysms: Technique and applications. The British Journal of Radiology 2009:82:56-61

Elastoplasticity analysis of the nails used in long bone fractures

R Paterson[1,2,3], G Del Sel[3], A Paterson[1,2,3], M T Miralles[1,4] and N Del Sel[4]

[1] Centro de Investigación en Diseño Industrial de Productos Complejos, Fac. de Arquitectura, Diseño y Urbanismo, Universidad de Buenos Aires (CIDI-FADU-UBA), Intendente Güiraldes 2160, Pabellón III, Ciudad Universitaria, C1428EGA, Buenos Aires, Argentina.
[2] Departamento de Hidraúlica, Facultad de Ingeniería, Universidad de Buenos Aires (FI-UBA), Paseo Colón 850, C1063ACV, Buenos Aires, Argentina.
[3] Sanatorio Mater Dei, San Martin de Tours 2952, C1425DNB, Buenos Aires, Argentina.
[4] Facultad de Ciencias Físicomatemáticas e Ingeniería, Pontificia Universidad Católica Argentina "Santa María de los Buenos Aires", Av. A. Moreau de Justo 1300, C1107AAZ, Buenos Aires, Argentina.

E-mail: mmiralles@gmail.com

Abstract. Elastoplastic endomedullary nail-insertion system in long bone fractures is a method which allows generating forces, moments, frictions and stress states in parts of the bone previously selected. Adding previous knowledge on the distribution of forces and acting moments over the bone fragments to be threaded, as well as on the elastoplastic and mechanical properties of the nails to be inserted (designing a special, specific structure for each case), the results, coming from 26 years of applying this technique, have been promising. It is from this perspective that we call this threading and anchoring procedure "Selective Tension System" (STS). Physicochemical analyses and mechanical trials on elastoplastic nails used in the osteosynthesis in long bone fractures are presented. The magnitude of the forces produced by flattening the nails and the reacting forces at both ends are measured. It is expected that the evidence provided on the elastic variability of these nails will be useful as guidance on the availability and choice of the elastoplastic combinations that best fit each patient.

Key Words: Elastoplastic nails, long bone fractures, plastic deformation, elastic force, nail bone interface.

1. Introduction

The Selective Tension System (STS) method enables reducing bone fragments applying elastic forces that remain in time, from which permanent strain-stress states are generated in the bones, promoting their consolidation.

This procedure facilitates selecting the magnitudes, direction and application points of the elastic forces applied to the fractured bone fragments.

In summary, the advantages of the STS technique are: low aggression to the bone tissue during insertion of the nails, the formation of a spatial structure that replaces the loss of resistance in the area of the bone fracture and, finally, the simultaneous stimulation for the formation of new bone over the area of the fracture.

The osteosynthesis method (STS) is based on pre-stressed concrete beams technique, but with the opposite aim: the exerted forces are not meant to prevent a fracture, but to repair a previously fractured structure.

These nails are inserted into the medullar canal from an area as distant to the fractured region as possible. Bone fragments are strung together. The nails are flattened as the result of the insertion in the bone, thus modifying their initial level of curvature determined from preoperative imaging studies in order to provide forces in the fracture zone.

To illustrate the technique, the repair of a distal third femur shaft fracture for a woman (age 83, weight 750 N) is presented. The figure 1 and figure 2 corresponds to the pre-surgery radiographs in the frontal and sagittal plane of mentioned fracture. In this case four elastoplastic nails were introduced. Nails are secured with screws (figure 3), joining and releasing the superior ends (figure 4). Figure 3 shows the additional screw in order to fix the femoral intercondylar fissure present on this fracture.

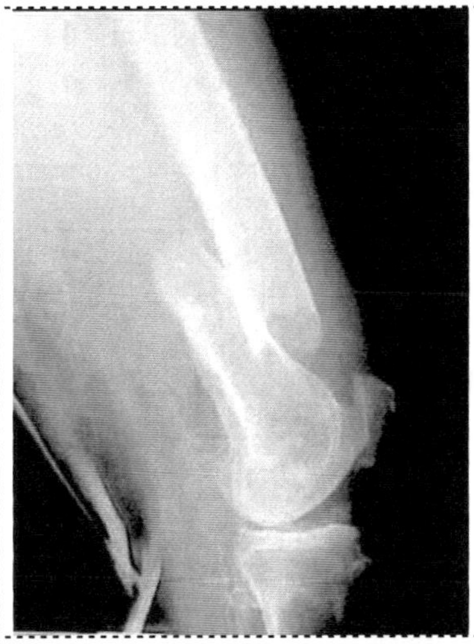

Figure 1. Pre-surgery radiographs showing the sagittal plane of the human femur.

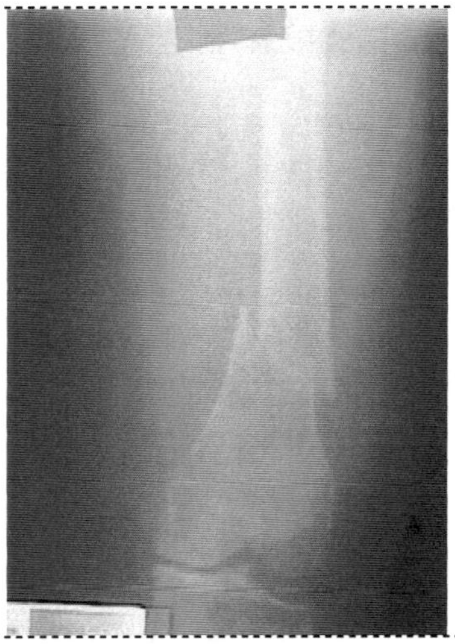

Figure 2. Pre-surgery radiographs showing the frontal plane of the human femur.

The nails that have been inserted are joined together in bundles and fixed to one of the fractured fragments in order to make sure stress forces remain unchanged. This allows to balance the forces, and the flexor-compression, torsion and shearing momentum tending to displace bone fragments.

The forces exerted by the nails on the bone fragments produce stress fields (tension-deformation) on the contact areas.

The distribution of these forces is the result of the elastic properties of the system, which depends on the number of nails used, the elastic properties of the bone, the friction forces between nails and between the nails and the bone, and the kind of fixation between the nails. It consists, from a structural stand point, a hyperstatic system, being the nail preconformation determinant for the generation of stresses in the osseous matrix.

An elastoplastic nail is a straight nail with a bending made before their introduction to obtain an appropriate arch and level of curvature that flattens when the nail is inserted into the medullar canal of a long bone. The remaining elastic deformation of the nail acts exerting forces on the contact areas of the intramedular cortex.

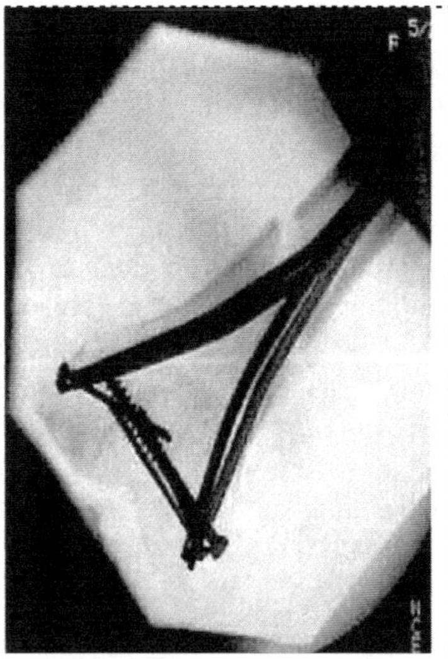

Figure 3. Detail of the anchor and threaded nails.

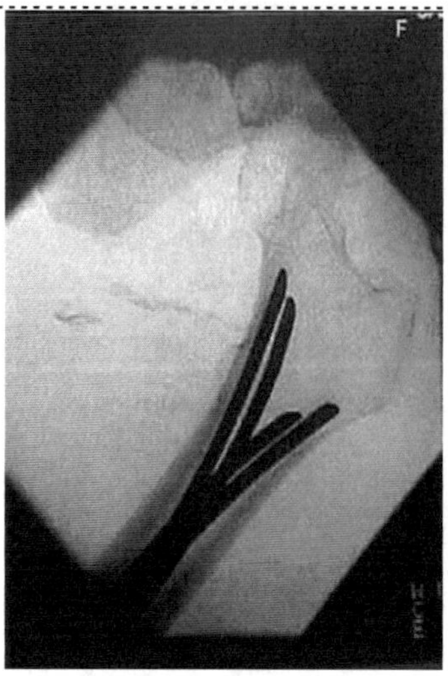

Figure 4. Proximal femoral metaphysic showing the spatial distribution of beam nails.

The prior conformation of the arches and the crushing of the nails during the introduction to the bone are essential in the STS, because this will depend on the three-dimensional state of stress and consequent bone formation. Figure 5 shows a repaired fractured tibia outlining the comments above.

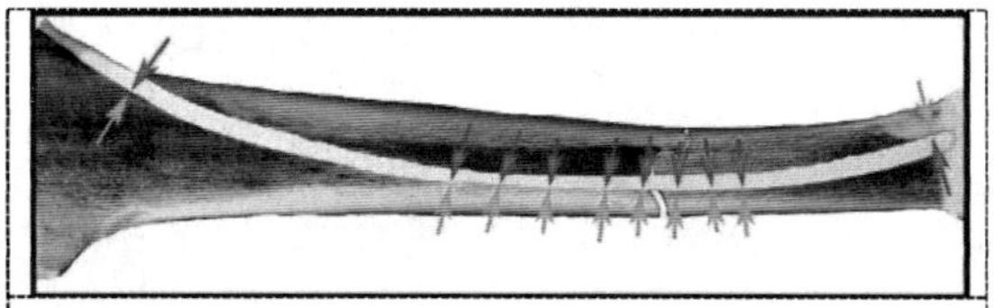

Figure 5. Diagram showing action and reaction forces in the nail-bone system around the fractured area.

By modifying the number of nails, their diameters, shapes, constituting materials and superficial treatments, elastoplastic properties and their mechanical and tribological characteristics can be modified. This fact is widely well known [1-4], and it is why knowing the elastoplastic properties of the nails is of great relevance for the right deployment of this technique.

This paper focuses on two aspects related to the elastoplastic properties of the nails. First determine the values for the percentage residual elastic coefficient (%REC) from a set of 23 elastoplastic nails.

A protocol of progressive flattening for the arched preformed nails allowed to determine the values of maximum force and the %REC. The estimate of the maximum magnitude of the forces exerted by the nails is indicative of the reactions exerted by the bone, once the nail is inserted. Second, in order to study the diversity and behavior of the anchoring, symmetrical and asymmetrical configurations showed the order of magnitude of the forces exerted on the fixing points.

2. Methods

2.1. Chemical and metallurgical properties of elastoplastic nails.

Average chemical composition of the studied nails (AISI 316 stainless steel) was as follows: 0.07% carbon, 16.4% chrome, 10.7% nickel, 1.98% molybdenum. Complementary metallographic studies and uniaxial traction studies were published [5-6]. An austenitic structure appearing from a strong deformation in cold was observed [5].

2.2. Study of the elastoplastic properties

The need to understand the elastoplastic properties was announced from the beginning of the use of the technique of inserting nails [7-13].

Elastoplasticity studies shown along this work belong to 23 nails (nationals and imported), from different manufacturers (see column 6, Table 1).The lengths and diameters of each nail are shown in columns 2 and 3, Table 1. The diameters were measured with a micrometer Palmer ($\pm 1 x 10^{-5}$ m) and the length was measured with a ruler ($\pm 1 x 10^{-3}$ m).

In order to perform these studies, a special device was designed. This device allows placing preformed nails in a vertical plane and measure the value of the arch sag (see figure 7). A flexible metal bracket was made in order to keep the arches of the tested nails on a vertical plane. A screw system allows applying force, in a controlled manner, on the middle of the nails. The free nail ends were separated 300 mm from each other. The forces were registered en Newton ($\pm 2x10^{-1}$ N). In this arrangement the maximum bending moment occurs in only one section of the sample.

Table 1. The 23 elastoplastic tested nails. A sub-index in column 6 indicates the procedence of the elastoplastic nails used for this measurements: (1)Argentina, (2)Austria, (3)United States. L: nail length (m), D: nail diameter (m), %REC: percentage residual elastic coefficient, as defined in equation (1).

No.	L ($\pm 1 \times 10^{-3}$ m)	D ($\pm 1 \times 10^{-5}$m)	Max. F ($\pm$ 0.2 N)	%REC	Origin
1	0.401	0.00471	16.5	51.8	IOA(1)
2	0.339	0.00464	12.7	53.0	IOA
3	0.390	0.00438	13.3	61.5	FICO (1)
4	0.427	0.00433	15.6	63.5	FICO
5	0.341	0.00438	14.3	49.1	FICO
6	0.358	0.00424	11.5	54.0	FICO
7	0.386	0.00421	10.9	55.0	FICO
8	0.353	0.00450	13.0	51.8	FICO
9	0.433	0.00450	15.1	66.8	UKH (2)
10	0.420	0.00440	13.5	59.0	FICO
11	0.409	0.00433	12.2	53.4	FICO
12	0.398	0.00445	13.3	56.2	FICO
13	0.430	0.00451	13.1	50.1	FICO
14	0.395	0.00470	14.7	46.9	FICO
15	0.430	0.00450	15.6	68.6	De PUY (.3)
16	0.375	0.00446	13.3	57.7	FICO
17	0.375	0.00448	14.7	63.5	FICO
18	0.370	0.00438	16.2	72.0	FICO)
19	0.370	0.00400	9.5	65.9	De PUY
20	0.375	0.00422	15.3	62.8	FICO
21	0.285	0.00359	19.3	85.5	De PUY
22	0.375	0.00422	11.7	54.4	FICO
23	0.495	0.00449	16.8	70.8	De PUY

The flattening force is applied on the central point of the arch as shown in figure 6 a). Progressive flattening in 17 steps of 3 mm was applied on the arches. The value of the progressively flattened arches was measured, as well as the corresponding force values. A graph (figure 11), was made based on this data (see section 3).

The deflection arrived at its maximum force, see figure 6 b), then the force is released gradually to cancel and measure the residual deflection from where the percentage of residual elastic coefficient of the nails (%REC) is determined.

This percentage was calculated from the relationship of the arches sag as shown in equation (1):

$$\%REC = \frac{A_r - A_M}{A_0 - A_M} \quad (1)$$

Where A_r is the residual arch sag (without external force); A_M the minimum arch sag (maximum force) and A_0 the initial arch sag (without force).

The values of maximum force and the %REC obtained, in each case, are presented in the fourth and fifth column of Table 1, respectively.

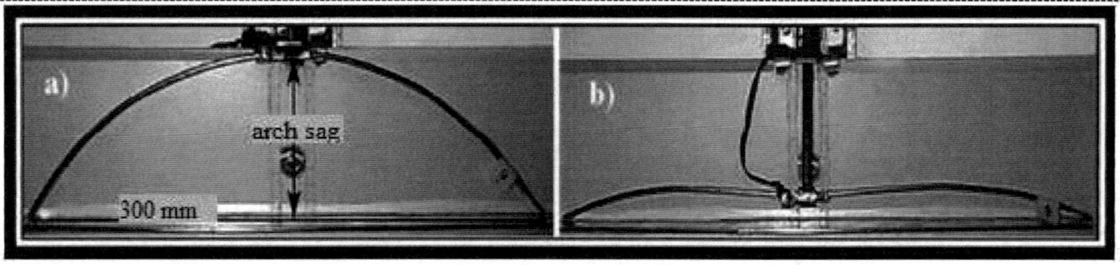

Figure 6. a) Initial arch sag of the nail before starting the test; b) maximum flattening corresponding to the maximum force applied.

As a consequence of the arch sag of the nail, nail-bone contact areas increase in number; this allows the nails to act as a bridge, exerting elastic forces on both ends of the fractured bone.

2.3 Characteristics of the forces on the extremes of the nails

Two digital force platforms range 0-1500 N (± 0.2 N), Soehnle (Germany), were used to measure the reaction force on each of the extremes of the nails. A weight of approximately 50 N was suspended from the upper area of the arches, (symmetric, partial skew and maximum skew). The rough, undulated platform surfaces prevent the nails from sliding (fixed extremes). The horizontal and secant straight lines between the extreme of the nail and the weight exertion point were used as reference for the measures of the angles of the semi-arches (figure 8, figure 9 and figure 10).

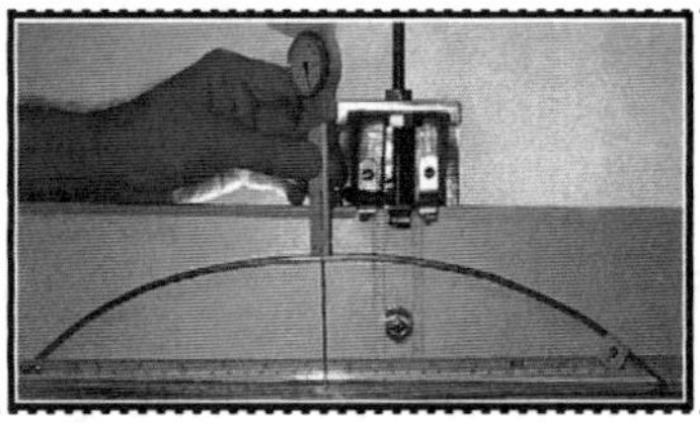

Figure 7. Arch sag measurement of an elastoplastic nail.

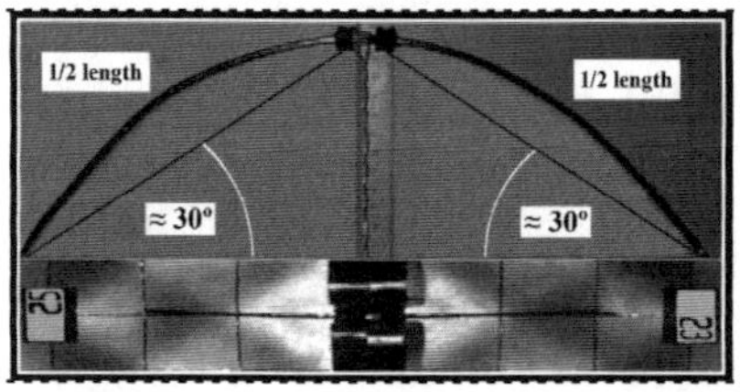

Figure 8. Nail under symmetric condition. The reaction forces in the extremes of the nail are equivalent.

When the nail curvature is symmetric, (figure 8) reaction forces in the extremes of the nail are equal. When the arch is skewed towards one of the extremes (figure 9) the reaction force is skewed in the same direction. Under extreme skew conditions, the reaction force is concentrated on the skewed extreme of the nail, with the other extreme barely uncharged (figure10).

Figure 9. Nail under partial skew condition. The major reaction force is in this extreme skew arch of the nail.

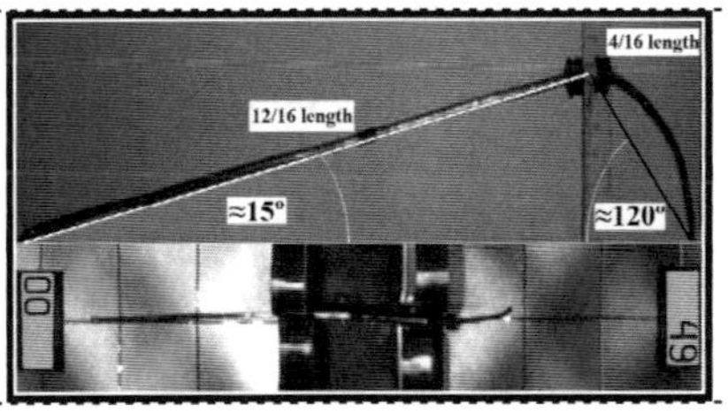

Figure 10. Nail under extreme skew condition. The reaction force is concentrated on the right with the other extreme uncharged.

3. Results

3.1. Experimental determination of elastic forces

Figure 11 shows the flattening forces versus arch sag of the various studied nails (Table 1). As can be observed on the upper curves, the highest values of force are between 95 N and 162 N.

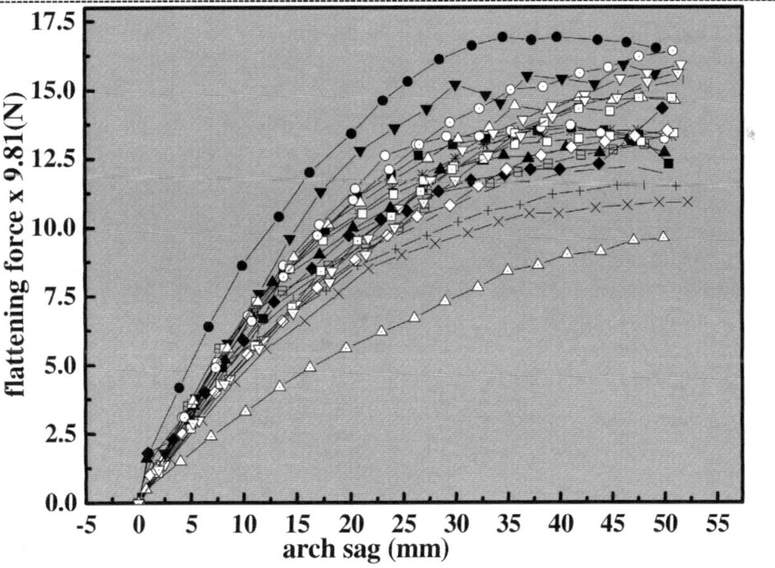

Figure 11. Flattening force [N] vs. arch sag [mm] of the nails tested (separation of the support points 300 mm). The lines between points are only to guide the eye.

The mean percentage of residual elastic coefficient (%REC), regardless of nail geometry and origin, was 59.71%.

Following the method described in section 2.3 the magnitude of the forces at the ends of the nails is determined.

Journal of Physics: Conference Series **332** (2011) 012002 doi:10.1088/1742-6596/332/1/012002

Based on the images shown on figure 9 and figure 10, the force values corresponding to different anchoring methods can be observed.

This study allows highlighting the magnitude of the internal forces that would take place inside the bone given its interaction with the nail. The shape of the arch is critical in this way. The chosen nail has a mean elasticity value.

Since the properties of the bone matrix will most likely differ from those of any artificial system as the one used for this experiment, the results must be interpreted as a study of the differences between the forces at the extremes according to the curvature, instead of the values themselves.

This qualitative knowledge is fundamental for the design of the architecture of the reparation using STS, given the fact that hyperostosis will depend on it.

4. Discussion and Conclusions

This paper presents the percentage residual elastic coefficient (%REC) measurements from 23 elastoplastic nails, maximum force values for each case (Table 1) and force values for each arch sag (figure 11), showing the variability of these properties and the qualitative behavior and order of magnitude for the reaction forces on the anchors.

These studies aim to achieve a wide understanding of the elastoplastic properties of the nails used on the repair of long bones using STS.

The ostheosynthesis by Selective Tension Systems method is the result of the theorical and experimental proof of the materials used, and the effects they can produce in the repair of long bone fractures. This task has been carried out thanks to the constant and permanent interaction between the medical and clinical-surgical.

An accumulated experience of 26 years of follow-ups on patients who had fractures repaired using STS is assessed. People with fractures also present great variations in size, cortical thickness, collagen and mineral matrix densities, skeletal tissue and body mass. These characteristics can be used as selection criteria in the usage of elastoplastic nails, according to the kind of elastic capabilities and forces that best agree with on each case.

All treatments were conducted by the same group of doctors specialized in traumatology and orthopedics. This article is intended for guidance in the selection criteria of the nails.

Osteosynthesis using elastoplastic nails -of which their plastic deformation and reaction elastic force characteristics- are already known, allows the surgeon to act in both aspects, statically and dynamically. Statically, through the number of nails to be used, the sum of their individual structural endurance and the quality of the fixation of the nails, and dynamically, as a result of the directions, magnitudes and application points of the forces exerted by the nails in the fixed extreme at one of the osseous fragment and over the bone area in contact with the flattened nail.

This double approach, both dynamic and static, is considered to be a new and promising prospect in future treatments of long bone fractures.

References

[1] Firica A, Troianescu P and Petre M, 1978L'Osteosintesi delle fratture di femore con chiodi flessibile endomidollari *G.I.O.T.***IV** I pp 23-29.

[2] Firica A, Popescu R, Scarlet M, Dimitriu M, Ionescu V, Protopescu C, Buga M, Constantinescu, I and Iliescu N 1981 L'ostheosynthèse stable élastique, nouveau concept biomécanique *Suppl. II Revue de Chirugie Orthopédique et Réparatrice de l'Appareil Moteur* **67**pp 82-91.

[3] Paterson R and Del Sel G 2001 Estudio de las fuerzas que entregan los clavos intramedulares elastoplásticos rectificados y confinados *A.A.O.T.* **66** pp 135-142.

[4] Eveleigh R 1995 A review of biomechanical studies of intramedullary nails *Medical Engineering & Physics* **17** 5 pp 323-331.

[5] Miralles M, Paterson A, Alonso P, Natale E, Paterson R, Del Sel G, Denegri R and Mariño M 2004 Caracterización del comportamiento de clavos elastoplásticos intramedulares utilizados en la osteosíntesis of La Serena La Serena Chile pp 303-308.

[6] Paterson R, Galvele J and López R 1972 Problemas Metalúrgicos en Materiales de Osteosíntesis de Uso *Programa Multinacional de Metalurgia PMM/I-105* Departamento de Metalurgia, Comisión Nacional de Energía Atómica (CNEA) Buenos Aires Argentina.
[7] Pollock F, Hall R Jr and Davenport K 1983 Distal third tibial shaft fractures: treatment with flexible intramedullary rods *Orthopaedic Transactions The Journal of Bone and Joint Surgery* **7** 3 p 528.
[8] Wiss D 1986 Flexible medullary nailing of acute tibial shaft fractures *Clinical Orthopaedics* **212** pp 122-132.
[9] Pankovich A Goldflies M and Pearson R 1979 Close Ender nailing of femoral shaft fractures *TheJournal of Bone & Joint Surgery* (Br), **61-A** pp 222-232.
[10] Hall, R Jr. and Pankovich A 1987 Ender nailing of acute fractures of the humerus *The Journal of Bone & Joint Surgery (Br)***69-A** pp 558-567.
[11] Walters J, Sheperd-Wilson W, Lyons T and Close R 1990 Femoral shaft fractures treated by Ender nails using trochanteric approach *The Journal of Bone & Joint Surgery* **72-B** pp 14-18.
[12] Merianos P, Cambouridis P and Smyrnis P 1985 The treatment of 143 tibial shaft fractures by Ender's nailing and early weigth bearing *The Journal of Bone & Joint Surgery (Br* **67-B**pp 576-580.
[13] Brumback R, Bosse M J, Poka A and Burgess A 1986 Intramedullary stabilisation of humeral shaft fractures in patients with multiple trauma *The Journal of Bone & Joint Surgery (Br)* **68-A** pp 960-970.

Acknowledgments
The authors wish to thank MSc. Ing. Ricardo Vecchio for his valuables comments of this manuscript. This work was partially supported by Universidad de Buenos Aires (Prot. UBACyT-A435) and Universidad Católica Argentina "Santa María de los Buenos Aires".

Acoustic pollution in hospital environments

J.M. Olivera[1], L.A. Rocha[1,2], V.I. Rotger[1] and M.C.Herrera[1]
[1] Gabinete de Tecnologías Médicas, Departamento de Bioingeniería, FACET-UNT
[2] Instituto de Maternidad y Ginecología "Nuestra Señora de las Mercedes", Tucumán

E-mail: jolivera@herrera.unt.edu.ar

Abstract. There are many different services within a hospital. This means different types of noise which can be considered as acoustic pollution. Knowing that preterm infants exposed to high amounts of noise in the NICU are at a much higher risk because of their neurologic immaturity and physiologic instability, that excessive levels of noise also affect the persons and it can also impede some studies on patients, it was proposed to evaluate the Sound Pressure Level in some services of the Instituto de Maternidad, Tucumán, Argentina. There were evaluated the Level III NICU, the laundry service, a physical space destined for a service of evoked potential and a neonatal incubator under working conditions. The measurements were performed with a type II sonometer (CENTER 322) and it was also used an incubator analyzer (FLUKE INCU) for the incubator. The average values obtained were of 63.6 dBA for the NICU, 82.5dBA for the laundry room, 52.7 dBA for the evoked potential room and 62.8 dBA in the inside of the incubator under 64 dBA in the outside. The reports were documented in compliance with the appropriate standards.

1. Introduction

It can be said that Preventive Maintenance (PM) is a set of tasks performed by personnel for the purpose of maintaining equipment and facilities in satisfactory operating condition by providing systematic inspection to avoid failures either before they occur or before they develop into a major one, according to selected guidelines or standards [1]. The professional that takes care of these policies in a health institution is the Clinical Engineer.

The American College of Clinical Engineering (ACCE) defines the Clinical Engineer as *A professional who supports and advances patient care by applying engineering and managerial skills to healthcare technology.*"-ACCE Definition, 1992-. As clinical medicine has become increasingly dependent on more sophisticated technologies and the complex equipment associated with it, the clinical engineer, as the name implies, has become the bridge between modern medicine and equally modern engineering. The clinical engineer is, by education and training, a problem solver, working with complex human and technological systems [2].

One of the tasks that should be performed in a regular PM schedule, at any health institution, is the control of the acoustic pollution in different indoor environments. There are many sounds present in hospital environments, including those from beepers, alarms, machines, air-conditioning, staff activities, conversations, and medical equipment, among other sources. These can be severely irritating and at times harmful to patients, depending on their current conditions like age, hearing ability, medication intake, cultural background, and some psychological variables [3, 4]. But, not only indoor sources are present. In some cases outdoors sources like traffic, especially in big cities, should be analyzed. Therefore, the acoustic pollution in healthcare environments is complex and requires a careful, strategic design and continuous control. Actually, acoustic pollution, or simply NOISE, is a component of new healthcare design guidelines and there are many studies trying to identify strategies to improve this matter.

But what would be considered as noise? Sound arises when fluctuations in air pressure give rise to pressure waves which travel through the atmosphere. As they travel they interact in various ways with their surroundings. Noise is a word which is normally applied to unwanted sound. Another way to understand Sound is by defining it as any pressure variation that the human ear can detect. Compared to the static air pressure (10^5 Pa), the audible sound pressure variations are very small ranging from about 20 μPa (20×10^{-6} Pa) to 100 Pa, being 20 μPa the average threshold of normal person´s hearing. A sound pressure of approximately 100 Pa is so loud that it causes pain and is therefore called the threshold of pain. The ratio between these two extremes is more than a million to one. Because of this, plus the ear logarithmic frequency response, the scale to represent acoustics parameters should be logarithmic. This logarithmic ratio is called a decibel or dB. The frequency range for a young, healthy human ear, is from 20 Hz to 20 kHz. In terms of sound pressure level, audible sounds range from the threshold of hearing at 0 dB to the threshold of pain which can be over 130 dB. There are many tables that represent different levels of sound pressure and human response, one of the most popular is table 1 [5].

Table 1. Sound Levels and Human Response

Common sounds	Noise Level [dB]	Effect
Rocket launching pad (no ear protection)	180	Irreversible hearing loss
Carrier deck jet operation Air raid siren	140	Painfully loud
Thunderclap	130	
Jet takeoff (200 ft) Auto horn (3 ft)	120	Maximum vocal effort
Pile driver Rock concert	110	Extremely loud
Garbage truck Firecrackers	100	Very loud
Heavy truck (50 ft) City traffic	90	Very annoying Hearing damage (8 Hrs)
Alarm clock (2 ft) Hair dryer	80	Annoying
Noisy restaurant Freeway traffic Business office	70	Telephone use difficult
Air conditioning unit Conversational speech	60	Intrusive
Light auto traffic (100 ft)	50	Quiet
Living room Bedroom Quiet office	40	
Library Soft whisper (15 ft)	30	Very quiet
Broadcasting studio	20	
	10	Just audible
	0	Hearing begins

Although an increase of 6 dB represents a doubling of the sound pressure, a human ear requires an increase of about 10 dB before the sound subjectively appears to be twice as loud. The smallest change that can be perceived is normally about 3 dB. This is a direct consequence of the human frequency response.

The sound level meters can acquire the sound level pressure and perform several different types of processing. It is well known that our hearing is less sensitive at very low and very high frequencies. In order to account for this, weighting filters can be applied when measuring sound. The signal may pass through a weighting network which is relatively simple to build by applying an electronic circuit whose sensitivity varies with frequency in the same way as the human ear. This has resulted in three different internationally standardized characteristics termed the "A", "B" and "C" weightings. The most common frequency weighting in current use is "A-weighting". It provides results often denoted as dB(A) which conform approximately to the response of the human ear. A "C-weighting" curve is also used, particularly when evaluating very loud or very low-frequency sounds. The Noise can be classified as continuous, intermittent or impulsive. When measuring noise, we need to know the type of noise in order to set the parameters to measure, the equipment to use, and the duration of the measurement. Some sonometers may have an octave or one-third octave band filter attached or integrated into the instrument; in this case, they are called Octave Band Analyzers [5, 6].

The objective of this paper is to measure Sound Pressure Levels at some services of the Instituto de Maternidad y Ginecología Nuestra Señora de las Mercedes, Tucumán, Argentina. In particular, in this preliminary study, the selected services were the level 3 neonatal intensive care unit, laundry, evoked potential facilities and a neonatal incubator.

Why were these services selected?

Neonatal Intensive Care Unit (NICU): Numerous studies have documented the continuous noise exposure of infants associated with NICUs. Noise levels are often related to the new modalities of respiratory therapy such as high-frequency oscillatory ventilation and high-frequency jet ventilation, devices alarms, visitors (patient's family), among other sources. The exposure of medically fragile and low birthweight infants to NICU noise has been related to anoxia and bradycardia and has a negative impact on blood pressure, heart rate, perfusion, oxygen saturation and cerebral blood flow. Consequently, preterm infants exposed to high amounts of noise in the NICU are at a much higher risk because of their neurologic immaturity and physiologic instability. The Committee on Environmental Health of the American Academy of Pediatrics recommend a maximum noise pressure of 45 db(A) in daytime and 35 dB(A) at night [7, 8].

Laundry: This is a typical case of noise at work and an occupational point of view can be used for its analysis. The Argentine´s standards and laws to be applied are the IRAM 4079-1//2 (Maximum admissible levels at occupational environments to prevent hearing impairment. Relationship between noise exposure and hearing threshold permanent displacement), IRAM 4113-1//2 (Description, measurement and assessment of environmental noise), IRAM 4060-2 (Estimation of Equivalent Continuous Sound Pressure Level A-weighted with hearing protectors), National Law of Health and Safe at Work No 19587 and its complementary resolutions [9, 10]. Occupational noise-induced hearing loss is one of the causes responsible for numerous occupational ailments and injuries. When the loss or reduction of hearing ability is not the result of an accident or an illness, it is usually an easily recognized process, indicated by basic symptoms such as the need for words to be repeated, misunderstandings in conversation, difficulty in understanding a conversation in certain circumstances, etc. But it should be noted that when these symptoms are present, there is a very high probability of an irreversible process.

Unfortunately, there are no low-cost solutions to reduce this problem, although the modern medicine has demonstrated that injuries appear at sound pressure exposure levels lower than those accepted by standards few years ago. In fact, in Argentina since 2003, the maximum Equivalent Continuous Sound Pressure Level (LAeq), for an 8 h working-day or 48 h working-week, has been reduced from 90 dB to 85 dB according to international standards. In other words, this means to recognize that the ear is affected at levels four times lower than the ones accepted few years ago.

Noise evaluation of a physical space destined for a service of evoked potential. In this case, the hospital authorities required a noise evaluation in order to implement a new Visual Evoked Potential service. Evoked potentials are used to measure the electrical activity in certain areas of the brain and spinal cord. Electrical activity is produced by stimulation of specific sensory nerve pathways. These tests are used in combination with other diagnostic tests to assist in the diagnosis of neurological disorders. Evoked potentials are used because they can indicate problems along nerve pathways that are too subtle to show up during a neurologic examination or to be noticed by the patient. The disruption may not even be visible on MRI exam. There are different types of potentials, i.e *Visual Evoked Potentials (VEP)*: This test requires that the patient observes a flashing checkerboard pattern projected on a monitor, this procedure takes up approximately 60 minutes to be completed; *Auditory Evoked Potentials (AEP)*: This examination involves listening to clicking noises generated in a set of headphones, this exam may take 60 minutes to complete. There are many other types of evoked potential [11]. In this case, the evaluation was performed according to IRAM 4026 (auditory cabin) and IRAM 4113-1 (Description, measurement and assessment of environmental noise. Basic quantities and assessment procedures) standards.

Neonatal Incubator: This is a major challenge because of the complex conditions proper of level III NICUs. The Instituto de Maternidad de Tucumán, is one of the 5 main hospitals in the province, it is Reference Center for the north-west region in Argentina and also the last step before deriving the most complex cases to Hospital Juan Garrahan in Buenos Aires, Argentina. In this case, the authorities require the implementation of a protocol to evaluate the performance of the incubators in working conditions One of the parameters to be evaluated is sound pressure within the baby compartment. Other parameters available to evaluate are air flow, temperature –both radiant and convection-, relative humidity. The tests were performed in compliance with IEC 60601-2-19:2009 and IRAM 4220-2-19:1995 standards.

2. Materials and Methods

In all cases, the environment sound pressure level was measure with a type II sonometer (CENTER 322) with data-logger. It was calibrated with a Sound Level Calibrator (TES-1356) at 94dB according with the manufacturer´s instructions. The instrument autonomy is about 48 h when it is powered by a 9V alkaline battery.

The sonometer parameters were set to:

- Level Range: Auto 30-120 dB
- Frequency Weighting: A
- Time Weighting: Fast
- Sampling time was programmed depending on the environment under evaluation

The recorded data were downloaded to a notebook, via an USB-RS232 cable adapter, by the TestLink SE322 software for Sound Level Meter, provided by the manufacture's sonometer. Furthermore, they can be also saved as a "csv" file format to be processed in any spreadsheet.

Neonatal Intensive Care Unit (NICU): The sound level pressure was recorded during 24 h, the sampling time was set to 5 sec and the sonometer was placed in a corner between 2 incubators, with patients, trying to get data under the worst conditions.

Figure 1. General view of the level III NICU and the sonometer placed in the corner

The personnel were asked "to ignore the presence of the new instrument" in the service and it was explained that it does not record voices or conversations. Figure 1 shows a view of the level III NICU and the sonometer placed in the corner.

Laundry: The sound level pressure was recorded during 5 min, the sampling time was set to 1 sec. The personnel were instructed "to ignore the presence of the instrument" in the service and it was guaranteed that the device does not record voices or conversations. The sonometer was placed in the centre of the engine room, inside a foam rubber bag to avoid vibrations, with the microphone pointing to the noise sources. Figure 2 shows two pictures of the engine room, Figure 3 shows the sonometer inside the foam rubber bag.

Figure 2. Two pictures of the engine room of the laundry service

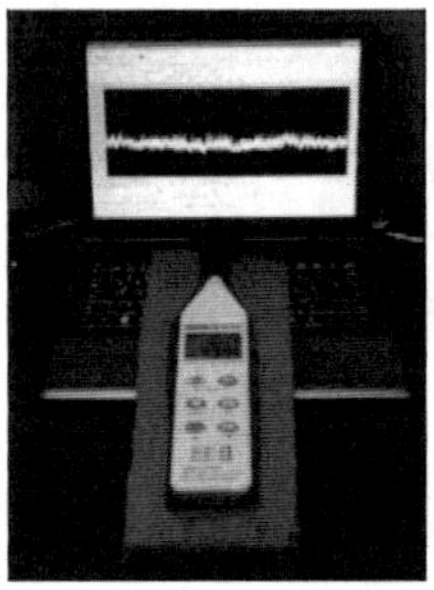

Figure 3. Sonometer inside the foam rubber bag

Noise evaluation of a physical space destined for a service of evoked potential: The sound level pressure was registered during approximately 40 min; the sampling time was set to 1 sec. The sonometer was placed in the centre of the room, inside a foam rubber bag to avoid vibrations. The door was closed and there were no personnel during the record. The data were recorded during a high level of people circulation in the hall at approximately 10 am. Figure 4 shows the hall, the entry and the room interior.

Figure 4. Hall, the entry and interior of the room destined for VEP

Neonatal Incubator: This test was performed in the hall between Level III & Level II NICUs as the service had no incubator available. Thus, it was decided to evaluate the incubator under worse conditions than those within the room, the ambient temperature being 24°C. This is showed in Figure 5. In this case, the test was performed with both, a sonometer and an incubator analyzer (Fluke Incubator Analyzer-INCU). The INCU is a portable device designed to verify the proper operation and environment of infant incubators. This unit focuses on the record in parameters important to the care of infants over time, such as airflow, sound level, temperature (four individual measurement probes), and relative humidity. It can operate stand-alone and is programmed by a PC via an RS232 port, the user selects the desired record time/interval via the software and then initiates the start of the test from

the INCU. After the test is completed, the user uploads the data collected into the PC for display and analysis. The recorded data can be saved in a file or printed in a report.
The INCU parameters were set to:
Interval Time: 1 h 20 min
Sample interval: 1 min

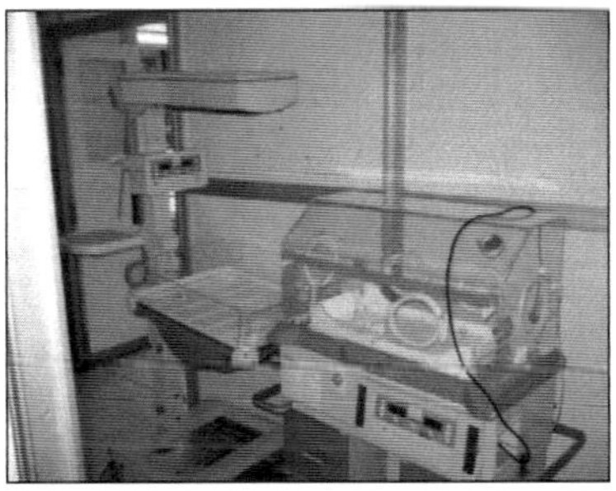

Figure 5. Incubator under test

3. Results

Neonatal Intensive Care Unit (NICU): It is well known that the sound pressure levels in level III NICUs are higher than in the other levels I & II. A previous study, performed in the same service in 2009, showed an average noise of 64.8 dBA [12]. In this case, we try to carry out the evaluation under the worst possible conditions. Figure 6 graphs the recorded data

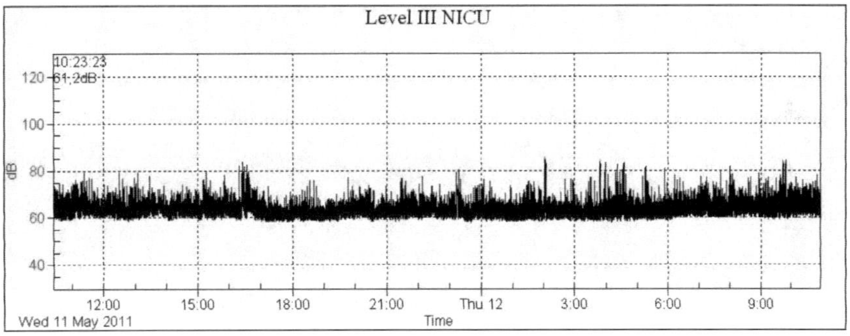

Figure 6. Level III noise measurement

The statistics values obtained are the following (24 h format):

Start Time: 11/05/2011 10:23:23 End Time: 12/05/2011 10:55:13
Sampling Rate: 3 sec Data: 29449
Maximum: 85.4 dBA@12/05/2011 2:01:59 Minimum: 58.0 dBA@11/05/2011 23:41:08
Average value: 63,6 dBA

If we compare these results with those obtained in 2009, even though the present values seem to be lower, it is possible to conclude that the working conditions should be improved if we take into account the American Academy of Pediatrics recommendations. But, when these results are compared to other real values, obtained in other institutions in our country and all over the world, it can be said that this is an acceptable result [13, 14].

Laundry: This is the first time that this type of evaluation is performed in this service and the personnel were pleased for being taken into account. Figure 7 graphs de recorded data.

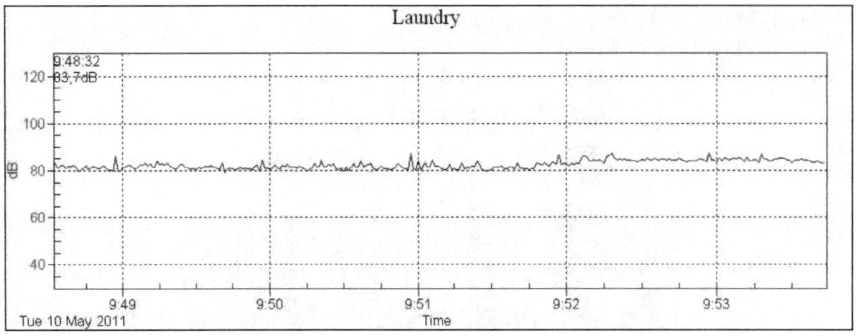

Figure 7. Laundry noise measurement

The statistics values obtained are the following (24 h format):
Start Time: 10/05/2011 9:48:32 End Time: 10/05/2011 9:54
Sampling Rate: 1 sec Data: 312
Maximum:83.8@10/05/2011 9:52:56 Minimum:81.8@10/05/2011 9:51:34
Average value: 82.5 dBA

Even when the average value seems to be high, it is lower than the maximum fixed by the law which is 85 dB.

Noise evaluation at a physical space destined for a service of evoked potential: The hospital needs a new Visual Evoked Potential service. Currently, the room destined for this matter is in a hall with an important people circulation. Even when the sound-noise is not the stimulus for the patient under evaluation, the environmental noise could interfere. Figure 8 graphs the recorded data.

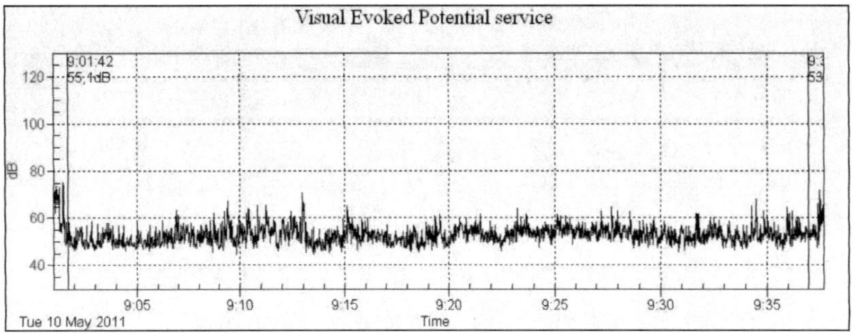

Figure 8. Visual evoked potential room measurement

The statistics values obtained are the following (24 h format):
Start Time: 10/05/2011 9:01:01 End Time: 10/05/2011 9:38
Sampling Rate: 1 sec Data: 2200
Maximum:72.0@10/05/2011 9:01:02 (opened door)
Cursor A: 55.1@10/05/2011 9:01:42 (closed door)
Cursor B: 53.5@10/05/2011 9:36:59 (closed door)
Max.Between A and B: 70.7@10/05/2011 9:12:59 (closed door)
Min.Between A and B: 44.6@10/05/2011 9:09:50 (closed door)
Avg. Between A and B: 52.7 (closed door)

It should be noted that, for a correct noise evaluation of this service, the noise has to be divided into its frequency components. Unfortunately, when this evaluation was performed, the octave band analyzer was not available and the room did not have any type of acoustic sound conditioning; therefore, only a general environmental noise evaluation was performed. Figure 9 shows the maximum values expected in compliance with IRAM 4026 standard.

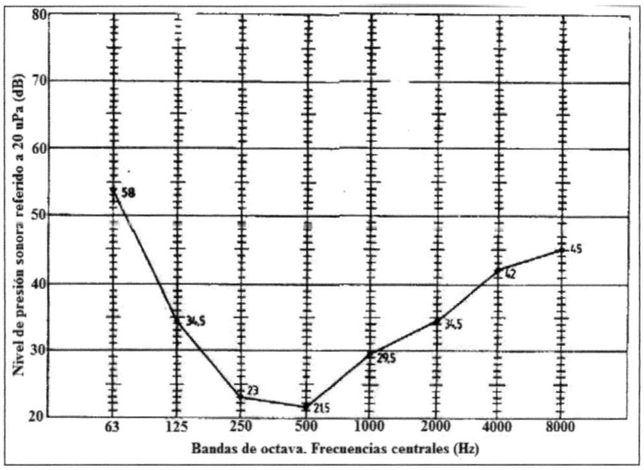

Figure 9. Maximum admissible Sound Pressure Level in audiometric cabin. Reproduced from IRAM 4026 standard

Neonatal Incubator: The real intention of this test is to check if the cabinet cover has any noise attenuation factor. The actual model of incubators at level III NICU has a technical specification of "own noise" less or equal to 50 dBA, but there is no specification for cover attenuation factor. This matter is under analysis in many institutions [15]. Our preliminary results show that the attenuation factor is very poor; for an external noise of 64 dBA, the inside shows 62.8 dBA. These values are shown in Figure 10.

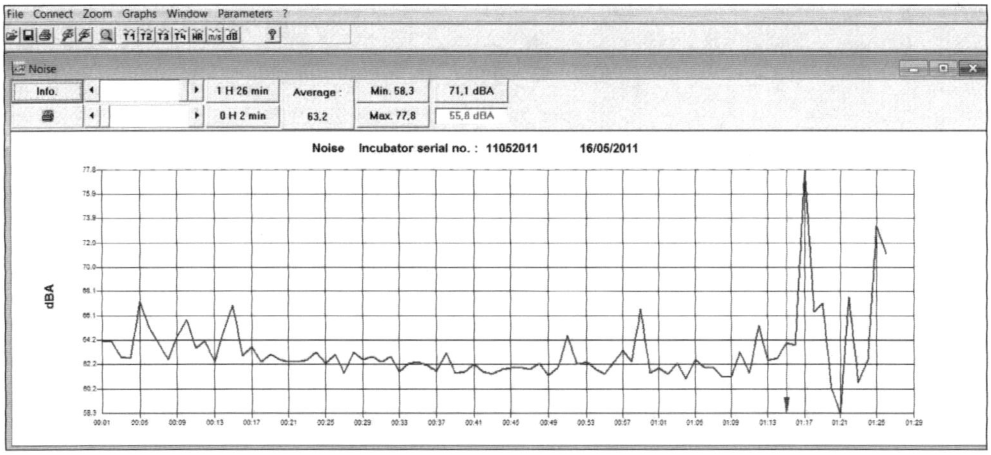

Figure 10. Noise inside the cabinet. The mark points the instant when the instrument is put outside over the incubator

The INCU also performed 4 temperature records according the diagram showed in Figure 11 and in compliance with IEC 60601-2-19:2009 and IRAM 4220-2-19:1995 Standards. Figure 12 shows the graphic report of the INCU air sensors. Temperature Sensor T1 is used for convection measurements and could correspond to point B in Figure 11; Temperature Sensor T2 is used for Convection or Radiant measurements and corresponds with point M in Figure 11; Temperature Sensor T3 is used for convection measurements and could correspond with point C in Figure 11; Temperature Sensor T4 is used for Mattress Temperature measurements (made by conduction) and it does not have any correspondence with any point in Figure 11.

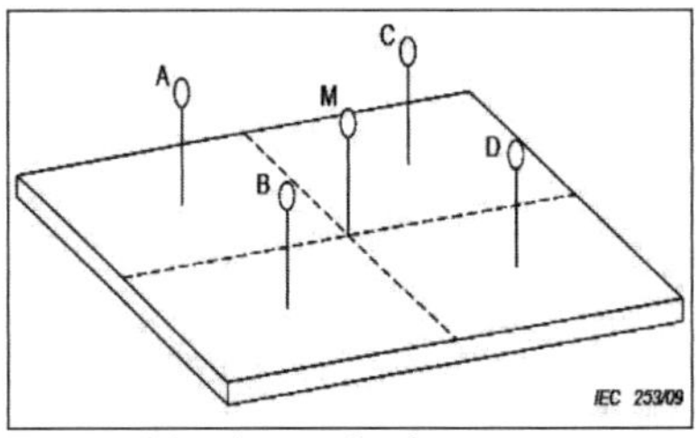

Figure 11. Positioning of air temperature sensors. Reproduced form IEC 60601-2-19:2009

The standard sates that:

- *During steady state condition, the INCUBATOR TEMPERATURE shall not differ from the AVERAGE INCUBATOR TEMPERATURE by more than 0.5°C (1°C transportable) during at least 1 hour at the control temperatures of 32°C and 36°C*
- *... in each point A,B,C,D,E, shall not differ from the AVERAGE INCUBATOR TEMPERATURE (test at a set T of 32°C - 36°C) by more than ±0.8°C (±1.5°C transportable).*
- *The warm-up time of the equipment shall not differ by more than 20% from the warm-up time specified in the instructions for use. INCU: Check the difference between the time stated by the manufacturer and the time to raise the temperature by 11 C, starting at the environmental conditions with a setting temperature 12°C above the ambient.*
- *After adjusting the temperature from 30°C to 34°C or Transit 32°C to 36°C, the overshoot in the incubator temperature shall not exceed 2°C.*

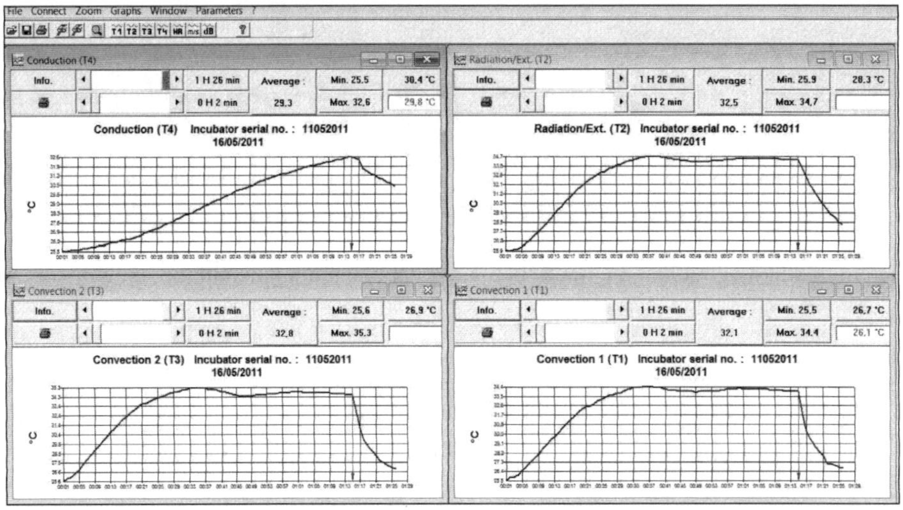

Figure 12. Temperature values inside the baby compartment. The mark points the instant when the instrument is put outside over the incubator.

The results showed in Figure 12 verify the standards requirements. Air flow and its average value 0.22 m/s, lower than the max fix by the standard which is 0.35 m/s were measured as well.

4. Conclusion

Environmental noise data is often used in legal proceedings. In fact, we should always collect data with the assumption that it may someday have to be examined in court testimony. Therefore it is of outmost importance that the conditions of the measurement be carefully documented in a formal measurement report. A comprehensive and carefully documented formal has the best chance in any legal proceedings or noise control negotiations settled out of court. All reports should be documented according ISO 1996 that states the information must be recorded. It is also important to write the report in an easy-to-understand, readable style. Depending on the target audience, the use of graphics, sketches, and illustrations can sometimes help to explain the data. In other cases, text and figures will suffice [5].

We do not conclude anything new if we said that a PM schedule is a major task in any health institution, but we believe that a very significant contribution is to include in the schedule a regular control of the acoustic pollution all over the health institution.

Acknowledgments

Authors want to deeply thank to the following students for their collaboration in collecting the data: Farias E, Damonte ME, Gerónimo JL, Herrera Argiró M, Herrera JP, Jimenez Pereyra FM, Lazarte Cabrera A, Orellana D, Paez C, Varela C, Viscido Castro A. Also thanks to all the hospital personnel in the person of the hospital director Dra. Rossana Chala. This work was financial supported in part by grants from the UNT Research Council (CIUNT), 26/E422 Research Program and INSIBIO-CONICET.

References

[1] Industrial Accident Prevention Association 2007 A Health and Safety Guideline for Your Workplace: Preventive Maintenance IAPA guidelines www.iapa.ca
[2] American College of Clinical Engineering *http://www.accenet.org*
[3] Ceilings & Interior Systems Construction Association (CISCA) 2010 White Paper: Acoustics in Health Care Environments *http://www.cisca.org*
[4] Goines L & Hagler L 2007 Noise Pollution: A Modern Plague *Southern Medical Journal* 100(3):287-294
[5] Brüel&Kjær 2001 Environmental Noise–booklet *Brüel&Kjær Sound & Vibration Measurement*
[6] Tim South 2004 Managing Noise and Vibration at Work *Elsevier Butterworth-Heinemann*
[7] Levy G, Woolston D, Browne J 2003 Mean Noise Amounts in Level II vs Level III Neonatal Intensive Care Units *NEONATAL NETWORK* **22:2** 33-38
[8] Committee on environmental health 1997 Noise: A Hazard for the Fetus and Newborn Pediatrics *American Academy of Pediatrics* **100(4):**724-7
[9] Instituto Argentino de Normalización y Certificación (IRAM) *www.iram.org.ar*
[10] Ley Nacional de Higiene y Seguridad en el trabajo Nro 19587
[11] Evans A, MD 2010 Clinical Utility of Evoked Potentials *http://emedicine.medscape.com*
[12] Brandán R, Halloy N, Sanchez MA, Sappia LD, Sueldo J, Rocha LA, Herrera MC, Rotger VI y Olivera JM 2009 Contaminación Acústica en salas de neonatología *Anales del XVII Congreso Argentino de Bioingeniería y VI Jornadas de Ingeniería Clínica, SABI 2009* **1:**1-4
[13] Alvarez Abril A, Boschi C, Gomez M 2007 Revisión de ruido en unidades de cuidados intensivos neonatales–Análisis regional *Anales del XVI Congreso Argentino de Bioingeniería y V Jornadas de Ingeniería Clínica, SABI 2007* **1:**1-4
[14] Werner AF (MD) Evaluación del ruido ambiental en las Unidades de Cuidados Intensivos de Recién Nacidos 2008 Anales *del VI Congreso Iberoamericano de Acústica-FIA 2008* **1:**1-11
[15] Wubben SW, Brueggeman PM, Stevens DC, Helseth CC, Blaschke K 2011 The sound of operation and the acoustic attenuation of the Ohmeda Medical Giraffe OmniBed *Noise and Health Journal* **13:50** 37-4

Proposal for periodic verifications of electromedical devices integrated to terrestrial Technical Ambulance Inspection (TAI).

R Del Aguila Heidenreich, O Vanella, R Bruni, R Taborda

LIADE (Laboratorio de Investigación Aplicada y Desarrollo), Facultad de Ciencias Exactas Físicas y Naturales – Universidad Nacional de Córdoba. Velez Sarfield Avenue 1611, Córdoba, Argentina.

e-mail: liade@efn.uncor.edu Web page: www.liade.efn.uncor.edu

Summary. In Argentina, electromedical devices may only be commercialized if they meet safety and performance requirements established by current regulations, ensuring their safety and intended performance when leaving the Factory. However, during usage, natural wearing and overloading may change this condition, especially if used in extra hospital services performed by ambulances, which are likely to be subjected to rough handling conditions and hitting. This proposal explains the chosen methodology to address the periodic verification activities of electro medical devices within the process of terrestrial Technical Ambulance Inspection (TAI). Among the results stand out the set of methods for verification and the lists used to record the outcome of this evaluation. Outstanding conclusions include that the operations meet the conditions of an analogous mechanism to that of a Technical Vehicle Inspection (existing for other vehicles), and that the same working structure can be used as a basis for making a manual of procedures for a TAI.
Key words: Inspection, ambulances, electromedical equipment

1. Introduction

In Argentina, the National Administration of Drugs, Food and Medical Technology (ANMAT – Administración Nacional de Medicamentos, Alimentos y Tecnología Médica) is the government entity which regulates the activities related to medical products. The ANMAT, through its disposition 4306/99, approves the MERCOSUR regulation "Essential Requirements of safety and efficacy for medical products", in order to minimize or reduce failures, the appearance of adverse events and/or malfunction caused by medical devices usage and, at the same time, provide patients, users and other people a high level of protection and provide the functions that the manufacturer has assigned to these devices [1]. It also establishes the enforcement of these regulations for the manufacturers/importers who want to commercialize products of its kind in the country.

In the case of electrical appliances, this means that the equipment must meet safety and performance requirements established by current regulations [2] [3].

The "basic security" and "essential performance" of the electrical appliances are part of the overall security situation, which comprises the safety of: equipment, installation to which it is connected to and the application [4].

The compliance of the regulatory requirements only guarantees that the equipment is safe and performs according to its intended use at the moment it leaves the Factory. However, during usage, natural wear, overloading, misuse or repairs can change its security condition or performance. Thus arises the need for a regular check to ensure that the equipment maintains the security status and initial performance [5].

Ambulances are recognized as a component of pre hospital (or extra hospital) systems and emergency response [6], which is of vital importance for the future of the victim because the probability of death or disability depends on the attention received, at the scene or on route, during the first hour (golden hour) after the event, with special emphasis on the first ten minutes (10 platinum minutes) [7]. Consequently, ambulances are a critical factor in any health system, and the security and proper functioning of the units and equipment must be ensured throughout its lifetime.

Reflecting on the overall security situation of the electromedical devices mentioned above, we see that an ambulance is an environment with singularities which affect the aforementioned condition in areas as diverse as transportation, assembly or positioning, connection, commissioning, operation (often in harsh and adverse circumstances) and position of the operator, his assistants, and the patient, related to electromedical device during use.

Regulations are abundant [8] at both the national [9] and international level [10] which establishes different requirements for both the base vehicle as well as for the elements used in the complex installations of an ambulance. As in the case of medical products, compliance with these standards only assures that the "ambulance" system is safe and works properly prior to its commercialization. Furthermore, the application of these standards is voluntary.

In the regulated field, our country has legislation specifying the minimum conditions to be met by emergency and urgency services but, in practice, there isn´t effective enforcement due to jurisdictional problems [11] and the absence of control mechanisms. Moreover, the scope of these laws is generally limited to listings and input devices to be included within the provided vehicles for the delivery of health care and patient transport.

Taborda and Vanella [12] warn about the need for a TAI, defined as a periodic verification process to confirm the compliance with the safety and performance of these vehicles, considering that it should include the different devices and facilities of an ambulance; based on the general applicable rules to these and other specific regulatory entities when insufficient.

For the effective implementation of TAI, it is necessary to determine which features to test and how to test them, to provide written procedures for each device and installation to evaluate and determine the acceptance criteria for each [13].

In the general context pointed out here, one of the systems to control are the electrical and electronic equipment used for medical practice [12] [14].

The current paper explains the chosen methodology to approach the periodic verification activities of electromedical devices in the process of Technical Ambulance Inspection.

2. Methods

Overall, the verification proposal consists on doing an inspection which involves:

- Visual assessment.
- Functional Operation test.
- Measurements and safety tests.
- Verification of the documentation accompanying the ambulance.
- Systematic recording of previous items' results.

Even though the activities described are valid to evaluate the electrical appliances of an ambulance, they can also be applied to other subsystems that comprise it.

The operations arising from this study should be conducted under boundary conditions imposed by TAI's general process [15], such as:

- Being time-efficient (the total duration of ITA does not exceed the seven hours).

- Avoid dismantling equipment and components (except for those essential to test the installation of medical gases).
- Avoid destructive or potentially destructive testing (devices must remain in use after inspection).
- Simplify the tests.

The electromedical devices to check are those established for each ambulance type (A1, A2, B o C) in clause 4.10.4 of the current standard IRAM 16030. For instance, given a type B ambulance:

- Cardioverter Defibrillator.
- Electrocardiograph Monitor.
- Transcutaneous pacemaker.
- Vacuum pump.
- Respirator.
- Pulse Oximeter.

The first task is to verify that each device is registered in ANMAT, in case it fails to comply with this legal requirement, it should not be used.

Then a subset of the total tests defined by the relevant standards must be selected, observing the restrictions imposed by the boundary conditions previously mentioned. The selection criteria contemplate the associated risks due to the use of medical devices and are detailed below:

a) Recommendations issued by Emergency Care Research Institute (ECRI) [16].
b) S.O.T. Analysis (Severity, Occurrence, Trends).
c) Experience of the Multidisciplinary team from the Faculty of Physical and Natural Sciences, National University of Cordoba.

The S.O.T. analysis is a qualitative tool for decision-making, which can be used to prioritize the resolution of problems. The application can be tailored for the task of selecting the set of tests to reflect that the problem in this case, is the risk that affects the patient, operator, other people, other equipment and the ambulance environment; from this perspective we consider that this type of analysis is relevant to achieve our goals because, from its three components, two (Gravity = Severity and Occurrence) are the elements which constitute the risk concept mentioned by the ISO 14971 standard [17].

The S.O.T. consists on describing for each dangerous situation its gravity or severity (severe, serious, not serious), occurrence probability (high, medium, low) and trend or evolution forecast (worsens over time, extinguishes or solves itself, remains stable).

The level for the classification in which each S.O.T. element is divided, is obtained by using a three point Likert Scale [18], modified to express the expert's opinions about each item. The weight of each point on the scale leads to a representative value of each of the components (S-O-T) of the dangerous situation. The final product of the three values (S x O x T) obtained defines the priorities.

To the selected set of tests, are added those which are established by the IRAM 16030 standard in its clauses 4.10.3.1 to 4.10.3.8 inclusive (except for the "free fall" test and electromagnetic assessment compatibility), which ensure the safety and performance of the devices in mobile situations and field applications. That is, including tests to ensure that, even when the devices withstand rough handling and hitting for being employed on portable and extra hospital conditions, they remain safe and effective.

The resulting tests from the previous selection are classified as: a) common or general (for all devices) and b) particular to each device.

The common set of tests are subdivided into qualitative observations and evaluation of general electrical safety requirements, while the individual tests are done in qualitative observations and safety assessments as well as essential performance for each device.

Finally, verification lists are generated, in which are registered test results.

3. Results

The application of the described methodology results in the following:

Common qualitative observations

The items to check are:

1. **Case/Chassis:** examining the outside of the unit in response to the cleanliness and overall physical condition. The case must be intact and the elements between frames (screws, nuts, etc…) must be present and tight. There should be no signs of liquid leakage, or misuse.
2. **Assembly:** if the equipment is fixed to the body of the ambulance or inside a piece of furniture, the supports or fixing elements must be firm and in proper condition to ensure the adequate fastening.
3. **Power connector:** inspect it searching for any type of damage. Exert force on the legs to see if they are firm. Shake the connector in search of noises that may indicate loose screws or broken pieces.
4. **Power Cable:** inspect it for signs of damage. Also check the battery charging cable (if applicable).
5. **Equipment Connectors:** Check its general state; check that it firmly holds the cable connector when connected.
6. **Other device cables:** inspect them searching for possible breaks in the insulation and damage of any kind. Check whether they are mechanically secured on fastening points to the device, so that mechanical requirements during usage are not supported by the conductors within.
7. **Master Switch:** activate it and check that it can be activated freely and that the on/off positions switch forcefully.
8. **Fuses:** check that the fuse in the device is of the type and value indicated on the chassis. Also check for the presence of a spare fuse.
9. **Controls, switches:** check the physical condition of all controls and switches, check if they are securely mounted and that their movement is correct. During the inspection course, check whether each control and switch perform the correct function. If there are any redundant controls, make sure they perform their proper function; then activate them individually and check that the function doesn´t work.
10. **Battery/charger (applies to battery-operated equipment):** check the physical conditions of all batteries. Verify that the charger is plugged into AC current, and that it´s connected to the device; and also that it has remained like that for long enough. Next, leave the device working using the battery for several minutes to verify it retains the charge. After that check the remaining charge by using the battery status test of the device. An alternative way to evaluate the battery state is by performing the entire inspection with the device disconnected from any other power source.
11. **Indicators/displays:** During the inspection, confirm the correct operation of all leds, indicators, gauges and displays. When appropriate, it should be checked that an adjustment on the display intensity control produces a change on the brightness of the stroke or other information on the screen.
12. **Alarms** [19]**:** Operate the equipment on a way to provoke the activation of every alarm (both visual and audible alarms. When in doubt, consult the documentation provided by the manufacturer). Check that both visual and audible alarms are produced properly.
13. **Audible Operation warnings** [19]**:** Operate the device so as to activate all its audible signals. Confirm that the sound level meets regulatory requirements and that the controls work properly.
14. **Labels and markings:** check that all the used units are in SI (International System of Units), with the possible exception of blood pressure. Check that all the labels or control markings are present and legible, as well as operating instructions and conversion tables provided by the manufacturer (in accordance with regulatory requirements).
15. **Accessories:** Check that all required accessories for the device's normal operation are present and in proper condition.

Evaluation of general electrical safety requirements
Required tools:
- Automatic electrical safety analyzer in accordance with IEC 60601-1.
- Or else, measuring elements as required by regulations.

The Parameters to be determined during the inspection are:
1. Protective ground resistance.
2. Patient auxiliary current.
3. Patient leakage current.
4. Contact current.
5. Ground leakage current.
6. Power insulation resistance.
7. Applied parts' insurance resistance.

The correspondent results for the individual tests are presented only for a defibrillator monitor, selected as an example.

Defibrillator/monitor
Qualitative observations
1. **Pallet/electrodes:** examine the pallets searching for physical damage and observing it´s cleanliness state. Check for dry electrolytic gel remnants, physiological fluids and dirt. Dirty electrodes cause poor electrical contact and can cause burns. Confirm the existence of an adequate number of spare ECG electrodes.
2. **Monitor Performance:**
   I. If the unit can monitor the ECG signal through the paddles, check if the ECG is displayed using as input both the pallets and the hoses.
   II. Inject an ECG simulated signal with known fixed frequency and amplitude, and observe the following:
      a) Baseline: must be horizontal; no significant deviations will be accepted.
      b) The ECG complexes, for a fixed frequency of the simulator, must appear regularly spaced. Check the amplitude and frequency calculated by the monitor.
      c) All portions of the ECG simulated signal must be clearly visible.
      d) Check whether there is interference of 50 Hz or any other significant noise overlapping with the baseline with the ECG simulator connected. Interference in the baseline may appear as a thick line at high gains, but shouldn´t be visible along the lower two thirds of the gain control range.
      e) Control monitor derivation selections, checking it's possible to obtain all available derivations.
   III. Response to a 1mV step. Press and hold the calibration 1mV button over a period of approximately 3 seconds (or apply externally a 1mV step if the unit does not have this feature available). Notice the following:
      a) The line should show a sharp rising edge, with well-marked and defined vertices not rounded or jagged.
      b) After 3 seconds the level should not drop significantly.
3. **Internal discharge of stored energy:** It must be possible to safely discharge the stored energy in case the medical technician decides not to use the defibrillator after being loaded. With the defibrillator tester's help, perform a load, activate the discharge button and measure the released energy by activating the discharge from the pallets.
4. **Electrodes:** confirm there is an adequate quantity of good state electrodes at hand (at least 20).

Evaluation of safety and essential performance
Required tools:

- Defibrillator tester in accordance with regulations (e.g.: Fluke 5000).
- Chronometer.

1. Measurement of released energy:
    a) Measure the energy released at 50 Ω load, selecting a max., min. and 2 intermediate energy levels (e.g.: 20, 100, 200 and 360 Joules).
    b) The deviation shall not exceed ±15% or ±4 J, whichever is higher.
    c) The released energy after 30 seconds elapsed since the device is loaded must be >85% from the selected level (check at 360 J).

2. Charging time control:
    I. Charging time at full output power <15 s.
    II. Check that this condition is met even after 5 consecutive discharges (frequency not less than 3 per minute), for E in the last third of the device's range.

3. Internal Discharge: when disconnecting the device's power, it should be discharged internally with a <10 seconds time constant.

4. Synchronism:
    I. Connect the patient cable to the ECG tester simulator. Select the synchronism option on defibrillator and perform a discharge.
    II. Check that the time frame from the R wave until the discharge is <80 milliseconds.

5. Defibrillator monitor:
    I. Recovery time after defibrillator discharge <10 s.
    II. Check that the EGC recording is not significantly interfered during charge.

The verification checklists used to register the results presented above are shown in figures 1 to 4 from the following page on.

ITA	*Inspección Técnica de Ambulancias Terrestres*		
	Estación de control:	*Fecha:*	*Cód. Interno de Inspección:*
Tipo de Amb.: A B C	*Tipo de Inspección: 1° 2° Otra*	*Fecha última inspección:*	
Pat. Vehículo:	*Pat. Part. Ambulancia:*	*Dueño o Responsable:*	
Marca y Modelo del vehículo:		*Empresa Integradora:*	

4.5.3 Instrumental biomédico		
Equipo Testeado:	*Marca y Modelo*	*Nº de Serie:*
Nº de Registro en ANMAT:	*Observaciones:*	

	1. Evaluación Cualitativa General			
	Condición evaluada	*Cumple S*	*Cumple N*	*Observaciones*
1.1	*Chasis o gabinete: estado físico general, limpieza, tornillos y tuercas firmes*			
1.2	*Montaje: soportes o elementos de fijación firmes*			
1.2	*Ficha de alimentación en buen estado*			
1.3	*Cable de alimentación en buen estado*			
1.4	*Conectores del equipo: estado general*			
1.5	*Otros cables del equipo: estado general, asegurados mecánicamente en los puntos de fijación*			
1.6	*Interruptor general: puede activarse libremente y su accionamiento es firme*			
1.7	*Fusibles: tipo y valor correcto. Cuenta con fusible de repuesto.*			
1.8	*Controles, interruptores: estado físico general, montaje firme, y funcionamiento adecuado*			
1.9	*Batería/cargador: la batería retiene carga*			

Figura 1. Example for a checklist of common qualitative observations.

	2. Evaluación Cuantitativa General (Requerimientos de Seguridad Eléctrica)					
2.1	*Resistencia de Tierra de Protección*					
		Cond. Normal	*Cond. 1° Defecto*			
2.2	*Corriente Auxiliar de Paciente*					
2.3	*Corriente de Fuga de Paciente*					
2.4	*Corriente de Contacto*					
2.5	*Corriente de Fuga a Tierra*					
2.6	*Resistencia de Aislación de Alimentación*					
2.7	*Resistencia de Aislación de Partes Aplicadas*					

Figure 2. Example for a checklist of general electrical security requirements evaluation.

Desfibrilador/Cardioversor				
Marca y Modelo:		Nº de Serie:		
Nº de Registro en ANMAT:		Observaciones		
1. Evaluación Cualitativa Particular				
	Condición evaluada	Cumple S	Cumple N	Observaciones
1.1	Paletas en buen estado de conservación. Limpieza adecuada.			
1.2	Electrodos para monitoreo de señal cardíaca (20 como mínimo), en buen estado.			
1.2	Gel electrolítico: fácilmente accesible y en cantidad suficiente.			
1.3	Si la señal de ECG puede monitorearse desde las paletas, verificar que lo haga			
Inyectando señal de ECG simulada, de frecuencia y amplitud conocidas				
1.4	Línea de base horizontal, sin desviaciones significativas			
1.5	Complejos de ECG aparecen regularmente espaciados			
1.6	Todas las porciones de la señal se ven claramente			
1.7	No existe ruido de 50 Hz u otra interferencia en grado significativo			
1.8	Selección de derivaciones: pueden obtenerse todas las derivaciones disponibles			
Respuesta al escalón de 1mV: oprimir y mantener presionado el botón de calibración durante 3 s o aplicar la señal externamente				
1.9	Flanco ascendente abrupto, con vértices bien marcados, no redondeados y sin picos			
1.10	Luego de 3s el nivel no decae significativamente			
Descarga interna de energía almacenada				
1.11	La función se realiza en forma correcta al activar control de descarga interna			
1.12	La Energía se descarga internamente antes de los 10 s posteriores a la interrupción de la alimentación			

Figure 3. Example for a checklist of qualitative observations for a defibrillator.

2. Evaluación Cuantitativa Particular					
	Condición Evaluada		Cumple S	Cumple N	Observaciones
Medición de Energía Liberada					
	Valores seteados	Valores medidos			
2.1	Emáx=................	Emáx=................			
2.2	Ei1=....................	Ei1=....................			
2.3	Ei2=....................	Ei2=....................			
2.4	Emín=................	Emín=................			
2.5	La Energía liberada luego de transcurridos 30 s de terminada la carga es > 85% del valor seteado				
Tiempo de Carga					
2.6	Tiempo de carga para máxima Energía de salida menor que 15 s				
2.7	Condición anterior, pero luego de 5 descargas consecutivas				
Sincronismo					
2.8	El intervalo entre la onda R y la descarga es inferior a 80ms				
Monitor del desfibrilador					
2.9	El tiempo de recuperación luego de una descarga es inferior a 10 s				
2.10	Durante la carga el registro de la señal cardíaca no se ve significativamente afectado				

Figure 4. Example for a checklist of Essential safety and performance evaluation for a defibrillator

4. Conclusions

- The procedures and the checklists proposed in this work are used to assess the compliance of safety conditions and performance of the electrical devices included in terrestrial ambulances.
- The inspection meets the conditions of a periodic verification mechanism, analogous to a Technical Vehicle Inspection as the one already existing for the rest of the vehicles.
- The Inspection is performed by executing observational assessment tests or measurements and performance verifications, satisfying concepts and security definitions and effectiveness regulations.
- Even though the described activities are valid for assessing electromedical devices in an ambulance, they can also be applied to other subsystems that comprise it.
- The structure of work can be used as a basis for the preparation of a procedures manual for a terrestrial Technical Ambulance Inspection in compliance with the existing boundary conditions.

References

[1] Administración Nacional de Medicamentos, Alimentos y Tecnología Médica (ANMAT) 1999 *Requisitos esenciales de seguridad y eficacia de los productos. Disposición ANMAT N° 4306/99* [http://www.anmat.gov.ar/webanmat/Legislacion/ProductosMedicos/Disposicion_ANMAT_4306-1999.pdf] (consultado el 06 de Mayo de 2011).

[2] Instituto Argentino de Normalización (IRAM) 2002.Norma IRAM 4220-1: 2002 *Aparatos electromédicos. Requisitos generales de seguridad*

[3] International Electrotechnical Commission (IEC) 2005 Norma IEC 60601-1: 2005 *Medical Electrical Equipment. General Requirements for basic safety and essential performance*

[4] International Electrotechnical Commission (IEC) 1994 Norma IEC 60513: 1994 *Medical Fundamental aspects of safety standards for medical electrical equipment*

[5] International Electrotechnical Commission (IEC) 2007 Norma IEC 62353: 2007 *Medical electrical equipment - Recurrent test and test after repair of medical electrical equipment*

[6] Outes A, Sanguinetti H y Salinas J 2003 Historia de la atención de la emergencia en la Ciudad de Buenos Aires *107 Emergencia* (Publicación de la Asociación de Profesionales del SAME). **1** 1: pp. 6 – 10

[7] Hospital de Urgencias de la Municipalidad de Córdoba Cátedra de Emergentología 2010 Seminario "Emergencias". Conferencia "Atención de víctimas en masa". Dictado para la Carrera de Ingeniería Biomédica Facultad de Ciencias Exactas, Físicas y Naturales Universidad Nacional de Córdoba Córdoba, Argentina

[8] Bruni R G, Vanella O R, Del Aguila Heidenreich R y Taborda R A M 2011 *Análisis comparativo de normativa aplicable a ambulancias terrestres en América Latina* SABi 2011 XVIII Congreso Argentino de Bioingeniería y VII Jornadas de Ingeniería Clínica Mar del Plata, Argentina

[9] Instituto Argentino de Normalización (IRAM) 2006 Norma IRAM 16030: 2006 *Vehículos de transporte sanitario y su equipamiento. Ambulancias Terrestres. Requisitos.*

[10] Asociación Española de Normalización y Certificación (AENOR) 2007 Norma Europea UNE-EN 1789: 2007 *Vehículos de transporte sanitario y sus equipos. Ambulancia de Carretera*
Associaçao Brasileira de Normas Técnicas (ABNT) 2000 Norma NBR 14561: 2000 *Vehículos para atendimento a emergências e resgate*
Instituto Nacional de Normalización (INN) 1998 Norma NCH 2426 Of.98 *Ambulancias para transporte terrestre de pacientes – Clasificación y requisitos de transformación y equipamiento*
General Services Administration (GSA) 2007 KKK-A-1822F *Federal Specification for the*

Star-of-Life Ambulance

[11] Dirección de calidad de los servicios de salud 1997 Resolución 794/97 *La normativa para móviles de traslado sanitario – Servicios terrestres* Ministerio de Salud Buenos Aires, Argentina

[12] Taborda R y Vanella O 2007 Seguridad y normatización en ambulancias terrestres *Revista argentina de Bioingeniería.* **13** 3

[13] Pedemonte F A 2009 *Programa de Inspección Técnica de Ambulancias Terrestres* Proyecto integrador de la Carrera de Ingeniería Biomédica Facultad de Ciencias Exactas Físicas y Naturales Universidad Nacional de Córdoba Córdoba, Argentina

[14] Vanella O 2011 Seminario de la IEEE "Seguridad y Normalización en Ambulancias". Conferencia "Seguridad de Aparatos Electromédicos de Ambulancias Terrestres" Córdoba, Argentina

[15] Del Aguila Heidenreich R 2011 Seminario de la IEEE "Seguridad y Normalización en Ambulancias" Conferencia "Inspección Técnica de Ambulancias Terrestres" Córdoba, Argentina

[16] Emergency Care Research Institute (ECRI) *Health Devices. Inspection and Preventive Management*

[17] International Organization for Standardization (ISO) 2007 Norma ISO 14971: 2007 *Productos Sanitarios. Aplicación de la gestión de riesgos a los productos sanitarios* Genova, Suiza

[18] Likert R 1932 *A Technique for the Measurement of Attitudes* Archives of Psychology **140** (1932): 1-55

[19] International Electrotechnical Commission (IEC) 2006 Norma IEC 60601-1-8: 2006 *Medical electrical equipment. Part 1-8: General requirements for basic safety and essential performance. Collateral Standard: General requirements, tests and guidance for alarm systems in medical electrical equipment and medical electrical systems*

SABI 2011 IOP Publishing
Journal of Physics: Conference Series **332** (2011) 012005 doi:10.1088/1742-6596/332/1/012005

Low flow measurement for infusion pumps: implementation and uncertainty determination of the normalized method

J Cebeiro, A Musacchio, and E Fernández Sardá

Departamento de Tecnología Médica, Hospital de Pediatría Prof. Dr. Juan P. Garrahan. Combate de los Pozos 1881, Buenos Aires, Argentina.

E-mail: jcebeiro@garrahan.gov.ar

Abstract. Intravenous drug delivery is a standard practice in hospitalized patients. As the blood concentration reached depends directly on infusion rate, it is important to use safe devices that guarantee output accuracy. In pediatric intensive care units, low infusion rates (i.e. lower than 10.0 ml/h) are frequently used. Thus, it would be necessary to use control programs to search for deviations at this flow range. We describe the implementation of a gravimetric method to test infusion pumps in low flow delivery. The procedure recommended by the ISO/IEC 60601-2-24 standard was used being a reasonable option among the methods frequently used in hospitals, such as infusion pumps analyzers and volumetric cylinders. The main uncertainty sources affecting this method are revised and a numeric and graphic uncertainty analysis is presented in order to show its dependence on flow. Additionally, the obtained uncertainties are compared to those presented by an automatic flow analyzer. Finally, the results of a series of tests performed on a syringe infusion pump operating at low rates are shown.

1. Introduction

Continuous drug therapy is generally administered via the intravenous pathway with a delivery system regulating drug concentration in blood in order to achieve and maintain a desired result. In order to achieve the adequate therapeutic dose, the device is set at a specific flow rate. Although underinfusion may not provide sufficient therapy, overinfusion can produce serious toxic side effects. The therapeutic range and risks associated with under- and overinfusion are highly drug and patient dependent [1]. In pediatric patients, low infusion rates, often less than 10 ml/h, are used. This fact significantly reduces the admissible range of flow deviations of the device that should reach the highest required performance standards for neonatal care.

Several reports on the failure of infusion pumps have been published [2], [3] and the FDA (Food and Drug Administration) has initiated recalls of different infusion devices [3]. In order to prevent accidents or failure to accomplish clinical expectancies, periodic trials have been proposed. As it stands for other medical devices, it is necessary to include infusion pumps in a scheduled testing program aimed at risk reduction [4], [5]. Procedures, guidelines, and recommendations written by manufacturers and medical technology agencies exist to investigate if factory specifications are achieved over the whole useful life of the device. The Biomedical Benchmark™© published by ECRI (Emergency Care Research Institute) introduces testing techniques and establishes intervals for assessing flow accuracy [6]. In this study, we have followed the method specified in the IEC/ISO60601-2-24 standard: Particular requirements for the safety of infusion pumps and controllers [7]. It provides a detailed description of the gravimetric procedure, including instrument set up, equations for performance parameters calculation, and graphical representation. An in-depth analysis of some interesting aspects of the implementation, such as weight loss due to evaporation, stability settings, temperature drift, environmental vibration, balance level, time capture, and load symmetry can be found in the study performed by Clarkson [8]. Although this study states that balance resolution is the dominant error (i.e. uncertainty) source in testing infusion devices using the gravimetric method, there are other components, such as balance linearity and repeatability [9], that can increase the final uncertainty of the method. Furthermore, these uncertainties can individually exhibit higher values than the resolution and be different among balances with the same resolution but from different manufacturers. Definition of the weighing uncertainty is the basis for the final method uncertainty calculation necessary for method selection and a mandatory requirement of the ISO 17025 standard for testing and calibration laboratories [10], [11].

Different techniques exist for flow measurement: gravimetric [6], [7], [8], [12], volumetric [6], and automatic [13], [15]. The automatic methods rely on optical-volumetric analyzers that simply require connecting the infusion line to an inlet port and show flow and volume readings on their screens [13], [14]. Some of them can simultaneously test several infusion devices and transmit samples to the PC via an RS-232 interface, thus allowing digital storage and further numeric analysis [15].

The present study describes the implementation of a normalized method for infusion-pump-accuracy determination for low flow delivery (1.0-25.0 ml/h). Uncertainty sources are revised together with a numeric and graphic analysis performed in order to show its dependence on flow rate. Additionally, the obtained uncertainties are compared to those presented by an automatic flow analyzer. Finally, as an example, the results of a series of tests performed on a syringe infusion pump operating at low rates are shown.

In the present study we use the concept of uncertainty as defined by the GUM (Guide to expression of uncertainty in measurement) [16], [17] whereas the word error may either refer to the numerical parameters indicated in section 2.1 (equation 3 and 4) or to a spurious distortion on a measurement process. The uncertainty specification in weighing for balances is referred to as precision.

2. Materials and methods

2.1. Method description

The procedure indicated for syringe infusion pumps in the ISO/IEC 60601-2-24 standard was used [7]. However, the analysis discussed here is also valid for other types, such as rotary peristaltic, linear peristaltic, and cassette infusion pumps. Being a gravimetric method, it consists of the recording of the weight contained by a collector recipient placed on a precision balance which receives the flow delivered by the infusion pump being tested. Sample values are transmitted to a PC at a fixed time interval. The assay lasts two hours and is conducted for a fixed flow set in the infusion pump. While the first-hour samples are rejected, the second-hour samples are used to calculate different parameters of interest, such as mean and errors. The weight difference between consecutive samples is converted to flow (F_i) through the density value and time interval according to equation (1).

$$F_i = \frac{60(w_i - w_{i-1})}{Sd} \tag{1}$$

Where F_i is flow expressed in *ml/h* units, d is the infused liquid density, S is the time interval in minutes, and w_i-w_{i-1} is the weight difference between consecutive samples separated by time interval S.

Figure 1shows the assay scheme for the syringe infusion pump: flow coming from the pump arrives in the collector recipient on the balance through the plastic tube – infusion line – connected to the syringe and the needle. The pump and the liquid must have the same level to prevent pressure differences that hinder or facilitate flow because of hydrostatic factors. The needle size should be 18G (gauge) in order to generate standard pressure drops depending on diameter, flow, and liquid viscosity. For example, an 18G, 6-9 cm catheter generates a mean pressure drop of 14.1–17.9 mmHg h l^{-1} for flows in the 100-300 ml/h range. For low infusion rates, drop formation in the needle may cause a stepwise increment (steps of approximately 0.05 ml). In order to avoid this increase, the tip of the needle should be immersed in the liquid, allowing it to flow in a continuous fashion.

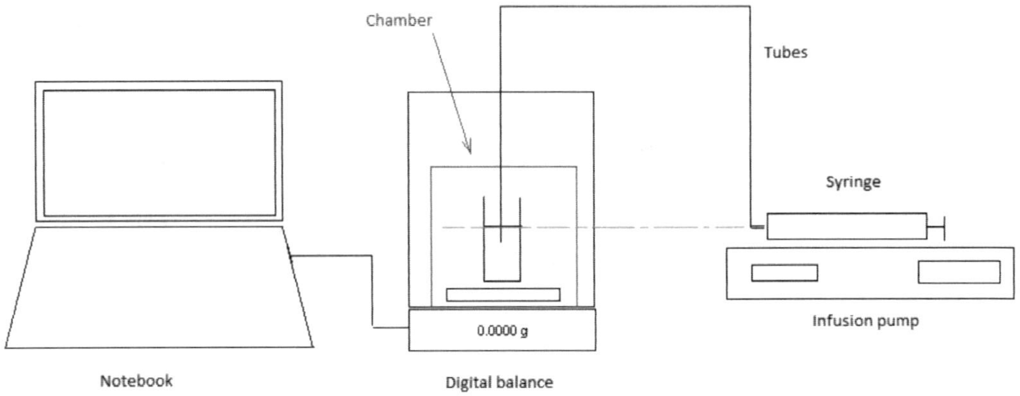

Figure 1: The ISO 60601-2-24 procedure setup.

A Pioneer 214 analytical balance (Ohaus, USA) with a full capacity of 210 g was used. Table 1 shows its specifications according to the manufacturer's certification. Distilled water was used as infusion liquid (Braun, Argentina) with a density of 0.998 g ml^{-1} ($\pm$0.001 g ml^{-1}) at 20° C. Weight samples were collected in a notebook at 30-second intervals [7] via an RS-232 serial port protocol using a graphical user interface developed with Borland Builder® C++. Data transferred to the PC were stored in text files and later imported into the Matlab® numerical computing environment to develop numeric and graphic analyses including error calculation in several observation windows.

Table 1. Sources of uncertainty.

Variable	u_c	Units
Resolution	0.1	mg
Linearity	0.3	mg
Repeatability	0.05	mg
Interval (S)	130	msec
Density (d)	0.001	g ml^{-1}

Interval time uncertainty was not specified in the balance data sheet; therefore, it was assessed by measuring the time between consecutive samples received by the software. The histogram of these time intervals over a 2-hour acquisition showed a maximum deviation of 130 ms. Additionally, this software-assessed uncertainty was instrumentally confirmed by measuring the time interval between consecutive data sets sent by the balance through the DB9 Tx pin using a digital oscilloscope TD2 2012B (Tektronix, USA).

The high sensitivity of precision balances may produce distortions caused by tension and movement of the components shown in figure 1. Hoses made of soft and elastic materials, such as PVC, are often subject to movements and exert forces when under pressure. In order to provide physical isolation we used a cylindrical (diameter 9 cm, height 12 cm, volume ~800 cm^3) acrylic mechanical isolation chamber covering the central volume of the balance case, including the weighing pan (figure 1). Every mechanical effort is supported by the chamber and discharged over the balance floor, thus reducing the weighing deviation. The present study has been developed under ISO/IEC 60601 atmospheric conditions (table 2).

Table 2. Environmental conditions.

Variable	Value	Unit
Atmospheric pressure	1013±10	hPa
Temperature	22.0 ± 2.0	°C
Relative humidity	60 ± 10	%

In high precision weighing it may be necessary to consider the force exerted by air buoyancy, even though in our working conditions and balance precision the effect was less than 0.1 %.

2.2. Evaporation

A recipient containing water on the weighing pan has a weight loss due to water evaporation. The effect depends on the size of the exchange area between water and air and relative humidity of the peripheral atmosphere. The magnitude of this exchange is represented by the evaporation rate; it is expressed in units of µl/min. In order to quantify this magnitude, we conducted preliminary trials with different recipients and conditions. A total number of six combinations were studied (table 3). Trials consisted of the recording of weight samples of a recipient containing distilled water at two samples per second during two hours, the ISO/IEC standard assay time for testing flow accuracy [7]. The duration selected allows an accurate estimate of the water loss during the entire period of the assay. The aim was to investigate how the results may be influenced by recipient shape and relative humidity. Ideally, evaporation rates should be independent of environmental conditions; therefore, we used an engine oil film covering the recipient area in order to avoid water evaporation [8]. An additional trial in a saturated atmosphere was performed using a wet cloth inside the chamber (described in 2.1) in order to evaluate the behaviour in a saturated atmosphere (a relative humidity of 100%).

Table 3. Evaporation assays

Recipient	Volume (cm^3)	Exchange area (cm^2)	Mechanical Isolation Chamber	Humidity at onset
Beaker	100	15.2	No	Ambient
Beaker	100	15.2	Yes	Ambient
Beaker	100	15.2	Yes	100%
Beaker	250	34.2	No	Ambient
Graduated cylinder	25	5.0	No	Ambient
Beaker + Oil film	100	0	Yes	Ambient

2.3. Uncertainty estimation

Section 2.1 describes an indirect method where weight is measured to subsequently calculate flow through equation (1). The GUM of the BIMP (Bureau International des Poids et Mesures) states that combined standard uncertainty for indirect measures is calculated as follows [16]:

$$u_c(F) = \left(\sum_{k=1}^{N}\left(\frac{\partial F}{\partial x_k}\right)^2 u^2(x_k)\right)^{\frac{1}{2}} \quad (2)$$

Where F is the indirectly measured flow, $u_c(F)$ is the combined standard uncertainty of the indirect measurement, x_k represents every single variable in equation (1), $u(x_k)$ is the corresponding standard uncertainty to every single variable, and $\frac{\partial F}{\partial x_k}$ is the partial derivate of equation (1) with respect to each given variable x_k. The value N is the number of variables combined to calculate the flow. The following four direct measure values are involved: density, time, and two weights. The weighing standard uncertainty may be considered twice in equation (2), so a value of $N=4$ was used.
Individual standard uncertainties were calculated according to the data in table 1. Standard weight uncertainty is the result of the quadratic addition of the three uncertainties: resolution, linearity, and repeatability adjusted according to the type of distribution and coverage factors [16]. For time uncertainty calculation, a square distribution function with a half-width of 130 msec was assumed. For density uncertainty calculation a square distribution was also used.

2.4. Trumpet curves

Performance parameters were calculated for several observation windows according to:

$$E_P(\max) = \underset{j=1}{\overset{m}{MAX}}\left[\frac{S}{P}\sum_{i=j}^{j+\frac{P}{S}-1} 100\left(\frac{F_i-r}{r}\right)\right] \quad (3)$$

$$E_P(\min) = \underset{j=1}{\overset{m}{MIN}}\left[\frac{S}{P}\sum_{i=j}^{j+\frac{P}{S}-1} 100\left(\frac{F_i-r}{r}\right)\right] \quad (4)$$

where F_i is the flow calculated according to equation (1), P the observation window size in minutes, r the set flow rate on the infusion pump, S the sampling period, and m the maximum number of observation windows for this window size P [7].

These errors represent the maximum and minimum of the mean flow error for a given observation window, usually between 2 and 31 minutes. These values are helpful for the assessment of pump performance and may be the basis to determine if an infusion pump is indicated for an intended use according to the infused drug half life.

3. Results

3.1. Evaporation assay

Figure 3 shows the weight loss for the different recipients and environmental conditions listed in table 3. Evaporation rates are indicated in table 4 for each combination together with a linearity estimator: the correlation coefficient. Additionally, the percentage represented by the evaporation rate with respect to three infusion rates is shown in order to illustrate how the result is influenced by this effect.

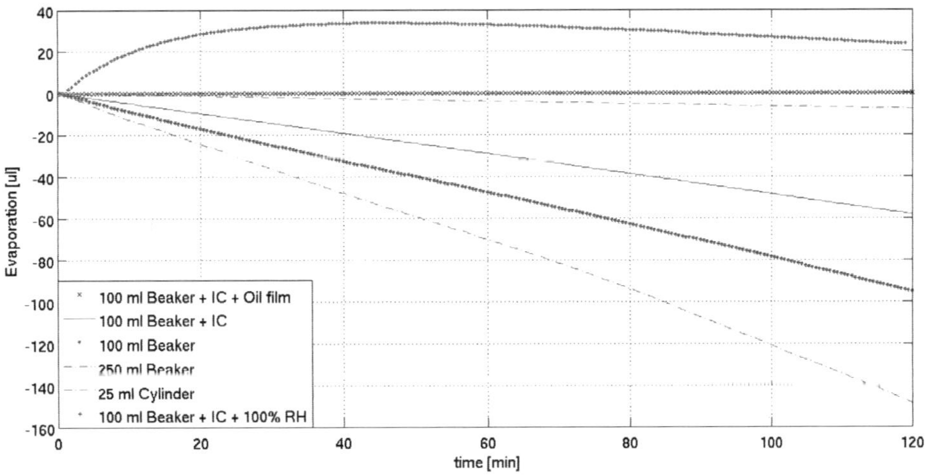

Figure 3: Water loss for the different recipients and environmental conditions. IC: with isolation chamber. RH: relative humidity.

3.2. Uncertainty

Figure 4 plots the combined standard uncertainty (calculated according to equation 2) together with the individual contribution of each component as a function of flow for a range between 0.1 and 25.0ml/h. Figure 5 plots the percentages represented by these magnitudes as a function of the measured flow. Table 5 shows the percentage uncertainty for particular flow rates covering the mentioned range for the studied method and the widely-used IDA4-Plus Flow Analyzer. Additionally, the table shows the time required by the IDA4 to reach the volume necessary in order to achieve the uncertainty specified by the manufacturer at the corresponding flow rate [14].

Table 4. Evaporation loss for different containers. The percentage of error shows the portion that evaporation loss represents for different flow rates.

				Percentage of error (%)		
				Flow rate (ml/h)		
Recipient	Capacity (ml)	Evaporation rate (μl/min)	R^2	0.50	1.0	10.0
Beaker	100	0.77	0.9997	9.3	4.6	0.5
Beaker [a]	100	0.48	1	5.8	2.9	0.3
Beaker	250	1.21	0.9987	14.5	7.2	0.7
Beaker + oil film [a]	100	$<8.3\ 10^{-4}$	1	<0.1	<0.05	<0.005
Graduated cylinder	25	0.06	0.9992	0.7	0.4	0.04

[a] With mechanical isolation chamber, the others without.

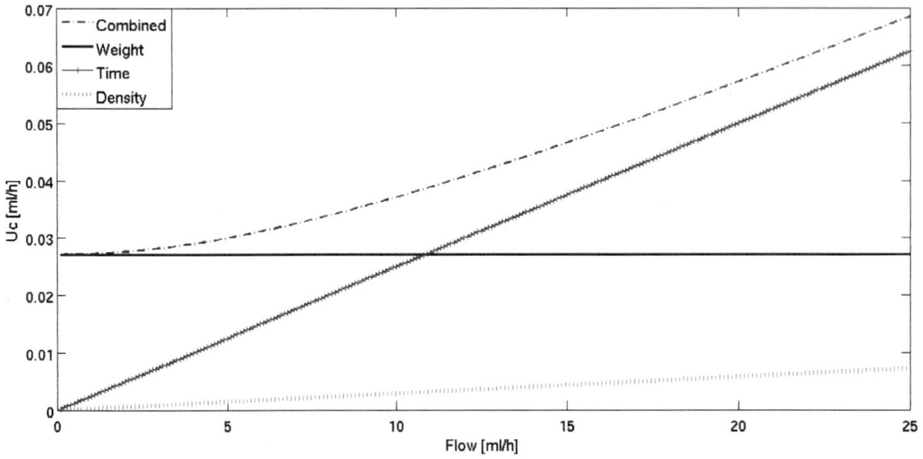

Figure 4: Individual source contributions (time, weight, and density) and combined standard uncertainty.

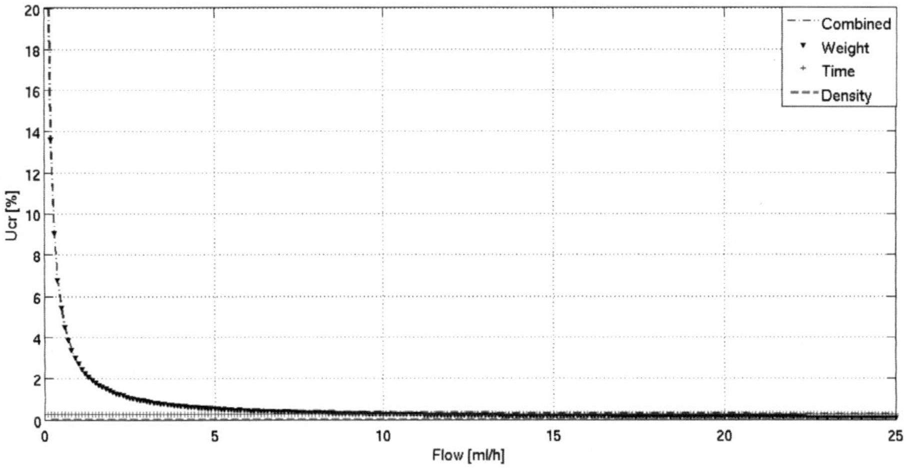

Figure 5: Percentage individual source contributions (time, weight, and density) and percentage combined standard uncertainty.

Table 5. Percentage uncertainties at different flow rates calculated for the studied method (uc) and the IDA4 flow analyzer (uc IDA4). Time values expressed in hours represent the measuring time needed by the IDA4 to reach the specified uncertainties.

	Flow (ml/h)						
	0.5	1.0	1.5	2.0	5.0	10.0	25.0
uc (%)	7	3	2	2	1	1	1
uc IDA4 (%)	4	3	3	2	2	2	1
IDA4 time (h)	20.0	10.0	6.7	5.0	2.0	1.0	0.8

3.3. Performance assays

In order to show the results presented by the method, a series of trials was conducted. An Activa22 syringe infusion pump (Adox, Argentina) was tested with set infusion rates of 2.0, 5.0, 10.0, and 20.0 ml/h. More than 8 hours of data were recorded. Figure 6 plots trumpet curves for Activa22 at different rates. Collected data were used for error calculation for different observation windows according to equations (3) and (4). Trumpet curves are shown for observation windows in the range of 2 to 31 minutes together with the mean flow.

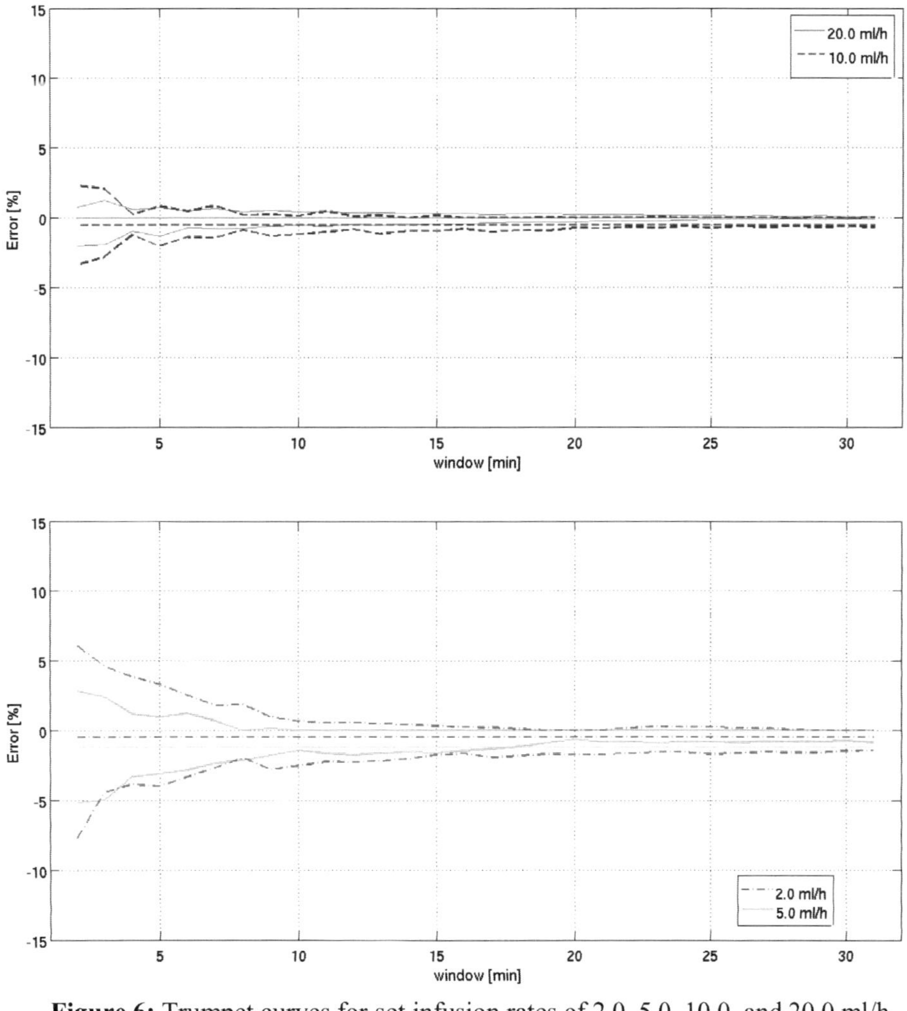

Figure 6: Trumpet curves for set infusion rates of 2.0, 5.0, 10.0, and 20.0 ml/h

4. Discussion

Evaporation test results present an increased water-loss rate in recipients with wider transversal sections (tables 3 and 4, figure 3). Peripheral atmosphere saturation produces erratic behaviour with moisture condensation over the beaker walls increasing weight at the first hour but exhibiting a low decrease towards the second. For the other cases, water evaporation takes place with constant rates strongly related to the area of liquid in contact with ambient air. For some rates, the percentage represented by evaporation is significant, consisting of an important portion of overall flow (up to 7%

for flows of 1.0 ml/h or smaller); therefore, water loss cannot be neglected. Correlation coefficients presented indicate a linear dependence between time and volume, supporting the idea of a constant evaporation rate for the whole period of two hours. The introduction of a mechanical isolation chamber only partially reduces water loss by evaporation, which still remains significant and above some currently sold infusion pump specifications. The addition of a thin (2-3 mm) engine oil film over water surface reduced the evaporation making it undetectable (<0.1 mg in 2 hours), and the evaporation curve remained zero.

Uncertainty of the normalized method proved to be lower than 1% for flows of 5.0 ml/h and higher, decreasing as flow increased. For flows of 1.0 ml/h uncertainty was 3% (table 5). The differences in balance resolution, time interval, and period of assay make it difficult to compare our results with those obtained by Ilfeld et al. [12].

The method performance is comparable to, or even better than some commercially-available infusion pump analyzers [13], [14]. At low flow rates, the normalized method presents the advantage of requiring less trial times as shown in table 5. While a 10-hour test is needed to reach a specification of 3% with a flow of 1.0 ml/h [14] using an IDA4 Plus flow analyzer, two samples are enough for the normalized method to obtain similar or better performance. Moreover, although the ISO/IEC standard recommends a two-hour test, for low flow measurement this time continues to be shorter than some IDA4 times. In any case, times longer than 31 minutes will be required by IDA4 to obtain trumpets curves for the longest window. This analyzer presents other kinds of advantages, such as demanding less work from the department staff, and some models allow testing of four infusion pumps simultaneously.

The sampling time used (30 sec) is that indicated by the standard. It can be increased to 2 minutes in order to reduce relative uncertainty without affecting the trumpet curves, even though some distortion may appear in flow-time curves due to aliasing effects.

The uncertainty analysis performed is also valid for higher flows, although in this case some previous considerations become less important, such as evaporation and balance precision. Some others, such as balance capacity, appear as limiting parameters.

Performance assays show a classic syringe pump result with smaller differences between $E_P(max)$ and $E_P(min)$ towards longer observation windows and higher flows: as flow rate increases error reduces and $E_P(max)$ and $E_P(min)$ tend to each other in the classic trumpet shape (figure 6).

5. Conclusion

A normalized method based on the ISO/IEC 60601-2-24 standard was used. Flow measurement is affected by several factors, such as air currents, movements, and evaporation.

Covering water with an engine oil film allowed to make the assay independent of environmental conditions, reducing evaporation rate until making it undetectable.

Method uncertainty turned out to be equivalent to other alternatives presenting an important additional advantage: considerable shorter testing times for low infusion rate analysis. Another interesting feature of the normalized method is cost. Even when using two balances of different capacities and precision to cover the full range of currently-available infusion pumps (0.5-1500 ml/h) it is still significantly less expensive than a flow analyzer. Calibration services for analytical balances are available everywhere and cheaper than those for infusion pump analyzers.

The normalized method proved to be an interesting alternative for testing accuracy in infusion pumps used for low flow delivery in hospital intensive care units, anesthesiology, and on wards.

For lower flows, such as those delivered by implantable infusion pumps, even more precise balances should be used (0.01 mg). Such balances are more sensitive to environmental-mechanic perturbations than the one used for this study. These flow rates cannot be measured by IDA4 according to specifications [13].

For higher infusion rates it is necessary to use higher-capacity balances, but less precision is required. In these cases, the uncertainty analysis presented in section 2.3 and 3.2 may be applied.

6. Future studies

The next step is investigating the method repeatability by performing series of trials and comparing trumpet curves obtained in several measurements under the same conditions. In order to contrast the performance of the method against flow analyzers, an experimental study including statistical tests and comparison among trumpet curves should be performed.

Acknowledgments

We would like to thank engineers Cristian Khourian and Marcelo Silvosa, from Tecnologías para la Salud y la Discapacidad and Energía departments at Instituto Nacional de Tecnología Industrial (INTI) for reading the draft of this work and providing useful suggestions.

References

[1] Bronzino J D 2006 *Biomedical Engineering Handbook* (Boca Raton: CRC Taylor & Francis Group) chapter 68.
[2] Geddes L A 1998 *Medical Device Accidents* (Boca Raton: CRC Press)
[3] Federal Drug Administration 2009 *Baxter Colleague Single and Triple Channel Volumetric Infusion Pumps, Class I Recall*
[4] Shepherd M, Frank R P, Dyro J F and Baretich M F 2004 Identification of Human Errors During Device-Related Accident Investigation *IEEE Engineering in Medicine and Biology* **23** 3 66-72
[5] Dyro J F 2004 *Clinical Engineering Handbook* (California: Elsevier Academic Press).
[6] ECRI Institute 2008 *Inspection and Preventive Maintenance for General Purpose Infusion Pumps* Biomedical Benchmark™©
[7] ISO/IEC 60601-2-24 1998 *Medical electrical equipment: Particular requirements for the safety of infusionpumps and controller*
[8] Clarkson M D 2002 Accuracy estimations of testing of infusion devices using weighing balances *Medical Engineering & Physics* **23** 3 229-235
[9] Nater R, Reichmuth A et al. 2009 *Dictionary of Weighing Terms - A Guide to the Terminology of Weighing* (Springer)
[10] ISO/IEC 17025 2005 *General requirements for the competence of testing and calibration laboratories*
[11] Rubio D, Ponce S, Álvarez A, and Fascioli E 2008 *Proc. IV Lat. Am. Cong. On Biom. Eng. (Margarita Island, Venezuela)* vol. 18 (Berlin: Springer Heidelberg)
[12] Ilfeld B, Morey T and Kayser Enneking F 2002 The delivery rate accuracy of portable infusion pumps used for continuous regional analgesia *Anesthesia and analgesia* **95** 1331-1336
[13] Fluke Corporation 2005 *Operators Manual IDA4 Plus Infusion Pump Analyzer*
[14] Bio-Tek Instruments INC 1997 *IDA2 Plus: Operators Manual*
[15] Fluke Corporation 2005 *Hydrograph Software Operators Guide, Rev.01*
[16] Bureau International des Poids et Mesures 2010 *Evaluation of measurement data- Guide to expression of uncertainty in measurement (GUM)*
[17] Bureau International des Poids et Mesures 2008 *International vocabulary of metrology — Basic and general concepts and associated terms(VIM)*

RTOS kernel in portable electrocardiograph

C A Centeno, J A Voos, G G Riva, C Zerbini, E A Gonzalez

Grupo de Ingeniería Clínica, Facultad Regional Córdoba, Universidad Tecnológica Nacional, Maestro López esquina Cruz Roja, Ciudad Universitaria, X5016ZAA Córdoba, Argentina.

ccenteno@gmail.com

Abstract. This paper presents the use of a Real Time Operating System (RTOS) on a portable electrocardiograph based on a microcontroller platform. All medical device digital functions are performed by the microcontroller. The electrocardiograph CPU is based on the 18F4550 microcontroller, in which an uCOS-II RTOS can be embedded. The decision associated with the kernel use is based on its benefits, the license for educational use and its intrinsic time control and peripherals management. The feasibility of its use on the electrocardiograph is evaluated based on the minimum memory requirements due to the kernel structure. The kernel's own tools were used for time estimation and evaluation of resources used by each process. After this feasibility analysis, the migration from cyclic code to a structure based on separate processes or tasks able to synchronize events is used; resulting in an electrocardiograph running on one Central Processing Unit (CPU) based on RTOS.

1. Introduction

Due to the increasing integration of peripheral devices within a chip or integrated circuit, it is possible to implement more functionality in electronic designs based on these devices. As a result of this technological advancement, an extensive line of microcontrollers is available, which provides special features to platforms based on them.

For the our previously designed and implemented portable electrocardiograph [1] [2] the microcontroller PIC16F877A from the Microchip family was used.

At the current stage of development, the electrocardiograph CPU uses a microcontroller (uC) PIC18F4550. This PIC model includes more memory, which is a prerequisite for the possible implementation of RTOS kernel. An important feature, necessary for such a change, is the possibility of using the available RAM as stack space.

The availability of these capabilities allows to consider the use of an RTOS to manage all functions associated with control of the electrocardiograph stages, whether analog or digital. The migration from cyclic executive to multitasking mode requires a previous analysis of each process and its interaction with the other ones.

2. Materials and Methods

2.1. Analysis of a Cyclic Executive

The initial system is based on a cyclic executive, in which control processes execute one after another in an infinite loop. In the following, we make a description of each hardware stage and how they interact with the uC. Finally, a summary of this information is presented in Table 1.

2.1.1. Matrix Keyboard.
The keyboard is made up of eight keys in a linear arrangement connected to 8-to-3 BCD encoder lines. The encoder is connected to the uC through three programmable input/output lines. Debouncing is implemented by software and special keys are not defined. The keyboard function is to allow the user to select various modes and functions available to the electronic device, such as lead selection, gain, notch filter, etc.

2.1.2. Input Multiplexer.
In an electrocardiogram it is necessary to obtain different leads implemented by using a multiplexer that allows the combination of patient's biomedical signals which must then be amplified. This block is controlled by four input / output port lines of the uC.

2.1.3. Conditioning module.
In an electrocardiograph a conditioning module is necessary for amplification of biopotential signals. After this process we obtain an appropriate signal level to be digitized. Because biopotential signals are different from one person to another respecting frequency and spectral composition, this electronic stage must provide comparable signals.

2.1.4. Filtering Stage.
The spectral composition of the conditioned signals allows us to determine certain diseases. If this signal is affected by noise, however, a correct determination is difficult to be reached. Among the spectral components that affect the obtained signal are the common mode signals that are induced on the device and the connections between it and the patient. In particular, the frequency at the common mode signal is 50Hz. For this reason, a Notch filter is introduced which is intended to mitigate the adverse effects of this frequency component. This stage is controlled through a 2-to-4 decoder connected to two input/output port lines of the uC.

2.1.5. Programmable gain amplifier.
Due to the mentioned variability of the signal amplitude after signal conditioning, it is necessary to use a variable-gain amplifier. For that purpose, a programmable gain amplifier can select between x1, x2, x4, x5 and x10 gain steps. For communication with this device, a synchronous serial communication via Serial Peripheral Interface (SPI) is used, which allows sending configuration commands from the uC.

2.1.6. Programmable offset voltage.
One of the design conditions poses the need to work with amplified signals with offset voltage to obtain the maximum excursion of the input voltage applied to the Analog/Digital Converter (ADC). This programmable voltage is obtained from a dedicated chip which is controlled by the SPI interface from the uC.

2.1.7. Analogic to Digital Converter ADC.
The conversion of analog signals to digital ones is made by a 12-bit ADC, which delivers the digital value by using the SPI interface. The maximum conversion rate is up to 100kbps.

2.1.8. Decoder selector

The use of multiple stages in the design of the electrocardiograph creates the need for more input/output control lines than available at our uC. In consequence, a 3-to-8 decoder for extension to five additional lines is included.

2.1.9. Digital stage.
The control of all above detailed stages is performed by a 8-bits PIC18F4550 uC, which includes 2048 bytes of RAM, 32768 bytes of ROM, 3 timers, SPI interface, USB interface and USART, among other parameters.

2.1.10. Liquid crystal display.
The digitized information is presented on a graphic 240X64 pixels LCD. On it, the biopotencial signal and data status of the electrocardiograph are displayed, such as gain, applied filters, reference levels, etc. The LCD uses 10 input/output lines for communication with the uC.

2.1.11. Communication interface.
Due to the future addition of digital post-processing, storage of studies in a database and increasing of the resolution in the graphical presentation, a communication interface must be provided to transmit the acquired information to a personal computer. For this purpose, we take advantage of the available USART interface. Through the use of an optocoupler, galvanic isolation required for this type of device is obtained at a RS232 port. An additional device can be included in order to interface to USB sockets.

Table 1. Connection between the modules and microcontroller

Module	**Connection**
Matrix keyboard	3 lines I/O
Input multiplexer	4 lines I/O
Filtering	1 lines I/O
Programable gain amplifier	SPI by software
Programable offset voltage	SPI by software
Analogic to digital converter	SPI by software
Decoder selector	3 lines I/O
Liquid cristal display	10 lines I/O
Communication interface	Interfaz USART

2.2. Module interactions
Each module has an associated specific function, which must be executed in a timely, synchronized manner in order to achieve the desired result. Table 2 presents the cyclic executive that controls each of the modules of the electrocardiograph.

Each of the steps/tasks must be properly synchronized in time. This way, a whole cycle can be completed within the set time. The conversion frequency is 500Hz, so this time is 2mSeg.

Additionally, it should be noted that the full cycle of on-screen display is 10mSec. Table 3 lists the minimum and maximum time needed to complete each phase/task.

Table 2. Cyclic Executive – Pseudocode – Flowchart

Descripcion	Flowchart
1- Scan Keyboard a. Change adquisition parameters i. Filtering ii. Gain iii. Offset iv. Derivation	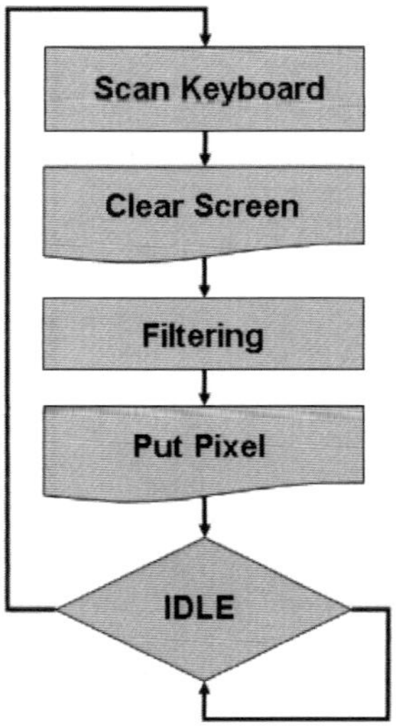
2- Clear Screen	
3- Processing – Digital Filtering	
4- Put pixel – Line Draw	
5- Wait for a new cycle (including 5 conversions). End of the IDLE state by using a flag.	
6- Repeat step 1-	

Table 3. Time required for each process.

Process	Minimum Time	Maximum Time
Leer_teclado	0.2 mSeg	0.2 mSeg
Borrar_pantalla	8.6 mSeg	10.47 mSeg
Dibujar línea	0.55 mSeg	3.55 mSeg
Conversión_ADC	0.320 mSeg	0.320 mSeg
Envío de datos vía USART	0.100 mSeg	0.100 mSeg
Total cycle time	9.77 mSeg	14.64mSeg

The study of the run time shows that the process *Borrar_pantalla* requires more than 10 mSec, so it is necessary to divide it in two stages, that is run completely every two cycles of screen printing. Additionally, due to the available screen resolution, it is only possible to print a dot for every five acquired samples by the ADC.

Other processes, and their additional times, must be completed within the established 10mSec as well.

If sending the converted information to a PC is considered, a transmission rate must be accordingly defined. On the other hand, it should be mentioned that 12 bits AD convertion takes at least 2bytes for each acquired value. Using asynchronous transfer without flow control, we introduce a marker byte in each packet transmission to ensure proper reception on the PC. In short, for every transmitted data 3 bytes must be sent, so the baud rate must be greater than or equal to 19200bps.

The necessary synchronization between each process in order to avoid data loss is shown in Figure 1. It must be kept in mind that start control of each repetition is performed using one of the available timers.

Figure 1. Temporal diagram

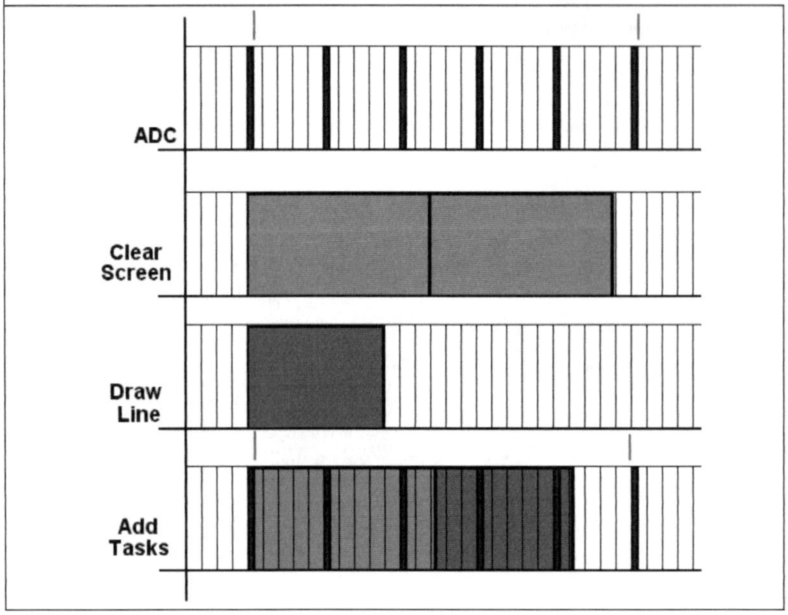

2.3. Using a Real Time Operating System (RTOS)

The uCOS-II is a Real Time Operating System [4] (RTOS kernel) that allows preemptive multitasking processing (i.e., the execution of a process is interrupted when a higher priority process is ready to run). It also offers elements called "events", such as mailbox, semaphores, queues, and others for synchronization and/or data transfer between processes. In addition, it has functions for time control and it may suspend or restart processes, as well as remove any of them.

We will highlight as part of the following analysis the hardware requirements for the RTOS port to PIC18 to be successfully embedded in the development platform [5] [6].

The RTOS, through the priority assigned to each process, determines which process runs first. This priority is set between zero and a maximum configurable value. That capability would allow the process "conversión_ADC" to have the highest priority to be executed as frequently as required, during time remaining from the other defined processes.

One of the advantages of using an RTOS is to exploit time previously wasted to consume clock cycles (e.g., to introduce delays), for the execution of a process / task.

2.4. System requirements

An RTOS typically requires RAM and ROM memories to implement its functionality, so it is very important to know the minimum required resources.

In the following, we analyze software modules that compose the RTOS and our specific hardware.

For each task that integrates the system, the RTOS requires a Task Control Block (TCB); and each element needs an Event Control Block (ECB) for synchronization. The total amount of TCBs and ECBs depends on how the final system is structured.

One of the features of the adopted RTOS is the ability to modify its configuration, and therefore its intrinsic capabilities, namely to enable and / or disable functions to obtain the smallest size of the resulting executable. A proper configuration allows us to save memory, both RAM and ROM. On the other hand, it is also possible to define how to manage the available stack for different tasks. This can be either Minimum or Extended mode. In the last one, the user is able to know the stack required by the tested process.

In Table 4, we present details of the amount of RAM and ROM associated with the different possible configurations of our system.

Table 4. Memory needed to use RTOS

Process	*Mode*	*ROM*	*RAM*	bytes
uCOS II – PIC18	Minimum mode	4712		Bytes
	Extended mode	5363		Bytes
TCB	Minimum mode		14	Bytes
	Extended mode		26	Bytes
ECB	Both modes		8	Bytes
Statistics Task	Minimum mode	322	124	Bytes
	Extended mode	973	220	Bytes
Use of SEMAPHORES	Both modes	1870	24	Bytes
Use of MAILBOX	Both modes	2108	30	Bytes
MAILBOX + SEMAPHORES	Both modes	3139	30	Bytes

Each task needs, in order to interact with the other ones, a priority and an associated stack. The definition of the amount of stack is very important, and depends entirely on factors such as the CPU overhead due to the kernel (i.e., how many times per second the scheduler is called), the number of tasks that make up the entire system, and their duration times, among others.

Each time the scheduler puts a task in *Waiting* state, it saves its context into its associated stack, so that when that task has to return the *Run* state, it does it from the point where it left. An Incorrect stack allocation of each process can cause the system to become unstable. Figure 2 shows the possible states that each process can adopt [5].

Priority and stack size are saved in the associated TCB in the Minimum mode; while in the Extended mode, the points where the stack starts and ends are also saved. Thus it is possible, by using functions defined for this purpose, to know at any instant the percentage of stack used. This feature defines the optimal stack value for each task, with the premise of full use of RAM resources. The TCB of each task also changes its configuration depending on the mode used, either Minimal or Extended.

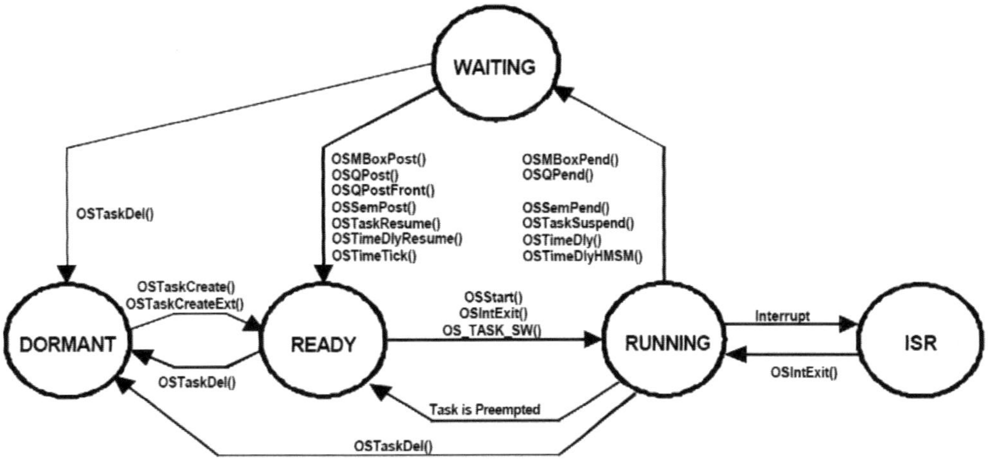

Figure 2. States defined in the system

It is also necessary to know when the CPU is overloaded. Overloading can cause the CPU not being able to complete a process before it must be executed again. This may be because more tasks are required than can be processed, or because their times are shorter than the minimum CPU time. In order to minimize the overhead, we need to know what the percentage use of the CPU is ; if this value reaches 100%, the system may become unstable. In order to know this value, the system has a statistical task, which also consumes valuable resources. This task should only be compiled at the time of evaluation, and not in the final product.

After weighting each of the capabilities of the system, we conclude that 7600 bytes of ROM and 925 bytes of RAM are necessary for use of the RTOS uCOS-II on a PIC18. This amount of resources is about 50% of available RAM and ROM on the selected microcontroller.

3. Results

3.1. Definition of Processes/Tasks

After analyzing each of the modules of the electrocardiograph, tasks (also called processes) that should compose the general system were determined. From this, four tasks were defined, which are described below.

3.1.1. Tarea_Conversion process

This process makes use of an ADC to convert the analog signal into a digital value, and an SPI interface to link the uC with the ADC. It should be noted that the SPI is implemented by software, due to pinout constraints. This task runs with a period of 2 mSeg. The conversion result is sent to Tarea_LCD and Tarea_COM tasks by means of events.

3.1.2. Tarea_Leer_Teclado process

This process runs with a period of 200 mSeg, reading the value from the 10-to-1 encoder of the key pressed. This process must be run after the ADC completes a conversion, because a result of keystrokes, changes settings of the programmable gain amplifier (PGA) and programmable voltage reference. Both modules make use of the SPI interface.

3.1.3. Tarea_LCD process

This process is executed after the *Tarea_Conversion* process, and takes as input the value sent by that process through a *mailbox* event.

3.1.4. Tarea_COM process

This process is executed after the *Tarea_LCD* process, with the value received through a *mailbox* from the *Tarea_Conversion* process. This value is transmitted to a PC through the available USART hardware interface.

3.2. Process interaction

In Figure 3 the interactions of each of the described processes are presented.

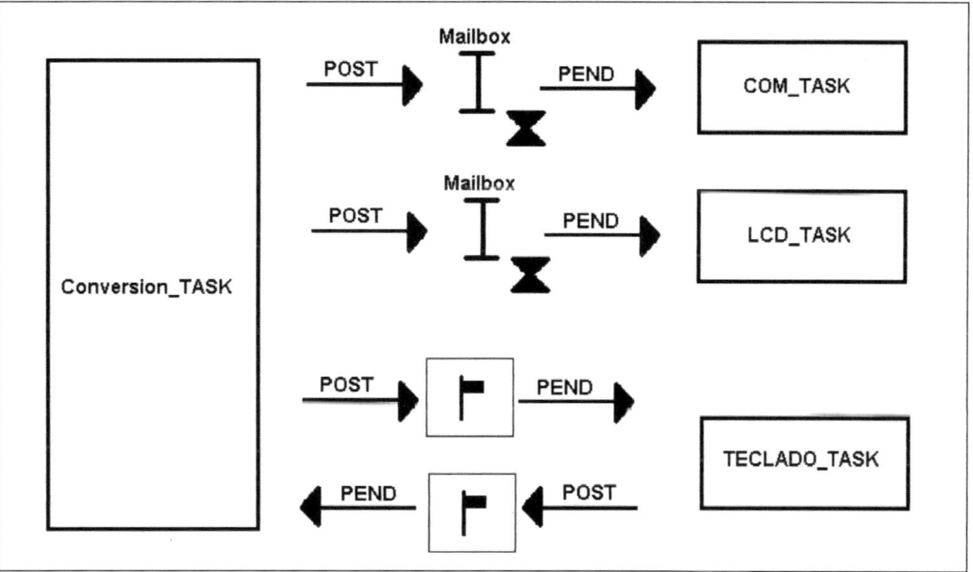

Figure 3. Interaction between all processes

We can summarize that at least 40 bytes are needed by the scheduler for the control for this configuration using 5 processes. At the same time, the use of events, i.e., mailbox, flags, etc, requires reserving RAM memory for each of them.

The system makes use of events for synchronization of processes or tasks for optimal use of available physical hardware resources. In Table 4, the *Conversión_Task* and *Task_LCD* processes are presented as an example.

The most critical module for work in multitasking mode is the SPI serial communication interface. Due to the impossibility of using specific included SPI hardware, which was used for other purposes, this interface has to be managed by software. The sharing of the same channel by two diferent and independent processes, to syncronize these tasks requiered the use of semaphores. The Teclado_Task process and the Task_Conversion process can access to the same resource, so was used two independent events for the synchronization between them.

Table 4. Process Task_LCD and process Conversión_TASK

```
void Task_LCD(void *pdata)
{
#if OS_CRITICAL_METHOD == 3
   OS_CPU_SR  cpu_sr;
#endif

char  *rxmsg;
INT8U  err;
INT16U valor_LCD;
x=0;inicio = 0;fin = 25;

for(;;)
{
rxmsg = (char *)OSMboxPend(COM_Mbox, 0, &err);
// Espera mensaje de Conversion_TASK

if(err == OS_ERR_PEND_ISR)
            valor_LCD = 0;
else if(err == OS_ERR_PEVENT_NULL)
            valor_LCD = 0;
else if(err == OS_ERR_EVENT_TYPE)
            valor_LCD = 0;
else if(err == OS_ERR_PEVENT_NULL)
            valor_LCD = 0;
else
            valor_LCD = *rxmsg;

cuenta++;
if(cuenta>=7)
{
            bloquex++;
            cuenta=0;
            if(bloquex>=29)
            {
                        bloquex = 0;
                        x=0;
            }
}
```

```
void Task_Conversion(void *pdata)
{
#if OS_CRITICAL_METHOD == 3
    OS_CPU_SR  cpu_sr;
#endif
INT8U cant_datos;
INT16U dato_ADC, maximo_ADC;
for(;;)
{
cant_datos++;

if(Bits1.bit1_velocidad_25_50mm){
            // si es 1 barro 25 puntos por segundo
   if(cant_datos>=10) {
            Bits1.bit1_interrupcion=1;
            // salgo del loop infinito e imprimo punto
            cant_datos=0;
   }
} else
{
   if(cant_datos>=5)
   {
      Bits1.bit1_interrupcion=1;
      cant_datos=0;
   }
}
dato_ADC = Leer_Adc();

if(dato_ADC >= 4096)
   dato_ADC = 4096;
if(dato_ADC >= maximo_ADC)
   maximo_ADC = dato_ADC;

if(Bits1.bit1_interrupcion){
   Bits1.bit1_interrupcion = 0;
   if( OSSemAccept( ADC_Sem ) )
   {
            OSSemPost( TECLADO_Sem );
   }
   OSMboxPost(LCD_Mbox, (void *)&maximo_ADC);
   OSMboxPost(COM_Mbox, (void *)&maximo_ADC);
   maximo_ADC = 0;
}
```

A summary of the resources after the migration complete cyclic executive to RTOS platform is presented in Table 5. It shows a comparison of the resources cost both in the executive cyclic version and the multitasking version.

Table 5. Summary of resources

Resource	*Cyclic executive*	*%*	*RTOS*	*%*
RAM	614 bytes	29,98%	1101 bytes	53,76%
ROM	4268 bytes	26,05%	11544 bytes	70,46%
System composition	* Process interrupt-driven conversion *Polling of other processes		* 5 TASK * 4 EVENT * Mail Box Enabled * Minimum MODE	

4. Conclusions

The use of an RTOS for CPU tasks on an electrocardiograph requires changes in the definition of operations to be performed by modules in the system. On the other hand, the inclusion of multitasking allows for much better exploitation of CPU capacity to process tasks in the system. Factors to be considered while deciding about using an RTOS on the developed platform are: requirement of minimum amount of RAM and ROM, eventual migration to another microcontroller, and possibility of implementation on a license-free RTOS platform or adding license costs to the product.

For stability analysis of the software version based on RTOS versus the cyclic executive version, two electronic equipments would be necessary. In this work, the analysis of a possible migration was performed on an existing platform; therefore it was necessary to adapt existing software modules in order to minimize the use of RAM and ROM. As part of these operations, we suppress variables which were previously used and are not further necessary in the new platform in order to reduce memory consumption.

Even if the electrocardiograph is able to run all the previously mentioned tasks with an RTOS, in the present it is not posible to add more functionalities. This is due to the lack of remaining ROM after the implementation of the RTOS system on the PIC18F4550. However, it may be considered using the PIC18F4675 uC in order to increase functionality without modifying the physical hardware.

The final application interacts properly with the associated hardware, and allows for further enhancements to the system. We therefore conclude that it is suitable and convenient to use an RTOS on the developed electrocardiograph.

References

[1] Monitor de pulsos cardíacos con interfaz a PC. SABI 2003. Córdoba. Argentina. 2003

[2] Procesamiento digital sobre monitor cardíaco. CLAIB 2007. IV Congreso Latinoamericano de Ingenieria Biomédica. Caracas. Venezuela. 2007

[3] Sistema para el análisis de la morfología de la señal de ECG de ratones de laboratorio infectados con mal de chagas. CBEB 2010. Crongreso Brasileiro de Ingenieria Biomédica.

[4] Medical Data Transmission System for Remore Healthcare Centres, Gonzalez E, Cagnolo F, Olmos C, Centeno A, Riva G and Zerbini, J. Phys.: Conf. Ser., Vol. 90, No. 1. (2007), 012029. doi:10.1088/1742-6596/90/1/012029 Key: citeulike:2069632

[5] uCOSII, The Real-Time Kernel , Jean J. Labrosse, R&D Technical Books, 1998,ISBN 0-87930-543-6.

[6] MPLAB-uCOS-II-PIC18 Version 2.51 www.micrium.com

FPGA implementation of a ZigBee wireless network control interface to transmit biomedical signals

M A Gómez López[1,2,4], C B Goy[1,2,3], P C Bolognini[2] and M C Herrera[2,3]

[1] Departamento de Electricidad, Electrónica y Computación, Facultad de Ciencias Exactas y Tecnología, Universidad Nacional de Tucumán, Av. Independencia 1900, Tucumán, Argentina
[2] Laboratorio de Investigaciones Cardiovasculares Multidisciplinario, Facultad de Ciencias Exactas y Tecnología, Universidad Nacional de Tucumán, Av. Independencia 1900, Tucumán, Argentina
[3] Instituto Superior de Investigaciones Biológicas (INSIBIO), Consejo Nacional de Investigaciones Científicas y Técnicas (CONICET), Chacabuco 461, Tucumán, Argentina

E-mail: mgomezlopez@herrera.unt.edu.ar

Abstract. In recent years, cardiac hemodynamic monitors have incorporated new technologies based on wireless sensor networks which can implement different types of communication protocols. More precisely, a digital conductance catheter system recently developed adds a wireless ZigBee module (IEEE 802.15.4 standards) to transmit cardiac signals (ECG, intraventricular pressure and volume) which would allow the physicians to evaluate the patient´s cardiac status in a noninvasively way. The aim of this paper is to describe a control interface, implemented in a FPGA device, to manage a ZigBee wireless network. ZigBee technology is used due to its excellent performance including simplicity, low-power consumption, short-range transmission and low cost. FPGA internal memory stores 8-bit signals with which the control interface prepares the information packets. These data were send to the ZigBee END DEVICE module that receives and transmits wirelessly to the external COORDINATOR module. Using an USB port, the COORDINATOR sends the signals to a personal computer for displaying. Each functional block of control interface was assessed by means of temporal diagrams. Three biological signals, organized in packets and converted to RS232 serial protocol, were sucessfully transmitted and displayed in a PC screen. For this purpose, a custom-made graphical software was designed using LabView.

1. Introduction

In recent years, the increasing development of wireless communication protocols has allowed the inclusion of these technologies to all types of medical devices [1]-[4]. The information, obtained from various sensors or systems located on/inside patients, is transmitted through communication networks to remote stations where a medical specialist or an intelligent system takes or process the data. From these data, the physician-patient interaction will be established, depending on the application, either

[4] To whom any correspondence should be addressed.

immediately or in the near future.

Cardiac hemodynamic monitors are some examples of these technologies. Some cardiac variables - such as the right ventricular pressure, right ventricle pre-ejection time and systolic time interval- are sensed in/on the body of patients -heart, arteries, veins- and transmitted by wireless systems to a monitoring station [5]-[9].

Recently, our research group has developed a cardiac hemodynamic monitor based on conductance catheter system (CCS). More precisely, a digital conductance catheter system (DCCS), completely implemented in a FPGA (Field Programmable Gate Array), has been designed and tested experimentally [10]. It should be a platform to optimize the conductance catheter system in order to develop an implantable device [11]. This platform adds a wireless communication module with which the researcher can access the patient's hemodynamic status noninvasively.

In general, the technical features required of the wireless communication protocols used in medical equipment are the short range and low power consumption. ZigBee is a protocol that ensures these specifications under the IEEE 802.15.4 Standard [12].

With this background, this paper describes the FPGA-embedded logical design of a control interface wireless network that allows transmission of cardiac hemodynamic signals from the FPGA internal memory to a personal computer for visual presentation.

2. Design

Briefly, newly conductance-catheter equipment DCCS is a development system based on left intraventricular volumen (LIVV) determination by recording multiple intracavitary conductances, left intraventricular pressure (LIVP) using a miniature ceramic transducer and a simultaneous electrocardiographic signal (ECG) [10]. All of them are digitized with 8 bits and sampled at 432Hz.

Fully implemented in a FPGA, DCCS stores these signals in a 1005x8 RAM memory and adequates the data to the ZigBee protocol communication.

FPGA is an embedded programmable logic device FLEX10k70RC240 with SRAM (Static Read Access Memory) technology [Altera Corp, 2003. FLEX 10K. Embedded Programmable Logic Device Family] which is part of an educational board (UP2 Education Board, Altera). Inside FPGA, the logic equations are implemented using MAX+PLUS II development system. Schematically, the design is realized by means of Altera´s hardware language description (AHDL).

In turn, FPGA is connected to the wireless transmission kit (Kit XBee ZB, ZigBee-PRO) which has two modules, the first called END DEVICE and the second called COORDINATOR. The first module can operate in low power consumption mode and the second receives and sends the data, using an USB2UART port, to a PC where, finally, data are displayed on a screen. These modules are reconfigurable modems that implement standard ZigBee communication protocols. Figure 1 shows a whole schematic diagram of the system.

Furthermore, the ZigBee kit includes two USB2UART boards which, in turn, could have a serial port from an USB port. Moreover, both modules are operated in API mode (Application Programming Interface) that allows a better management of network properties.

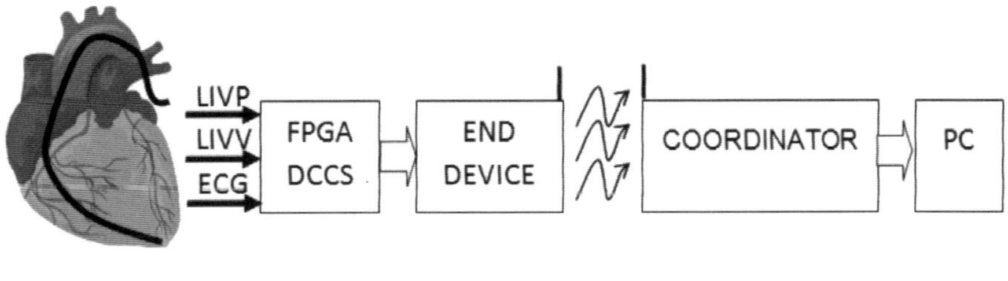

Figure 1. Block diagram of the proposed system. See text for explanations.

3. API message structure

Figure 2 presents a complete message configuration necessary to transmit by the wireless system.

	Frame Fields		Offset (Byte)	Example	Description
API Packet	Start Delimiter		0	0x7E	
	Length		MSB 1	0x00	Number of bytes between the length and the checksum
			LSB 2	0x30	
	Frame-specific Data	Frame Type	3	0x10	
		Frame ID	4	0x01	
		64-bit Destination Address	MSB 5	0x00	Set to the 64-bit address of the destination device.
			6	0x13	
			7	0xA2	
			8	0x00	
			9	0x40	
			10	0x3A	
			11	0x4A	
			LSB 12	0x39	
		16-bit Destination Network Address	MSB 13	0xFF	Set to the 16-bit address of the destination device, if known. Set to 0xFFFE if the address is unknown, or if sending a broadcast.
			LSB 14	0xFE	
		Broadcast Radius	15	0x00	
		Options	16	0x00	
		RF Data	17	...	Data that is sent to the destination device
			...	...	
			48	...	
	Checksum		49	...	0xFF - the 8 bit sum of bytes from offset 3 to this byte.

Figure 2. Message configuration including API-specific structure. Detail Data Structure (in API mode). Reproduced from IEEE 802.15.4 Standard [12].

Each API packet consists of a header with 17 memory address or bytes -from position 0 to 16- (see figure 2) and a set of data –RF Data– (LIVP, LIVV and ECG signals) that are located from position 17 to 48 (33 bytes) and, finally, one byte for the Checksum (see details later in the text) .

4. System description

Figure 3 shows the logical diagram incorporated in the FPGA. Digital data -corresponding to biological signals- are stored in ROM memory (1 byte per data); in turn, this memory delivers the data to a read/write RAM memory where the API packet is built. The control needed to build each signal packet is implemented by the MESSAGE CONTROL block that creates both, CLK1 and CLK2, from the FPGA main CLOCK.

Also, COUNTER MOD 1001 provides the address lines to ROM memory while COUNTER MOD 56 provides 56 address lines to RAM memory, both activated by CLK1. These blocks allow to create the information packets with a specific format. Furthermore, ADDER block generates a control signal that defines the header and the position of the data and sets an interval between packets to avoid overlap. Finally, TRANSMITTER converts the data to RS232 serial protocol required to access to the END DEVICE module (see figure 1).

When the transmission is requested (START in HIGH), the bytes related to each API message are extracted from the RAM, one by one. Thus, these bytes are converted to RS232 asynchronous serial protocol and are buffered at the END DEVICE module until they are transmitted wirelessly to the destination address specified in the API message header. Besides, when START is in LOW; the END DEVICE module is “sleeping” (LOW state), so it has very low power consumption.

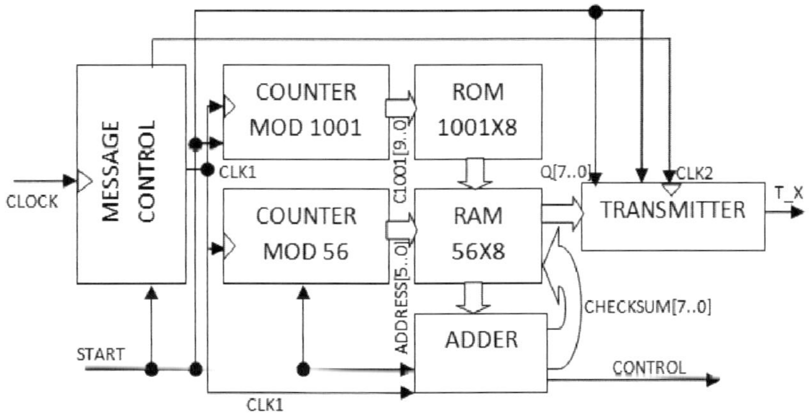

Figure 3. FPGA Logic Design.

4.1. MESSAGE CONTROL

MESSAGE CONTROL is a logical block that sets a counter (28847 counts). It has two input signals; one is the 25.175MHz CLOCK (from the UP1 Education Board) and the other is the START that enables the transmission. The MESSAGE CONTROL outputs are two clocks of 872Hz (CLK1) and 9597Hz (CLK2), respectively, both implemented with lpm_Counter function (MAX+PLUS II, Altera). CLK1 sets the loading/downloading frequency of ROM and RAM memories, whereas that CLK2 sets transmission speed of UART port.

Figure 4 shows the MESSAGE CONTROL temporal response including CLK1 and CLK2 signals. If START is in HIGH, the counters do not count; if it is in LOW, they count normally. START has a major role as it is the signal that initiates the transmission and "freezes" the whole system when it is enabled. Note that every eleven pulses of CLK2, there is only one pulse of CLK1. During these eleven pulses, one data (one byte) is converted according to RS232 protocol.

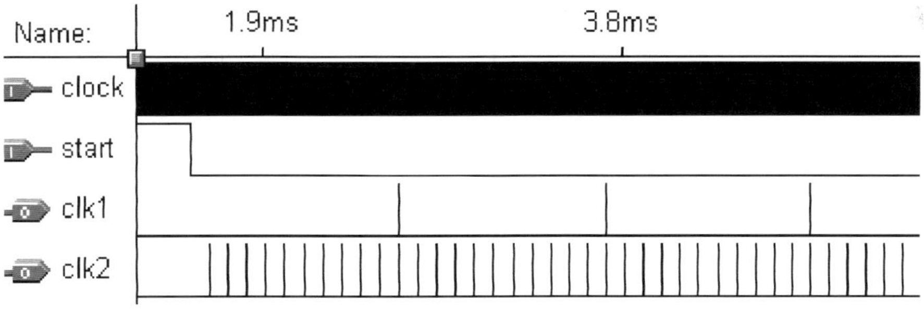

Figure 4. MESSAGE CONTROL Inputs and Outputs versus Time.

4.2. COUNTER MOD 1001

Figure 5 presents the COUNTER MOD 1001 functional block; it provides the address lines of the ROM memory (10bits bus called C1001 [9..0]). ROM memory stores the 34 bytes API packet. Implemented with lpm_Counter function (MAX+PLUS II, Altera), COUNTER MOD 1001 counts from 0 to 1000 where START enables the count.

Besides, COUNTER MOD 1001 has two clocks. CLOCK1 is the counter "pacemaker" with a frequency of 872Hz. The first 17 accounts match with the 17 bytes of the HEADER and one additional pulse necessary for synchronization. From the 19 account, CLOCK1 counts 34 pulses for 34 bytes (data). The following 4 pulses are used to generate a minimum waiting time interval called GUARD that the ZigBee module specifies for a successful transmission. The other clock, CLOCK_EN, is designed to generate the timing as described above; that is, CLOCK_EN disables CLOCK 1 account both during the 17 pulses of HEADER and during the 4 pulses of GUARD.

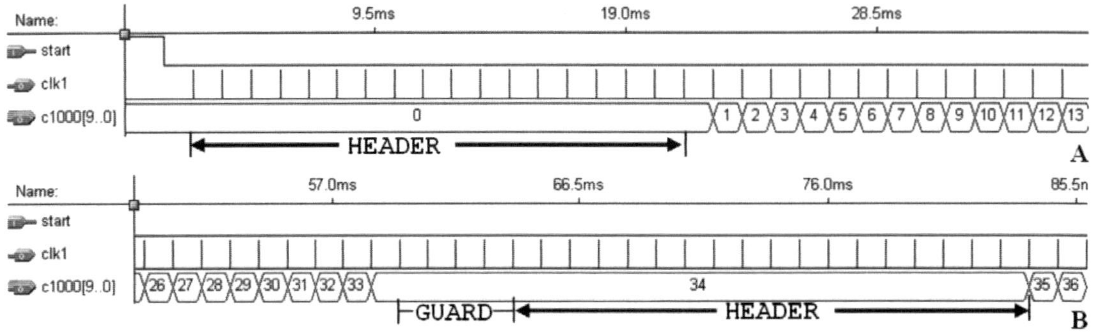

Figure 5. COUNTER MOD 1001 inputs and outputs (function of time). **A**- 0 to 13 counts. **B**- 26 to 36 counts. This behavior is periodically repeated every 56 pulses of CLK1.

4.3. ROM 1001x8

Figure 6 shows the ROM 1001x8 functional block implemented with lpm_rom function (MAX+PLUS II, Altera). The input is 10 bits-bus C1001 [9..0] or the ROM memory address. The output SIGNAL [7..0] has 1001 bytes of ECG, LIVP and LIVV, in that order. It should be noted that during the GUARD and HEADER intervals, logical block output retains the last biological signal value.

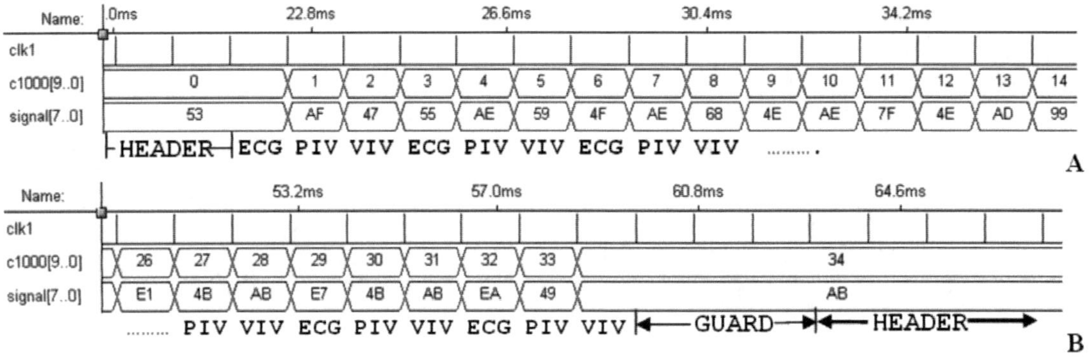

Figure 6. ROM 1001X8 Output vs Time. **A**- 0 to 14 address. **B**- 26 to 34 address.

4.4. COUNTER MOD 56

Figure 7 presents the COUNTER MOD 56 functional block implemented with lpm_Counter function (MAX+PLUS II, Altera). Its inputs are START and CLK1 signals and its output is a 6 bits-bus address line (ADDRESS [5..0]). COUNTER MOD 56 counts from 0 to 55 and provides 56 address lines to the RAM memory. Again, START enables the count.

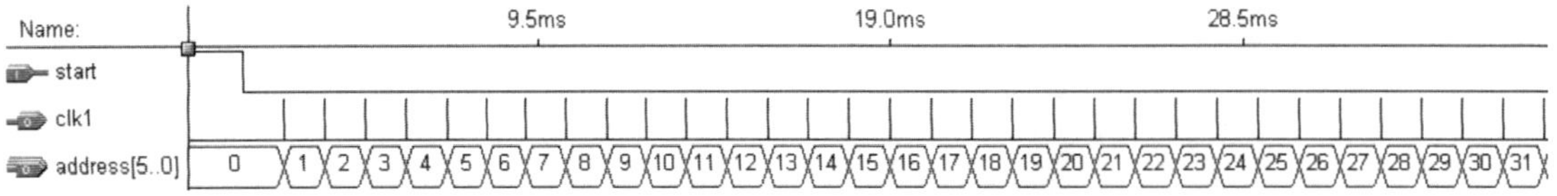

Figure 7. COUNTER MOD 56 temporal signals. It only displays 0 to 31 counts. Identical behavior is observed for 32 to 55 counts.

4.5. RAM 56x8

Figure 8 describes the RAM 56 functional block implemented with lpm_ram_dq function (MAX+PLUS II, Altera). It prepares the data for the TRANSMITTER logical block that sends them to the ZigBee module. RAM 56`s inputs are SIGNAL [7..0], ADDRESS [5..0] and CHECKSUM [7..0] while RAM 56`s outputs are Q [7..0], TX and CLKE.

Q [7..0] has the contents of the addresses listed in ADDRESS [5..0]. If ADDRESS [5..0] is minor than 18, RAM 56 block reads the contents of a memory initialization file (.mif) that includes the 17 values of message-header (HEADER). However, if ADDRESS [5..0] varies between 18 and 51, RAM 56 block writes the input data from SIGNAL [7..0] in these addresses. Finally, if ADDRESS [5..0] is equal to 52, RAM 56 block writes in the 52 memory address the value present in CHECKSUM [7..0] which is obtained at the output ADDER block described below.

TX output is in HIGH level when the data are transmitted; otherwise, TX output is in LOW. Also, CLKE output is in LOW during the GUARD time interval; otherwise, CLKE output is in HIGH. In addition, during GUARD interval, 0x00 hexadecimal value is placed at Q [7..0] output.

TX transition –from HIGH to LOW level- allows to control the GUARD time interval necessary to avoid saturating the END DEVICE buffer. In turn, GUARD interval time is obtained putting the TRANSMITTER output in HIGH level when TX is in LOW level.

TX, CLKE and START signals are entered into a three-input AND gate and its output is CLK_EN, the COUNTER MOD 1001 enable line. If any of these input signals are in LOW, CLK_EN too; hence, COUNTER MOD 1001 holds the account.

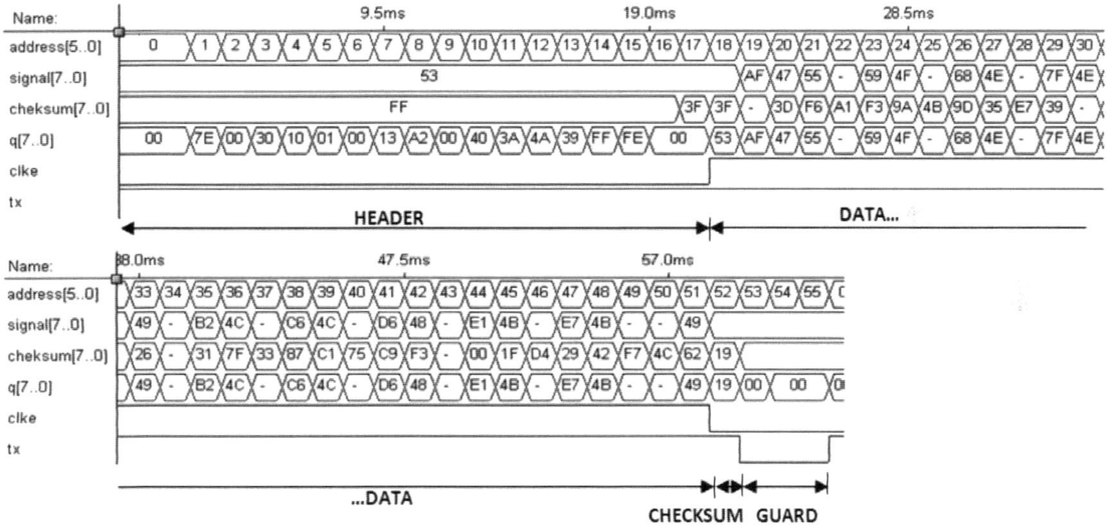

Figure 8. RAM56X8 temporal signals. Note that during the first 17 values of ADDRESS [5..0], Q [7..0] has the 17 bytes of HEADER. Then, Q [7..0] copies the values of data from SIGNAL [7..0] during the following 34 bytes. Finally, the CHECKSUM [7..0] and GUARD interval time are displayed.

4.6. ADDER

Figure 9 shows the ADDER functional block implemented with lpm_add_sub function (MAX+PLUS II, Altera). ADDER`s inputs are Q[7..0], START and CLK1 while ADDER`s outputs are the bus CHECKSUM [7..0] and CONTROL.

ADDER block main function is to implement the CHECKSUM [7 .. 0] byte (see byte 49, figure 2); it is calculated as:

$$\text{CHECKSUM [7..0]} = \text{HEX “FF”} - \Sigma \text{ offset 3 to 48 bytes} \tag{1}$$

$$\Sigma \text{ offset 3 to 48 bytes} = adder_{i+1}\,[7..0] = adder_i\,[7..0]+Q_i\,[7..0] \quad (2)$$

where the subscript i indicates the current value.

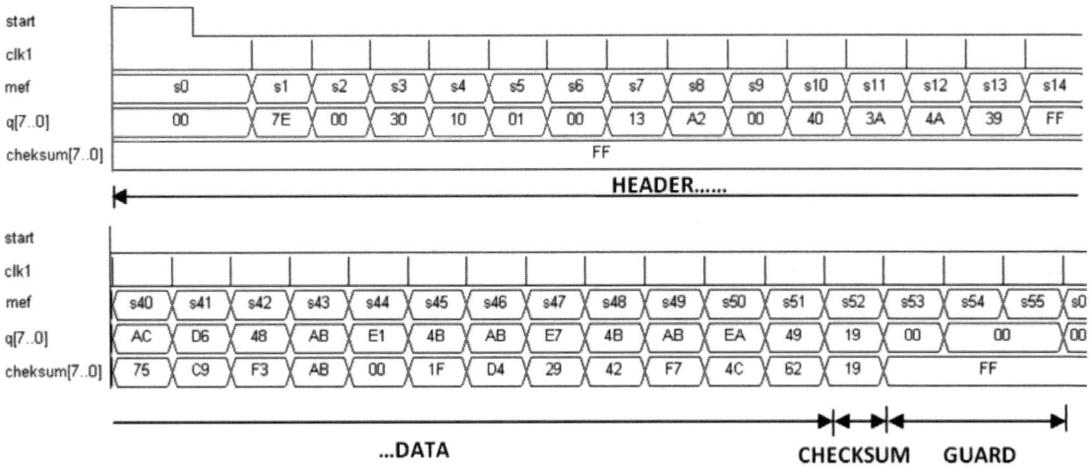

Figure 9. ADDER temporal signals. mef , internal variable that identifies the finite state machine.

Implemented with a 56 finite state machine, *adder* [7..0] is an 8 bits internal variable accumulating the partial sums. START input is connected to the RESET of the finite state machine starting it. First, the cumulative sum of 3 to 48 bytes of data and then, the difference indicated in equation (1) are calculated. In both cases, the results are truncated keeping only the lowest 8 bits. In HEADER time interval (first 18 states), *adder* [7..0] is initialized with HEX "00"; therefore, applying equation (1), CHECKSUM [7..0]=HEX "FF". The same occur during GUARD interval (54 to 56 pulses). During 19 to 53 states, the cumulative sum *adder* [7..0] is calculated with the bytes that enter sequentially from *Q* [7..0] applying equation (2); CHECKSUM [7..0] is determined using (1). Finally, CONTROL signal function is to send a HIGH level pulse when ADDER is in the first state; it serves to measure the time interval between packets (not included in figure 9).

4.7. TRANSMITTER

Figure 10 presents the TRANSMITTER functional block implemented with HDL (Finite State Machine, MAX+PLUS II, Altera). Its inputs are Q[7..0], START and CL while its output is the T_x bit.

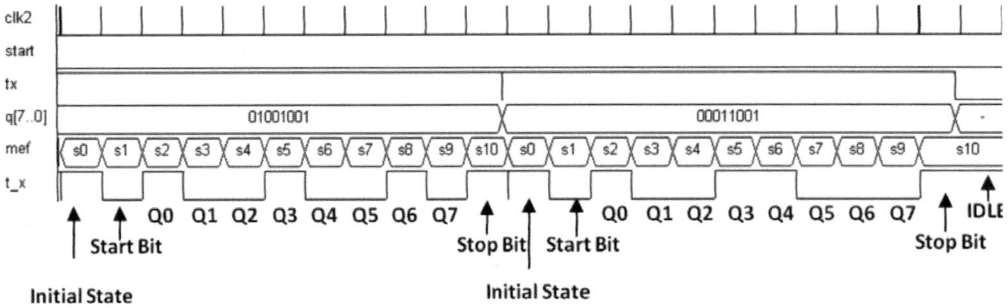

Figure 10. TRANSMITTER temporal signals. mef, internal variable that identifies each state of the finite state machine.

In TRANSMITTER, Q[7..0] bytes are converted to RS232 asynchronous serial protocol adopting a LOW level for the "Start Bit" and a HIGH level for the "Stop Bit" without parity bit.

It is implemented with an 11 finite state machine synchronized with CLK2 (9597Hz). The sequence begins at Initial State (see figure 10). This finite state machine is activated if the TX signal (from the RAM 56x8 block, see item 4.5) is in HIGH level; otherwise, T_x output is in IDLE state (inactive). This control assures that the TRANSMITTER block is not sending information during the GUARD interval. Besides, this machine begins when the START signal is in LOW; this control assures that the TRANSMITTER block is synchronized with the global enable signal.

5. Results

5.1. About the FPGA Logic Resources

After the FPGA design was completed, the resource utilization indices were calculated and presented in Table 1. A low percentage of used FPGA resources is observed except for embedded memory blocks (EABs) that store the data. However, ROM memory contents are refreshed when 1001 words were transmitted, therefore, does not affect the availability for future improvements. The few resources used by the ZigBee interface allows successfully incorporate the design to the DCCS.

Table 1. Device Used Resources

Parameter	Used Resources
Total dedicated input pins used	1/6 (16%)
Total I/O pins used	3/183 (2%)
Total logic cells used	241/3744 (6%)
Total embedded cells used	56/72 (77%)
Total EABs[a] used	7/9 (77%)
Total input pins required	2
Total output pins required	2
Total flipflops required	121
Synthesized logic cells	55/3744 (1%)
Memory bits	9312 (50%)
Registered performance	13.33MHz

[a] Embedded Array Blocks.

5.2. About the transmitted data

Figure 11a shows the oscilloscope screen (software supplied by Tektronix). Clearly, it shows one packet of data (Channel 1) for each positive CONTROL pulse (Channel 2). The Channel 1 signal is the COORDINATOR received messages or PC input as a function of time.

To display the transmitted information, a graphical interface was designed using LabView software and VISA library (Configure Serial Port). It is able to extract only data of interest for each received message rejecting the headers and transmission technical information. In this particular design, 50 bytes messages are received via the RS232 port, where the relevant data are located from the fifteenth byte (Figure 11b). Note that the transmitted packets size (END DEVICE) differs from the received packets size (COORDINATOR) depending of the ZigBee protocol.

Figure 11b presents three transmitted biological signals from the FPGA-embeded DCCS, through the ZigBee modules, to the personal computer. Thus, the ECG, LIVP and arterial pressure were successfully displayed on a personal computer screen. These signals were compared with the signals in the ROM memory; missing data were not detected (loss of information) neither distortion of them.

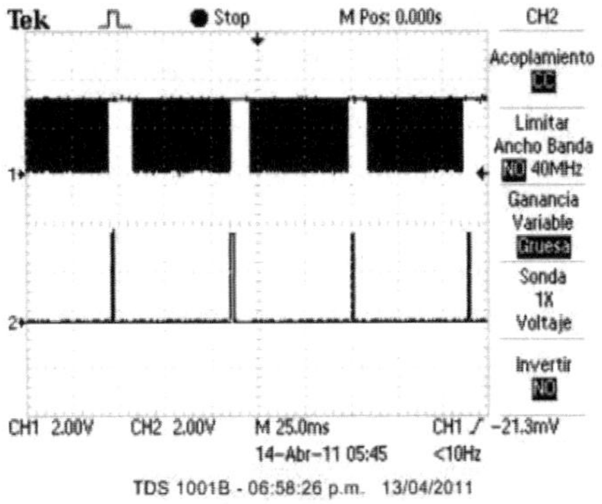

Figure 11a. TEKTRONIX oscilloscope screen (viewed on the computer). Channel 1: COORDINATOR received messages or PC input as a function of time. Channel 2: FPGA CONTROL output.

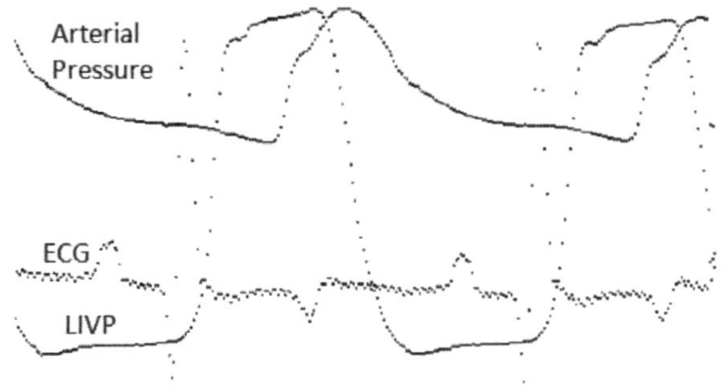

Figure 11b. Wirelessly transmitted biological signals (from the FPGA to the PC via ZigBee) visualized by custom-made software (developed in LABVIEW). LIVP, ECG and Arterial Pressure are presented (superimposed).

6. Conclusion

A control interface for a ZigBee wireless network using FPGA logical design has successfully implemented. This interface receives 3 simultaneous cardiac signals from a digital conductance catheter system previously developed, generates data packets with a specific format and sends them to the transmitter device (END DEVICE) of a ZigBee module. This device wirelessly transmits the data to a COORDINATOR module. The latter is connected to a personal computer where data are displayed. Results demonstrate that the control interface performance is correct as it successfully recovers the transmitted signals.

Acknowledgments

This work was supported in part by grants from the UNT Research Council (CIUNT), 26/E422-3 Research Program and INSIBIO (CONICET).

References

[1] Cai K and Liang X 2010 A Zigbee based mesh network for ECG monitoring system *4th Int.Conf. on Bioinformatics and Biomedical Engineering (iCBBE)* (Guangzhou: China Inf. Coll. Zhongkai Univ. of Agric. & Eng) 1-4

[2] Jung J and Lee J 2008 Device access control and reliable data transmission in zigbee based health monitoring system *IEEE 10th Int. Conf. on: Advanced Communication Technology (*Seoul: Electron. & Telecommun. Res. Inst.) 795-97

[3] Junnila S, Defee I, Zakrzewski M, Vainio A and Vanhala J 2008 UUTE home network for wireless health monitoring *IEEE Int. Conf. on: Biocomputation, Bioinformatics, and Biomedical* (Tampere: Dept. of Signal Process. Tampere Univ. of Technol.)125-30

[4] Kuryloski P, Giani A, Giannantonio R, Gilani K, Gravina R, Seppa V, Seto E, Shia V, Wang C, Yan P, Yang A, Hyttinen J, Sastry S, Wicker S and Bajcsy R 2009 Dexternet: an open platform for heterogeneous body sensor networks and its applications *6th IEEE Int. Workshop on Wearable and Implantable Body Sensor NetworK* (New York: Ithaca Dept. of ECE Cornell Univ.) 92-7

[5] Bennett T, Kjellstrom B, Taepke R and Ryden L 2005 Development of implantable devices for continuous ambulatory monitoring of central hemodynamic values *Heart Failure Patients Rev.* **28** 573–84

[6] Steinhaus D, Reynolds D, Gadler F, Kay G, Hess M and Bennett T 2005 Implant experience with an implantable hemodynamic monitor for the management *Symptomatic Heart Failure* **28** 747-53

[7] Yu C, Wang L, Chau E, Hon-Wah Chan R, Kong S, Tang M, Christensen J, Stadler R and Lau C 2005 Intrathoracic impedance monitoring in patients with heart failure. Correlation with fluid status and feasibility of early warning *Preceding Hospitalization Circulation* **112** 841-48

[8] Tallaj J, Singla I and Bourge R 2011 Implantable hemodynamic monitors *Cardiology Clinics* **29** (2) 289-99

[9] Walton A and Krum H 2005 The heartpod implantable heart failure therapy system *Heart Lung and Circulation* **14S** S31–3

[10] Gómez López M 2008 Equipo para mediciones en el sistema cardiovascular: Tecnología FPGA aplicada a la determinación de volúmenes cardíacos. *Tesis de Maestría.* (Tucumán: Departamento de Bioingeniería FACET UNT)

[11] Herrera M, Olivera J and Gómez López M 2010 Implantable hemodynamic monitors: can be conductance catheter system successfully implemented? *32nd Anual Int. IEEE EMBS Conf.* Digital 3549-52

[12] Norma IEEE 802.15.4 2006 IEEE Standard for Telecommunications and Information Exchange Between Systems - LAN/MAN - Specific Requirements - Part 15.4: Wireless Medium Access Control (MAC) and Physical Layer (PHY) Specifications for Wireless Personal Area Networks (WPANs)

Mechanical properties of the aortic arterial wall during 24 hours: a preliminary study in conscious sheep

S. Graf[1,3], D. Craiem[1,3], M. Valero[1], M. Alfonso[1], J. G. Barra[1,2], R.L. Armentano[1]

[1] Fac. de Ingeniería y Ciencias Exactas y Naturales,Universidad Favaloro
[2] Dto. de Ciencias Fisiológicas, Universidad Favaloro
[3] CONICET, Buenos Aires, Argentina

E-mail: graf@favaloro.edu.ar

Abstract. Previous experiences in animals showed a different behavior between the variability of pressure, arterial diameter and elasticity when they were registered for a couple of hours. To better understand arterial mechanics variability, we propose to measure simultaneously aortic pressure and diameter during 24 hours in a sheep. For that purpose, we developed a portable prototype device. It allows continuously recording physiological signals throughout the day and storing them in a solid state memory for later analysis. Pulse wave velocity and Peterson modulus were assessed beat-to-beat as arterial stiffness indexes. We identified 53,762 heart beats during 24 hours that were separated into 2 groups: below or above median mean pressure (71 mmHg). Mean diameter, pulse wave velocity and Peterson modulus increased for higher pressure values ($p<0.05$) whereas heart rate slowed down ($p<0.05$). Pressure-diameter loops were successfully recreated all along the experience. This new methodology sets the basis for further experiences involving the estimation of 24 hours arterial mechanics variability.

1. Introduction

The alterations of the arterial system can be evaluated from a biomechanical perspective analyzing the function and structure of the arteries. Previous studies from our group applied mathematical models to assess the geometric, elastic and viscous properties of the arteries. They were carried out starting from pressure-diameter loops in chronically instrumented conscious animals [1, 2]. Although the heart rate variability was extensively studied, insufficient information is available about the variability of the arterial mechanical properties during the day and night. We showed that pressure, diameter and elasticity variability were dissimilar during short term (one hour) experiences performed in sheep [3, 4]. Particularly, elasticity did not followed pressure variations during all the time. On the basis of these observations, we propose to extend the analysis to 24 hours, measuring simultaneously arterial pressure and diameter and pulse wave velocity (PWV). The online assessment of these properties could help to better understand the different regulatory mechanisms that control blood flow and pressure during day and night. Recently, we developed an autonomous and portable prototype device, which allows continuous recording and digital storage of physiological signals for analysis. The device was validated in vitro and contrasted against synthesized (simulated) signals [5]. Using this device, the aortic blood pressure, diameter and consequently arterial stiffness evolution can be estimated beat to beat. To validate arterial stiffness measurements, PWV was also calculated using two arterial diameter sensors [6]. This work presents the methodology, the system implementation and the preliminary results from an animal experience.

2. Materials and methods

2.1. Prototype description

The portable prototype device is based on a dsPIC30F6014A (Microchip, USA) microcontroller (figure 1). Basically, it includes a signal conditioning stage (filtering and amplification), a 12-bit A-D converter stage and a data storage stage (4GB SD type flash memory card). The signals can be monitored in real time by a liquid crystal display, or via serial communication using a computer.

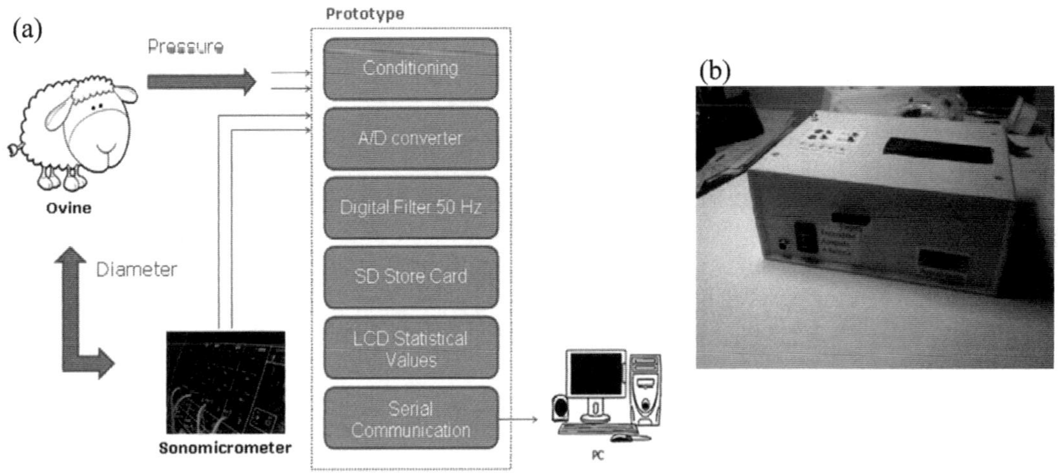

Figure 1. (a) Block diagram and (b) picture of the implemented prototype.

The equipment allows acquiring and storing up to 4 channels simultaneously, at a configurable sampling frequency of 0.5, 1, 2 and 4 KHz and at configurable time intervals. In order to verify the proper operation, several tests were conducted for at least 24 hours. The memory storage capability was tested using synthetic signals resulting in no loss of data. Keeping a constant signal on each input channel, it was demonstrated that the variation throughout a day was less than 1%. Settling time of the device was less than 5 minutes [5].

2.2. Animal instrumentation

A healthy Corrediale crossbreed sheep of 47 kg and 1.5 year-old was used for the experiment. Surgery was performed under general anesthesia premedicated with acepromazine (0.2 mg/kg) induced with thiopental sodium (20 mg/kg) and maintained with isoflurane (2.5 % in pure oxygen at 2 L/min) with assisted mechanical ventilation (Neumovent 910, Córdoba, Argentina). During surgery, blood pressure, electrocardiogram, capnography and heart rate were monitored (Siemens Sirecust 404-1). Two thoracotomies in the fourth and seventh left intercostal spaces were performed. The proximal and distal thirds of the aorta were exposed to implant pressure and diameter sensors. The solid-state pressure microtransducer of 2.5 mm (Konigsberg Instruments, Inc., Pasadena, CA) was implanted in the proximal third. The pressure microtransducer has been previously calibrated in vitro in saline at 37°C against a pressure caliper (Xcaliber, Viggo-Spectramed, Oxnard, CA). Both in proximal and distal regions of the aorta, ultrasonic crystals (5 MHz, 4 mm) were sutured with 6-0 silk on the adventitia in two places diametrically opposed, to measure outer diameter by sonomicrometry (Triton Technology Inc., San Diego, CA). The first pair was sutured to the proximal third of the vessel at 5 mm of the pressure sensor to avoid interference and the second pair was sutured in the distal third (the distance between the two sensors were 14.7 cm). The optimal placement of the crystals was verified by the digital oscilloscope screen (Tektronix TOS 220, Tektronix Inc. Beaverton, USA). The correct recording of the sensors was verified during operation by monitoring the signals in real time with the prototype connected to a PC before and after closure of thorax. After sensors implantation, cables were tunneled to emerge through the skin on the back of the sheep, which was then sutured. Finally, the incisions were closed by layers.

2.3. Experimental protocol

The experiment was carried out after 7 days of the surgical instrumentation to ensure proper postoperative recovery. The first channel of the prototype was adjusted to work with the pressure microtransducer, while the two following channels were conditioned to receive the sonomicrometer diameter signals. The prototype device was configured to acquire and store in memory three signals over 24 hours continuously for 5 minutes, separated by an interval of 5 minutes. Sampling frequency was set to 1 KHz. The activity of the sheep was monitored using digital images of 512×512 pixels taken with a digital camera (Logitech, USA), connected to a PC. The day after the study, data from the memory card was downloaded and signals were processed using proper software developed in our laboratory.

For each cardiac cycle, systolic, diastolic and mean values from pressure and diameter signals were automatically obtained. PWV was assessed as the ratio of the distance between the two diameter sensors and the resulting temporal delay between the onsets of the two diameter waveforms, identified by the tangent method [7]. At a sampling frequency of 1 KHz and considering a distance of 14.7 cm between sensors, the estimated error in PWV was ~3.4%. Heart rate was calculated from the pressure signal. The pressure-strain elastic (Peterson) modulus (E_p) was calculated as defined in equation (1):

$$E_p = \frac{\Delta P}{\Delta D} \cdot D_D , \qquad (1)$$

where ΔP is the difference between systolic and diastolic pressure, and ΔD is the difference between systolic and diastolic diameter (D_D).

2.4. Statistical analysis

Data are expressed as mean ± standard deviation. In order to characterize the mechanical behavior of the arterial wall, the total amount of identified beats along 24 hours were separated into 2 groups: below or above the median mean pressure. Differences were compared using Student *t* test. A value of $p<0.05$ was considered significant.

3. Results

The instrumented animal during the experience is shown in figure 2. Note the sensors wires emerging from the back of the animal. The sheep was free to move using a custom hose-spring system hanging from the ceiling. The prototype device and the sonomicrometer were located on the other side of the wall. The animal was isolated and remained calmed during the experience. All measurements were made without sedation with the animal in conscious state. It was free to eat and rest in any moment and did not present discomfort during the experience.

Aortic pressure, diameter and PWV temporal evolution for 24 hours are shown in figure 3. The pressure-diameter measures in the aorta are shown in figure 4, with the excursion of the diastolic, mean and systolic values during 24 hours. Pressure values ranged from 40 to 100 mmHg and diameter values from 19.5 to 23.5 mm.

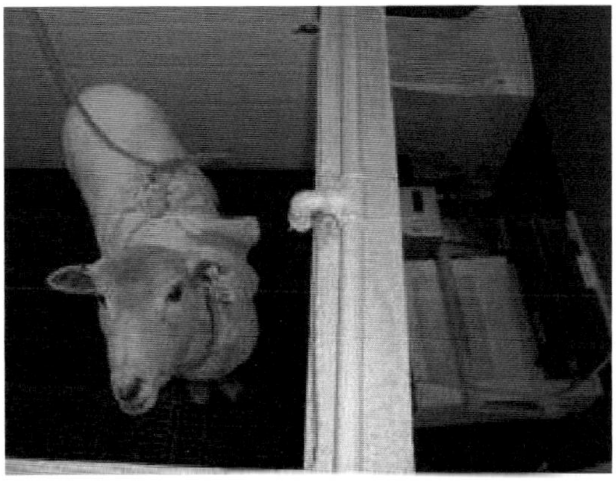

Figure 2. Conscious instrumented sheep and associated equipment used for signals recordings during 24 hours.

We identified 53,762 heart beats during 24 hours that were separated into 2 groups according to the median pressure value, which resulted in 71 mmHg. The hemodynamic parameters of each group are detailed in table 1. Mean diameter, PWV and E_p increased during high pressure group ($p<0.05$) whereas heart rate slowed down ($p<0.05$).

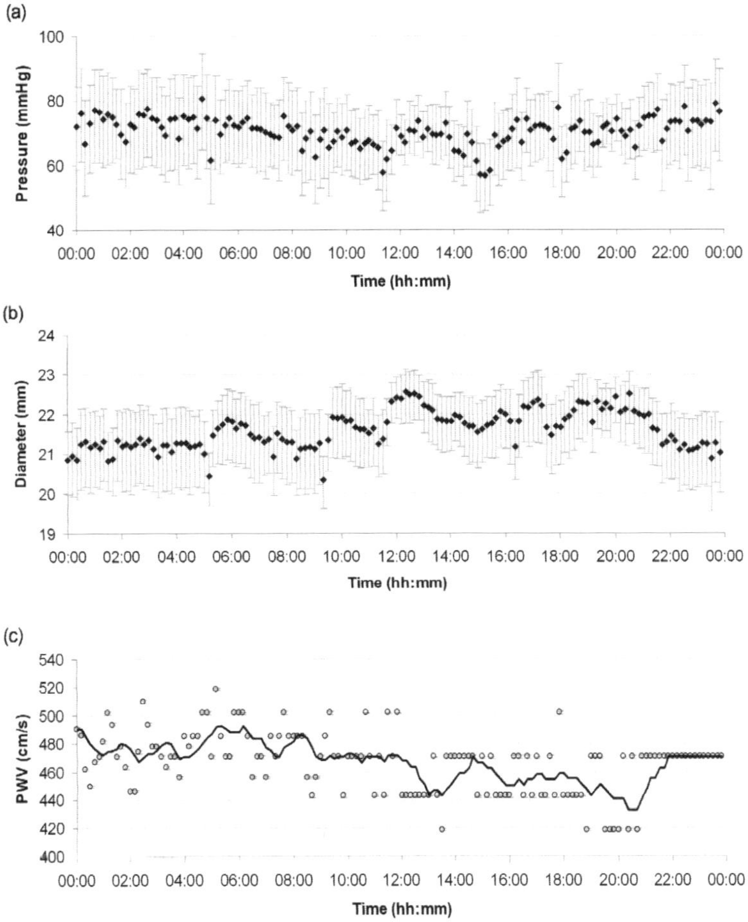

Figure 3. Temporal evolution of pressure, diameter and PWV every 10 minutes during 24 hours. The plotted values represent the median from 5-minutes continuous recordings. For pressure and diameter the systolic, mean and diastolic values are shown. For the PWV, an additional trend line is shown.

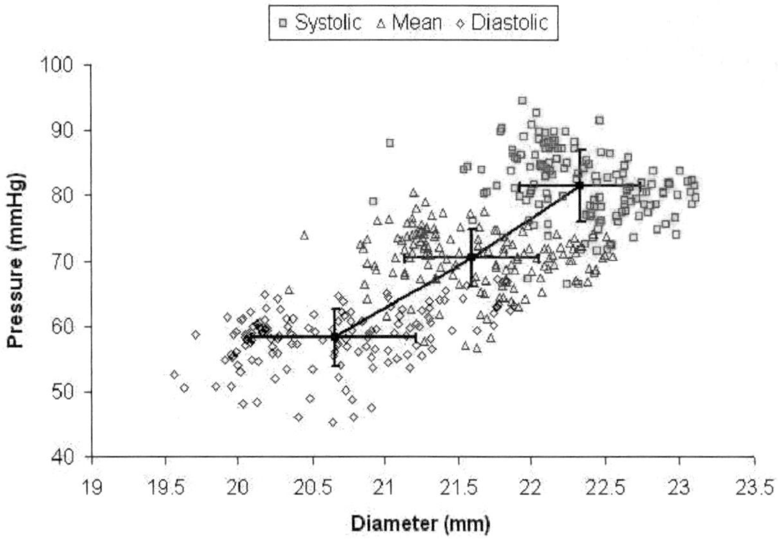

Figure 4. Diastolic, mean and systolic median values of pressure and proximal diameter registered during 24 hours, every 5 minutes, with their respective mean and standard deviation.

Table 1. Hemodynamic parameters during 24 hours separated by groups.

	Low pressure group n=26,881 beats	High pressure group n=26,881 beats
Mean pressure (mmHg)	66 ± 4	75 ± 4 *
Mean diameter (mm)	21.56 ± 0.51	21.67 ± 0.55 *
PWV (cm/s)	461 ± 34	478 ± 37 *
Heart rate (bpm)	98 ± 18	93 ± 15 *
E_p (mmHg)	285 ± 64	299 ± 70 *

* $p<0.05$ (Student t test) between low and high pressure group.

4. Discussion

In this work we show for the first time the continuous assessment of aortic elastic properties in a sheep during 24 hours. Aortic pressure and diameter were recorded using a custom portable prototype especially designed in our laboratory. The device successfully acquired and stored more than 50,000 heart beats throughout one day. In this particular experience, pressure varied ~30% and diameter only ~10% with respect to 24 hours mean values. Aortic stiffness was independently estimated using two indexes: the Peterson modulus E_p, calculated from pressure-diameter loops, and the PWV, estimated using 2 diameter sensors separated with a known distance. Although the animal was calm during the experiment, the arterial pressure showed an important variation (figure 3). As expected, aortic wall stiffness was positively associated to pressure changes. To study the association between pressure values and arterial mechanics, we divided all the measured beats in 2 groups using the median mean pressure as a threshold. Arterial stiffness was higher for pressure beats above the threshold: for a 15% of pressure change, PWV and E_p increased ~5%. However, pressure and stiffness were not permanently correlated during the experience. In fact, we found that the relationship between arterial pressure and diameter was not systematic, evidencing the complexity of the arterial system. As shown in figure 4, pressure-diameter values do not lie on the same unique elasticity curve during 24 hours. This reflects the natural biological variability of the arterial wall properties, where the active components of the wall (vascular smooth muscle) continuously regulate the vessel size. In previous works of our group, different beat-to beat elastic models were tested in vivo [8, 9]. So, we are encouraged to initiate a new set of experiments aimed to analyzing the variability of the mechanical properties in aorta. Analyzing the animal activity along the 24 hours,

we concluded that could not follow a circadian behavior. As can be seen from the figure 3, there is not any comprehensible day/night pattern. That might be related to the permanent availability of food. The digestive system of ruminants might generate recurrent pressure maneuvers that could modulate the arterial pressure. Also, the activity of the animal could be altered by the hose-spring system. Further experiments are necessary to explore these suggestions.
Some issues remain to be solved to ensure the full portability of the prototype. First, although the system supports the direct connection of the pressure microtransducer, it requires an external sonomicrometer to generate the diameter signals. A new portable sonomicrometry system is being designed in our laboratory and will be incorporated in the near future. Secondly, we are reducing the size of the prototype using surface mount components and small batteries that will allow a better portability. Then, the device will become completely autonomous without requiring any external connection, allowing the animal to carry the whole system on his back.

5. Conclusions

Through a beat to beat 24 hours analysis we studied the influence of pressure on the aortic mechanical properties of a sheep. In general, when pressure increased, diameter, PWV and E_P concomitantly increased whereas heart rate decreased. Pressure-diameter loops were successfully recreated all along the experience, confirming the capability of assessing true arterial wall behavior during 24 hours. These findings encourage us to program new experiments aimed to analyze the arterial mechanical properties throughout the day.

6. Acknowledgments

This work was in part subsidized by the project PIP number 112-200901-00734 (CONICET) and project PICTO 31355 (ANPCyT).

References

[1] Armentano R L, Barra J G, Levenson J, Simon A and Pichel R 1995 Arterial wall mechanics in conscious dogs: Assesment of viscous, inertial, and elastic moduli to characterize aortic wall behaviour *Circ. Res.* **76(3)** 468-78

[2] Barra J G, Armentano R L, Levenson J, Cabrera-Fisher E I, Pichel R H and Simon A 1993 Assesment of smooth muscle contribution to descending thoracic elastic mechanics in conscious dogs *Circ. Res.* **73** 1040-50

[3] Gamero L G, Armentano R L and Levenson J 2002 Arterial wall diameter and viscoelasticity variability *Compu. in Card.* **29** 513-6

[4] Craiem D, Graf S, Salvucci F, Chironi G, Megnien J L, Simon A, Armentano R L 2010 Physiological principles of the ambulatory arterial stiffness index explained by the non-linearity of arterial elasticity *Physiol. Meas.* **31(7)** 1037-46

[5] Alfonso M 2009 Sistema portable para el cálculo y registro de parámetros fisiológicos. Thesis. FICEN. Universidad Favaloro

[6] Laurent S et al. 2006 Expert consensus document on arterial stiffness: methodological issues and clinical applications *Eur. Heart J.* **27** 2588 – 605

[7] Chiu C Y, Arand P W, Schroff A, Feldman T and Carroll J 1991 Determination of pulse wave velocities with computarized algorithms *Am. Heart J.* **121** 126

[8] Armentano R L, Barra J G, Pessana F M, Craiem D, Graf S, Santana D B and Sanchez R A 2007 Smart smooth muscle spring-dampers. Smooth muscle smart filtering helps to more efficiently protect the arterial wall *IEEE Eng. Med. Biol. Mag.* **26** 62-70

[9] Gamero L G, Armentano R L, Barra J G, Simon A and Levenson J 2001 Identification of arterial wall dynamics in conscious dogs *Exp. Physiol.* **86** 519-28

Design and implementation of a sigma delta technology based pulse oximeter's acquisition stage.

E E Rossi[1], A Peñalva[1], F Schaumburg[1].
Laboratorio de Instrumental electromédico para diagnóstico y monitoreo. Universidad Nacional de Entre Ríos, Ruta provincial Nº 11, Argentina

E-mail: rossiesteban@gmail.com, schaumburg.f@gmail.com, elpenia@gmail.com

Abstract. Pulse oximetry is a widely used tool in medical practice for estimating patient's fraction of hemoglobin bonded to oxygen. Conventional oximetry presents limitations when changes in the baseline, or low amplitude of signals involved occur. The aim of this paper is to simultaneously solve these constraints and to simplify the circuitry needed, by using ΣΔ technology. For this purpose, a board for the acquisition of the needed signals was developed, together with a PC managed software which controls it, and displays and processes in real time the information acquired. Also laboratory and field tests where designed and executed to verify the performance of this equipment in adverse situations. A simple, robust and economic instrument was achieved, capable of obtaining signals even in situations where conventional oximetry fails.

1. Introduction

Pulse oximetry is a technique used to estimate arterial oxygen saturation (SpO_2). It is widely used because of the advantages which presents compared to the other accepted methods to measure this variable, namely continuity of the record, non invasiveness, simplicity to use and lower cost. All this at the expense of an improvable accuracy. [1][2][3].

Transmittance pulse oximetry, consists of measuring the intensity of the light which an irrigated tissue (such as a finger or the earlobe) is capable of transmitting. The light used is composed by two different wavelengths, typically 660 nm and 940 nm. Time variation of the transmitted intensity for any wavelength is called the plethysmographic signal. [1][3][4].

These signals consist of a high amplitude DC level (due mainly to static components of tissue and ambient light) and a low amplitude AC level (due mainly to volumetric changes in vascular tissues, due to heart's activity). Numerically AC component represents between 0.1% and 2% of the plethysmographic signal. The plethysmography signal's bandwidth is 10 Hz. By comparing the AC and DC components for each wavelength is possible to obtain a highly SpO_2 correlated index (R). [3][5][7][9].

This particular waveform, together with the need for two different signals (one for each wavelengths) leads conventional approaches, which are those that use low resolution (less than or equal to 16 bits) digital-analog converters (ADC), to implement complex and expensive analog processing stages. The large difference in magnitude between AC and DC, requires to digitize them separately, so that they can receive different amplification, thus achieving good resolution for AC components without causing saturation. This implies that there should be four analog processing channels: one for AC due to 660 nm emission, one for DC due to 660 nm emission, and similarly for the AC and DC due to 940 nm emission. Figure 1 shows a block diagram of a conventional pulse oximeter. [8].

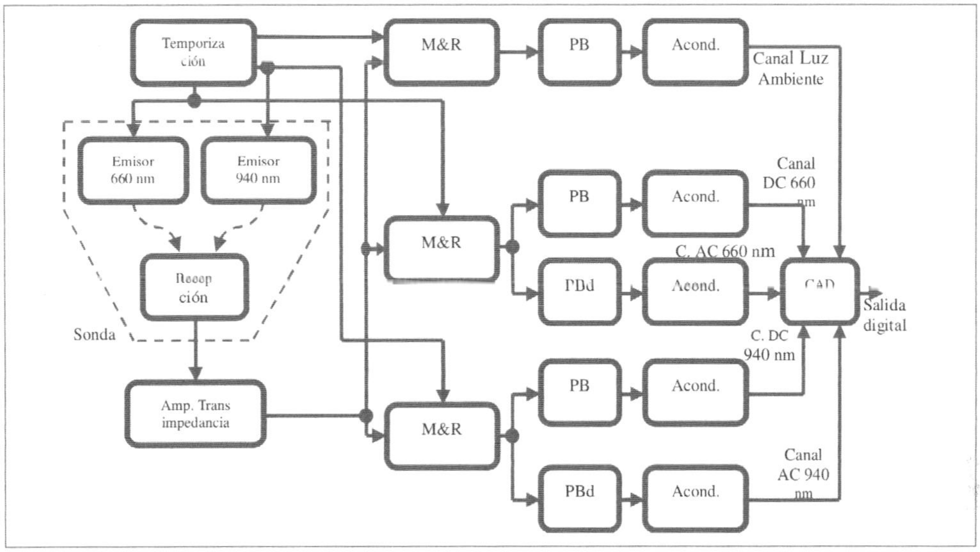

Figure 1. Conventional pulse oximeter's acquisition stage block diagram. M&R: Sample and Hold, PB: lowpass filter, PBD: bandpass filter.

On the other side, to get acceptable resolution with those ADC is necessary to amplify signals as much as possible. As a consequence, large variations in the DC signal imply overpassing conversion's range. This phenomenon is also called "saturation of the ADC input" or "ADC saturation", and implies the loss of information. This is why conventional oximeter readings are incorrect or are not valid in situations involving large variations in signal's DC level or baseline (such as patient or sensor movement, or intense light in operating room) or situations where signal's amplitude is poor (low blood differential pressure, hypothermia, hypotension, hyper-pigmentation, bad sensor anchorage). [1][2][4][9].

ΣΔ AD converters have three characteristics of great interest for biomedical signal acquisition: High resolution (up to 24 bits), allowing signal's low amplification without loss of useful information, together with large DC level variations without involving ADC saturation; Oversampling (which in some ΣΔ ADC is greater than the bandwidth of operational amplifiers) and; Intrinsic behavior as a lowpass filter for the input signal, solving the aliasing problem without the need for complex filters or directly without filters. [6].

2. Objective

To develop an oximetry's acquisition stage, which by the use of a ΣΔ ADC will give solution to the previously appointed problems of conventional pulse oximeters, and which also will minimize the analog processing stage, resulting in a more robust and economic design . This device must be able to be used in the context of a laboratory for teaching purposes.

3. Materials and Methods

3.1. Spiral development model.

The "spiral development model" was chosen in order to implement this acquisition module. In this model, the life of a development is divided into stages which are driven in a cyclical manner while the product progresses. When starting a spiral cycle, the risks associated with product in the state it is are assessed, then its aspects to be improved are determined, and a new design aimed at meeting them is made. The spiral has an angular component that indicates the progress of the project in a cycle, and a radial component indicating the cost and complexity of development. [10][11].

During design some of the IEC 60601-1, ISO 14971:2007 and the IRAM-FAAA 37226 AB requirements, where taken into account. The decision to leave out the rest of the guidelines imposed by these regulations, was based on the intended use of this device, which is within the scope of a laboratory, under controlled conditions.

3.2. Validation.

A validation process was followed, in which the result of the design and implementation process was assessed in order to ensure compliance with requirements and specifications defined. It consisted of laboratory and field testing..

The former consisted of measurements carried out in a laboratory in order to verify the correct functioning of the various blocks that compose the designed device, by simulating the conditions to which it would be exposed, in a controlled manner. These tests were carried out at the "Laboratorio de Ensayo y Calibración de Equipos Médicos" (LEyCEM, FI-UNER). The equipment used for this purpose was: A FLUKE 196C oscilloscope, a Fluke 271 DDS Signal generator, and a TES 1330 lightmeter. Three repetitions were performed for each laboratory test, with the aim of reducing the influence of the uncertainty of measurement process.

Field tests are tests performed on volunteers in order to check the overall operation of the device. Those tests were designed to verify compliance with the objectives set for this work (that is not losing information of plethysmographic signals, which is the first step for solving the problems they entail). All field tests were performed three times on two volunteers. The emission intensity of the emitters used, in each case was constant and equal to half the maximum attainable with the device.

3.2.1. *AD conversion.* In this laboratory test, in order to asses that the frequency response of the AD conversion block was the desired, signals of known amplitude and frequency were injected in this block, and records of the digitized signals were obtainded. This scanning was performed using 1 Vpp amplitude sinusoidal signals, and the following frequencies: 0.1, 0.2, 0.5, 1, 2, 5, 10, 20, 50, 100, 200, 500, 1 000, 2,000 and 5 000 Hz. Records obtained were analyzed using MATLAB®.

3.2.2. *Light emission.* Light emission. In this laboratory test, in order to evaluate if the relationship between emitted light intensity and the current driver control signal (see 4.1.1) was linear, a sweep of this signal was performed, and illuminance was measured for each wavelength using the light meter mentioned. Emitters were placed inside an opaque container. To achieve reliable measurements with the lightmeter, emissions were continuous, and not pulsed as it in the normal operation of the device.

3.2.3. *Light reception.* In this laboratory test in order to verify that the relationship between light intensity incident on the photodiode and the amplifier output voltage, is linear, this voltage was measured for increasing values of the current drivers control signal. For this purpose, the transmitter and receiver were placed facing each other, inside an opaque container. Prior to this linearity was observed between current drivers control signal, and the emitted light intensity.

3.2.4. *Record during movement.* In this field test, in order to verify the continuity of the records (i.e. no information loss) during patient movement (which also implies sensor movement), records were taken in the following format: 25 s in a quiet state, followed by 35 s with the volunteer doing a light jog on a treadmill. To prevent complete detachment of the sensor, its cable was fixed to the finger using tape (as it is done in clinical practice). Data collected was analyzed in order to determine the changes in the signal's baseline, and the absence of ADC saturation.

3.2.5. *Record with strong ambient light.* In this field test, in order to verify the ability of the designed device, to keep record in situations which involve intense ambient light, plethysmographic signals records (for both wavelengths) were obtained, with three different ambient light levels: 50, 500 and 5 000 lux. The lightmeter was used to measure those levels. Records obtained were analyzed for determining the absence of saturation, and the correct AC components acquisition.

3.2.6. *Record during low perfusion.* In this field test, in order to verify the ability of the device to acquire adequately low amplitude plethysmographic signals, records were obtained on voluntaries with low perfusion in the limb the sensor was placed on. To induce this, a cuff for non-invasive blood pressure measuring was placed in the same arm than the sensor was. The cuff remained inflated at a pressure between the volunteer's maximum and minimum (previously determined). Records obtained were analyzed in order to verify the correct acquisition of plethysmographic signals.

3.2.7. *Record of a SpO_2 variation.* Calibration of the device, as well as algorithms for R and SpO_2 calculation are issues beyond the scope of this work. Either way, a change in a volunteer's SpO_2 induced by a voluntary apnea episode, should cause the value reported by the device to drop after this maneuver. Also, after normal breathing is restored, the SpO_2 value should return to a value similar to the initial. In order to verify this, in this field test, records were taken in volunteers who followed this breathing pattern: Normal breathing for 10 s, Forced expiratory reserve volume expiration, Avoiding inspiration for 25 s, Normal breathe during 50 s. Records obtained were analyzed to asses if there was correspondence with the expected behavior.

4. Results

4.1. ΣΔ ADC based pulse oximeter's acquisition stage

4.1.1. *Hardware.* Sensor was manufactured by reusing the mechanical components of a commercial sensor. Narrow spectrum LED emitters were selected with emission peaks at 660 nm and 940 nm respectively. As a receptor, a photodiode with spectrum including both emission wavelengths was chosen. A transimpedance amplifier was included in the same sensor, so as to amplify and convert to voltage the photocurrent, and to decrease tribological noise effects due to cable movement. The only block included between this amplifier and ADC, was anti-aliasing filter. The ΣΔ ADC was chosen to allow sampling of the three signals (plethysmographic signals and ambient light signal) at a rate greater than or equal to ten times the maximum signal's frequency, with a resolution allowing at least 100% baseline variations, and 1% resolution for plethysmographic signal's AC component. For circuit's control and communication a microcontroller (μC) was used. A pulse width modulated signal (PWM) provided by the μC, was chosen for controlling emission intensity. Acquisition stage-PC connection was accomplished by the means of a USB 2.0 port. Emission's control was accomplished by using current drivers, which are circuits that generate currents that control when and for how long a LED emits, together with emission's intensity. Circuit was mechanically protected by the means of an ABS plastic cabinet. Figures 2, 3 and 4, show hardware achieved.

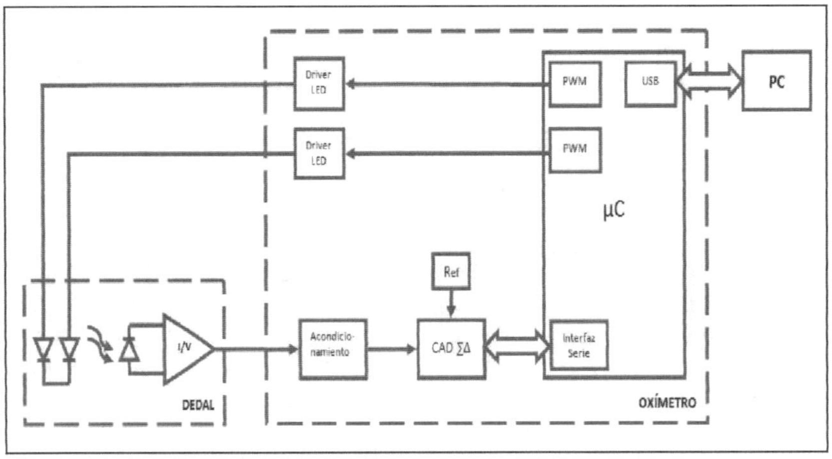

Figure 2. Block diagram for the circuit achieved.

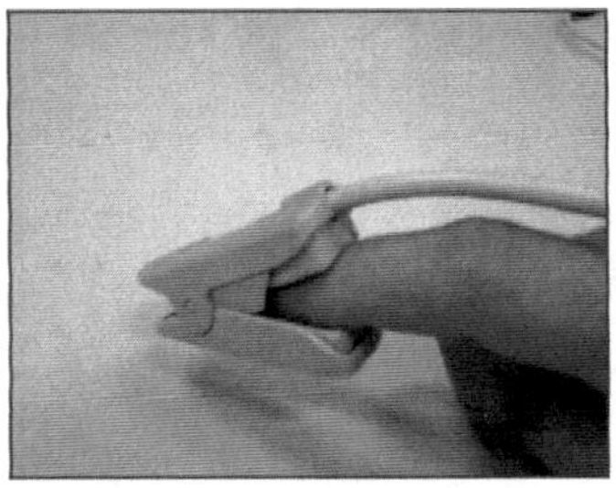

Figure 3. Thimble containing two emitters, a photo-receiver and a transimpedance amplifier.

Figure 4. Cabinet containing the circuits for the acquisition stage.

4.1.2. *Software.* As shown in Fig. 5, the PC managed software allows real-time visualization of the acquired signals and calculation of SpO_2 and R. The screen has two plotting areas: one for AC components, and one for DC components and ambient light. On these charts, amplitude and time scales can be changed using the zoom controls. There also is a bar menu which allows record operating (starting, pausing, completing, saving, etc.), selection of signal to be plotted, PWM duty cycle selection, among others.

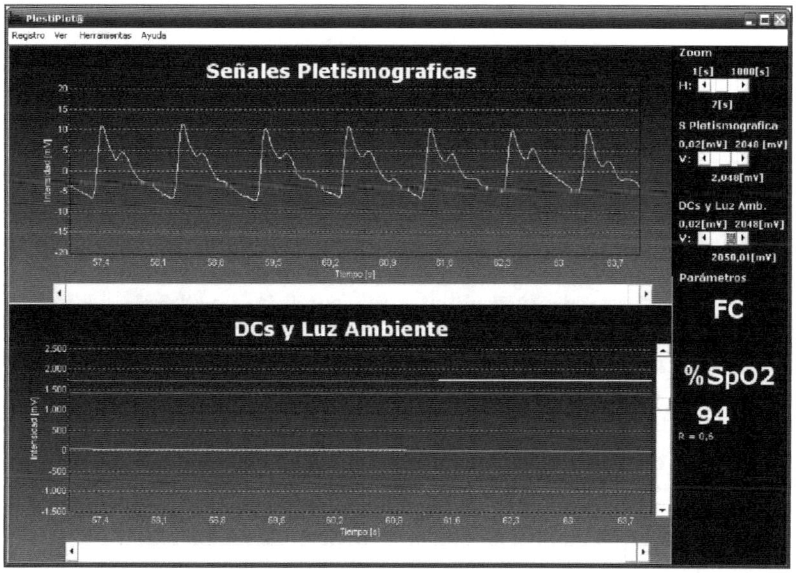

Figure 5. PC managed user interface, for processing, plotting and storing the acquired data.

4.2. Validation.

4.2.1. *AD conversion.* The obtained ADC's frequency response, matched its datasheet. No aliasing was found in the analyzed records. In figures 6 and 7 respectively, ADC's frequency response and laboratory setup used for this purpose, can be observed.

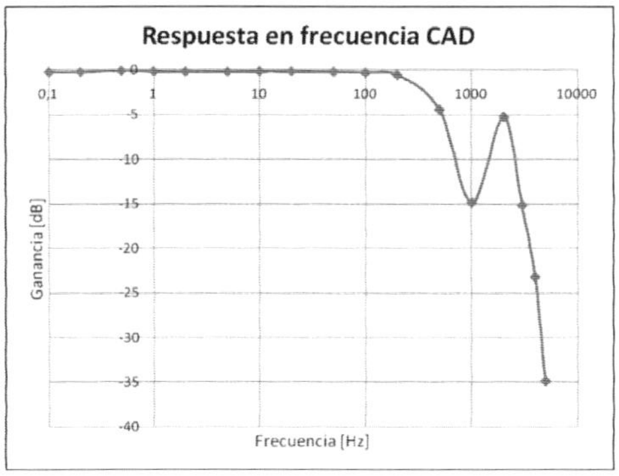

Figure 6. ΣΔ ADC's frequency response.

Figure 7. Laboratory setup. The device and the equipment used can be seen.

4.2.2. *Light emission.* Relationship between the emitted intensity (in lux) and PWM duty cycle (in %), was obtained. Linearity between those magnitudes was found. In figures 8 and 9 respectively, this relationship and laboratory setup, is shown.

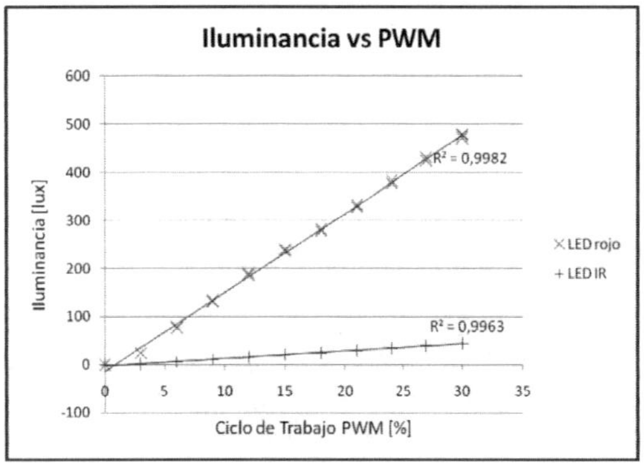

Figure 8. Relationship between light emitted intensity (for each LED) and PWM duty cycle for current driver controlling.

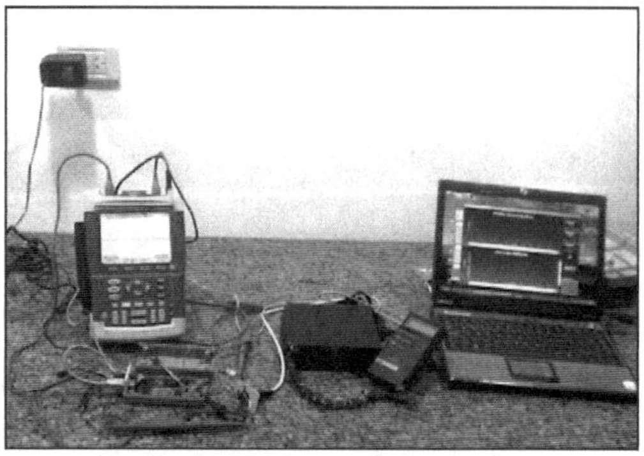

Figure 9. Laboratory setup. The device and the equipment used can be seen.

4.2.3. *Light reception.* Relationship between incident light intensity on the photodiode, and transimpedance amplifier's output voltage, was obtained for each LED. In both cases, linearity was found. Figure 10 shows this relationship.

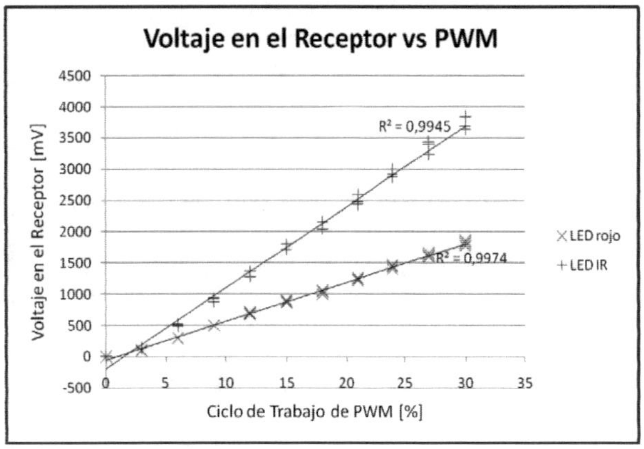

Figure 10. Relationship between light intensity received by the photodiode (for each wavelength) and PWM duty cycle. The regression lines used to evaluate the linearity between the measured variables, and a close to the unit r^2 parameter supports this linearity.

Journal of Physics: Conference Series **332** (2011) 012009 doi:10.1088/1742-6596/332/1/012009

4.2.4. *Record during movement.* Records were obtained during volunteer movement. It was found that there was no ADC saturation, i.e. ADC input signal's amplitude never exceeded ADC's voltage reference. In the worst of recorded cases, there was a near 100% baseline variation. Also, this variation could have been as large as 360% without causing problems. That record is shown in figure 11.

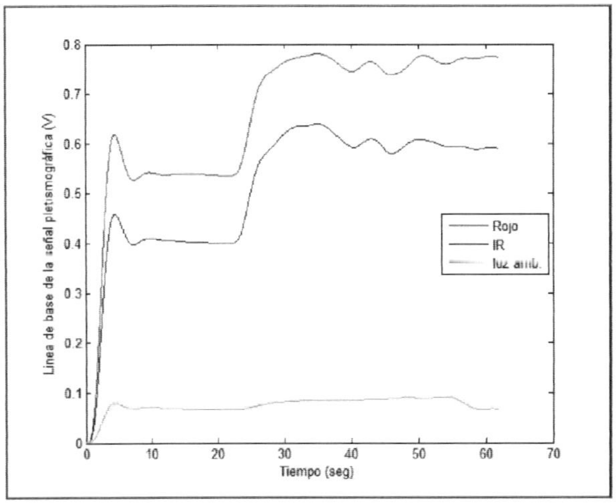

Figure 11. Worst case record obtained during movement. DC components and ambient light are shown. It can be observed that after the second 25, DC increases to nearly twice its previous value (ΔDC = 100%) as a result of movement. As the ADC's reference voltage used was 2.048 V, DC variation could have been close to 360% without causing saturation.

4.2.5. *Record with strong ambient light.* Records with three different levels of intense ambient light were obtained. In none of them ADC saturation occurred, and in every case it was possible to visualize AC components correctly. Figure 12 shows a record obtained in the worst lighting scenarios (5000 lux ambient ligth).

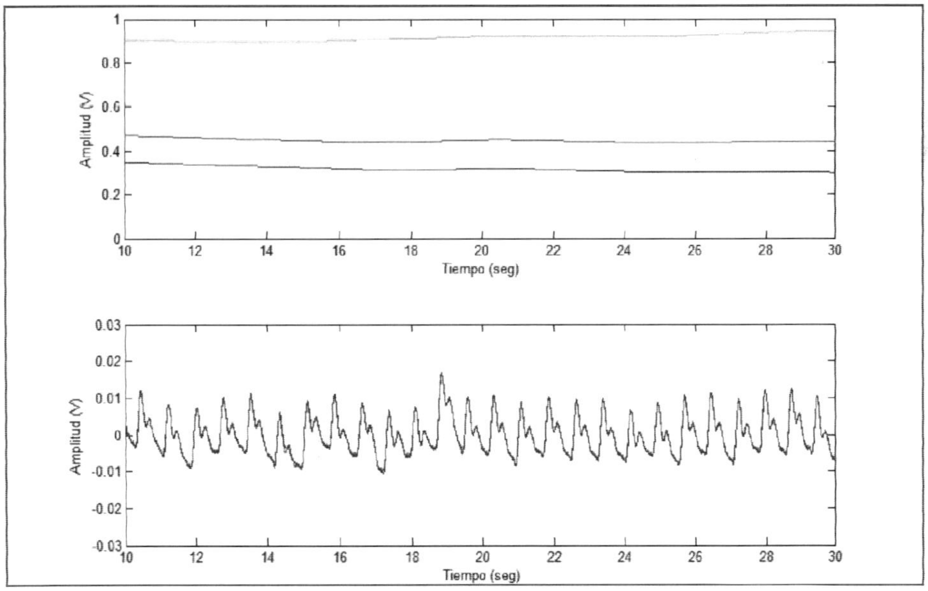

Figure 12. Above: ambient light (green) and DC components (blue for 940 nm, and red for 660 nm) in a measuring scenario with strong ambient light (5 000 lux). Below: AC components obtained correctly under these conditions.

4.2.6. *Record during low perfusion.* Records were obtained under the named conditions. In all cases plethysmographic signals were acquired properly, as shown in Figure 13. In the worst case, the AC component was acquired with a 0.5 resolution.

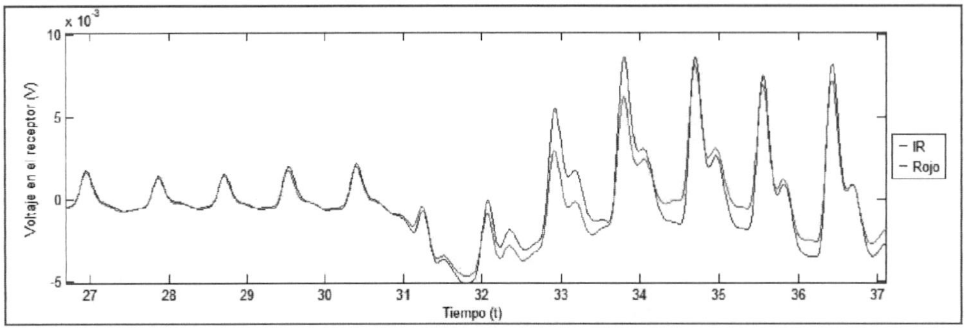

Figure 13. Low amplitude register due to a non invasive pressure cuff, used in the same limb were the oximetry sensor was placed. This condition was normalized after the second 30. The magnitude difference in the signal before and after this moment can be observed.

4.2.7. *Record of a SpO_2 variation.* SpO_2 records were obtained in volunteers who followed the breathing pattern mentioned. The behavior of this variable was the expected, as shown in Figure 14.

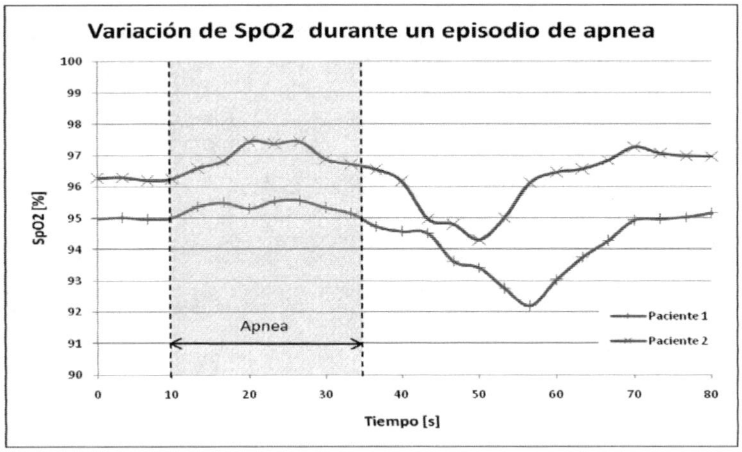

Figure 14. SpO_2 register in two volunteers. After the apnea episode, a SpO_2 decrease can be observed, followed a subsequent return to its initial value.

5. Conclusions

A ΣΔ ADC based pulse oximeter's acquisition stage prototype was achieved successfully. Since this prototype was designed for laboratory use, a PC managed software which allows hardware control and communication was implemented, also successfully.

During laboratory tests, compliance with requirements defined for each prototype's block was verified. In addition, during field tests and meeting this article's objectives, the device proved to be able to provide records during adverse situations, without loss of data due to ADC or sensor saturation, or low-amplitude signals. In all cases plethysmographic signals were recovered correctly.

Furthermore, although the device was not calibrated, the correlation between SpO_2 value changes thrown by the prototype, and the expected variations during apnea, is an indicative of acquired data validity.

Moreover, comparing block diagrams shown in Figures 1 and 2, it can be concluded that the reduced complexity of the proposed circuit, means component reduction and also cost savings.

Calibration of the device, microcontroller implementation of algorithms for R and SpO_2 calculation, improvement of those algorithms, isolation of the patient circuit, a PC independent user interface, and the expansion of the validation process (i.e. performing the field tests detailed in this article, but over more volunteers) remains for future work.

References

[1] Mardirossian, George; Schneider, Roland E., "Limitations of Pulse Oximetry" (*Anesthesia Progress*, v.39, pp.194-196, 1992)

[2] Niederbacher Velásquez, J., García Niño, M., Gómez Moya, G., "Valores de referencia de saturación arterial de oxígeno mediante pulso-oximetría en niños sanos de Bucaramanga", (*Med UNAB*, v.6, N°17, pp.63-69, ago 2003)

[3] Zamora, B.W.G, Pérez Barriga, R., "Oxímetro de pulso a PC con interface de Pic16F877", (*XV Congreso Argentino de Bioingeniería*)

[4] Sola, A.; Chow, L.; Rogido, M., "Oximetría de pulso en la asistencia neonatal en 2005. Revisión de los conocimientos actuales" (*Anales de Pediatría*, v.62, N°3, pp.266-281, 2005)

[5] Townsend, Neil, "Medical Electonics" (Michaelmas Term, 2001)

[6] Yao,J., Warren, S., Ph.D., "Stimulating Student Learning with a Novel "In-House" Pulse Oximeter Design", *(2005 American Society for Engineering Education Annual Conference & Exposition*)

[7] Jubran, Amal, "Pulse Oximetry: Review" (*Critical Care,* v.3, pp.11-17, 1999)

[8] Park, Sangil, "Principles of Sigma-Delta Modulation for Analog-to-Digital Converters" (*Strategic Applications: Motorola Digital Signal Processor Operation*)

[9] López-Herranz, Patricia G., "Oximetría de pulso: A la vanguardia en la monitorización no invasiva de la oxigenación" (*Revista Médica del Hospital General de México*, v.66, N°3, pp.160-169, jul-sep 2003)

[10] Boehm B.W., "A spiral Model of Software Development and Enhancement" (*Computer,* v.21, pp.61-72, 1988)

[11] Lerendegui, N. M., "Diseño de Equipos y Software"

Estimation of local pulse wave velocity using arterial diameter waveforms: Experimental validation in sheep

S. Graf[1,3], D. Craiem[1,3], J. G. Barra[1,2], R.L. Armentano[1]

[1] Fac. de Ingeniería y Ciencias Exactas y Naturales, Universidad Favaloro
[2] Dto. de Ciencias Fisiológicas, Universidad Favaloro
[3] CONICET, Buenos Aires, Argentina

E-mail: graf@favaloro.edu.ar

Abstract. Increased arterial stiffness is associated with an increased risk of cardiovascular events. Estimation of arterial stiffness using local pulse wave velocity (PWV) promises to be very useful for noninvasive diagnosis of arteriosclerosis. In this work we estimated in an instrumented sheep, the local aortic pulse wave velocity using two sonomicrometry diameter sensors (separated 7.5 cm) according to the transit time method (PWV_{TT}) with a sampling rate of 4 KHz. We simultaneously measured aortic pressure in order to determine from pressure-diameter loops (PWV_{PDLoop}), the "true" local aortic pulse wave velocity. A pneumatic cuff occluder was implanted in the aorta in order to compare both methods under a wide range of pressure levels. Mean pressure values ranged from 47 to 101 mmHg and mean proximal diameter values from 12.5. to 15.2 mm. There were no significant differences between PWV_{TT} and PWV_{PDLoop} values (451±43 vs. 447±48 cm/s, p=ns, paired t-test). Both methods correlated significantly ($R=0.81$, $p<0.05$). The mean difference between both methods was only -4±29 cm/s, whereas the range of the limits of agreement (mean ± 2 standard deviation) was -61 to +53 cm/s, showing no trend. In conclusion, the diameter waveforms transit time method was found to allow an accurate and precise estimation of the local aortic PWV.

1. Introduction

Cardiovascular disease is one of the most important causes of death around the world [1]. Increased arterial stiffness is associated with an increased risk of cardiovascular events [2]. Measurement of the changes in arterial diameter and pressure at the same site allows direct measurements of arterial stiffness [3, 4]. However, in clinical practice this is not easy to achieve. Alternative methods have emerged, being the pulse wave velocity (PWV) one of the most accepted [5].

From the Moens Korteweg relation, it is possible to relate the square of PWV, directly to the product of the elastic modulus and wall thickness, and inversely with the product of the arterial diameter and the density of blood [3]. PWV is commonly assessed over a given arterial segment length, such as carotid-femoral, carotid-radial or brachial-radial [5]. Many of these measurements are based upon simultaneous measurements of pressure or flow at two sites in the arteries and the determination of the time it takes a wave to propagate from one site to the other (transit time method) [5]. These measurements represent the average wave speed over the distance between the measurement sites. However, due to the non elastic and non geometric uniformity of the arterial tree, arterial stiffness is different in proximal and elastic aortas compared with distal and muscular arteries. Therefore, "local" PWV measurements, at specific arterial sites are desirable.

In clinical practice, ultrasound is frequently used to no invasively assess blood velocity, wall thickness and diameter waveforms [6]. Therefore, recently several methods emerged for the estimation of local PWV using ultrasound principles. Among others: a) the flow and area method (QA), in which the local PWV is estimated during a reflection free period of the cardiac cycle, as the ratio between the change in flow and the change in cross sectional area [7, 8], b) the 2D distension waveforms method, in which multiple adjacent distension waveforms are determined simultaneously along a short arterial segment, using a single 2D-vessel wall tracking system with a high frame rate (651 Hz) [9], c) tissue Doppler imaging method, in which a relatively new color Doppler modality for measuring tissue motion offers multiple recording sites along the artery [10]. However, none of these methods includes an exhaustive in vivo validation.

In this work we propose to estimate in an instrumented sheep, the local aortic pulse wave velocity using two sonomicrometry diameter sensors according to the transit time method. We simultaneously measured aortic pressure in order to determine from pressure-diameter loops (PDLoop method considered as gold standard) [4], the "true" local aortic pulse wave velocity. A pneumatic cuff occluder was implanted in the aorta in order to compare both methods under a wide range of pressure levels.

2. Material and methods

2.1. Animal instrumentation

A healthy corrediale sheep of 60 kg and 2 year old was used for the experiment. Surgery instrumentation was performed under general anesthesia premedicated with acepromazine (0.2 mg / kg) induced with thiopental sodium (20 mg/kg) and maintained with isoflurane (2.5% in pure oxygen at 2 L / min) with assisted mechanical ventilation (Neumovent 910). During surgery, blood pressure, electrocardiogram, capnography and heart rate were monitored (Siemens Sirecust 404-1). By two thoracotomies in the fourth and seventh intercostal left spaces, the proximal and distal thirds of the aorta were accessed to implant two pairs of ultrasonic crystals (5 MHz, 4 mm), for subsequent instantaneous diameter determination by sonomicrometry (Triton Technology Inc. San Diego, CA). Each pair of crystals was sutured in two places diametrically opposite in the adventitia of the aorta with 6-0 silk suture, with a separation of 7.5 cm between pairs. The optimal placement of the crystals was verified by the digital oscilloscope screen (Tektronix TOS 220, Tektronix Inc. Beaverton, USA).

A pneumatic cuff occluder made from silicon rubber was implanted around the descending thoracic aorta, distal to the ultrasonic crystals. This ensured that no artifacts appeared in diameter measurements during aortic occlusions. After sensors implantation, wires were tunneled to emerge through a small hole in the skin at the back of the sheep, which was then sutured. Finally, the thoracotomies were closed by layers.

2.2. Experimental protocol and recordings

The study comprised the recording of instantaneous aortic pressure and aortic diameter signals. A high-fidelity pressure catheter (Millar Instruments Inc, Texas, USA) connected to an amplifier (Gould 6600 Series Transducer) was used to measure the pressure in the proximal third of the aorta. The catheter was advanced from the left brachial artery through a little incision, and was positioned near the proximal ultrasonic crystals. The pressure transducer had been previously calibrated in vitro in physiological saline at 37 °C with a pressure caliper (Xcaliber, Viggo-Spectramed, Oxnard. Cal.). The diameter signals were calibrated in millimeters using the 1-mm step calibration facility of the sonomicrometer. Aortic pressure and diameter signals were digitized at 4 KHz on a computer (PC Pentium IV 300 MHz) using an analog to digital converter (National Instruments Lab PC) and digitally stored for later analysis.

The acquisition started with approximately 10 consecutive beats in basal state (CTL) and then the pneumatic occluder was inflated in order to obtain high-pressure values for a couple of seconds (OCL). Finally, the occluder was released in order to obtain low-pressure values (REL). In all cases, the instantaneous pressure-diameter loops were monitored and registered until stabilization was evidenced. This procedure was repeated several times.

2.3. Data analysis

Signals were processed using a proprietary software developed in our laboratory. Twenty eight pressure-diameter loops, corresponding to baseline, mechanical occlusion and release conditions were considered. For each cardiac cycle, systolic, diastolic and mean (average value computed from numerical integral of the waveform) values from pressure and diameter signals were obtained. Concomitantly, local arterial stiffness indexes were computed with the following indexes.
Local arterial distension was calculated as:

$$Distension = \frac{(D_s - D_d)}{D_d} x100 \qquad (1)$$

where D_s and D_d are systolic and diastolic proximal diameters. Peterson modulus (E_P) was calculated as:

$$E_p = \frac{P_s - P_d}{D_s - D_d} \cdot D_d \cdot 0.1334 \qquad (2)$$

where P_s and P_d are systolic and diastolic blood pressures.
For the assessment of PWW using pressure and proximal diameter waveforms (PWV_{PDLoop}) we considered the Bramwell-Hill equation [11],

$$PWV = \sqrt{\frac{V \cdot dP}{dV \cdot \delta}} \qquad (3)$$

where dV represent the change in volume (V) due to a change in pressure (dP), and δ blood density (1.06 g/cm3). For each cardiac cycle, the diastolic phase was separated from the pressure-diameter loops and the resulting diameter was adjusted to a logarithmic model, according to the following formula:

$$D = \alpha + \beta \cdot \ln P \qquad (4)$$

where α and β are two constant determined by the fitting procedure, and P the arterial pressure (See Figure 1). Assuming a circular section for the artery, changes in volume and diameter are related according to [12]:

$$\frac{dV}{V} \approx \frac{2 \cdot dD}{D} \qquad (5)$$

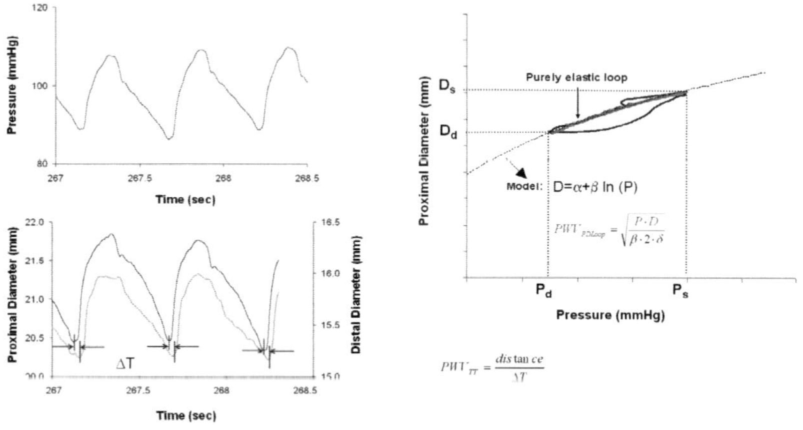

Figure 1. Left: Typical example of aortic pressure waveform (top) and aortic proximal (bottom, in black) and distal (bottom, in grey) diameter waveforms. Arrows indicate temporal delay (ΔT) between diameter waveforms. Right: Pulse wave velocity assessed from transit time and from pressure-diameter loop methods

Finally, after deriving equation (4) respect to pressure (dD/dP), using (5) and replacing in (3), pulse wave velocity was computed as:

$$PWV_{PDLoop} = \sqrt{\frac{P_d \cdot D_d}{\beta \cdot 2 \cdot \delta}} \qquad (6)$$

In the same cardiac cycle, PWW was assessed using the transit time method (PWV_{TT}), computed as the ratio of the distance between the two diameter sensors (7.5 cm) and the temporal delay (ΔT) between the two arterial diameter waves (see figure 1), as:

$$PWV_{TT} = \frac{7.5cm}{\Delta T} \qquad (7)$$

We considered the foot (point of the onset of the wave) on each waveform as the reference point, and was determined using the intersecting tangent algorithm [13]. Pulse wave velocity computed using these reference points correspond to the diastolic PWV.

Data are expressed as mean ± standard deviation. Comparisons between variables were analyzed by paired Student-t test or ANOVA as appropriate. A value of $p<0.05$ was considered statistically significant.

3. Results

Mean pressure values ranged from 47 to 101 mmHg and mean proximal diameter values from 12.5. to 15.2 mm. Figure 2 shows mean pressure values plotted against mean diameter values during CTL, OCL, and REL conditions.

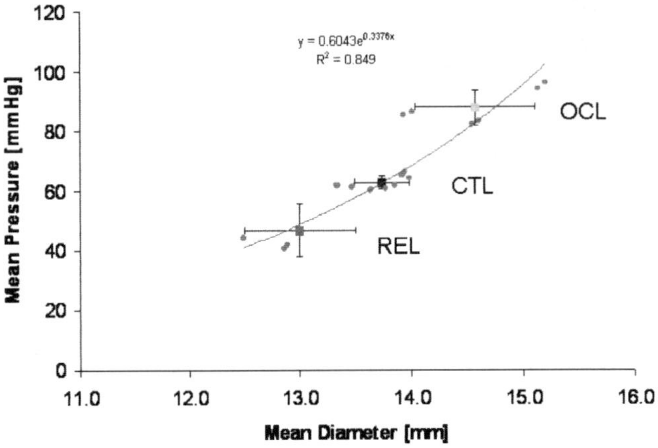

Figure 2. Mean aortic pressure values plotted against mean aortic proximal diameter values during control (CTL), cuff occlusion (OCL) and cuff release (REL) conditions with their respective mean (squares box) and standard deviation, and the non linear regression curve.

Values of aortic pressure, aortic proximal diameter and aortic stiffness indexes during control, occlusion and release conditions are given in Table 1.

Table 1. Hemodynamic parameters during control (CTL), cuff occlusion (OCL) and cuff release (REL) conditions. E_p: Peterson modulus. Values are expressed as mean ± standard deviation. [a]p<0.05 vs. CTL; [b]p<0.05 vs. REL; ANOVA and Neuman- Keuls post hoc tests.

	CTL (n=14)			REL (n=8)			OCL (n=6)		
Aortic Pressure									
Systolic, mmHg	69	±	2	53	±	8 [a]	94	±	5 [ab]
Diastolic, mmHg	57	±	3	41	±	9 [a]	82	±	6 [ab]
Mean, mmHg	63	±	2	47	±	9 [a]	88	±	6 [ab]
Pulse, mmHg	11.7	±	2.9	12.4	±	1.9	11.5	±	1.5
Aortic Proximal Diameter									
Systolic, mm	14.01	±	0.23	13.31	±	0.47 [a]	14.75	±	0.52 [ab]
Diastolic, mm	13.52	±	0.28	12.74	±	0.56 [a]	14.38	±	0.58 [ab]
Mean, mm	13.74	±	0.24	13.00	±	0.50 [a]	14.57	±	0.54 [ab]
Pulse, mm	0.48	±	0.11	0.57	±	0.12	0.37	±	0.07 [ab]
Local Stiffness Indexes									
Distension, %	4	±	1	5	±	1 [a]	3	±	1 [b]
E_p, kPa	37	±	3	32	±	3 [a]	52	±	7 [ab]
PWV_{TT}, cm/s	450	±	27	417	±	34 [a]	498	±	45 [ab]
PWV_{PDLoop}, cm/s	442	±	15	402	±	35 [a]	518	±	25 [ab]

Mean pressure increased by 40 % during OCL and decreased by 26 % during REL, compared to control values. Concomitantly, mean proximal diameter increased by 6% during OCL and decreased by 5% during REL. Aortic distension decreased by 25% during OCL and increased by 25% during REL. E_p increased by 41% during OCL and decreased by 14% during REL compared to CTL. PWV increased by 17% for PWV_{PDLoop} and 12% for PWV_{TT} during occlusion, and decreased 9% for PWV_{PDLoop} and 7% for PWV_{TT} during cuff release, compared to control values. All these changes were statically significant ($p<0.05$)

There were no significant differences between PWV_{TT} and PWV_{PDLoop} values (451±43 vs. 447±48, p=ns, paired t-test). Figure 3 (left panel) shows PWV estimated by the transit time method plotted against PWV estimated by the PDLoop method. The methods correlated significantly ($R=0.81$, $p<0.05$). In figure 3 (right panel), the difference between the two methods is plotted against the average of the two methods. The mean difference between both methods was only -4±29 cm/s, whereas the range of the limits of agreement (mean ± 2 standard deviation) was -61 to +53 cm/s, showing no trend.

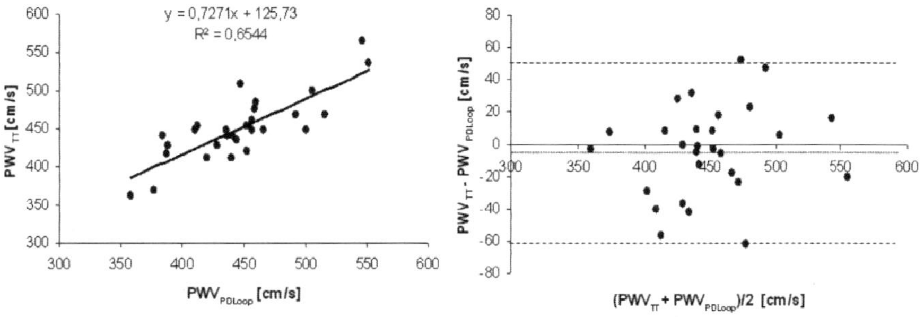

Figure 3 Left: PWV by the transit time method plotted against PWV by the PDLoop method. Right: The difference between the two methods plotted against the average of the two methods. The dotted line is the mean difference and the dashed lines represent the mean ± two times SD.

Considering systolic pressure and systolic proximal diameter values in equation 6, we found that PWV_{PDLoop} was 13±5 % higher ($p<0.05$) than the corresponding diastolic PWV_{PDLoop}, assessed with diastolic pressure and diameter values. A linear regression analysis showed that PWV_{TT} was related to mean pressure ($R=0.73$; $p<0.05$), mean diameter ($R=0.80$; $p<0.05$), Peterson modulus ($R=0.72$; $p<0.05$) and distension ($R= -0.78$; $p<0.05$). Similarly, PWV_{PDLoop} was related to mean pressure ($R=0.95$; $p<0.05$), mean diameter ($R=0.92$; $p<0.05$), Peterson modulus ($R=0.86$ $p<0.05$) and distension ($R= -0.82$; $p<0.05$).

4. Discussion

In this study we demonstrated the feasibility of estimating local PWV from two diameter waveforms. Undoubtedly, for a complete characterization of the arterial wall mechanics, a simultaneous assessment of arterial pressure and arterial diameter is required [4]. Therefore, we instrumented the aorta of a sheep with a pressure sensor and two diameter sensors. The use of a pneumatic cuff occluder, allowed a variation of the mean aortic pressure of ~56 mmHg. This is essential for comparison due to the pressure dependence of the mechanical properties of arteries. Figure 2 shows the typical nonlinear behavior of the arterial wall.

The procedure to determine PWV from pressure-diameter loops is based on the Bramwell-Hill equation. We assumed a circular section for the artery and that changes in volume and diameter are related according to equation (5). This simplification is allowed because in unimpeded arteries length is constant so that change in volume during the cardiac cycle is caused by a change in luminal cross-sectional area alone [12]. For practical purpose this relation can be expressed as a function of diameter. This is allowed if it is assumed that the artery lumen is circular in cross-section and that the change in diameter (dD) is small relative to the diameter. For the transit time method, considering a separation distance of 7.5 cm between sensors, a maximum PWV value of 556 cm/s and a sampling frequency of 4 KHz, the uncertainty in determining the time delay was less than 2%.

We found no significant differences between PWV_{TT} and PWV_{PDLoop} methods. The mean difference was only 4 cm/s, and 95% of the differences were less than two standard deviations, showing no trend. By other side, using the systolic and diastolic pressure and diameter values in equation (6), we determined a systo-diastolic variation of PWV of 13% during the cardiac cycle.

It is well known that PWV is strongly related to blood pressure [3]. An increase in blood pressure level will increase PWV. Our data confirmed this relationship. However, changes in PWV can not be only explained by changes in pressure. For example, comparing CTL and OCL condition, an increase of 40% in mean pressure caused a rise of only 17% in PWV. For the same increase in pressure, the Peterson modulus increased 41%, the mean diameter increased only 6% and the distension decreased 25%. By other side, comparing CTL and REL condition, a decrease of 26% in mean pressure caused a fall of only 9% in PWV.

Finally, further studies involving structural changes in the wall, and not just functional changes, will be necessary to demonstrate that the methodology remains valid.

5. Conclusion

The diameter waveforms transit time method was found to allow an accurate and precise estimation of the local aortic PWV in a wide range of pressure levels.

Acknowledgments

This research was supported in part by grant from PIP No 112-200901-00734 (CONICET) and PICTO 31355 (ANPCyT).

References

[1] Naghavi M, Falk E, Hecht HS, Jamieson MJ, Kaul S, Berman D, et al 2006 SHAPE Task Force. From vulnerable plaque to vulnerable patient- Part III: Executive summary of the Screening for Heart Attack Prevention and Education (SHAPE) Task Force report *Am J Cardiol* **98** 2

[2] Mitchell GF, Hwang SJ, Vasan RS, Larson MG, Pencina MJ, Hamburg NM, Vita JA, Levy D, Benjamin EJ 2010 Arterial stiffness and cardiovascular events: the Framingham Heart Study *Circulation* **2;121(4)** 505-11

[3] Nichols WW, O'Rourke MF. Vascular impedance. In: McDonald's Blood Flow in Arteries: Theoretical, Experimental and Clinical Principles. 4th ed. London, UK: Edward Arnold; 1998:54–97, 243–283, 347–395.

[4] Armentano R L, Barra J G, Levenson J, Simon A and Pichel R 1995 Arterial wall mechanics in conscious dogs: Assesment of viscous, inertial, and elastic moduli to characterize aortic wall behaviour *Circ. Res.* **76(3)** 468-78

[5] Laurent S, Cockcroft J, Van Bortel L, Boutouyrie P,Giannatasio C, Hayoz D, et al 2006 Expert consensus document on arterial stiffness: methodological issues and clinical applications *Eur Heart J* 2006 **22** 2588–2605

[6] Reneman RS, Meinders JM, Hoeks AP. Non-invasive ultrasound in arterial wall dynamics in humans: what have we learned and what remains to be solved 2005 *Eur Heart J.* **26(10)** 960-966

[7] Hermeling E, Reesink KD, Reneman RS, Hoeks AP. Measurement of local pulse wave velocity:

effects of signal processing on precision 2007 *Ultrasound Med Biol* **33(5)** 774-781
[8] Hermeling E, Reesink KD, Kornmann LM, Reneman RS, Hoeks AP. The dicrotic notch as alternative time-reference point to measure local pulse wave velocity in the carotid artery by means of ultrasonography 2009 *J Hypertens* **27(10)** 2028-2035
[9] Meinders JM, Kornet L, Brands PJ, Hoeks AP 2001 Assessment of local pulse wave velocity in arteries using 2D distension waveforms *Ultrason Imaging* **23(4)** 199-215
[10] Eriksson A, Greiff E, Loupas T, Persson M, Pesque P. Arterial pulse wave velocity with tissue Doppler imaging 2002 *Ultrasound Med Biol* **28(5)** 571-80.
[11] Bramwell JC, Hill AV. The velocity of pulse wave in man 1922 *Proc Soc Exp Biol Med* **93** 298-306
[12] Reneman RS, Hoeks APG, Westerhof N. Non-invasive assessment of artery wall properties in humans: methods and interpretation 1996 *Journal of Vascular Investigation* **2(2)** 53-64
[13] Chiu C Y, Arand PW, Shroff A, Feldman T, Carroll J. Determination of pulse wave velocities with computerized algorithms 1991 *Am Heart J* **121** 1460

The effect of Functional Electric Stimulation in stroke patients' motor control – a case report

Denise Pripas, Allan Rogers Venditi Beas, Caroline Fioramonte, Pedro Claudio Gonsales de Castro, Daniel Gustavo Goroso, Maria Cecília dos Santos Moreira

Instituto de Medicina Física e Reabilitação - HCFMUSP

de.pripas@gmail.com

Abstract. Functional Electric Stimulation (FES) has been studied as a therapeutic resource to reduce spasticity in hemiplegic patients, however there are no studies about the effects of FES in motor control of these patients during functional tasks like balance maintenance. Muscular activation of gastrocnemius medialis and semitendinosus was investigated in both limbs of a hemiparetic patient during self-disturbed quiet stance before and after FES on tibialis anterior, by surface electromyography. The instant of maximum activation peak of GM and ST were calculated immediately after a motor self-disturbance, in order to observe muscular synergy between these two muscles, and possible balance strategies used (ankle or hip strategy). At the preserved limb there occurred distal-proximal synergy (GM followed by ST), expected for small perturbations; however, at spastic limb there was inversion of this synergy (proximal-distal) after FES. It is possible that intervention of electricity had inhibited synergical pathways due to antidromic effect, making it difficult to use ankle strategy in the spastic limb.

1. Introduction

Stroke is a disease with vascular origin, in which there occurs a focal perturbation of encephalic function, due to ischemia or bleeding on encephalic nervous tissues. It is the second principal cause of death in the world, and among the survivors, there is a large ratio of disabled people that need constant support from their families and from social and health institutions [1].

Hemiplegia is the most prevalent sequel in patients that suffered stroke [2], which is commonly characterized by spasticity, motor control deterioration and consequently changes in gait pattern [3]. Usual motor alterations are decrease of triple flexion of lower limb [4], reduction on dorsiflexion range of motion, equinus foot, and knee hyperextension [5]. Thus, these subjects have higher energy expenditure in walking [6] and also balance deficit [4].

The main factor that contributes to these alterations is spasticity [7], due to loss of central inhibition of the muscle stretch reflex by injury to upper motor neurons, which predominates in extensor muscles at the lower limbs [2].

Spasticity reduces the ability to regulate voluntary movement. The reason that this occurs is that patients with stroke sequels have their reciprocal inhibition (RI) decreased [8], because the predominance of hypertonic muscles blocks the interaction between synergistic and antagonistic muscle groups, resulting in a static fixation of the joint – in this case, the equine foot – instead of dynamic stability during motor tasks [9].

At post-stroke rehabilitation, neuromuscular facilitation is commonly used, which emphasizes the inhibition of abnormal reflex patterns and tone normalization before training standing and gait [10]. Therefore, a possible treatment would be interfering in the mechanism that modulates RI and presynaptic inhibition [11]. It is known that FES has the immediate effect of increasing RI of stimulated muscles' antagonists [12], so it is possible that, by stimulating the dorsiflexor muscles, a decrease in spasticity of plantar flexors can be caused, favoring a more efficient pattern of muscle activation in spastic hemiplegic patients.

Since 1961, when Liberson et al. [13] described a method of FES on hemiplegics' fibular nerve, induce RI in spastic muscles by FES has been studied by several researches [14,15], and all of them found significant results such as increase of RI and, consequently, spasticity decrease in these patients. However, it is important to highlight that, with the exception of Alfieri's research [14], none of them measured the duration of RI on the spastic muscles. On the other hand, Alfieri showed that there was a reduction of spasticity after FES with duration of effect ranging from 10 minutes to 3 hours, with an average of 1 hour; but this was measured by the Ashworth Scale, which is subjective and unreliable to measure spasticity objectively.

Moreover, most of the cited studies has evaluated spasticity of plantar flexors in a static manner, but did not assess how these muscles behaved in functional motor tasks such as balance and gait.

With this background, we found that FES can be a useful therapeutic approach in reducing the spasticity of hemiplegic patients, but it is necessary to study the effects of FES in motor control during functional tasks such as maintaining balance.

2. Materials and methods

The present study investigated muscle activation of the lower limb in hemiparetic patients during self-disturbed stance before and after FES of tibialis anterior, by surface electromyography. However, we used kinematic variables to enable the division of the electromyography signal in two phases: disturbance and post-disturbance. The post-disturbance, when the subject is recovering from the perturbation, is the phase of interest in this study.

The research was developed in the *Laboratório de Movimento* of the *Instituto de Medicina Física e Reabilitação* of the *Hospital das Clínicas* of the *Faculdade de Medicina da Universidade de São Paulo* (IMREA – HC / FMUSP). The laboratory has advanced technological resources such as (1) Imaging System, (2) Power Platforms System and (3) Surface Electromyography System. These systems are integrated with the Acquisition and Processing System of the collected data.

2.1. Subject selection

We assessed a male volunteer, aged 51, who was in a rehabilitation program in the model of Light Hemiplegia in the *Estação Especial da Lapa* unit of IMREA.

Inclusion criteria were: diagnosis of chronic stroke (more than one year of injury), level of spasticity in lower limbs = 2 (Modified Ashworth), level of dorsiflexion muscle strength ≥ 3, and the presence of functional ambulation without aids or orthoses.

Exclusion criteria were: age above 60 years, previous use of FES as a treatment, botulinum toxin injection in the plantar flexor muscles in the last six months, orthopedic disorder or equinus foot deformity, cognitive disorders, other neurological disorders, and hemodynamic instability.

The volunteer was included in the study after he agreed to participate in the research and signed the INFORMED CONSENT FOR RESEARCH PARTICIPATION, making him aware of the procedures related to the collection of data. Additionally, all of his doubts were clarified as to the procedures to be performed. Participation was voluntary and unpaid.

The experimental methods used were non-invasive and pose no hazard to the health of the volunteer. We did not use any type of substance, and no medical/ therapeutic treatment was prescribed.

2.2. Subject Preparation

After the explanations of testing procedures, the individual wore appropriate clothing (swimsuit or shorts), and then he was properly prepared for the attachment of reflective markers at specific anatomical points, defined by the anthropometric model of Dempster (1995) [16]. This was necessary because through these markers, body segments were determined in order to calculate the total center of mass (COM) using the Ortho Trak 6.2 software [16], which was subsequently utilized for the separation of task intervals by the method of Costa et al. [17].

To capture the electromyography signal, after the abrasion of the skin with alcohol, surface electrodes were fixed on the following locations: the gastrocnemius medialis muscles (at 32% of the distance between the medial popliteal fossa and the medial side of the Achilles tendon), the semitendinosus (at 50% of the distance between the medial epicondyle and the ischial tuberosity), and the tibialis tuberosity (ground electrode). These points were chosen following the recommendations of SENIAM [18] and Sacco [19].

2.3. Experimental protocol

The experimental sequence of this study was composed of the following stages: (1) system calibration, (2) execution of the task, (3) intervention using the FES, and (4) post-intervention biomechanical analysis, every 45 minutes during a period of 3 hours.

The task consisted of the recovery of balance after a self-perturbation caused by the subject. The volunteer was instructed to position himself standing on the force platform, so the axis of mediolateral movement was represented by the x coordinate, and the anteroposterior by y, in relation to the axis of reference adopted on the laboratory platform. The initial position was 60° of trunk flexion (relative to the ground) with arms resting on the back of a chair - to avoid possible falls, since the subject had a deficit of balance - and feet comfortably parallel to the body (Figure 1).

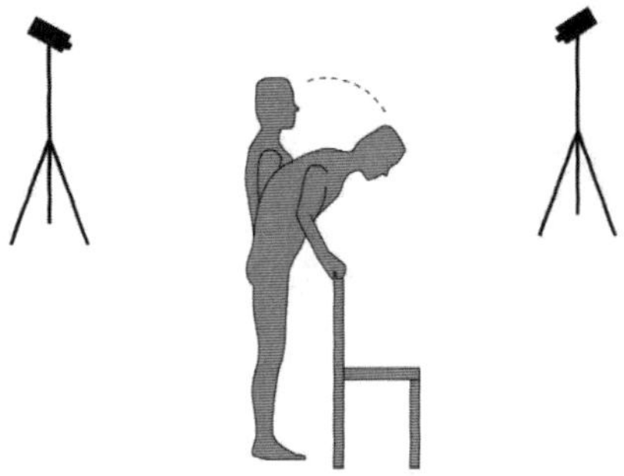

Figure 1: Schematic representation of the task: the volunteer's initial position (60° of trunk flexion) and final position (already in the upright posture).

Data collection began after an initial sound stimulus, when the volunteer was asked to make a sudden extension of the trunk to the upright position (Figure 1), causing him a motor disturbance. Afterwards, he was told to keep his eyes open and fixed on the horizon and stay in the upright quiet position for 60 seconds, when the second sound stimulus indicated the end of the first data collection. After 1 minute of rest, the second collection was performed, and then the third, respecting the same resting interval. Of these three collections, the first was considered the task's adjustment period and was discarded. The second was chosen as valid for analysis, but in cases when the subject had made some unwanted movement like shaking his head or arms, we used the third collection. This task has been used in other experiments with the intention of causing a motor disturbance, and was described in more details by Castro [20,21] and Costa [17].

The intervention with FES was initiated immediately after the initial biomechanical analysis. For this, the volunteer was positioned in a chair with his feet flat on the floor, with hip ischial support, keeping the hip, knee, and ankle in 90°.

For the motor point mapping, two adhesive electrodes were used: the first, with 5 cm of diameter, was used as the reference electrode; the second, with 3.2 cm of diameter, was used to map the motor point. It was considered an indication of proper motor point placement when there was muscle stimulation with FES, gradually increasing the amplitude and causing a pure dorsiflexion. After locating the motor point, two electrodes of the same size (5 cm) were applied for electrical stimulation.

Electrical stimulation parameters were: frequency = 20 Hz [22], pulse width = 300US, ramp-up = 2s, ramp down = 2s, contraction time = 6s, rest time = 10s, and the intensity was regulated until the occurrence of a minimum degree of dorsiflexion contraction (TA action) without joint movement [14]. The application time was 40 minutes.

After the intervention with FES, the volunteers were re-evaluated by the biomechanical system according to the procedures described above, and these were repeated four more times with time intervals of 45 minutes, totalizing a period of 3 hours after FES application.

Thus, each subject was analyzed in six trials, ordered from 1 to 6, namely: trial 1 - prior to FES, trial 2 - immediately after FES, trial 3 - 45 minutes after FES, trial 4 - 1hour and 30 minutes after FES, trial 5 - 2 hours and 15 minutes after FES, and trial 6 - 3 hours after FES.

2.4. Acquisition and processing of data

After pre-processing of the raw data by Evart software [23], the system provides the files with kinematic and EMG data. OrthoTrak 6.2 software [16] integrates the signals from the cameras and creates a three dimensional model of the studied subject by photogrammetry, after manual processing by a program operator.

Before processing the EMG signals, it was necessary to determine the following intervals performed by the volunteer: disturbance interval (when the subject performed trunk extension for upright posture) and post-disturbance interval (when he reached the upright position and kept oscillating to maintain postural control). For the division of these intervals, a pre-processing of kinematic data was carried out.

2.4.1. *Pre-processing of data.* For pre-processing of data, kinematic and EMG variables captured by the software Evart were used.

The imaging system generates a raw data file with information of markers fixed on the subject during the movement of self-perturbation, with the coordinates of the joints, the COM and the centers of mass of each segment at coordinates x, y and z. This file can be read in an Excel spreadsheet and has the extension *.TRBCoord. Another file with extension *.xls contains the values of the electromyography signals of ST and GM muscles of both legs during the collection interval [16].

In order to identify the end of self-disturbance, a criterion was established to select the task intervals, and these were calculated by a mathematical routine created in Matlab, which also provides the frame in which each interval began.

The criteria used were, in this order: a) speed of center of mass position vector, b) speed in z, and c) acceleration of center of mass position vector. Both the criteria and the equations used for calculation can be observed in detail in the study of Costa [17].

2.4.2. *Electromyography signal processing.* After the *.xls file was created, the linear envelope of the EMG signal was generated by the Excel software, and the first peak time of gastrocnemius medialis (GM) and semitendinosus (ST) muscles was calculated in the post-disturbance interval by the same software.

3. Results and discussion

The instant of maximum peaks of GM and ST muscles activation allows us to observe the synergism between these two muscles in an attempt to maintain balance after a motor disturbance, and possible strategies used to keep balance (ankle or hip strategy) on each trial (1-6), and are shown in Figure 2.

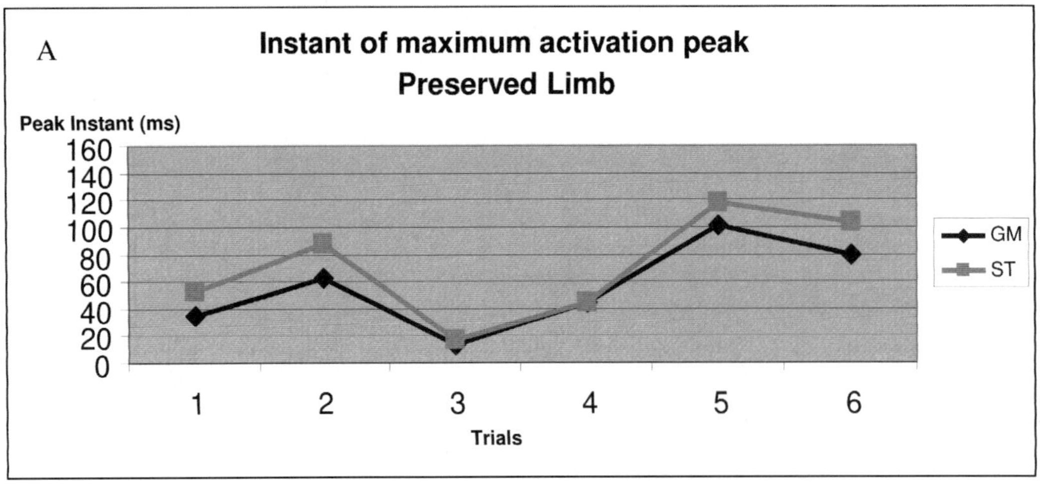

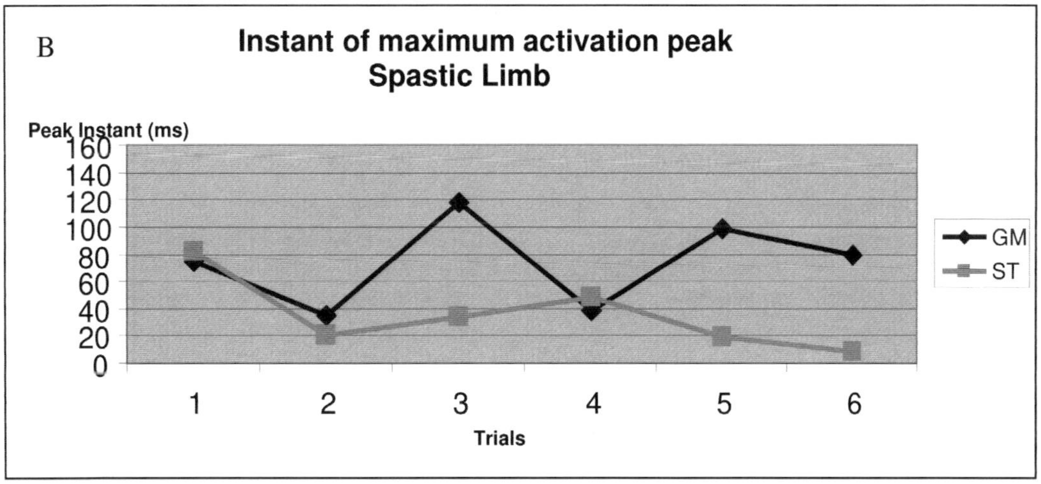

Figure 2: Instant of maximum peaks of the gastrocnemius medialis (GM) and semitendinosus (ST) muscles at trials 1 (prior to FES), 2 (immediately after FES), 3 (45min after FES), 4 (1h30min after FES), 5 (2h15min after FES), and 6 (3h after FES), in the preserved limb (A) and spastic limb (B).

One can observe that in the preserved limb (Figure 2A), there is a proximal-distal synergy (ankle strategy), in which activation of the GM occurs followed by ST activation for all trials, as expected for small perturbations [24]. In the spastic limb (Figure 2B), this pattern occurs only in the first trial (prior to FES); but in all trials after the completion of the FES (with the exception of trial 4) this pattern is reversed, and proximal-distal activation occurs. This synergistic pattern is related to the hip strategy, which relates to major disturbance of balance [24]. One can imagine that after the use of FES the spastic member showed greater difficulty in maintaining balance, since the same task was considered a major disturbance.

In 1986, DiFabio [25] performed a study with EMG in hemiparetic subjects and found latencies in distal muscles of their paretic limb that were longer than the contralateral limb, and he commented that patients with hemiparesis become fixed on stereotyped patterns of movement, showing a loss of flexibility and adaptability, as well as delayed responses and disruption of synergistic muscle coordination of the hemiparetic side, therefore they need to activate the proximal muscles before the distal muscles.

It is possible that FES did not cause an effective RI in the study subject's GM muscle, or it may even have caused an opposite effect to that which was expected: the action potential propagated by the motoneurons can also have passed along collateral axons making synapse with spinal inhibitory interneurons called Renshaw cells. Antidromic activation of these interneurons, in turn, inhibits the activity of agonists and synergistic hyperactive motoneurons [26]. Thus, instead of facilitating the synergy between GM and ST, there may be an inhibition of these synergistic pathways leading to a proximal-distal synergy in the spastic limb.

Another plausible hypothesis is that, due to spasticity, the subject presented greater weight-bearing at the preserved limb (despite having been instructed to distribute his weight equally between both legs), and therefore, this member had responded to self-disturbance of balance with little assistance of the spastic limb.

However, this research is a case report, so we cannot discard the possibility of measurement error or a non-sensitive method to detect changes caused in the synergy of GM and ST by RI.

More studies are needed, with a larger sample of subjects to verify the effectiveness of the RI by FES in the synergy of hemiplegic patients, and even different methodologies to detect its effect.

4. Conclusion

Spastic hemiplegic stroke patients present altered proximal-distal synergy in the spastic limb during maintenance of the balance after a motor self-disturbance. FES is a useful tool in reducing the spasticity of GM by RI, but there is the possibility that the antidromic effect, caused by an electrical intervention, can inhibit certain muscle synergies such as the one that is used in the ankle strategy of balance, so it can be more difficult to maintain balance after a disturbance.

Therefore, further studies are needed with larger samples of subjects and different methodologies in order to determine the benefit or harm caused by FES in these patients' motor control.

References

[1] Organização Mundial da Saúde 2006 *Manual STEPS de Acidentes Vascular Cerebrais da OMS: enfoque passo a passo para a vigilância de acidentes vascular cerebrais* (Genebra: Organização Mundial da Saúde)
[2] Lianza S 2001 *Medicina de Reabilitação* 3 ed. (Rio de Janeiro: Guanabara-Koogan)
[3] Woolley SM 2001*Top Stroke Rehabil.* **7**(4) 1-18
[4] Anderson TP 1994 *Reabilitação de pacientes com derrame cerebral completo* (São Paulo: Manole)
[5] Corrêa FI *et al.* 2005 *Arq. Neuropsiquiatr.* **63**(3-B) 847-51
[6] Bowden MG, Balasubramaniam CK, Neptune RR and Kautz AS 2006 *Stroke* **37** 872-6
[7] Knutson E and Martensson A 1980 *Scand. J. Rehab. Med.* **12** 93-106
[8] Okuma Y, Mizuno Y and Lee RG 2002 *Clin. Neurophysiol.* **113**(2) 292-7
Leonard CT, Sandholdt DY, McMillan JA and Queen S 2006 *J. Child. Neurol.* **21**(3) 240-6
Morira H, Crone C, Christenhuis D, Petersen NT and Nielsen JB 2001 *Brain* **124** 826-37
[9] Maisson J 1984 *Human neurology* **2** 261-7
[10] Barbeau H, Norman K, Fung J, Visintin M and Ladouceur M 1998 *Ann. N. Y. Acad. Sci.* **860** 377-92
[11] Larsson LE 1994 *Scand. J. Rehabil. Med. Suppl.* **30** 63-72
Levin MF and Hui-Chan CW 1992 *Electroencephalogr. Clin. Neurophysiol.* **85**(2) 131-42
[12] Agne JE 2004 *Eletroterapia: teoria e prática* (Santa Maria: Orium)
[13] Liberson WT, Holmquest HJ, Scot D and Dow M. 1961 *Arch. Phys. Med. Rehabil.* **42** 101-5
[14] Alfieri V 1982 *Scand. J. Rehab. Med.* **14** 177-82
[15] Visser SL and Zilvold G 1976 *EMG analysis in hemiplegic patients treated by means of FES of the peroneal nerve* 3rd Int. Cong. ISEK (Pavia, Itália, 30 Aug – 4 Sept 1976)
Carnstam B and Larsson LE 1975 *EEG and Clin. Neurophys.* **38** 214
Pompeu JE, Mattos ECT e Kohn AF 2009 *Rev. Fisiot. e Pesq.* **16**(3) 258-62
Kagamihara Y and Masakado Y 2005 *J. Clin. Neurophysiol.***22**(2) 136-47
Kesar TM *et al.* 2009 *Stroke* **40**(12) 3821-7
[16] *Manual Orthotrak 6.2* 2004 (Santa Rosa, CA, USA)
[17] Costa RMCL, Goroso DG and Lopes JAF 2008 *Criterion to identify the interval auto-perturbation in quiet position* 5ta Conf. Int. de la Facultad de Ingeniería Eléctrica de la Universidad de Oriente (Santiago, Cuba, July 2008)

[18] Hermens HJ, Freriks B, Disselhorst-Klug C and Rau G 2000 *J. Electromyogr. Kinesiol.* **10**(5) 361-74
[19] Sacco IC, Gomes AA, Otuzi ME, Pripas D and Onodera AN 2009 *J. Neurosci. Methods.* **180**(1) 133-7
[20] Castro PCG 2008 *Quantificação do trabalho mecânico em função de perturbações motora e visual.* [Dissertação]. (Mogi: Universidade de Mogi das Cruzes)
[21] Castro PSG, Goroso DG, Coelho DB, Lopes JAF e Moreira MCS 2009 *Acta Fisiátrica* **16**(4) 179-85
[22] Colling R. 1997 *Brief view: Distribution of Human Muscle Fibre Type Exercise Physiology* Available from: www.physiotherapy.curtin.edu.au/community/educational_resources/ep552_97/fibretypedist.htmlMuscle
[23] *Manual EVaRT 5.0* 2006 (Santa Rosa, CA, USA)
[24] Shumway-Cook A and Woolacott MH 2003 Controle *Motor- Teoria e Aplicações práticas* 2 ed. (Barueri: Manole)
[25] DiFabio RP, Badke MB and Duncan PW 1986 *Brain Res.* **363** 259
[26] Baker LL and Chandler SH 1987 *Brain Res.***420**(2) 340-50

SABI 2011 IOP Publishing
Journal of Physics: Conference Series **332** (2011) 012012 doi:10.1088/1742-6596/332/1/012012

Assessing the progress of rehabilitation in patients with ACL reconstruction using the International Knee Documentation Committee Subjective Knee Form.

J H Leguizamon, A Braidot, P Catalfamo Formento[1]
Laboratorio de Biomecánica, Facultad de Ingeniería, Universidad Nacional de Entre Ríos, Ruta 10, Km 11, 3101 Oro Verde, Entre Ríos.

Abstract. There are numerous assessment tools designed to provide information on the results of reconstructive surgery of anterior cruciate ligament (ACL). They are also used for monitoring progress and facilitating clinical decision-making during the rehabilitation process. A brief summary of some existing tools specifically designed to evaluate knee ligament injuries is presented in this article. Then, one of those outcome measures, the International Knee Documentation Committee Subjective Knee Form (IKDC) was applied to a group of patients (N = 10) who had undergone surgery for ACL reconstruction. The patients attended the same physiotherapy service and followed a unified rehabilitation protocol. The assessment was performed twice: four and six months after surgery. The results showed an improvement in the rehabilitation of most patients tested (verified by a difference equal to or greater than 9 points on the IKDC outcome between measurements 1 and 2). The IKDC probed to be an instrument of quick and easy application. It provided quantitative data about the progress of rehabilitation and could be applied in everyday clinical physiotherapy practice. However, the results suggested considering the IKDC as one component of an evaluation kit to make decisions regarding the progress of the rehabilitation treatment.

1. Introduction

After surgical interventions and during the early stages of rehabilitation programs it is necessary to measure and evaluate the evolution of the patient. Often tools for generic classification and diagnostic evaluation that rely on the initial pathology are implemented. A classification is used to describe the disease and the health status of the patient. An evaluation is utilized to understand the status and decide how to proceed.

1.1. WHO classifications

The World Health Organization (WHO) is responsible for issuing the diagnostic standards related to health conditions, using a classification system widely accepted. The objectives of the standards are to establish a common language on issues related to health and disease and to contribute to the epidemiological management and use of terminology in clinical practice. This family of classifications, as they are called, assists with the analysis and interpretation of data, information storage and retrieval, temporal and spatial comparison of population groups, incidence and prevalence of disease and identification of issues, characteristics and circumstances that affect the individual.

[1] Author to whom any correspondence should be addressed. E-mail: pcatalfamo@bioingenieria.edu.ar

Therefore it provides a consistent source of data to the responsible institutions in each country of the management and administration of health policies.

The International Classification of Function and Disability, ICF, part of the WHO Family of International Classifications, focuses on the description of health and health-related domains. These domains are classified from body, individual and societal perspectives by means of two lists: a list of body functions and structure, and a list of domains of activity and participation. Since the functioning and disability of an individual occurs in a context, the ICF also includes a list of environmental factors [1]. The domains in turn are organized into components. The domain of functioning and disability contains components such as body structures, body functions, activities and participation. The domain of environmental factors incorporate not only environmental but also personal factors, although these are not included in the classification itself but as part of the framework. The organization structure of the ICF is designed to help identify factors that prevent or improve the participation and integration of people within social, recreational or work environments [2]

The ICF guides a medical treatment towards a deficiency in the body structure, it guides the rehabilitation or training program towards certain activities and it also guides social policies and public health towards welfare and environmental factors. Hence the position within the classification relates with the development of structured assessment tools for evaluating results of intervention procedures and rehabilitation programs. Tools such as subjective evaluation questionnaires, functional tests and rating scales among others, go beyond the description of a condition. They are intended to quantify and qualify through the physical examination and diagnosis.

1.2. Evaluation instruments

Depending on the pathology to be evaluated different instruments are used. Especially in pathologies associated with the knee and given the structural and biomechanical complexity associated with it, many types of assessment instruments have been proposed. Overall, the instruments related to knee ligament can be divided in three categories: questionnaires, clinical tests and functional tests. The main between questionnaires and clinical tests is that the former measure disability while the latter measure the level of deterioration [3]. The high prevalence of anterior cruciate ligament injuries (ACL) [4] has bias the assessment tools toward the evaluation of functional state before and after injury. The tools are also designed to understand the processes of rehabilitation after ACL reconstruction (R-LCA), including other factors such as the level of activity at the time of intervention that can alter the outcome of treatment [5]. The requirements to be fulfilled by any evaluation before being accepted for administration are reliability, validity and responsiveness, the latter recently added as an important psychometric property to consider. Responsiveness is defined as the ability to detect clinically relevant changes, i.e. to detect improvement or deterioration in the patient over time [6, 7].

At the beginning of the 80s the need to compare the results of different surgical techniques or different methods of treatment was recognized. However, this was not possible due to the lack of standards for evaluation and the fact that different subjective rating scales provided different results. It was not easy to measure the evolution after surgery. Some efforts had been made using questionnaires that were administered by the surgeons themselves and presented in percentage scales [8]. Although the scales did not distinguish between objective, subjective and functional criteria, they provided the basis for the development of many evaluation frameworks [9].

The Lysholm score originally published in 1982 [10] has been one of the formats most widely used since its publication. It was intended to qualify knee instability in young patients and it was designed to be applied by qualified personnel in order to avoid bias. The scale evaluates among other things, different activities of varying intensities (generally, low demand activities). Since the activities do not necessarily represent a challenging effort to the patient, the final scores are often quite high. This affects the final analysis, given that in some cases the scale results in higher than expected values.

Another feature of the scale is that it does not include a physical examination, but it mainly focuses on the perception of the patient regarding their activities of daily living (ADL). The scale introduced the term instability, defined as a symptom [5], and reported as a sense of instability in the knee. After a first version, this scale was further developed and organized in 8 parameters with a maximum of 100 points in total [11]. This new version was used to assess the patient's level in sports and work activities, highlighting important variables in the process of rehabilitation. The combination of these two scales continues to be used in following ACL reconstruction surgery [12, 13] and rehabilitation programs [14].

The Cincinnati Scale was designed specifically to evaluate ACL injuries with emphasis on symptoms and the perception of the patient regarding the functional knee [15]. Since its publication it has been subjected to numerous modifications and has been developed to assess symptoms and functional limitations in sports and everyday life. The evaluation system consists of 11 components, including physical examination and radiographic stability, which is useful in evaluation of multiple ligament injuries. Further studies were performed to determine its validity, reliability and responsiveness [16]. Although widely used, it takes some time to be completed and presents certain level of complexity.

1.3. Use of the International Knee Documentation Committee Subjective Knee Form (IKDC)

The creation of the International Knee Documentation Committee (IKDC) emerged from the need to implement a standard assessment of ligaments and to provide common and universal terminology. Despite the existence of previous tools to assess ligament injuries and treatment outcomes, the systems often assign numerical values to factors that are difficult to quantify, which led to assigning arbitrary values that hampered comparisons with other tools. The initial objectives of the evaluation system would be to facilitate the publication of results of treatments and interventions of the knee [17], to include the essential criteria for evaluating results in a reproducible manner and to allow the use of the system in clinical practice with or without assistance of an experienced professional [18].

The original assessment includes a section on general information (documentation), a section of qualifications and a section of evaluation. The evaluation, the largest part, consists of four areas: subjective assessment, symptoms, range of motion and ligament examination. It was essentially designed to be used before the reconstruction and for the subsequent monitoring of patients until 10 years after surgery.

The Subjective Knee Form IKDC (S-IKDC) [18] emerged as an appendix to the original format and focused on the evaluation of improvement or worsening of symptoms, daily function and level of sports activity experienced by patients who had a variety of conditions, including injuries to ligaments, meniscus, cartilage and / or patellofemoral pain. The format consists of various subjective factors such as symptoms, functional abilities and sporting activities grouped into 18 questions. The questions on activities of daily living include walking on different surfaces, climbing stairs, standing still, kneeling, squatting, sitting with knees bent and rising from a chair. And with regard to sport activities, the format assesses abilities to run, jump using involved leg, move and stop and turn quickly. In summary, the S-IKDC was designed to detect symptoms and deterioration in function and sports activities due to knee injury and to maximize the reliability, validity and responsiveness of the test. Based on the evidence, the S-IKDC has proven to be a good indicator for evaluating the course of rehabilitation programs in subjects from different functional activity levels [19-21].

According to the criteria for certifying one such test, the validity and reliability was assessed as satisfactory by Irrgang in 2001. Also from the study, it was concluded that a true change is represented by a total change equal to or more than 9 points in the final score of the S-IKDC [18].

While the use of scales is recommended in the literature and supported by investigations regarding their reliability, validity and sensitivity, they are not used in everyday clinical practice. The aim of this study was, therefore, to assess the implementation of the S-IKDC in order to evaluate the rehabilitation progress of patients with ACL reconstruction.

2. Methods

2.1. Subjects

For this study, 10 patients (9 M, 1 F), age 16 to 40 years (mean 29.6 ± SD 9.5), weight 82.8 ± 16.2 kg, height 1.73 ± 0.05 m (Table 1) attending a physiotherapy service in Paraná, Entre Rios were recruited. All of them had been diagnosed with unilateral ACL injury and all had undergone semitendinosus-gracilis graft reconstructive surgery. The surgeries were performed between June and October 2010. All subjects attended the same physiotherapy service and followed a unified rehabilitation protocol that began two weeks after surgery and included physical activities of closed and open kinetic chain exercises.

Table 1. Characteristics of the participants

Subject	A	Ag	RK	TIS	TA1	TA2	CI
FS	2	22	R	4	5	7	Contact sport
DA	2	40	L	6	4	6	Occupational accident
FF	1	40	L	1	4	NA	Occupational accident
R	1	20	R	12	NA	6	Non-contact sport
EP	2	23	L	7	4	6	Contact sport
AC	2	22	L	2	4	6	Contact sport
O	1	UD	UD	UD	NA	6	Occupational accident
CG	2	39	L	~60	4	6	Contact sport
DY	1	31	L	4	4	NA	Occupational accident
V	2	16	R	<1	4	6	Non-contact sport

A: Number of assessment
Ag: Age
RK: Reconstructed Knee (R: right, L: left)
TIS: Time between injury and surgery (months)
TA1: Time between surgery and the first assessment (months)
TA2: Time between surgery and the second assessment (months)
CI: Cause of injury
UD: Unavailable Data
NA: The patient did not attend the evaluation session

All patients were informed of the purpose of the study and were invited to sign an informed consent before starting the procedure.

2.2. Procedure

The Subjective Knee Form IKDC was administered twice: 16 ± 1 weeks after surgery and 24 ± 1 weeks after surgery. The questions were read by the evaluator. For the interpretation of the results only data from patients who attended both sessions was analysed.

3. Results

Ten patients were recruited. Of those, six completed both assessments, two did not attend the second session and two did not attend the first session (Table 1). For processing the data obtained from the S-IKDC, Microsoft Excel® software was used, following the instructions of the evaluation form [18]. The results are presented in Table 2.

For the six patients analyzed, the leading cause of injury was contact sports activity (N = 4). The predominant age group was 16 to 23 years (N = 4), 39 and 40 years (N = 2).

All six patients reported continuing the rehabilitation program 3 times per week for at least 3 months, as offered by the health insurance systems. Five of them continued with individual training programs afterwards, and one reported having taken a one month break after the three months period and having started a physical program on his own after that.

The patients reported their pre-injury physical activity as practice of sports a few days a week (N = 3) and receive training and sports frequently (N = 3). At the time of the second assessment four patients had resumed or initiated training program every day (N = 2), 3 or 4 times a week (N = 2) or had started jogging (N=2).

Of the six patients who completed both evaluations, five showed an improvement in the second visit. This improvement was demonstrated by a difference in the results of the S-IKDC equal to or greater than 9 points between measurements 1 and 2 (Table 2). One patient presented a negative difference (of -1.1 points).

Table 2. Subjective Knee Form IKDC results for the first and second assessment and the difference (Diff) between assessments.

	IKDC		
Subjects	**1°**	**2°**	**Diff.**
FS	70,1	79,3	9,2
DA	69,0	88,5	19,5
FF	51,7	-	-
R	-	81,6	-
EP	64,4	74,7	10,3
AC	66,7	80,5	13,8
O	-	59,8	-
CG	65,5	77,0	11,5
DY	54,0	-	-
V	92,0	90,8	-1,1

4. Discussion

The results obtained are in the same range of those reported in the literature for one year [22], 5 years [20] and 6 years [23] after reconstructive surgery.

The administration of this subjective assessment proved to be fast (completing the questionnaire took no longer than 10 minutes) and easy procedure. However, it is not used in everyday clinical practice. Several factors could contribute to this state of affairs. One could be related to the healthcare coverage, which limits the visits for physical therapy to a fix number and it is unalterable by the opinion of the clinicians regarding the state of the patient. Hence, the assessment would not have a direct effect on the visits covered. Also, clinicians need to be familiar with the range of measures available and they should be able to evaluate the suitability of the measures for their clinical practice.

In general, there was no difficulty in interpreting and answering the questions of the form. The exception was shown in the section on sports activities. Some patient had not tried at the time of the first assessment some tasks such as jumping on one foot (N = 5) or kneeling (N = 2). The patients explained that either they were afraid of undertaking the activity which could affect the graft or rehabilitation process or that the physical therapist advised against performing the activity. For the second evaluation, only two patients had not attempted jumping and while all reported having kneeled (or attempted to).

One patient showed a change of -1.1 point from first to second assessment. This difference can not be considered a real change in the condition of the patient. However, the result is in partly explained by the high level of activity that the patient performed already at the fourth month of rehabilitation, i.e. the first assessment session (which is shown in the high value of the first session result of the S-IKDC (92.0)). This particular patient performed intense sports activities before the injury and he was trying to quickly return to the same activities. For this particular patient it would have been difficult to define his readiness to return to regular sports activity at the fourth month using the S-IKDC as the only evaluation procedure. In this case, other outcome measurements (such as assessment of muscle strength) would be useful to increase the information and for the decision – making process.

The general characteristics of the patients recruited for this study correlate well with the findings of previous studies, where the higher percentage of ACL injuries and reconstruction are found in recreational athletes and basketball players and only a small percentage on competitive athletes [20]. The subjects of this study participated in contact team sports, such as football (N = 3) and rugby (N = 1), with regular participation in sport games and competitive pre-injury level.

The S-IKDC has been submitted to several revisions in order to reduce the percentage of unanswered questions in the test groups. While this objective was in part achieved, the present study shows that there are still questions that patients find difficult to answer. The criteria to be used in evaluation of activities that patients do not performed (either for fear or pain), which are therefore responses left blank in the questionnaire, should be investigated in future studies.

Within the set of instruments designed to measure outcomes from the patient´s perspective, a recent study supports the S-IKDC as a measuring instrument of general but not universal applicability [24].

In conclusion, this study showed that the Subjective Knee Form IKDC is a tool of quick and easy application that provides quantitative data about the rehabilitation stage of the patient and could be applied in everyday clinical physiotherapy practice. However, its value as a status assessment of patient recovery is limited, which was also manifested by other researchers [25]. Hence, it would be sensible to use it as one component of an evaluation set. The whole set should be used then for decisions about the progress of the rehabilitation treatment.

5. Bibliography

1. Dahl TH 2002 International classification of functioning, disability and health: an introduction and discussion of its potential impact on rehabilitation services and research. *J Rehabil Med* **34** 201-4.
2. WHO 2001 ICF International Classiffcation of Functioning, Disability and Health World Health Organization.
3. Neeb TB, Aufdemkampe G, Wagener JH and Mastenbroek L 1997 Assessing anterior cruciate ligament injuries: the association and differential value of questionnaires, clinical tests, and functional tests. *J Orthop Sports Phys Ther* **26** 324-31.
4. Fu FH, Bennett CH, Lattermann C and Ma CB 1999 Current trends in anterior cruciate ligament reconstruction. Part I: Biology and biomechanics of reconstruction. *Am J Sports Med* **27** 821-30.
5. Tilley S and Thomas N 2010 What knee scoring system? *The Journal of Bone and Joint Surgery* 004.
6. Irrgang JJ, Anderson AF, Boland AL, Harner CD, Neyret P, Richmond JC and Shelbourne KD 2006 Responsiveness of the International Knee Documentation Committee Subjective Knee Form. *Am J Sports Med* **34** 1567-73.
7. Marx RG 2003 Knee rating scales. *Arthroscopy* **19** 1103-8.
8. O'Donoghue DH 1955 An analysis of end results of surgical treatment of major injuries to the ligaments of the knee. *J Bone Joint Surg Am* **37-A** 1-13; passim.
9. Labs K and Paul B 1997 To compare and contrast the various evaluation scoring systems after anterior cruciate ligament reconstruction. *Arch Orthop Trauma Surg* **116** 92-6.
10. Lysholm J and Gillquist J 1982 Evaluation of knee ligament surgery results with special emphasis on use of a scoring scale. *Am J Sports Med* **10** 150-4.
11. Tegner Y and Lysholm J 1985 Rating systems in the evaluation of knee ligament injuries. *Clin Orthop Relat Res* 43-9.
12. Bottoni CR, Liddell TR, Trainor TJ, Freccero DM and Lindell KK 2008 Postoperative range of motion following anterior cruciate ligament reconstruction using autograft hamstrings: a

prospective, randomized clinical trial of early versus delayed reconstructions. *Am J Sports Med* **36** 656-62.
13. Mitsou A, Vallianatos P, Piskopakis N and Maheras S 1990 Anterior cruciate ligament reconstruction by over-the-top repair combined with popliteus tendon plasty. *J Bone Joint Surg Br* **72** 398-404.
14. Bynum EB, Barrack RL and Alexander AH 1995 Open versus closed chain kinetic exercises after anterior cruciate ligament reconstruction. A prospective randomized study. *Am J Sports Med* **23** 401-6.
15. Noyes FR, Mooar PA, Matthews DS and Butler DL 1983 The symptomatic anterior cruciate-deficient knee. Part I: the long-term functional disability in athletically active individuals. *J Bone Joint Surg Am* **65** 154-62.
16. Barber-Westin SD, Noyes FR and McCloskey JW 1999 Rigorous statistical reliability, validity, and responsiveness testing of the Cincinnati knee rating system in 350 subjects with uninjured, injured, or anterior cruciate ligament-reconstructed knees. *Am J Sports Med* **27** 402-16.
17. Hefti F, Muller W, Jakob RP and Staubli HU 1993 Evaluation of knee ligament injuries with the IKDC form. *Knee Surg Sports Traumatol Arthrosc* **1** 226-34.
18. Irrgang JJ, Anderson AF, Boland AL, Harner CD, Kurosaka M, Neyret P, Richmond JC and Shelborne KD 2001 Development and validation of the international knee documentation committee subjective knee form. *Am J Sports Med* **29** 600-13.
19. Higgins LD, Taylor MK, Park D, Ghodadra N, Marchant M, Pietrobon R and Cook C 2007 Reliability and validity of the International Knee Documentation Committee (IKDC) Subjective Knee Form. *Joint Bone Spine* **74** 594-9.
20. Lee DY, Karim SA and Chang HC 2008 Return to sports after anterior cruciate ligament reconstruction - a review of patients with minimum 5-year follow-up. *Ann Acad Med Singapore* **37** 273-8.
21. Vitkus L, Fiodorovas M and Pocius G 2002 [Knee joint evaluation after anterior cruciate ligament plasty]. *Medicina (Kaunas)* **38** 510-4.
22. Moksnes H, Snyder-Mackler L and Risberg MA 2008 Individuals with an anterior cruciate ligament-deficient knee classified as noncopers may be candidates for nonsurgical rehabilitation. *J Orthop Sports Phys Ther* **38** 586-95.
23. Ferretti A, Monaco E, Giannetti S, Caperna L, Luzon D and Conteduca F 2011 A medium to long-term follow-up of ACL reconstruction using double gracilis and semitendinosus grafts. *Knee Surg Sports Traumatol Arthrosc* **19** 473-478.
24. Wang D, Jones MH, Khair MM and Miniaci A 2010 Patient-reported outcome measures for the knee. *J Knee Surg* **23** 137-51.
25. Strehl A and Eggli S 2007 The value of conservative treatment in ruptures of the anterior cruciate ligament (ACL). *J Trauma* **62** 1159-62.

Muscle-skeletal model of the thigh: a tool for understanding the biomechanics of gait in patients with cerebral palsy

Emiliano Pablo Ravera, Paola Andrea Catalfamo Formento, Marcos José Crespo, Ariel Andrés Braidot

Laboratory of Biomechanics, National University of Entre Ríos, 3101, Oro Verde, Entre Ríos, Argentina.

E-mail: emilianoravera@bioingenieria.edu.ar

Abstract. Cerebral Palsy represents the most common cause of physical disability in modern world and within the pediatrics orthopedics units. The gait analysis provides great contributions to the understanding of gait disorders in CP. Giving a more comprehensive treatment plan, including or excluding surgical procedures that can potentially decrease the number of surgical interventions in the life of these patients. Recommendations for orthopedic surgery may be based on a quantitative description of how to alter the properties probably muscle force generation, and how this affects the action of the muscle to determine how these muscles, impaired by disease or surgery, contributing to the movement of the segments of the limb during crouch gait. So the causes and appropriate treatment of gait abnormalities are difficult to determine because the movements generated by the muscular forces of these patients are not clearly understood. A correct determination of the etiology of abnormal patterns of the knee is the key to select the appropriate therapy, presenting a major challenge at present since there is no theoretical basis to determine the biomechanical causes of abnormal gait of these patients. The potential and necessity of using correct biomechanical models that consistently study the abnormalities becomes clear. Reinforcing and correcting a simple gait analysis and eliminating the unknowns when selecting the appropriate treatment is crucial in clinical settings. In this paper a computer muscle-skeletal model is proposed. The model represents a person's thigh simulating the six most representative muscles and joints of the hip and knee. In this way you can have a better understanding of gait abnormalities present in these patients. So the quality of these estimates of individual muscle dynamics facilitate better understanding of the biomechanics of gait pathologies helping to reach better diagnosis prior to surgery and rehabilitation treatments.

1. Introduction

Normal gait is a functional movement involving the most efficient energy transfer. Similarities between subjects are common but also small individual variations, so any deviation detected by comparing gait patterns are of great value [1].

Cerebral Palsy (CP) describes a group of developmental disorders of movement and posture causing activity restriction or disability, which are attributed to disturbances occurring in the fetal or infant brain [2]. It represents the most common cause of physical disability in modern world and within the pediatrics orthopedics units [3] [4].

Gait analysis provides great contributions to the understanding of gait disorders in CP [5]. It allows an objective and accurate quantification of the causes of gait disorders originating from musculoskeletal abnormalities in patients with CP and it demonstrates superiority to the visual or observational analysis of movement [5] [6]. It also provides a mean for a more comprehensive treatment plan, including or excluding surgical procedures that can potentially decrease the number of surgical interventions in the life of a patient with CP [7].

One of the most common abnormalities in children with CP is crouch gait, characterized by excessive knee flexion during the terminal swing phase and initial loading response and increased dorsiflexion during the whole gait cycle. Hip flexion may be similar to the normal pattern but often presents increased flexion and internal rotation during the complete gait cycle [8] [9] [10] [11].

Spasticity or static contracture of the hamstrings, are thought to cause the excessive knee flexion in many cases. Therefore crouch gait is commonly treated by surgical lengthening of the hamstrings, typically in combination with other orthopaedic procedures with the objective of diminishing the excessive knee flexion and improving the walking pattern [9] [11] [12] [13].

When crouched gait is caused primarily by contracture of the hamstrings, surgical lengthening may improve knee extension by decreasing the passive tension in the muscles [13]. When spasticity of the hamstrings contributes to the crouched gait pattern, surgical lengthening of the hamstrings may improve knee extension by attenuating the exaggerated response of muscles to stretch [14]. However, there are patients for whom the surgical procedure did not improve the knee flexion and it is thought that the cause of crouch gait on these patients may have been different to short or slow contraction of the hamstrings (related to spastic or otherwise affected hamstrings).

Studying the length and velocity of the muscle–tendon may help to identify patients who walk with abnormally short or slow hamstrings and would possibly benefit from surgery from those who walk with normal hamstring´s length and velocity of contraction [15].

Unfortunately, it is difficult to obtain a measure of the length and velocity of the hamstrings by simple inspection of kinematic data [16]. Hence, muscle-skeletal models are needed for calculation.

The models could improve recommendations for orthopaedic surgery, based on a quantitative description of how to modify muscle force generation and how these modifications affect the action of the muscles during crouch gait [17] [18].

However, the causes of and the appropriate treatment for gait abnormalities are difficult to determine because the movements generated by muscular forces of these patients are not clearly understood yet [19] [20]. Hence the explicit identification of the anomaly does not necessarily present a direct solution [21].

Due to the mechanical redundancy displayed by the human locomotor system, activities such as walking can be performed in different ways and in many combinations of muscle forces which can generate the same net joint moment. However, there are similarities in the muscle activation patterns (MAP) of different people performing the same well-learned task [22]. Experimental measurements of muscle forces during locomotion of animals showed that the MAP are in general stereotyped [23].

These consistencies suggest that the central nervous system uses specific principles for the control of individual muscle forces, and that the principles are the same for different persons. That leads to the hypothesis that the central nervous system selects optimal MAP with a set of criteria that are, as yet, unknown [22].

Assuming that the central nervous system minimizes stress on the muscles and the body, leads to the hypothesis that muscle strength may be found as the solution to an optimization problem [24].

However, it is thought that this crouch gait may be due to a variety of neuromuscular disorders, including abnormalities at the structure of hamstrings, which are not easily detectable.

A treatment plan is further complicated because there is no scientific basis to determine how the neuro-muscular-skeletal impairment contributes to the abnormal movement [25]. Abnormalities of the knee may be related to dynamic conditions such as spasticity, secondary static muscle contractures of the hip, knee and ankle, or a combination of short and spastic muscles.

A correct determination of the etiology of abnormal patterns of the knee is the key to select the appropriate therapy [26], presenting a major challenge at present since there is no theoretical basis to determine the biomechanical causes of abnormal gait of these patients [27] [28] [29].

At this point, the potential and necessity of using correct biomechanical models that consistently study the abnormalities become clear. Reinforcing and correcting a simple gait analysis and eliminating the unknowns when selecting the appropriate treatment are crucial in clinical settings [30].

In this study a computer model based on a simple rescaling method using subject-specific anthropometric data is proposed. The model is used to estimate the muscle tendon length and velocity of unimpaired subjects and patients treated for crouched gait with the objective of identifying those patients with abnormally short or slow hamstrings. It also estimates muscle forces of the six most representative muscles of the thigh using static optimization. It is believed that this will improve the understanding of gait abnormalities in patients with CP.

2. Methods

The musculoskeletal mechanical system proposed consists of the pelvis, femur, tibia and six muscles. Four of the muscles were considered two-joint muscles: Tensor Fascia Lata (TFL), Semimembranosus (SM), Sartorius (S) and Rectus Femoris (R) and two were modeled as monoarticular: Gluteus (G) and Iliac (I). The model has six degrees of freedom (DoF), three for the hip joint and three for the knee.

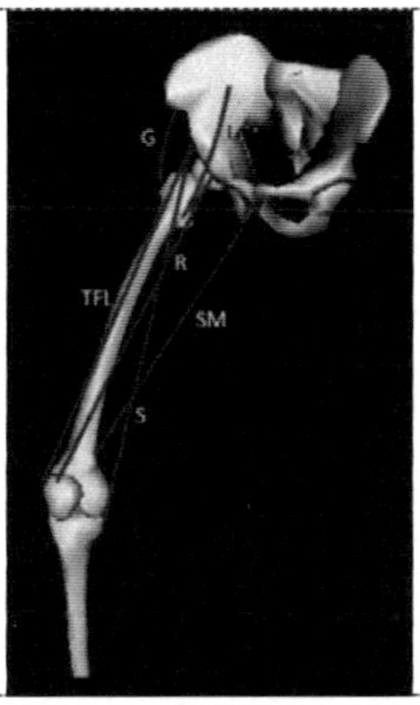

Figure 1. Mechanical System Modeling Scheme [31].

- Six muscles: Tensor Fascia Lata (TFL), Semimembranosus (SM), Sartorius (S) and Rectus Femoris (R), Gluteus (G) and Iliac (I).
- Six degrees of freedom (DoF).

Estimations of the patches of origin and insertion and directions of the muscles, assumed as the linear direction between the origin and insertion, were used by the muscle-skeletal model for dynamic evaluation of the muscle behavior throughout the gait cycle. The estimations were obtained from a complete three-dimensional muscle–skeletal model of the lower limb table 1.

Table 1. PCSA, origin (XO, YO, ZO) and insertion (XI, YI, ZI) of the muscle model [32]

Muscle	PCSA (cm^2)	XO (%)	YO (%)	ZO (%)	XI (%)	YI (%)	ZI (%)
TFL	5,9	11,4	26,6	12,5	3,0	15,2	20,8
SM	28,9	22,21	18,2	0,0	21,3	17,7	3,7
S	8,8	11,4	26,6	12,5	31,3	17,7	3,7
R	17,1	11,4	13,3	6,3	35,2	0,0	0,0
G	13	(***)	(***)	(***)	31,18(*)	6,3(**)	15,5(*)
I	60,8	(****)	(****)	(****)	31,18(*)	8,3(**)	21,5(*)

(*) Our estimates.
(**) Proportions of the patient's height.
(***) Middle position between the anterior superior iliac spines and posterior.
(****) Position estimated 7% of the width of the pelvis behind the origin of the gluteus.

The position of the patch ($\bar{p}_{origin}$, $\bar{p}_{insertion}$), origin and insertion of the six muscles, were calculated using axis systems embedded on and linked to the movement, $[\bar{u}, \bar{v}, \bar{w}]_{segment}$, of the pelvis and the leg respectively. Subject–specific anthropometric measures, $W_{segment}$, such as pelvis width and knee width were used for calculation.

The origin of the muscles was then determined using equations (1) and (2).

$$\bar{p}_o = \left(\bar{u}_{pelvis} \quad \bar{v}_{pelvis} \quad \bar{w}_{pelvis}\right) \cdot \begin{pmatrix} x_o \\ y_o \\ z_o \end{pmatrix} \cdot W_{pelvis} \qquad (1)$$

$$\bar{p}_{origin_i} = \bar{p}_{hip} + \bar{p}_o \qquad (2)$$

Similarly, the position of the insertions was calculated from the position of knee joint. The calculation used a weighted displacement linked to the movement of the leg. In this case, the weighting factor was the value of knee width of the patient (3).

$$\bar{p}_I = \left(\bar{u}_{leg} \quad \bar{v}_{leg} \quad \bar{w}_{leg}\right) \cdot \begin{pmatrix} x_I \\ y_I \\ z_I \end{pmatrix} \cdot W_{knee} \qquad (3)$$

Finally (4),

$$\bar{p}_{insertion_i} = \bar{p}_{knee} + \bar{p}_I \qquad (4)$$

Kinematic data from thirty trials from unimpaired subjects (10–25 years of age, 1.50–1.94 m in height and 49–105 kg in mass) and twenty trials from patients with CP that presented crouch gait (8–23 years of age, 1.08–1.83 m in height and 28–79 kg in mass) were used for calculation. Data from both, healthy and CP populations were provided by the Gait and Movement Laboratory at FLENI Institute for Neurological Research. All subjects analyzed were older than eight years, then it is considered that everyone had a mature gait pattern [33].

2.1. *Calculation of the length and velocity of shortening of hamstrings*

Since changes in length and velocity of the semimembranosus and the semitendinosus muscles are similar during the gait cycle, semimembranosus was considered as representative of medial hamstrings and used for this analysis [13].

Hamstrings length is then calculated as the distance between muscle position of origin and insertion (as line segments [34]). Shortening velocity is calculated as the first derivative with respect to time of the length of the muscle [13] [17], using the 25–parameter coefficients proposed by Savitsky and Golay [35]. Finally, the length of the semimembranosus is normalized to the length of the femur [17].

2.2. *Static optimization model*

The static optimization problem is given by (5) where the activation and / or muscle strength are calculated for each sample independently [22]. Thus, muscle forces are constrained by two physiological limitations: muscles may provide only contraction forces and these forces are limited to a maximum [36] [22].

$$\begin{aligned} &\underset{f}{Mín} \quad G(f_i^{(M)}) \\ &restrict \quad to \\ &C \cdot f = r \\ &y \\ &f_i^{(M)} \geq 0 \quad i \in \left\{1,\ldots,n^{(M)}\right\} \end{aligned} \qquad (5)$$

Where $C \cdot f = r$ is given by the net total torque of the six muscles actuating on the joints of the knee and hip (6) [36] [37]. Among the objective functions used in static optimization in gait, the use of polynomials criteria stands up (7).

$$\vec{M} = \sum_{i=1}^{6} \vec{r}_i \times \vec{F}_i = \sum_{i=1}^{6} \vec{r}_i \times \vec{d}_i F_i = B\vec{F} \quad (6)$$

These describe physiological conditions only if they are complemented by restrictions that prevent individual muscle forces from exceeding their physiological maximum when the external loads increase [24].

$$G(\vec{f}^{(M)}) = \sum_{i=1}^{n^{(M)}} \left(\frac{f_i^{(M)}}{N_i} \right)^p \quad (7)$$

In this paper we used a polynomial optimization function of order $p = 3$ [38] [31]. Where $f_i^{(M)}$ represents the amplitude of force of the $n^{(M)}$ muscles involved in the model and N_i is the normalization factors of the function, in this case the Muscle physiological cross-sectional areas (PSCA). The physiological behavior shown by this criterion is to minimize muscle fatigue [22].

3. Results

3.1. Model length and velocity of shortening of hamstrings

A *t-Student* analysis ($p < 5\%$) shows that the proposed muscle-skeletal model presented a better differentiation between patients and healthy populations in areas of increased hamstrings activation during walking.

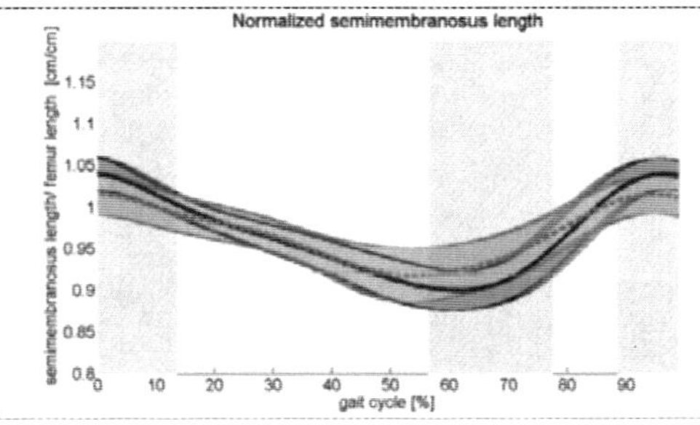

Figure 2. Semimembranosus length. Normal (line), crouch gait (dash) and area of difference $p > 95\%$ (gray bars) [30].

Figure 2 shows the mean and standard deviation of hamstring length for patients and healthy subjects. The figure shows larger differences between groups in the area of final swing and initial stance. Mean while there is a high probability of similarities ($p > 95\%$) between groups during the rest of stance phase.

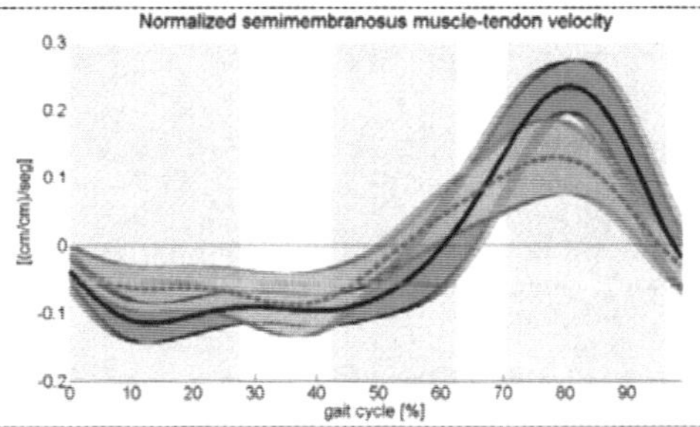

Figure 3. Semimembranosus shortening velocity. Normal (line), crouch gait (dash) and area of difference $p > 95\%$(gray bars) [30].

Figure 3 shows the mean and standard deviation of the velocity of the hamstrings for both groups. The *t-Student* test showed a statistically significant difference ($p > 95\%$) between populations in a large part of the gait cycle. The difference between groups is more notorious during middle swing, when the hamstrings reached the highest velocity of shortening.

3.2. Static optimization model

Figure 4 shows the forces of each of the six major muscles crossing the hip and knee joints for two gait cycles of three healthy subjects. It is possible to see that the patterns of force show a relationship between different trials of the same subject but there are not common patterns among different subjects. Force patterns found are similar to electromyographic recordings of the respective muscles reported in the literature [39].

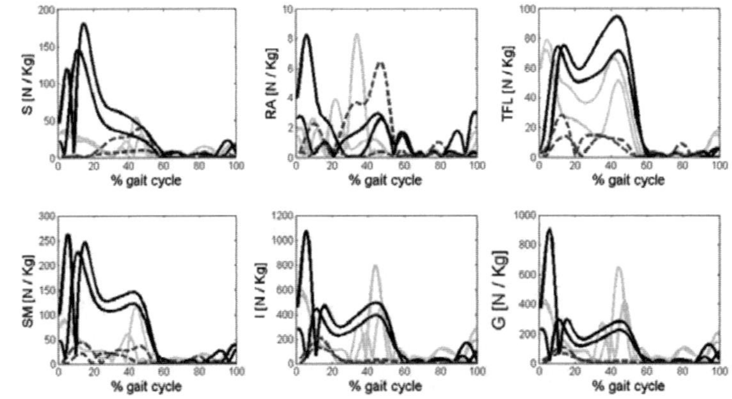

Figure 4. Module force patterns of two gait cycles of three subjects. Subject 1 (black line), 2 (gray line), 3 (dash).

The net joint moments at the hip obtained using articulated Link Segment Model (LSM) and recalculated by estimating the torque of the individual muscle forces simulated by the model presented in this work are shown in figure 5.

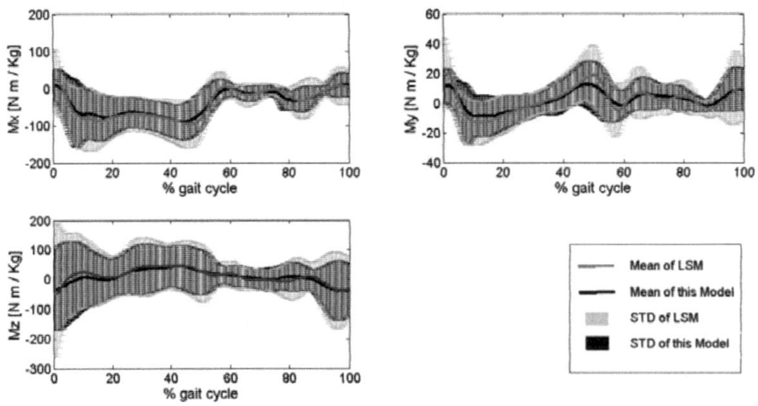

Figure 5. Net muscle moments of the hip in global coordinate system.

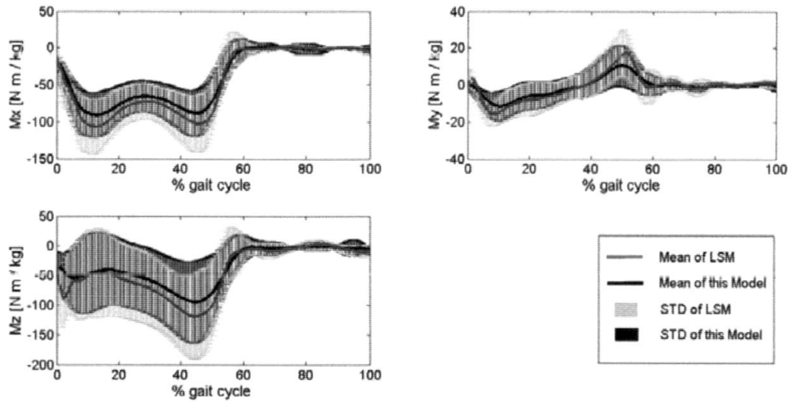

Figure 6. Net muscle moments of the knee in global coordinate system.

Figure 6 shows net joint moments of knee using articulated Link Segment Model (LSM) and recalculated values by estimating the torque to the individual muscle forces simulated by this model.

It is possible to see that the net joint moments for both joints calculated using the proposed model are very similar to the calculated using LSM. The small differences seen may be due to the filtering process applied to muscle forces prior to the reconstruction of torque. Also, the proposed model does not consider the effect of meniscis and ligaments, which may have influenced the differences found at the knee.

4. Discussion

The hamstrings length curves (figure 2) allowed for an optimal discrimination between groups (patients and healthy subjects) especially considering the differences at the end of swing and initial stance. However, there is a high probability of similarity between groups ($p > 95\%$) during mid and late stance. These findings were expected since previous work [40] showed that hamstrings exhibit little activity during mid and late stance while they are stretched to their maximum length at the end swing and initial stance.

In the case of velocity of shortening (figure 3) the difference between groups was more notorious during middle swing, when the hamstrings reached the highest velocity of shortening [40].

It is possible to see that the use of hamstrings length as a standalone parameter may not be sufficient for studying this group of patients, since some of them developed crouch gait, with spastic but not short muscles. This is consistent with previous results [12] that showed that from a population of patients with crouch gait, 30% had short and spastics muscles while 35% had spastic but not short hamstrings.

In the present study, 40% of patients walked with short and spastics hamstrings and 45% presented only spastic hamstrings. The minor differences in percentage may be due to the relatively small sample size used in this study, though it is sufficient to outline the trend.

In the evaluation of a movement that requires muscles submaximal efforts, the use of optimization problems makes it possible topredicted: (i) the reciprocal coactivation of monoarticular antagonists, (ii) synergistic co-activation of a two-joint articulation with their antagonists, (iii) simultaneous activation of muscles crossing the same joint, and (iv) the strong relationship between force and muscle activation as seen in figure 4 [31].

It also notes that achieved high concordance reconstruct the resulting net moments acting on the joints, leaving with the conclusion that the forces estimated by this model consistently approximates the actual forces acting. In this model, we observed that muscle forces for each person have a stereotyped MAP as reported in the literature [22] [23].

5. Conclusion

Patients with crouch gait normally walk with excessive flexion of the knee and hip [10] [12] [16]. It is thought that this abnormality may be due to a variety of neuromuscular disorders, including pathology of the hamstrings, which are not easily detectable. One possible solution is the implementation of biomechanical models.

The main obstacles faced by these models are that they are anatomically and physiologically incomplete, they may provide insufficient accuracy of the parameters and the appropriate validation remains a challenge.

This paper presents a model with six muscles and six DoF for the estimation of parameters of interest in the biomechanics of cerebral palsy gait such as length and velocity of shortening of hamstring and muscle strength with high concordance with existing models [24] [22] [23] and muscle activation.

A multi segment model was proposed in this study based on rescaling and adjusting for each patient using subject-specific anthropometric data. The model allowed for the discrimination between subjects with short, slow and normal hamstrings. It has the potential to provide information about muscle forces involved in walking. However, for a correct discrimination of forces in cerebral palsy patients further analysis regarding the most appropriate objective functions to use will be required since it is unlikely that minimization of metabolic cost would represent a real scenario for those patients. Ultimately the quality of these estimates of muscle dynamics facilitate better understanding of the biomechanics of gait pathologies helping to reach better diagnosis prior to surgery and rehabilitation treatments.

The main limitation of the proposed model is that the patient must not have undergone any orthopedic intervention in the leg which might affect the origin and insertion of the hamstrings.

The main advantage of the model, on the other hand, is that it does not require costly and time consuming magnetic resonance images for estimation of anatomical parameters.

References

[1] B Lofterød, T Terjesen, and I Skaaret, "Gait analysis: a new diagnostic tool," vol. 125 (15), no. 2014-2016, 2005.

[2] Unni G. Narayanan, "The role of gait analysis in the orthopaedic management of ambulatory cerebral palsy," *Current Opinion in Pediatrics*, vol. 19, pp. 38-43, March 2007.

[3] Alwyn Abraham and Janice M. Quinby, "Musculoskeletal problems in cerebral palsy," *Current Paediatrics*, vol. 15, pp. 9–14, 2005.

[4] Zubayer Karim, Andreas Kranzl, Margit Gfoehler, and Marcus G. Pandy, "Biomechanics of a crouched gait caused by spastic cerebral palsy in children," in *8th. World Congress on Computational Mechanics and 5th. European Congress on Computational Methods in Applied Sciences and Engineering*, Venice, Italy, 2008.

[5] Unni G. Narayanan, "The role of gait analysis in the orthopaedic management of ambulatory cerebral palsy," vol. 19, no. 38-43, 2007.

[6] Bjørn Lofterød, Terje Terjesen, Ingrid Skaaret, Ann-Britt Huse, and Reidun Jahnsen, "Preoperative gait analysis has a substantial effect on orthopedic decision making in children with cerebral palsy: Comparison between clinical evaluation and gait analysis in 60 patients," vol. 78 (1), no. 74-80, 2007.

[7] Mauro César De Morais Filho, Rafael Yoshida, Wander Da Silva Carvalho, Helena Elisa Stein, and Neil Ferreira Novo, "Are the recommendations from three-dimensional gait analysis associated with better postoperative outcomes in patients with cerebral palsy?," vol. 28, no. 316-322, 2008.

[8] Z. Matjacic and A. Olensek, "Biomechanical characterization and clinical implications of artificially induce crouch walking: Differences between pure iliopsoas, pure hamstrings and

comination of iliopsoas and hamstings contratures," *Journal of Biomechanics*, vol. 40, pp. 491-501, 2007.

[9] Scott L. Delp, Allison S. Arnold, Rosemary A. Speers, and Carolyn A. Moore, "Hamstrings and psoas lenghts during normal and crouch gait: Implication for muscle-tendon surgery," *Journal of Orthopaedic Research*, vol. 14, no. 1, pp. 144-151, January 1996.

[10] Jennifer L. Hicks, M. H. Schwartz, A. S. Arnold, and S. L. Delp, "Crouched postures reduce the capacity of muscles to extend the hip and knee during the single-limb stance phase of gait," *Journal of Biomechanics*, vol. 41, pp. 960–967, January 2008.

[11] M. H. Schwartz, V. Gasparroni, and T. F. Novacheck, "Can a simplified hamstrings length model predict the outcome of hamstrings lengthening surgery for the correction of crouch gait?," *Gait & Posture*, vol. 24, p. S22, November 2006.

[12] Marjolein M. Van der Krogt, Caroline A.M. Doorenbosch, and Jaap Harlaar, "Muscle length and lengthening velocity in voluntary crouch gait," *Gait & Posture*, vol. 26, pp. 532-538, 2007.

[13] Allison S. Arnold et al., "Do the hamstrings operate at increased muscle–tendon lengths and velocities after surgical lengthening?," *Journal of Biomechanics*, vol. 39, no. 8, pp. 1498-1506, March 2006.

[14] Allison S. Arnold, May Q. Liu, Michael H. Schwartz, Sylvia Ounpuu, and Scott L. Delp, "The role of estimating muscle-tendon lengths and velocities of the hamstrings in the evaluation and treatment of crouch gait," *Gait & Posture*, vol. 23, pp. 273-281, March 2006.

[15] Allison S. Arnold, Silvia S. Blemker, and Scott L. Delp, "Evaluation of a Deformable Musculoskeletal Model for Estimating Muscle-Tendon Lengths During Crouch Gait," *Annals of Biomedical Engineering*, vol. 29, pp. 263–274, January 2001.

[16] Caroline Stewart, Ilse Jonkers, and Andrew Roberts, "Estimation of hamstring length at initial contact based on kinematic gait data," *Gait and Posture*, vol. 20, pp. 61-66, 2004.

[17] Allison S. Arnold, Silvia S. Blemker, and Scott L. Delp, "Evaluation of a Deformable Musculoskeletal Model for Estimating Muscle-Tendon Lengths During Crouch Gait," *Biomedical Engineering Society*, no. 29, pp. 263–274, 2001.

[18] Allison S. Arnold, Silvia S. Blemker, and Scott L. Delp, "Evaluation of a Deformable Musculoskeletal Model for Estimating Muscle-Tendon Lengths During Crouch Gait," vol. 29, no. 263–274, 2001.

[19] J. L. Hicks, M. H. Schwartz, A. S. Arnold, and S. L. Delp, "Crouched postures reduce the capacity of muscles to extend the hip and knee during the single-limb stance phase of gait," *Journal of Biomechanics*, no. 41, pp. 960–967, 2008.

[20] Allison S. Arnold and Scott L. Delp, "The role of musculoskeletal models in patient assessment and treatment," *Clinics in developmental medicine*, no. 164/165, pp. 165-179, 2004.

[21] Allison S. Arnold, May Q. Liu, Michael H. Schwartz, Sylvia Ounpuu, and Scott L. Delp, "The role of estimating muscle-tendon lengths and velocities of the hamstrings in the evaluation and treatment of crouch gait," *Gait & Posture*, no. 23, pp. 273-281, 2006.

[22] Boris I. Prilutsky and Vladimir M. Zatsiorky, "Optimization-Based Models of Muscle Coordination," *Exerc. Sport Sci.*, no. 1, pp. 30-43, January 2002.

[23] B. I. Prilutsky, W. Herzog, and T. L. Allinger, "Forces of individual cat ankle extensor muscle during locomotion predicted using static optimization," *Journal of Biomechanics*, vol. 30, no. 10, pp. 1025-1033, 1997.

[24] John Rasmussen, Michael Damsgaard, and Michael Voigt, "Muscle recruitment by the min/max criterion-a comparative numerical study," *Journal of Biomechanis*, vol. 34, pp. 409-415, 2001.

[25] A. S. Arnold and S. L. Delp, "Computer modeling of gait abnormalities in cerebral palsy:application to treatment planning," vol. 6, no. 305-312, 2005.

[26] H. G. Chambers, "Treatment of functional limitations at the knee in ambulatory children with cerebral palsy," vol. 8 (Suppl. 5), no. 59-74, 2001.

[27] Allison S. Arnold, F. C. Anderson, M. G. Pandy, and S. L. Delp, "Muscular contributions to hip and knee extension during the single limb stance phase of normal gait: a framework for investigating the causes of crouch gait," vol. 38, no. 2181-2189, 2005.

[28] A. S. Arnold, S. Salinas, D. J. Asakawa, and S. L. Delp, "Accuracy of muscle moment arms estimate from MRI-Based musculoskeletal of the lower extremity," vol. 5, no. 108-119, 2000.

[29] Marjolein M. Van der Krogt, Caroline A.M. Doorenbosch, Jules G. Becher, and Jaap Harlaar, "Walking speed modifies spasticity effects in gastrocnemius and soleus in cerebral palsy gait," vol. 24, no. 422-428, 2009.

[30] Emiliano Pablo Ravera, Marcos Crespo, Paola Andrea Catalfamo, and Ariel Andrés Braidot, "Model to Estimate Hamstring Behavior in Cerebral Palsy Patient: As a Pre_Surgical Clinical Diagnosis Tool," in *32th Annual International Conference of the IEEE Engineering in Medicine and Biology Society*, Buenos Aires, 2010.

[31] Emiliano Pablo Ravera, Paola Andrea Catalfamo, Marcos José Crespo, and Ariél Andrés Braidot, "Modelo Músculo-esquelético para Obtener Fuerzas Musculares Individuales," in *V Latin American Congress on Biomedical Engineering CLAIB 2011*, La Habana, 2011, pp. 1-4.

[32] M.D. Klein Horsman, H.F.J.M. Koopman, F.C.T. Van der Helm, L. Poliacu Prosé, and H.E.J. Veeger, "Morphological muscle and joint parameters for musculoskeletal modelling of the lower extremity," *Clinical Biomechanics*, no. 22, pp. 239-247, 2007.

[33] David Sutherland, Richard Olshen, Edmund Biden, and Marilynn Wyatt, *The Development of Mature Walking*.: Mac Keith Press, 1988.

[34] Silvia S. Blemker and Scott L. Delp, "Three-Dimensional Representation of Complex Muscle Architectures and Geometries," *Annals of Biomedical Engineering*, vol. 33, no. 5, pp. 661-673, May 2005.

[35] John C. Russ, *The Image Processing Handbook*.: CSC Press, 2000, vol. I.

[36] Alexandre Terrier, Martin Aeberhard, Yvan Michellod, and Philippe Mullhaupt, "A musculoskeletal shoulder based on pseudo-inverse and null-space aptimization," *Medical Engineering & Physics*, vol. xxx, no. xxx, pp. xxx-xxx, 2010, doi: 10.1016/j.medengphy.2010.07.006.

[37] D. A. Winter, *Biomechanics and Motor Control of Human Movement*, Second edition ed.: Wiley-Interscience, 1990.

[38] Guoan Li, Janine E Pierce, and James H Herndon, "A global optimization method for prediction of muscle forces of human musculoskeletal system," *Journal of Biomechanics*, vol. 39, pp. 522-529, 2006.

[39] Michael W Whittle, *Gait Analysis an Introduction*, 3rd ed.: Butterworth-Heinemann, 2003.

[40] Anthony G. Schache, Tim V. Wrigley, Richard Baker, and Marcus G. Pandy, "Biomechanical response to hamstring muscle strain injury," *Gait & Posture*, vol. 29, pp. 332–338, 2009.

Journal of Physics: Conference Series **332** (2011) 012014 doi:10.1088/1742-6596/332/1/012014

Evaluation of arterial propagation velocity based on the automated analysis of the Pulse Wave Shape

F M Clara[1], A G Scandurra[1], G J Meschino[1] and L I Passoni[1]
[1] Laboratorio de Bioingeniería, Departamento de Electrónica
Facultad de Ingeniería, Universidad Nacional de Mar del Plata
Juan B. Justo 4302, Mar del Plata, Argentina

E-mail: fmclara@fi.mdp.edu.ar

Abstract. This paper proposes the automatic estimation of the arterial propagation velocity from the pulse wave raw records measured in the region of the radial artery. A fully automatic process is proposed to select and analyze typical pulse cycles from the raw data. An adaptive neuro-fuzzy inference system, together with a heuristic search is used to find a functional approximation of the pulse wave. The estimation of the propagation velocity is carried out via the analysis of the functional approximation obtained with the fuzzy model. The analysis of the pulse wave records with the proposed methodology showed small differences compared with the method used so far, based on a strong interaction with the user. To evaluate the proposed methodology, we estimated the propagation velocity in a population of healthy men from a wide range of ages. It has been found in these studies that propagation velocity increases linearly with age and it presents a considerable dispersion of values in healthy individuals. We conclude that this process could be used to evaluate indirectly the propagation velocity of the aorta, which is related to physiological age in healthy individuals and with the expectation of life in cardiovascular patients.

1. Introduction

The analysis of the pulse waveforms (PW) measured at the wrist is a well known method in assessing arterial stiffness, as shown in bibliography [1]. The pulse wave shape is determined by the interaction of incident waves generated by the heart and a reflected wave travelling back toward the heart and, possibly, one or more reflected waves travelling in either direction [2]. Hence, pulse wave can also supply an assessment of arterial wave reflection, which is closely related to cardiovascular condition [3].

The shape of pulse wave measured using a non invasive method at the radial artery of an adult healthy male is shown in Figure 1. To facilitate the analysis, the wave amplitude is normalized, corresponding to 100% for systolic pressure and 0% for diastolic pressure, so that the amplitudes are expressed as percentages [4]. It can be observed the systolic wave (SW), the diastolic wave (DW) and between them, the reflected wave (RW). The RW is caused by a reflection of the aortic pressure wave at the abdominal area [5] The analysis of the RW provides information on the vasoconstrictor tone of the arterial tree and also from the propagation velocity through the aortic region. Also it is shown the Reflected Wave Arrival Time (RWAT), estimated between the maximum of SW and the beginning of the RW.

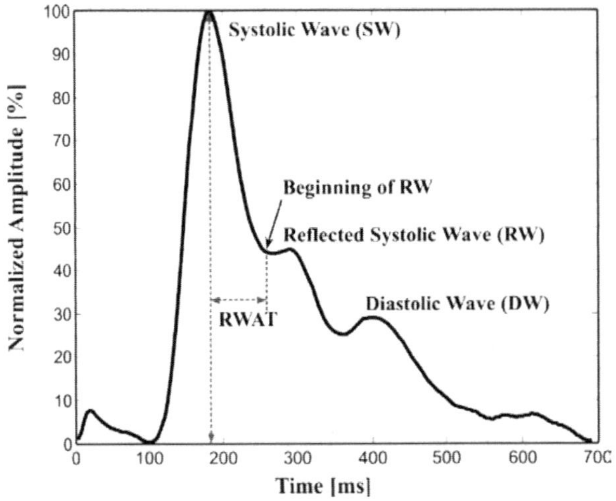

Figure 1. Pulse Wave. RWAT: Reflected Wave Time between the maximum of SW and the beginning of the RW. The minimum relative amplitude (0%) corresponds to diastolic pressure and the maximum (100%) to systolic pressure.

In young individuals, the RW has low amplitude and appears at the base of the systolic wave, when finalizing the systole. With increasing age, the increase of the propagation velocity of the pressure wave anticipates the arrival of the reflected wave to the wrist region where the pulse wave record is measured. Also, the augmentation of the vasoconstrictor tone in the reflection region increases its amplitude. For this reason, in adults, the reflected wave is observed in half of the roll-off of the systolic wave, and in older individuals its amplitude grows and is located near the apex of the systolic wave, sometimes forming a plateau [6].

In Figure 2, typical pulse waves corresponding to young, adult and elder people are shown, where RWAT is pointed. As it is noted, RWAT decreases with age and may be considered as a parameter of arterial aging exclusively related to the distensibility of the aortic walls, which determines the Pulse Wave Velocity (PWV). Aortic PWV increases when the aorta loses its elasticity by the natural aging process which is accelerated by hypertension (HT) [7].

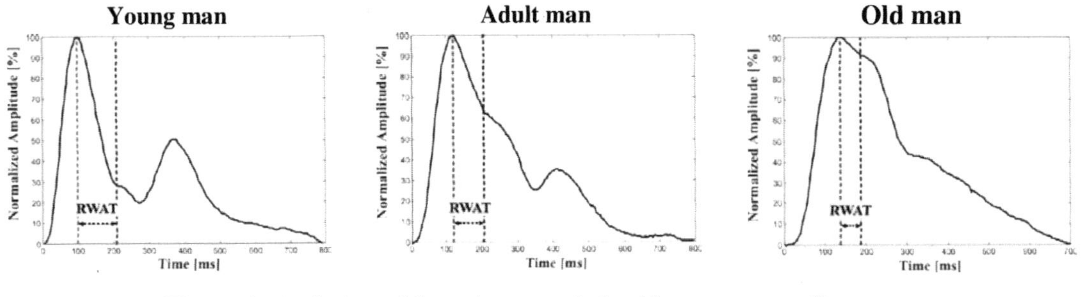

Figure 2. Evolution of the pulse wave in healthy men, according to age. SW maximum (red), RW (blue), and RWAT are detailed.

Many researchers attribute clinical significance to PWV; among them, Monnier found a close correlation between the aortic PWV and physiological age, claiming to consider PWV as an index of arterial aging [8]. Boutouyrie *et al* found that the increase of PWV is an independent predictor of coronary events production in hypertensive patients [9].

This paper proposes the automatic estimation of RWAT from PW raw records measured in the region of the radial artery. In order to avoid miscalculations in the RWAT, a functional approximation is obtained using a first order Takagi-Sugeno Fuzzy Inference System. The work of Ying [10] has shown that general SISO (Single Input – Single Output) Takagi-Sugeno Fuzzy Systems with linear rule consequent are universal approximators. In this process the parameters of the Fuzzy Inference System (FIS) are adapted using an Adaptive Neural Fuzzy Inference System (ANFIS) to optimize its performance. A similar approach has been used to process volumetric capnography records, to estimate derived variables of interest for the study of respiratory physiology [11, 12].

To evaluate the proposed methodology, we measured the RWAT and estimated PWV of PW records in a population of men from a wide range of ages. It has been found in these studies that PWV increases linearly with age and it presents a considerable dispersion of values in healthy individuals. We conclude that this automatic process of the PW records could be used to evaluate indirectly the aortic propagation velocity, which is related to physiological age in healthy individuals and with the expectation of life in cardiovascular patients.

2. Materials and Methods

2.1. Pulse Wave Acquisition

The acquisition system of the PW is based on a capacitive motion sensor, consisting of a small, rectangular insulated metal plate, (~2 cm^2), which is softly supported on the radial pulse zone without making any pressure. By connecting the patient to ground potential, a capacitor is formed between him and the metal plate. The pressure variations during the cardiac cycle causes variations in the radial artery diameter, transmitted to the surface of the skin, they generate the plate motion and, consequently, variations of its capacity. These capacity changes are used to generate a FM signal which is detected by the transducer and transformed into voltage variations, yielding a record of PW with morphological features similar to the radial pressure records obtained by applanation tonometry.

The transducer output is digitized and acquired in a conventional PC (5 ms, sampling period), for storage and processing.

PW recordings were obtained in a set of 201 healthy normotensive male volunteers aged between 2nd and 8th decade, according to bioethical procedures suggested by the Bioethics Committee of the National University of Mar del Plata.

2.2. The Pulse Wave Typical Cycle (PWTC)

The recorded raw sequence of PW contains many cycles of pulse waves. To estimate the RWAT, a pulse wave typical cycle (PWTC), which represents the entire sequence, must be selected. To perform this task a series of steps are carried out. The first one is the detection of each PW cycle: this process begins by estimating an approximate period using the Fourier transform; then the first maximum within the first period is localized. Thereafter, successive maximums are obtained considering each period. Once the maximums are obtained, the minimums are computed via the first derivative. Each PW cycle is neatly defined between consecutive minimums. Consequently a set of PW cycles, are overlapped to compute the statistical mode of the amplitude in each sample (PWM).

The PW cycle selected as typical (PWTC), is the one presenting the minimum Root Mean Square Error (RMSE) against the PWM.

2.3. Automated estimation of the RWAT

The RWAT ought to be computed, over the PWTC, as the lapse between the maximum of the systolic wave and the arrival of the reflected wave, identified by a slope change after the systolic peak. Given that requirement, this point must be precisely located; the PWTC must be smooth, free of noise, but it also must retain the information about the slopes changes. This property is clearly not found in the digitalized register without any smoothing process.

To ensure a precise location of the RWAT, we propose the use of a Takagi-Sugeno Fuzzy Inference System (FIS) as a data-driven model to get an analytic function of each PWTC. Hence, for each record, the estimation of the RWAT was automated, using a functional approximation of the PW.

A heuristic based on the change in the number of rules searches the best ANFIS model that approximates the PWTC, as follows:

(Step 1) Given a PWTC and the initial number of fuzzy rules n_0, obtain a FIS model using grid partition.
- Use ANFIS to optimize the parameters of the membership functions and the linear functions coefficients of the consequent of the FIS model.
- Compute the fitness function (*FF*) as a weighted linear combination of ANFIS RMSE and the range of the second derivative.

(Step 2) $n_i = n_{i-1} + 1$

(Step 3) Use ANFIS model to optimize the parameters as in step 1.

(Step 4) Compute the fitness function (FF_i) as in step 1.

(Step 5) If $FF_i \geq FF_{i-1}$, then stop and select the actual ANFIS model as the final one. Otherwise repeat steps 2 to 5.

To train the ANFIS model, the PWTC is sub-sampled at half alternately. Half of the data is selected for training and the other half is used for validation test. Hence, the ANFIS model is that which presents the minimum validation error, which avoids overtraining, and maximizes the generalization ability of the model.

The iterative process ends when the *FF* grows beyond the first local minimum which ensures an effective model with the minimum number of rules (parsimonious principle).

Once a PWTC good approximation is obtained with the ANFIS model, this analytical function (smooth and retaining its peak locations) is used to compute the RWAT. This parameter is computed as the lapse between the maximum of the systolic wave and the maximum of the second derivative.

Undoubtedly, RWAT depends on the height of the individual, because the path of the reflected wave increases with it. To become independent of it, Velocity Index [VI, m/s] is defined as an estimator of the aortic propagation velocity:

$$VI\left[\frac{m}{s}\right] = \frac{height[m]}{RWAT[s]}.$$

3. Results

In Figure 3, a raw sequence of PW is shown. In red is pointed the beginning of each PW and in green the systolic peak.

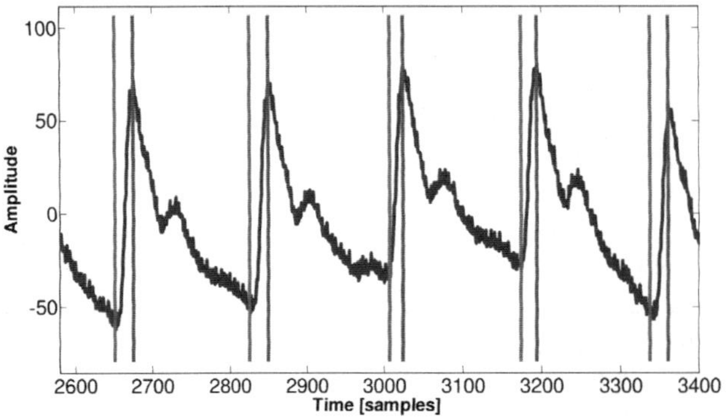

Figure 3. Typical raw PW record.

Figure 4 shows the set of overlapped PW cycles (blue) and the selected cycle (PWTC) according to the RSME criterion, is plotted in red.

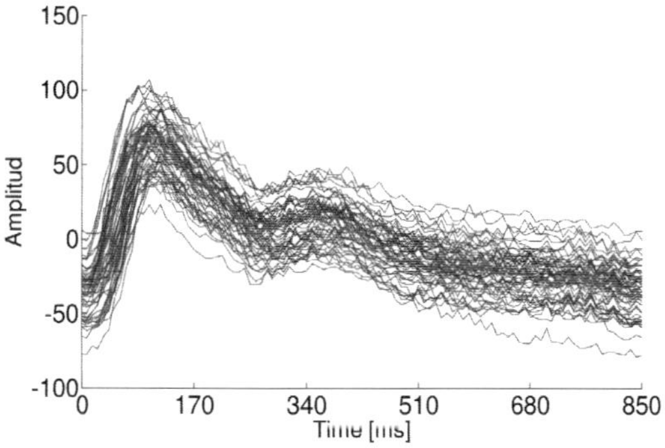

Figure 4. Set of overlapped PW cycles and the best cycle chosen automatically.

The result of the best ANFIS model, selected by the heuristic search, is shown in Figure 5: ANFIS output: in blue, PWTC in red. Figure 5 (a) and (b) show the PWTC of an adult and the PWTC of a young male, respectively.

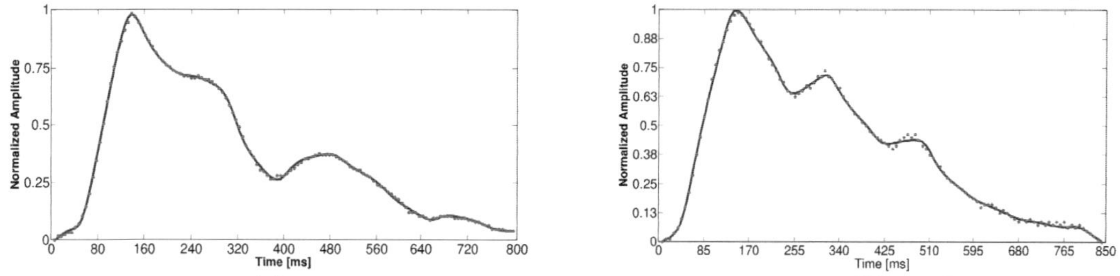

(a) (b)

Figure 5. Examples of ANFIS models (blue) and raw data (red).

Figure 6 shows the beginning of the RW and the second derivatives (black) computed of the ANFIS output (blue). The beginning of the RW, found as the first maximum of the second derivative after the systolic peak, is shown as a red line.

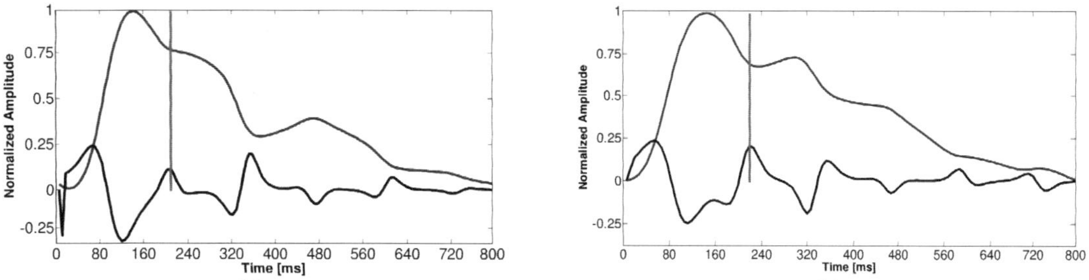

(a) (b)

Figure 6. ANFIS models (blue), second derivatives (out of scale) computed from the model (black) and beginning of Reflected Waves (RW) detected automatically (red).

In Figure 7, RWAT values are plotted against age of individuals between 2nd and 8th decades of life.

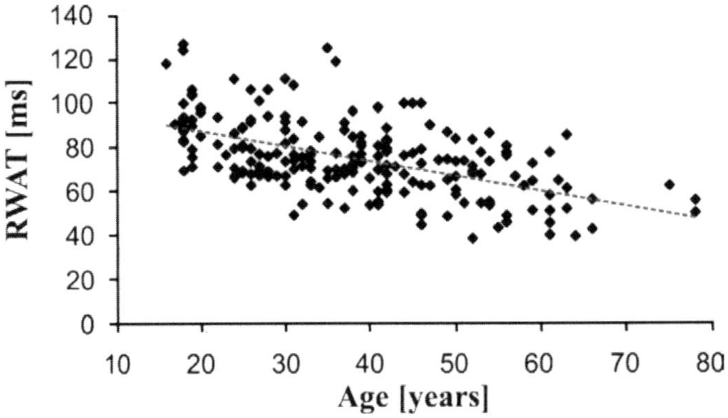

Figure 7. RWAT values related to age in a population of healthy men.

Figure 8 shows the VI values corresponding to the RWAT in Figure 7. It was found that the average value in normotensive VI increased reasonably linearly with age, a trend-line between the VI and normotensive age was $r = 0.57$, $p < 0.05$.

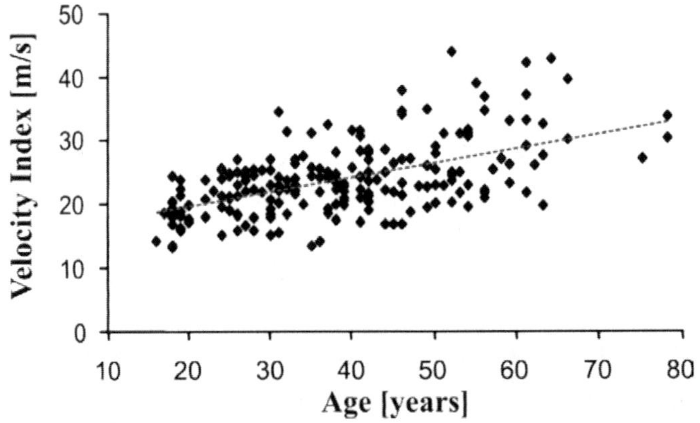

Figure 8. Velocity Index (VI) related to age of healthy men.

Table 1 shows the characteristics of the study population (mean values ± SD), and the index for each decade of life, highlighting their similar characteristics with regard to weight, heart rate and levels blood pressure. The low number of cases in the last two decades is due to the difficulty of finding healthy, non-medicated individuals, and clinically certified as such advanced ages.

In Figure 9, the Bland Altman [13] plot shows results of the comparison between the Velocity Index computed in the experimental case set with the proposed method and those obtained with the interactive process used so far. In the latter, the VI is computed on a set of pulse waves selected manually by the user, averaged and filtered with a FIR filter. As can be seen, the bias and the dispersion of the difference are mainly between the limits of agreement (bias ± 1.96 SD).

Table 1. Characteristics of the study population (Mean values ± SD)

Age Decade	N° of Cases	Systolic Pressure [mmHg]	Diastolic Pressure [mmHg]	Body Mass Index [kg/m^2]	Cardiac Frequency [bpm]	Velocity Index [m/s]
2ª	28	123 ± 27	75 ± 17	23 ± 2	73 ± 12	18,5 ± 2,7
3ª	36	118 ± 21	78 ± 7	24 ± 3	73 ± 12	21,2 ± 3,5
4ª	49	122 ± 21	82± 15	27 ± 3	72 ± 11	23,0 ± 4,6
5ª	46	123 ± 13	83 ± 9	26 ± 3	73 ± 11	24,6 ± 5,1
6ª	27	128 ± 14	85 ± 8	27 ± 3	71 ± 12	27,6 ± 6,2
7ª	12	133 ± 88	88 ± 7	27 ± 3	68 ± 10	31,1 ± 7,6
8ª	3	143 ± 3	85 ± 0	28 ± 3	75 ± 12	30,3 ± 3.3

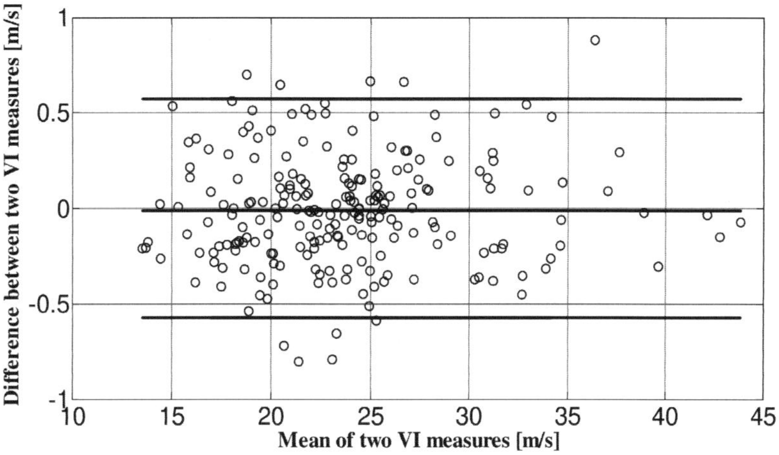

Figure 9. Bland-Altmann plot for Velocity Index (VI) measures.

4. Discussion

4.1. Methodological considerations

The proposed method to compute the RWAT is based on the requirements of a smooth pulse wave well tuned for each digitized register, providing low signal-to-noise ratio, while keeping the morphology of the PW.

While it is widely known that one of the most used filters for data smoothing and differentiation is the Savitzky-Golay method, it requires specifications of both, filter length and polynomial degree [14]. Results of the smoothing operation are therefore dependent on a priori choices. Unlike carrying out a heuristic search on the filter parameters, an adaptive neuro-fuzzy model is proposed to obtain an analytic function.

4.2. Physiological considerations

The distensibility of the wall of an artery depends on the amount, rate and state of its elastin fibers, collagen and vascular smooth muscle. Then an estimation of the propagation velocity provides an estimate of the state of the aortic walls, closely related to the functioning of the endothelium, which can be extrapolated to the rest of the arterial system [15].

In addition, during the course of aging are accumulating in collagen glycation end products and advanced glycation of proteins that generate cross-links between collagen or elastin fibers of the arterial wall, contributing to reduce their compliance [16, 17].

The elastic fibers begin to break down at early ages; until in old age have their influence on the properties of the aortic walls. Broken elastic fibers stretched to yield and be contained by the collagen fibers, much less distensible. This causes a gradual increase in diameter and decreased distensibility of the walls [18, 19].

The fragmentation progressive phenomena of elastic fibers impact disproportionately on the area of the ascending aorta, which has a relationship elastin/collagen= 3/1. In the abdominal aorta that relationship falls to 0.8 and the effect is less noticeable and more distant muscular arteries are hardly affected [20].

Meanwhile, other phenomena such as calcification alter the structure of the arterial walls in response to a multitude of interconnected causes such as dyslipidemia, hypertension, diabetes, smoking and oxidative stress, causing endothelial cell dysfunction [21, 22].

The above phenomena reduce the distensibility of the aortic walls and decrease the compliance of the artery, increasing the pulsatile component of blood pressure and ventricular afterload. These changes are highlighted by the increase in aortic velocity of propagation. High velocity of propagation implies high ventricular afterload and high pulse pressure, conditions that lead to the development of ventricular hypertrophy and to accelerated deterioration of elastic fibers by mechanical stress [23, 24].

In view of the above, it is explained that the decrease of the delay of arrival of the reflected wave, plotted in Figure 8, is due to increased propagation velocity of the aorta caused by natural aging. A wider delay corresponds to a healthy aorta, a situation typical of adolescents and youth. If it is pretended to compare the aortic state between different individuals, it should be noted that it also depends on the length of the path of the reflected wave, which is proportional to their height. The exact length of this route is uncertain, not only because the path is not straight but because the reflection does not originate from a particular point, but in various parts of the circulatory system, mainly in the area of the iliac bifurcation, affirming even that it would be a diffuse area and not a precise point [25, 26].

If we knew the exact length of the path of the reflected wave between the birth of the subclavian artery and the starting point for reflection, dividing twice that length by the time spent on the trajectory, we could get directly the value of the average propagation velocity in that area. This is currently not feasible, given the impossibility of accurately measuring the above mentioned circuit. For these reasons it is preferred to use the height of the individual to calculate the velocity, obtaining a velocity index with dimensions based on the identification of events in the radial pulse wave, which depends on the velocity of propagation in the aorta. The results shown in Figure 8 show that although there is wide variation in healthy individuals, its average value increases with age in a linear trend at least within the age range studied.

5. Conclusions

This paper proposes the automatic estimation of the arterial propagation velocity from the pulse wave raw records measured in the region of the radial artery. A fully automatic process is proposed to select and analyze typical pulse cycles from the raw data. An adaptive neuro-fuzzy inference system, together with a heuristic search is used to find a functional approximation of the PW.

The estimation of the pulse wave velocity is carried out via the analysis of the functional approximation obtained with the fuzzy model. The analysis of the fuzzy inference output was performed to compute the RWAT, and the aortic PWV was estimated with the height of the individuals.

The analysis of the PW records with the proposed methodology showed small differences compared with the method used so far, based on a strong interaction with the user. The automated process has a good performance also in those cases where the RW surpasses the level of the systolic peak, as in hypertensive patients.

The PW morphology gives information on the vasoconstrictor tone of the arterial tree and the propagation velocity through the aortic region. PW analysis also provides relevant information on the PWV. The PWV is useful to estimate the condition of the aortic walls, closely related to the functioning of the endothelium, which can be extrapolated to the rest of the arterial system.

The computed RWAT characterizes the arterial status, so a high delay corresponds to a healthy aorta, a situation typical of adolescents and youth. When comparing aortic differences between individuals, it should be noted that it also depends on the length of the path of the reflected wave, which is proportional to the individual height. There is a wide variation in healthy individuals, but its average value increases with age in a linear trend, at least within the age range studied. Such dispersion is not due to the inaccuracy of the measurement methods, but the variety of arterial anatomical features observed in practice. The aortic PWV assesses impairment of arterial walls in the area, but it is also an independent predictor of coronary events, which is currently providing a growing interest.

We conclude that the presented methodology shows encouraging results to indirectly evaluate the aortic propagation velocity, which is related to physiological age in healthy individuals and with the expectation of life in cardiovascular patients.

References

[1] Hsien-Tsai Wu, Chun-Ho Lee, An-Bang Liu, Wei-Sheng Chung, Chieh-Ju Tang, Cheuk-Kwan Sun, Hon-Kan Yip 2011 Arterial Stiffness Using Radial Arterial Waveforms Measured at the Wrist as an Indicator of Diabetic Control in the Elderly. *Biomedical Engineering, IEEE Transactions on* . **58**:2 243 – 252

[2] Greenwald S E, Carter AC, Berry C L 1990 *Circulation* **82** 114-122

[3] Xu L, Meng M, Qi X, Wang K 2010 Morphology Variability Analysis of Wrist Pulse Waveform for Assessment of Arteriosclerosis Status *Journal of Medical Systems* **34**: 3 331-339

[4] Clara F M, Cayrol M, Scandurra A, Meschino G, Garzillo M, Moyano E, Introzzi A 2005 La técnica de análisis de onda de pulso en la determinación del riesgo cardiovascular. *Rev Fed Arg Cardiol;* **34**: 213-220.

[5] Dart A, Kingwell B 2000 Pulse pressure. A review of mechanisms and clinical relevance. *JACC*; **37**: 975-984.

[6] Nichols WW 2005 Clinical measurement of arterial stiffness obtained from noninvasive pressure waveforms. *Am J Hypertens*; **18**:3 S-10S.

[7] Futterman L, Lemberg L. 2003 The effects of aging on arteries. *Am J Crit Care*.;**12** 472-475

[8] Monnier M 1987 Changes in pulse wave velocity with age in man: A longitudinal series over 20 years. *Experientia*; **43**: 378–81.

[9] Boutouyrie P, Tropeano AI, Asmar R et al. 2002 Aortic stiffness is an independent predictor of primary coronary events in hypertensive patients: A longitudinal study. *Hypertension*; **39**: 10–15.

[10] Ying, H. 1998 General SISO Takagi–Sugeno Fuzzy Systems with Linear Rule Consequent Are Universal Approximators. *IEEE Transactions On Fuzzy Systems* **6** 582-587

[11] Scandurra A, Maldonado E, Dai Pra A, Tusman G, Passoni L 2008 Modelo Híbrido para la aproximación funcional de registros de Capnografía Volumétrica In Annals of XIV Congreso Latino Ibero Americano de Investigación de Operaciones pp: 69.

[12] Tusman G, Scandurra A, Böhm S, Suarez-Sipmann F, Clara F 2009 Model fitting of volumetric capnograms improves calculations of airway dead space and slope of phase III *Journal of Clinical Monitoring and Computing* **23** 197-206.

[13] Bland JM, Altman DG 1986 Statistical methods for assessing agreement between two methods of clinical measurement *Lancet* **327**(8476): 307–10.

[14] Savitzky A, Golay M J 1964 Smoothing and Differentiation of Data by Simplified Least Squares Procedures *Anal. Chem.*, **36** (8), 1627–1639

[15] Anderson TJ. 2006 Arterial stiffness or endothelial dysfunction as a surrogate marker of vascular risk *Can J Cardiol*; **22**:72 B-80B.

[16] McVeigh GE, Bank AJ, Cohn JN. 2007Arterial Compliance. En: Cardiovascular Medicine. London: Springer, Cap. 8, p. 1811-1831.

[17] Zieman SJ, Melenovsky V, Kass DA. 2005Mechanisms, Pathophysiology, and Therapy of Arterial Stiffness *Arterioscler Thromb Vasc Biol* **25**: 932-943.

[18] Safar ME, Blacher J, Mourad JJ, London GM. 2000 Stiffness of carotid artery wall material and blood pressure in humans: application to antihypertensive therapy and stroke prevention *Stroke;* **31**(3):782-90.

[19] McVeigh GE, Bratteli CW, Morgan DJ, et al. 1999Age-Related Abnormalities in Arterial Compliance Identified by Pressure Pulse Contour Analysis *Hypertension*; **33**:1392-1398

[20] Rogers WJ. 2001Age-associated changes in regional aortic pulse wave velocity. *J Am Coll Cardiol*; 38:1123-1129.

[21] McEniery CM, McDonnell BJ, Aitken S, et al. 2009 Aortic Calcification Is Associated With Aortic Stiffness and Isolated Systolic Hypertension in Healthy Individuals; *Hypertension*; **53**(3):524-31.

[22] Fassett RG, D'Intini V, Healy H, Gowardman J, Gan JS, Sharman JE, Coombes JS. 2009Assessment of arterial stiffness, oxidative stress and inflammation in acute kidney injury. *BMC Nephrol*; 10:15.

[23] Viazzi F, Leoncini G, Parodi D, et al. 2002 Pulse pressure and subclinical cardiovascular damage in primary hypertension. *Nephrol Dial Transplant*; **17**: 1779 -1785.

[24] James M, Watt P, Potter J, Thurston H, Swales J 1995Pulse Pressure and Resistance Artery Structure in the Elderly. *Hypertension*; **26**: 301-306.

[25] Nichols W, O'Rourke M 2009 Aortic Pulse Wave Velocity, Reflection Site Distance, and Augmentation Index. *Hypertension*; **53**: e9.

[26] Pythoud F, Stergiopulos N, Westerhof N , 1996Meister J. Method for determining distribution of reflection sites in the arterial system. *Am J Physiol Heart Circ Physiol* **40** (5):1807- 1813

SABI 2011 IOP Publishing
Journal of Physics: Conference Series **332** (2011) 012015 doi:10.1088/1742-6596/332/1/012015

A novel time-domain signal processing algorithm for real time ventricular fibrillation detection

G E Monte[1], N C Scarone[1], and P O Liscovsky
Universidad Tecnológica Nacional, Facultad Regional Del Neuquén
P Rotter S/N
Plaza Huincul, ARGENTINA

E-mail: gmonte@frn.utn.edu.ar

Abstract. This paper presents an application of a novel algorithm for real time detection of ECG pathologies, especially ventricular fibrillation. It is based on segmentation and labeling process of an oversampled signal. After this treatment, analyzing sequence of segments, global signal behaviours are obtained in the same way like a human being does. The entire process can be seen as a morphological filtering after a smart data sampling. The algorithm does not require any ECG digital signal pre-processing, and the computational cost is low, so it can be embedded into the sensors for wearable and permanent applications. The proposed algorithms could be the input signal description to expert systems or to artificial intelligence software in order to detect other pathologies.

1. Introduction

Ventricular fibrillation (VF) is a serious class of arrhythmia which requires immediate and correct detection to save lives. In normal conditions, the electrical activities in the heart are synchronized. Under certain circumstances, this synchronism is lost and the ventricles enter in a state known as *fibrillation* [1]. As a result, there is a chaotic behaviour showing uncoordinated activity with poor or no blood pumping. Collapse and sudden cardiac death will follow in minutes unless medical help is provided immediately. If treated in time, VF can be converted into normal sinus rhythm. This requires shocking the heart with a device called a defibrillator [2].

VF detection algorithms have been studying for more than three decades. A large number of techniques of digital signal processing have been applied to the detection of pathologies in ECG signals. In the time domain, are highlighted; the threshold crossing intervals algorithm (TCI) [3] where decisions are based on the number and position of signal crossings through a certain threshold. The Tompkins algorithm based on QRS complex detection using slope, amplitude and width information [4]. The autocorrelation algorithms that use the periodicities within the ECG to infer signal parameters [5]. On the other hand, in the frequency domain, the spectral algorithm (SPEC) [6] analyses the change of patterns during VF in frequency bands by means of Fourier analysis. Wavelet transform also has been proposed [7], and more recently the Hilbert transform has been applied successfully [8]. Artificial intelligence techniques such as neural networks [9] and genetic algorithms [10] have been used with good results, but they require training stages or the computational cost is excessive.

[1] IDESIC research Group. UTN-FRN

Each method has advantages and disadvantages. Techniques applied in time domain are more efficient than others, but they are very sensitive to noise and artifacts. Signal transformations to other domains achieve detection of specific patterns but they involve a high computational cost.
Biomedical signals present a challenge to digital signal processing techniques. First, the signal to noise ratio is generally poor. Second, many artifacts are present like electromyographic noise, baseline drift, power line interference, radio frequency interference and skin-electrode interface noise. Such interferences can degrade the ECG signals; therefore the algorithms to analyze them must be smart enough to detect these artifacts or to avoid contaminated regions when the interference is excessively high. Figure 1 shows a simplified ideal ECG signal, where the important features are shape, relative position and duration of the expected waves. Figure 2 shows a real ECG waveform after filtering, sampling at the frequency of 128 Hz.

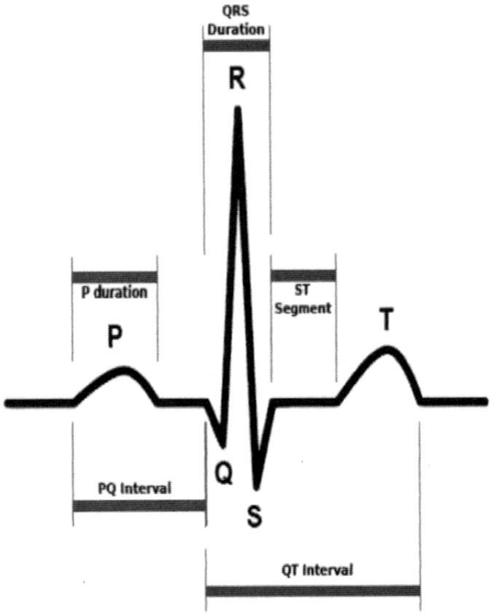

Figure 1. Ideal ECG waveform.

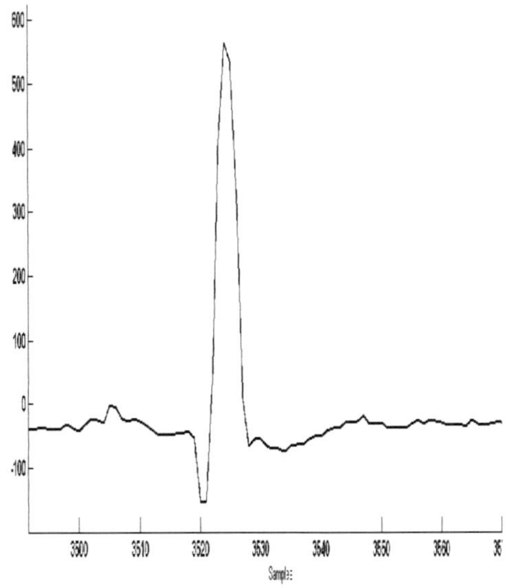

Figure 2. Real ECG signal from MIT-BIH Normal Sinus Rhythm Database record N° 16265.

The electrical signals produced by the heart in each beat are known since the mid-19th century. In 1901, William Einthoven published the document "String Galvanometer", which is considered the first electrocardiograph [11]. Throughout these years, Scientists, MDs and then Cardiologists have learned to infer the health state of the cardiovascular system analyzing the waveforms. ECG signals are time-variant; therefore almost any domain transformation requires windowing the signal. This window can include artifacts or noise mixed with normal signals. As a result, the signal should be pre-processed to discard regions of unwanted signals or to process regions of expected signals.
The aim for digital signal processing is to extract knowledge from data. Most of the algorithms start from data to get knowledge, e.g. to detect some pathology. Our proposal is to obtain an intermediate stage from data with a standardized structure. Then, from this platform, processes the information to get knowledge. This intermediate stage is important if we have multiple goals, because we do not have to go back to data to start a new algorithm. Figure 3 shows graphically these concepts.
This paper presents a time domain techniques for processing ECG signals that can be used to detect pathologies and to pre-process the signal to other more time-consuming processes. These techniques can be applied to any signal, as a result they conform a homogeneous platform for analysis.

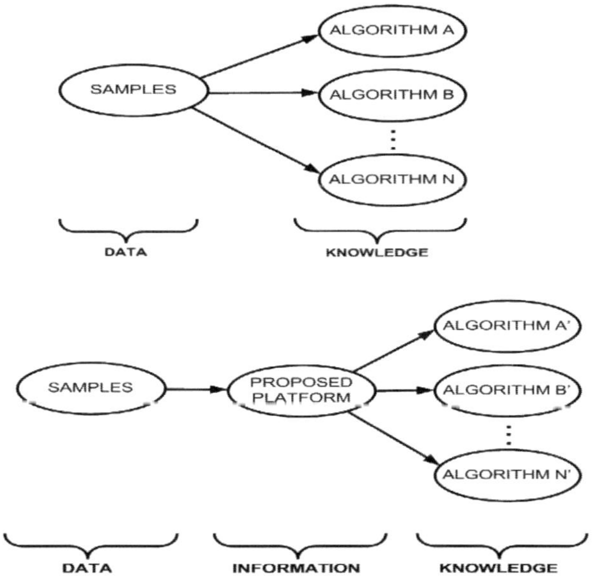

Figure 3. Above, classical approach: form data to knowledge. Below, our proposed scheme: from data to information and then to knowledge.

It starts in the sampling process to get a uniform platform with embedded information. The entire process imitates a human being observing the signal. We will see that the segmentation and labeling process is the raw information to infer global behaviour and to extract important features of the signal. In the following sections, these techniques are presented and applied to detect ventricular fibrillation as a particular example of signal feature extraction.

2. Materials and Methods

2.1 Segmentation description

If a human being watches the signal in figure 2, he can get information about the signal behaviour. We could conclude that the signal is "noisy" at the beginning. Then it reaches a maximum, a minimum and grows and falls rapidly with a spike like shape. After that, it starts to increase and then tends to a stable value. We can not measure the signal with precision but we are aware of the shape. We do not see isolated samples but the relationship among them. Therefore, we need a more complete sampling process in order to take into account samples sequences. The key process is to take into account how the signal trajectory between samples is. It is not necessary to do this for every pair of samples. Therefore, the first process is to segment the signal.

Let $x(t)$ the analog sensor signal band limited to **B** hertz and sampled at a frequency of **Fs** with uniform sampling, where $\mathbf{Fs} \gg \mathbf{2B}$. Since that the signal is oversampled, there will be redundancy between adjacent samples. The algorithm starts interpolating the sample between two samples, i.e. interpolates sample 2 from 1 and 3. Then, the error, as the difference between the real sample and the interpolated sample, is computed. If the error is less than a prefixed value, the sample has no relative importance because its value can be obtained from its neighbors. Now, the sample 3 is interpolated from 1 and 5 and so on. If the error exceeds a threshold, the sample is tagged as important and the process starts again, now from this sample. The interpolation is achieved by linear interpolation. If no important sample is found during N samples, the last sample is tagged and the process starts again from this sample. The tagged samples will be the left and right samples of the segment. The pseudo code of the algorithm called LIN1 is described in [12].

Journal of Physics: Conference Series **332** (2011) 012015 doi:10.1088/1742-6596/332/1/012015

As a result of the algorithm, two vectors are obtained, the tagged or essential samples vector *mark(n)*, and the temporal index vector *tempos(n)* .The algorithm is extremely simple; nevertheless it allows real time signal segmentation and a platform for the segment shape analysis. Only a division by two is involved in the average computation.

2.2 Labeling description

Each time the difference is computed in the interpolation process, the sign of the error is stored. When the segment ends, based on the majority of error signs accumulation, the behaviour inside the segment can be classified. The error sign is the clue to classify the segment. If the signal is oversampled, the trajectory of the real signal inside the segment is restricted by the interpolation error. Figure 4 shows an example of segmentation and labeling.

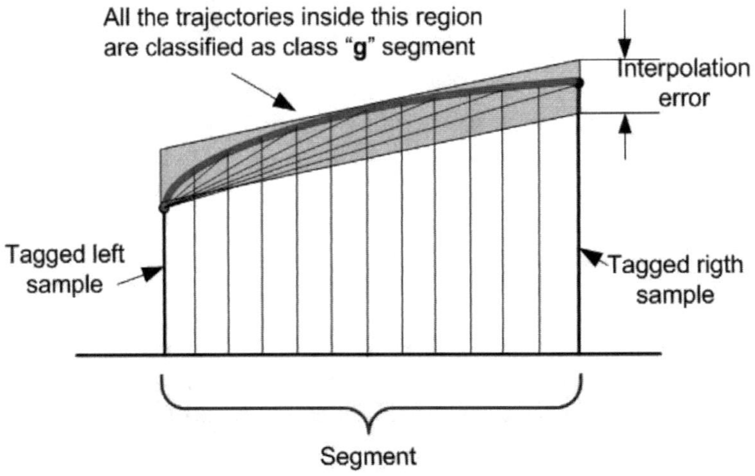

Figure 4. Example of segment class "g". In this case, the interpolated samples are below of the real ones.

Considering the value of the left and right samples and the sign of the interpolation error we have the simplified shapes shown in figure 5 and figure 6. In figure 5, the interpolation error does not reach a threshold so the segment ends due to it has reached the maximum size of N samples. In figure 6, the segments end due to interpolation error.

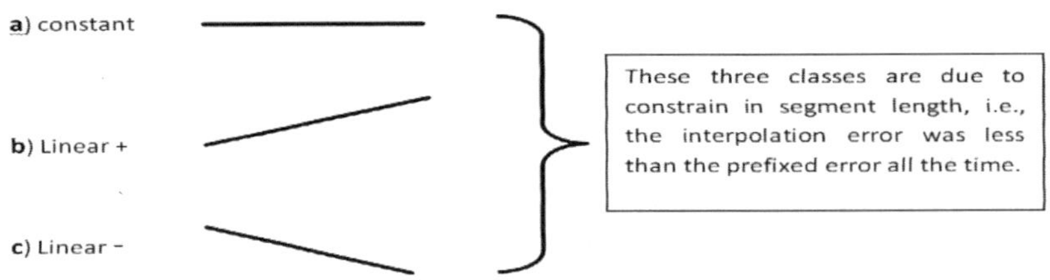

Figure 5. Class segments a,b and c. The interpolation error is less than the threshold and the segment ends, since it has reached the maximum length N.

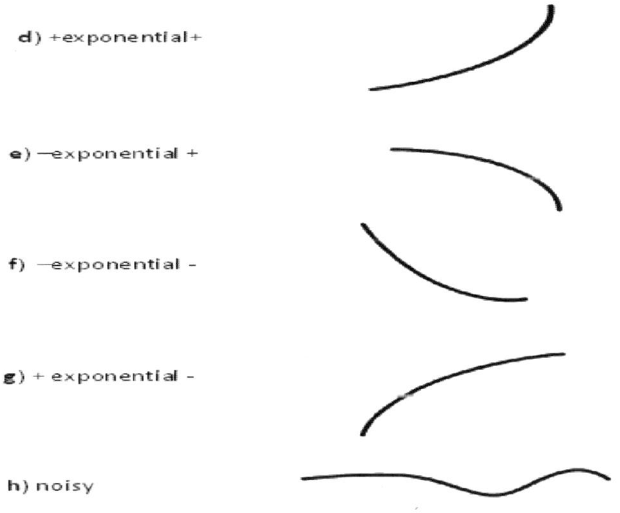

Figure 6. Class segments d,e,f,g and h. The segments are classified based on the error sign between real samples and the interpolated ones.

A new vector, *class(n)*, is added to the sensor signal description, whose elements are the segment class **a, b, c, d, e, f, g,** or **h**. The set of vectors *mark(n), class(n)* and *tempos(n)* in short **MCT**, conform the fundamental structure from which arises all the proposed algorithms and it is the proposed platform that captures information form data. The **MCT** vectors can be seen as a smart sampling process where the relationship among essential samples is clearly stated by the class vector. Why it is necessary an oversampling frequency? In order to get a uniform behaviour within the segment, i.e., most of the interpolation errors are all negative or they are all positive. Under conditions of oversampling and interpolation error approaching to zero, the probability of occurrence of segments **a,b,c** and **h** tend to zero. Therefore, only four classes of segments are important: **d,e,f,g** and the *sequence* of them indentifies the sensor signal. The vector *class(n)* is raw information in order to imitate a human observer. The vector *class(n)* is amplitude and time scale invariant. It captures only the shape information. For example, the pattern "**gefd**" or "**gggeeeeffffddd**", captures a sinusoidal waveform and the sequence "**ddddfffff**" codifies a positive spike like signal. The frequency and amplitude can be obtained from the **MCT** vectors. In addition, the *class(n)* vector provides precise localization of the signal local maximums and minimums. Maximums occur at the union of "**df**", "**de**", "**gf**" and "**ge**" segments. Minimums occur at the union of "**eg**", "**ed**", "**fg**" and "**fd**" segments.

We think that this vector is the "missing link" from the analog to the digital "world" in the traditional digitalization process, because it captures the shape information of the signal. Important signal features can be inferred from the sequence of segments. For example, a sequence of class "**g**" segments indicates that the signal has an exponential shape that tends to a steady state value if the length of the segments is increasing. A sequence of segments class "**d**" indicates that the signal is growing rapidly and class "**f**" that it is decaying quickly. Figure 7 shows some examples of waveform identification looking for sequences of segments.

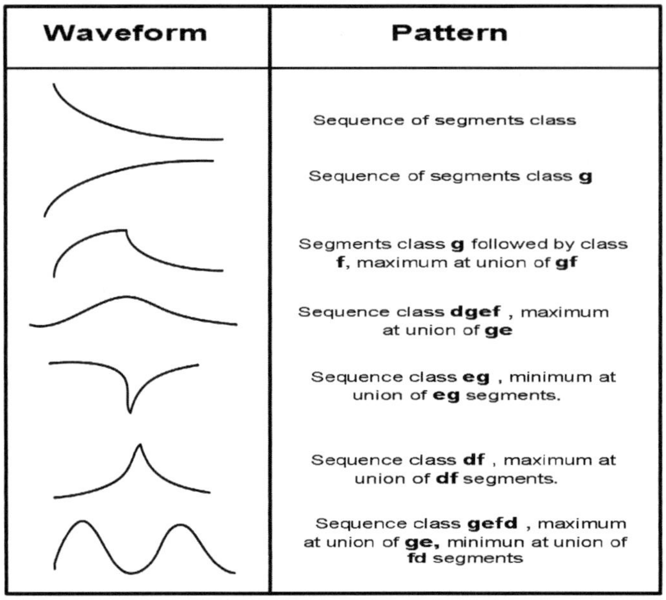

Figure 7. Examples of waveform identification through sequence of segments.

The MCT set of vectors is our normalized information platform from which arises all the proposed algorithms.
Matlab software was used to program the proposed algorithms. We tested the performance of pathologies detection using real data from the collection of recorded physiologic signals, freely available from PhysioNet [13].

3. Algorithm for fibrillation detection

One of the clues for detecting the chaotic behaviour of VF is the lack of regularity in the RR time interval. An additional characteristic is the increment in the heart beat frequency beyond normal limits. In episodes of ventricular tachycardia (VT) the heart beat frequency is high but there is regularity in the RR period with a spike like signals. Since that the RR time interval is a random variable, we will use the standard deviation as another parameter to infer VF. Therefore, we need a precise detection of the QRS complex which is presented in the next subsection.

3.1. QRS pattern detection

The algorithm is based on the information provided by the MCT vectors. The *class(n)* vector has the property that easily locates maximums and minimums of the segmented signal. Figure 8 shows a real QRS complex and the output of the segmentation and labeling processes. The red line is the linear interpolation from the tagged samples. This interpolation is not part of the process; it is shown only to remark the segments.

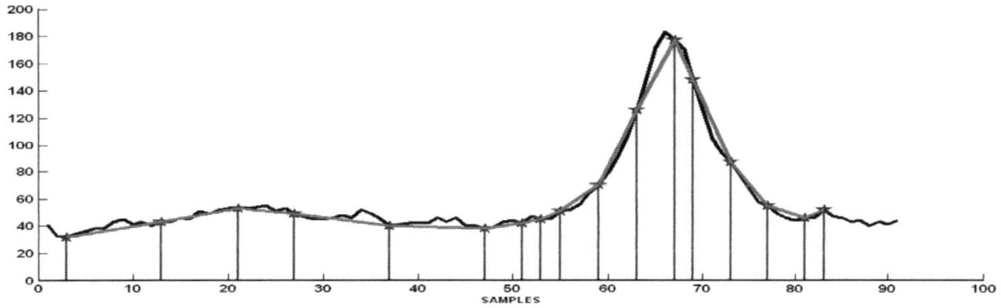

MCT VECTORS

3	13	21	27	37	47	51	53	55	59	63	67	69	73	77	81	83
32	43	53	49	40	38	42	45	51	70	126	177	148	87	55	46	52
F	G	D	E	F	E	G	G	D	D	D	G	E	F	F	F	D

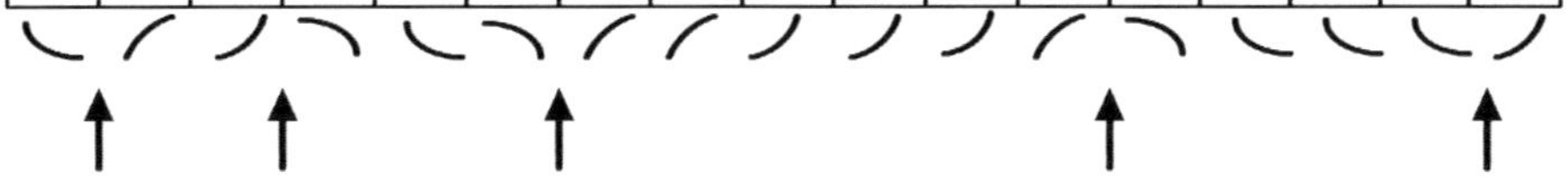

Figure 8. Segmentation and labeling applied to test signal (426m.dat from MIT-BIH Malignant Ventricular Arrhythmia database, interpolation error=3, Fs=250 Hz. Below, the resultant MCT vectors showing the maximums and minimums.

QRS pattern detection begins looking for the trio min-max-min. Then, five conditions must be satisfied in order to qualify as a normal pattern:

a) The raising section must be class "**d**" dominant segment.
b) The falling section must be class "**f**" dominant segment.
c) The difference of the min values must be a fraction of the difference between the max value and the average of the min values.
d) The difference between the height and the base should be greater than a preset value.
e) The energy of the impulse, estimated as the area of a triangle should be inside a range.

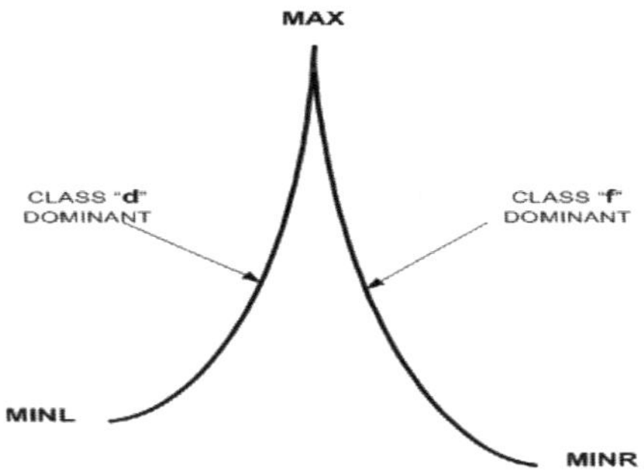

Figure 9. QRS pattern candidate and its nomenclature.

Specifically, for the PhysioNet database with a sampling frequency of 250 Hz and using the nomenclature of the figure 9, we employ:

a) The raising section should contain at least two class "**d**" segments.
b) The falling section should contain at least two consecutive class "**f**" segments.

c)

$$dif\_bases < \alpha[abs(mark(\max) - base\_average)] \quad (1)$$

Where, $dif\_bases = abs[mark(\min L) - mark(\min R)]$ (2)

$$base\_average = [mark(\min L) + mark(\min R)]/2 \quad (3)$$

d)

$$abs[(mark(\max) - base\_average)/base\_width] > \beta \quad (4)$$

Where, $base\_width = tempos(\min R) - tempos(\min L)$ (5)

e)

$$\delta 2 < Energy < \delta 1 \quad (6)$$

Where, $Energy = base\_width * (mark(\max) - base\_average)/2$ (7)

For the experimental results shown in this paper, we have used:

$$\alpha = 0.42; \quad \beta = 2.0; \quad \delta 1 = 1200; \quad \delta 2 = 6000.$$

Figure 10 shows experimental results of QRS detection. The proposed algorithm has an excellent performance even with base line drift and noise.

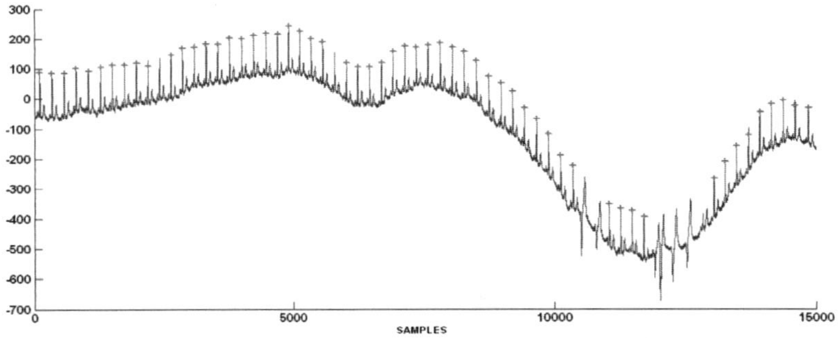

Figure 10. QRS detection applied to 426m.dat from MIT-BIH Malignant Ventricular Arrhythmia database, interpolation error=3, Fs=250 Hz. "+" implies detection.

If we analyze the reason of QRS pattern rejection, it is possible to detect other pathologies. For example, figure 11 shows a zoom of the lower part of figure 10 where pattern were rejected due to excessive difference between minimum values that could point out premature ventricular contraction (PVC).

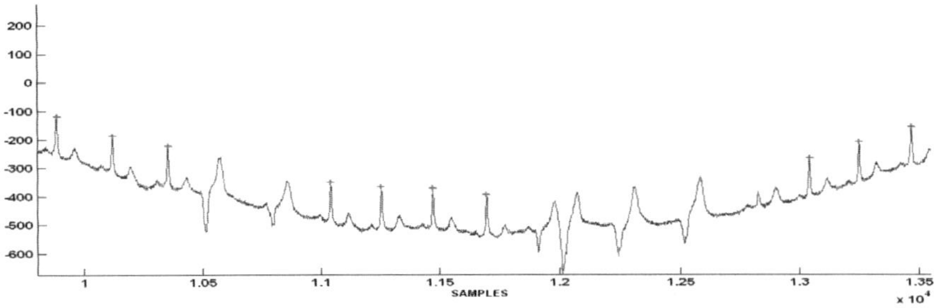

Figure 11. Zoom of the lower part of figure 11 showing rejected patterns that could indicate PVC (Premature Ventricular Contraction).

3.2. Ventricular fibrillation detection

Figure 12 shows the algorithm applied to an ECG signal with VF. We can observe the change of pattern showing increment of frequency content and an irregular shape.

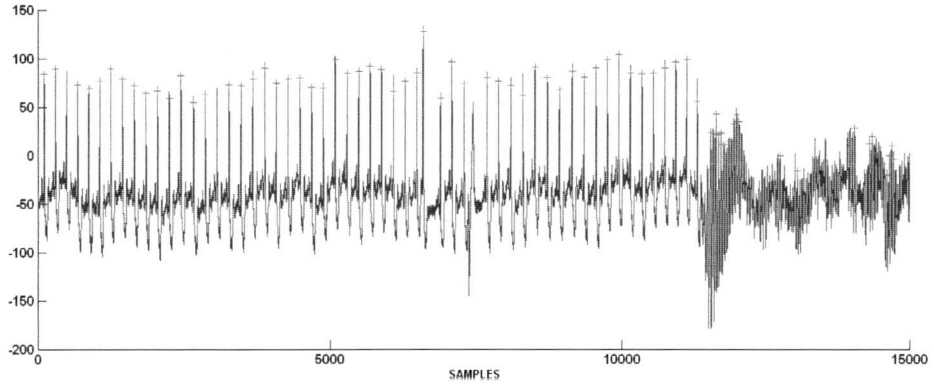

Figure 12. Transition to VF episode, from MIT-BIH Malignant Ventricular Arrhythmia database, interpolation error=3, Fs=250 Hz. Record 426. "+" implies QRS detection.

Figure 13 shows the instantaneous heart beat. During the VF episode the QRS detection follows the chaotic behaviour of the signal. If VF persists the detection usually tends to zero.

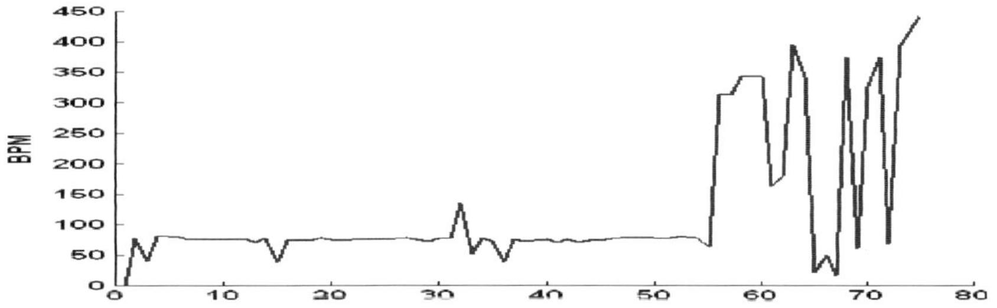

Figure 13. Instantaneous beat per minute for the signal in figure 13.

Therefore, the proposed algorithm for VF detection is shown in figure 14. There are two temporal windows, one for VF episode and one for back to normal state. Once we have detected a frequency in the range of, for example 200-500 BPM, the VF window is open to discard interference. In this state, VF is confirmed by two parameters. First, the estimated variance of the instantaneous heart beat should be high or the hear beat should tend to zero due to the algorithm is not detecting pulses. If none of these conditions are met, we open temporal window to go back to the normal state. These windows should be in the range from 2 to 5 seconds.

4. Conclusions

A normalized platform was proposed to analyze ECG signals. In this paper the focus was on ventricular fibrillation detection as an application of the proposed techniques. These algorithms can be seen as a complete sampling process followed of a morphological treatment. All the results shown in this paper were obtained from real ECG signals of the PhysioNet database without any prefiltering.

VF detection was used as an example of simplicity and versatility of the proposed techniques. This paper shows only preliminary results. Adding other algorithms based on MCT vectors would allow us to achieve sensory knowledge fusion to infer more accurately the different pathologies present in an ECG signal.

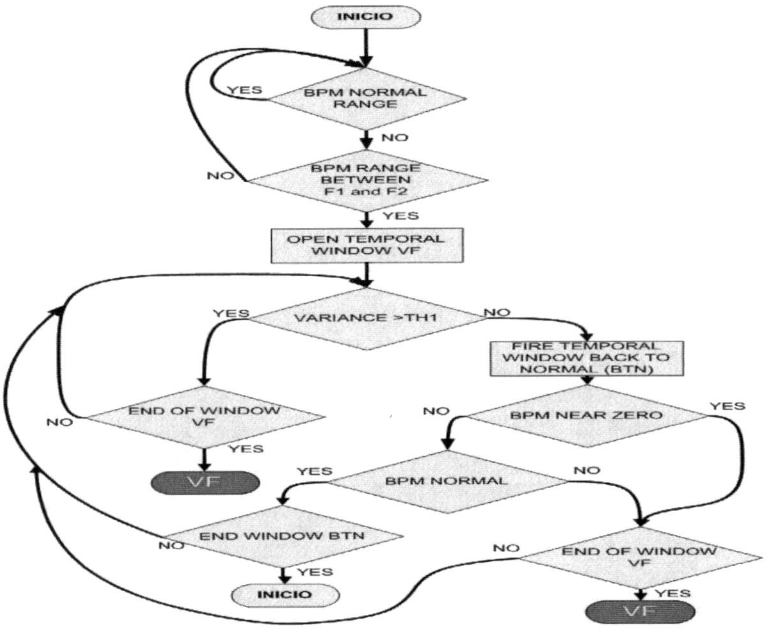

Figure 14. Algorithm for VF detection.

5. References

[1] Valentinuzzi M E 2010 *Cardiac Fibrillation-Defibrillation: Clinical and Engineering Aspects.* In Series on Bioengineering and Biomedical Eng. World Scientific Publishers, Singapore

[2] http://www.americanheart.org/presenter.jhtml?identifier=4784

[3] Thakor N V, Zhu Y S, and Pan K Y 1990 Ventricular tachycardia and fibrillation detection by a sequential hypothesis testing algorithm. *IEEE Transaction on Biomedical Engineering,* 37(9):837–43

[4] Pan J and Tompkins W J. 1985 A real-time QRS detection algorithm. *IEEE Trans Biomed Eng,* 32(3):230–6

[5] Chen S, Thakor N V, and Mower M M 1987 Ventricular fibrillation detection by a regression test on the autocorrelation function. *Med Biol Eng Comput,* 25(3):241–9

[6] Barro S, Ruiz R, Cabello D, and Mira J 1989 Algorithmic sequential decision-making in the frequency domain for life threatening ventricular arrhythmias and imitative artefacts: a diagnostic system. *J Biomed Eng,* 11(4):320–8

[7] S. M. Szilagyi, Z. Benyo, L. Szilagyi and L David 2003 Adaptive wavelet transform based ECG waveforms detection, *Proceedings of 25th Annual IEEE EMBS International Conference*, No. 24, pp. 12-15

[8] Amann A, Tratnig R and Unterkofler K S, 2005 A New Ventricular Fibrillation Detection Algorithm for Automated External Defibrillators, *Computers in Cardiology,* 32:559-4

[9] Clayton R H, Murray A and Campbell R W 1994 Recognition of Ventricular Fibrillation Using Neural Networks, *Medical and Biological Engineering and Computing,* Vol 32(2) 217-220

[10] Poli R, Cagnoni S and Valli, G 1995 Genetic design of optimum linear and nonlinear QRS detectors, *IEEE Transaction On Biomedical Engineering*, Vol. 42, pp. 1137-1141

[11] Geddes L A and Roeder R A 2002 The First Electronic Electrocardiograph *Cardiovascular Engineering* Vol 2 Number 2 pp 73-79

[12] Monte G Sensor Signal Preprocessing Techniques for Analysis and Prediction 2008 , *IECON 2008 34th Annual Conference of IEEE.* pp 1788-1793

[13] Massachusetts Institute of Technology, MIT-BIH arrhythmia database http://www.physionet.org/physiobank/database/mitdb

Journal of Physics: Conference Series **332** (2011) 012016 doi:10.1088/1742-6596/332/1/012016

Left Ventricular Hypertrophy: An allometric comparative analysis of different ECG markers

MP Bonomini[1*], **F Ingallina**[2], **V Barone**[2], **ME Valentinuzzi**[1], **PD Arini**[1,3]
[1] Instituto de Ingeniería Biomédica (IIBM), Facultad de Ingeniería (FIUBA), Universidad de Buenos Aires (UBA)
[2] Instituto de Investigaciones Médicas Dr. A. Lanari, Universidad de Buenos Aires (UBA)
[3] Instituto Argentino de Matemática, "Alberto P. Calderón", CONICET

E-mail: pbonomini@fi.uba.ar

Abstract. Allometry, in general biology, measures the relative growth of a part in relation to the whole living organism. Left ventricular hypertrophy (LVH) is the heart adaptation to excessive load (systolic or diastolic). The increase in left ventricular mass leads to an increase in the electrocardiographic voltages. Based on clinical data, we compared the allometric behavior of three different ECG markers of LVH. To do this, the allometric fit $AECG = \delta + \beta\ (VM)$ relating left ventricular mass (estimated from ecocardiographic data) and ECG amplitudes (expressed as the Cornell-Voltage, Sokolow and the ECG overall voltage indexes) were compared. Besides, sensitivity and specifity for each index were analyzed. The more sensitive the ECG criteria, the better the allometric fit. In conclusion: The allometric paradigm should be regarded as the way to design new and more sensitive ECG-based LVH markers.

1. Introduction

Left ventricular hypertrophy (LVH) is the heart way to adapt to overloads, either during diastolic or systolic periods. This adaptation consists of increasing the diameter of the cardiac fibers and, consecuently, of left ventricular mass. Such augmented mass directly affects the electrocardiographic signal by raising its voltage amplitudes. We address the following question here: does this increase in amplitude keep an allometric relationship with the increase in left ventricular (LV) mass? To figure this out, a comparison of the allometric adjustment of different LVH indexes was carried out.

Scaling of many biological processes can be described by the allometric equation, $Y = a(B_M)^b$, where Y is the biological process, B is the body or organ mass and a and b are scaling constants. In general, the weights of most individual organs scale as a constant fraction of body mass (i.e., the body mass exponent, b, equals 1.0). Biological rates (heart rate, respiratory rate) scale as b close to 0.25. Finally, volume rates (the product of volume and rate), such as cardiac output, ventilation and oxygen uptake, vary as b around 0.75. These emergent patterns provide insights into body-size dependent "principles of design" that seem to dictate several aspects of design and function across species among all mammals [1,2,3].

---

* Address for correspondence.

Noujaim *et al* [4] assumed that the heart behaves as a set of "fractal-like" networks tending to minimize propagation time across the conducting system while ensuring a hemodynamically optimal atrioventricular activation sequence. With the potential relationship given above and, subsequently, based on previously published values of PR interval, heart rate, and body masses of 541 mammals, they reported as best fit the equation $PR = 53(B_M)^{0.24}$.

The existence of many different criteria for diagnosing LVH makes clinical application more complex. The sensitivity of the LVH indexes based on ECG is generally quite low (usually less than 50%), while the specificity is quite high (often in the range of 85% to 90%) [5,6,7]. Published studies are currently insufficient to indicate whether any of the more recently proposed criteria are clearly superior to the others or are simply redundant. For these reasons, the aim of this work was to provide insight into new clues and theoretical support to help researchers find more sensitive markers of LVH.

1. Materials and Methods

1.1. Patient population

According to the Penn Convention and using the Deveraux equation [8], LV mass was assessed in 36 patients, from which, 23 out of 36 showed ecocardiographic mass index greater than 259 g and 166 g for men and women, respectively, leading to LVH diagnosis. The average age of the studied population was composed of 16 men (75.44±8.13 years old) and 20 women (72.75±13.54 years old). Pathologies were varied, since recruitment was done on outpatients of the General Hospital *Instituto de Investigaciones Médicas Dr. A. Lanari*, belonging to University of Buenos Aires (UBA). Patients with intraventricular conduction diseases (IVCD) were ruled out, since both LVH and IVCDs alter QRS patterns, therefore, the existence of an IVCD may impact the accuracy of ECG criteria for LVH [7].

1.2. ECG criteria for LVH

We compared the allometric adjustment of three ECG markers for LVH, i.e., Cornell index, ECG total 12-lead voltage and Sokolow index. They are widely used in clinical practice. These indexes were calculated as follows:

- **Cornell (voltage) index:** There are two versions of this index, one concerning voltage only and the other one combining QRS voltage and duration. We used here only the voltage version that combines the amplitude of the S wave in V_3 lead and the amplitude of R wave in aVL lead [7,9].
    - Men: $S_{V3} + R_{aVL}$ >2,8 mV (28 mm)
    - Women: $S_{V3} + R_{aVL}$ >2,0 mV (20 mm)
- **Total 12-lead voltage:** Total 12-lead voltage, measured as the sum of all S and R peaks of all 12 leads> 175 mm [7,10].
- **Sokolow index:** who in 1949 introduced the widely used criterion based on the sum of the amplitude of the S wave in V_1 lead plus the amplitude of R wave in V_5 or V_6 leads. The cut-off point for this index is >/= 3,5 mV (35 mm) [7,11]. We have chosen V_6 lead for our analysis.
    - $S_{V1}+R_{V6}$>= 3,5 mV (35 mm)

For the ECG studies, we have used a standard 12-lead ECG device and obtained 10 second-recordings. The device had a sampling rate of 400 Hz and 12 bits resolution. From the

ECG recordings, peak amplitudes were averaged out from all the beats contained in the entire recording.

1.3. Left ventricular mass calculation

In the late 80s, Levy and coworkers published a landmark paper evaluating a subset of individuals without known cardiovascular risk factors in the Framingham Cohort [8]. These authors calculated LV mass both with the ASE (American Society of Echocardiography) convention and Troy equation [12]:

$$\text{LV mass(Troy)} = 1.05\,([\text{LVIDD} + \text{PWTD} + \text{IVSTD}]^3 - [\text{LVIDD}]^3)\ \text{g}. \qquad (1)$$

Where:

LVIDD = Left Ventricular Internal Diameter in Diastole
PWTD = Posterior Wall Thickness in Diastole
IVSTD = Interventricular Septum Thickness in Diastole

and with the Penn Convention and Devereux equation [13]:

$$\text{LV mass(Deveraux)} = 1.04\,([\text{LVIDD} + \text{PWTD} + \text{IVSTD}]^3 - [\text{LVIDD}]^3)\ \text{-13,6 g} \qquad (2)$$

In this work, the Penn Convention and Devereaux equation were chosen. The authors proposed normal limits for LV mass for men and women, based on cut points at two standard deviations above the mean [8,14].

1.4. Allometric equation

Allometry, in general biology, measures the relative growth of a part in relation to the whole living organism. The term was first used by Snell, in 1891 [15], to express the mass of a mammal's brain as a function of the body mass. The growth velocity of a component *'y'* is related to the growth velocity of another component (or the whole organism) *'x'* in a constant way. This was clearly described by von Bertalanffy in 1957 [16]. Thus, the relative rate of change of a given event *'y'* is proportional to the relative rate of change of body mass or body weight *'x'*, i.e.,

$$\frac{dy/dt}{y} = B\frac{dx/dt}{x} \qquad (3)$$

After integration and some easy algebraic manipulation, equation (3) becomes

$$\ln y = A + B \ln x$$

or

$$y = Ax^B \qquad (4)$$

Originally, *'y'* was the weight of an organ (heart, stomach, other) and *'x'* was body weight or mass. The parameters *A* and *B* require numerical estimation by an appropriate procedure usually using empirical information. By the same token, let us say that the amplitude of the ECG (we use *Aecg* in a general form, since *Aecg* will be cuantified as the ECG criteria for LVH) follows a relationship with the number of ventricular hypertrophic fibers, and therefore, the ventricular mass (*Vm*). Thus, eq. (4) will be reformulated as in (5):

$$Aecg = \alpha(Vm)^{\beta} \tag{5}$$

After taking logarithms of both sides, the latter equation becomes

$$\ln Aecg = (\ln\alpha + \beta\ln\gamma) + \beta(\ln\Delta_{Vm}) \tag{6}$$

which can be reduced to

$$AECG = \delta + \beta(VM) \tag{7}$$

We define *AECG* as ECG voltage or amplitude, where $\delta = \ln\alpha + \beta\ln\gamma$, $VM = \ln\Delta Vm$ and $AECG = \ln Aecg$. The straight line, equation (5), in log-log plot with the parameters β and δ would represent the increase in ECG amplitude as function of the amount of hypertrophic fibers.

1.5. Numerical procedure

To calculate the two constants δ and β and later on apply the mathematical expression in equation (7), linear regression was implemented in order to evaluate the allometric fit on a log-log plot gathering the variables ventricular mass and LVH index. Notice that δ represents the intercept to the origin and β is the slope of the regression line when plotted on log-log coordinates. When $\beta < 1$, the relative weight of the ECG criteria (fraction of total ventricular mass) is greater in small than large hypertrophy. Also, sensitivity and specificity was estimated from our set of data by calculating the following equations:

$$Sensitivity = TP/(TP+FN) \tag{8}$$
$$Specificity = TN/(TN+FP), \tag{9}$$

where *TP* are the true positive, *FP* the false positive and *TN* the true negative cases, all of them confirmed by ecocardiographic analysis. More specifically, ecocardiographic mass calculation, as described above, and the criteria of a cardiologist were set as the gold standard for LVH diagnosis.

2. Results

Using the equation $AECG = \delta + \beta\ (VM)$ in log-log representation, the fitting procedure produced coefficients for every LVH index, displayed in Table 1. Moreover, graphic results for the allometric adjustments on log-log plots are showed in Figure 1. Linear regression and determination coefficients *r* are also displayed. Note that the ECG criteria for LVH that correlated best with ventricular mass in an allometric fit was the Voltage-Cornell index (r=0.7229), followed by the Total Voltage index (r=0.5292) and last, with a very low fit the Sokolow index (r=0.2769) with a slope deviation from zero not significant.

Table 1. Coefficients β and δ and goodness of fit *r* for all three LVH indexes.

	β	δ	r
Voltage-Cornell	0,89 ± 0,14	-0,90 ± 0,33	0,72
Total voltage	0,42 ± 0,11	1,1 ± 0,26	0,52
Sokolow	0,24 ± 0,14	0,72 ± 0,33	0,27

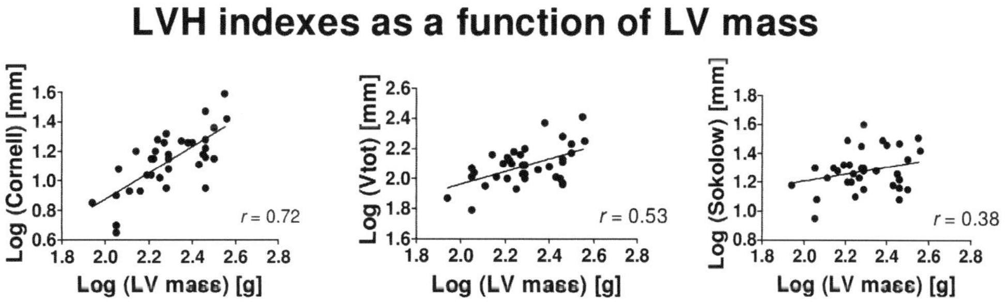

Figure 1: Linear regressions for the allometric fits of ECG-based LVH markers: Cornell index, total 12-lead voltage index and Sokolow index.

Notice as well, that all the slopes β in the adjustment are positive and smaller than 1. This means that indexes are inversely proportional to the amount of hypertrophy in the heart. This is, the greater the amount of hypertrophic fibers, the lesser the growth of the LVH index. Slopes greater or even equal to 1 would be more desirable in order not to lose sensitivity with the hypertrophy extent.

Table 2 shows the sensitivity and specificity as reviewed from the bibliography and as estimated from our set of data. In all cases, notice how the sensitivity suffered when calculated from a reduced set of samples.

Table 2. Sensitivity and specificity collected from bibliography and estimated from our pool of data, 23 patients with and 13 without LVH ecocardiographically confirmed.

	Sensitivity from bibliography	*Specificity from bibliography*	*Estimated Sensitivity*	*Estimated Specificity*	*References*
Voltage-Cornell	42 %	96 %	35 %	100 %	Casale 1985 [7]
Sokolow	22 %	100 %	13 %	100 %	Sokolow 1949 [9]
Total voltage	17.2 %	95 %	17 %	100 %	Siegel 1982 [8]

3. Discussion

The results show that sensitivity goes along with allometric behavior when searching for LVH markers based on ECG. However, certain constraints should be regarded. For instance, the population in study is quite homogeneous in terms of age, presenting mainly the same type of LVH. Thus, the analysis herein accomplished holds for the elderly only. It is important to notice that patients with intraventricular conduction diseases (IVCD) were excluded from the study in order not to confuse the symptoms, since both LVH and IVCD lead to similar changes in QRS.

Another limitation of the study is that not all the QRS-based indexes of LVH were analyzed. Nevertheless, the results found here encourage a more complete study including all electrocardiographic indexes of LVH over a larger and more heterogeneous sample population.

One point to mention is the slope β of the linear regressions in the allometric fit. All of them resulted positive and smaller than 1. This means that the greater the amount of hypertrophic fibers, the lesser the growth of the LVH marker. This leads to a loss of sensitivity with hypertrophy extent. Therefore, if more sensitive markers are required, β tending to 1 would be desired.

Interestingly, the Voltage-Cornell index, which showed the best allometric fit, is the one that best sensitivity achieved among the studied indexes. Voltage-Cornell index has a sensitivity of 42%, as collected from the bibliography [8] and 35% as estimated from our pool of data (see Methods, section 2.5. Numerical procedures).

Finally, even though the Cornell index was designed for different LVH cut-off points (132 g/m^2 for men and 109 g/m^2 for women) as used here, it adapted well to the new definition, postulating itself as the most robust ECG marker of LVH.

4. Conclusions

The voltage increase in ECG would follow an allometric relation with left ventricular mass. This fact should guide the search of new and more sensitive markers of hypertrophy by taking into account the allometric law. Moreover, slopes of the regression lines (or constants β) resulted all positive and smaller than one. This means that ECG-based indexes for LVH are inversely proportional to the amount of hypertrophy in the heart. Slopes greater or even equal to one would be more desirable in order not to lose sensitivity with hypertrophy extent.

References

[1] Lindstedt SL and Schaeffer PJ 2002 Use of allometry in predicting anatomical and physiological parameters of mammals *Laboratory Animals* **36** 1-19.

[2] Martin RD, Genoud M and Hemelrijk CK 2005 Problems of allometric scaling analysis: Examples from mammalian reproductive biology *J Experimental Biology* **208** 1731.

[3] Bonomini, MP, Arini PD, Valentinuzzi ME, Probability of ventricular fibrillation: Allometric model based on the ST deviation. Biomed Eng Online. 2011 Jan 13;10(1):2.

[4] Noujaim SF, Lucca E, Muñoz V, Persaud D, Berenfeld O, Meijler FL and Jalife J 2004 From mouse to whale: A universal scaling relation for the PR interval of the electrocardiogram of mammals *Circulation* **110** 2801.

[5] MacFarlane PW and Lawrie TD 1988 Comprehensive Electrocardiography: Theory and Practice in Health and Disease. Oxford, United Kingdom: Pergamon Press.

[6] Hsieh BP, Pham MX and Froelicher VF 2005 Prognostic value of electrocardiographic criteria for left ventricular hypertrophy *Am Heart J.* **150**(1) 161.

[7] Hancock W, Deal B, Mirvis D, Okin P, Kligfield P and Gettes L 2009 Recommendations for the standardization and interpretation of the electrocardiogram *Journal of the American College* **53**(11) 992.

[8] Levy D, Savage DD, Garrison RJ, Anderson KM, Kannel WB and Castelli WP 1987 Echocardiographic criteria for left ventricular hypertrophy: the Framingham Heart Study *Am J Cardiol* **59** 956.

[9] Casale P, Devereux R, Kligfield P, Eisenberg RR, Miller DH, Chaudhary BS and Phillips MC 1985 Electrocardiographic detection of left ventricular hypertrophy: development and prospective validation of improved criteria *J Am Coll Cardiol.***6** 572.

[10] Siegel RJ, Roberts WC 1982 Electrocardiographic observations in severe aortic valve stenosis: correlative necropsy study to clinical, hemodynamic, and ECG variables demonstrating relation to 12-lead QRS amplitude to peak systolic transaortic pressure gradient *Am Heart J* **103** 210.

[11] Sokolow M and LyonTP 1949 The ventricular complex in left ventricular hypertrophy as obtained by unipolar and limb leads *Am Heart J* **37** 161.

[12] Troy BL, Pombo J, and Rackley CE 1972 Measurement of left ventricular wall thickness and mass by echocardiography *Circulation* **45** 602.

[13] Devereux RB and Reichek N 1977 Echocardiographic determination of left ventricular mass in man. Anatomic validation of the method *Circulation* **55** 613.

[14] Foppa M, Duncan BB and Rohde LE 2005 Echocardiography-based left ventricular mass estimation. How should we define hypertrophy? *Cardiovasc Ultrasound* **3** 17.

[15] Snell O 1891 Die Abhängigkeit des Hirngewichtes von dem Körpergewicht und den geistigen Fähigkeiten *Archiv für Psychiatrie und Nervenkrankheiten* **110** 2801.

[16] Von Bertalanffy L 1957 Quantitative laws in metabolism and growth *Quaterly Review Biology* **32**(3) 217.

Journal of Physics: Conference Series **332** (2011) 012017 doi:10.1088/1742-6596/332/1/012017

Simple communication using a SSVEP-based BCI

Guillermo Sanchez, Pablo F. Diez, Enrique Avila, Eric Laciar Leber

Gabinete de Tecnología Médica, Facultad de Ingeniería – Universidad Nacional de San Juan, San Juan – Argentina.

E-mail: pdiez@gateme.unsj.edu.ar

Abstract. Majority of Brain-Computer Interface (BCI) for communication purposes are speller, i.e., the user has to select letter by letter. In this work, is proposed a different approach where the user can select words from a word set designed in order to answer a wide range of questions. The word selection process is commanded by a Steady-state visual evoked potential (SSVEP) based-BCI that allows selecting a word in an average time of 26 s with accuracies of 92% on average. This BCI is focus in the first stages on rehabilitation or even in first moments of some diseases (such as stroke), when the person is eager to communicate with family and doctors.

1. Introduction

According to statistics from the World Health Organization (WHO), there are in the world about 10% of people with permanent disabilities (motor, sensory, mental, and others), i.e., about 600 million people suffer some form of disability [1]. In particular, people suffering severe motor disabilities due to high cervical injury, cerebral palsy, multiple sclerosis or muscular dystrophy, have multiple consequences for all their neuromuscular system. These sequels prevent the person from proper communication with their environment.

A brain computer interface (BCI) is a communication system that can generate control signals from the brain activity. These are obtained directly by an electroencephalogram (EEG), without using any muscle [2].

Over the past two decades several studies have evaluated the possibility that EEG signals may create a new technology trend that does not require any effort. Thus, a subject can establish an interaction with the outside world, either by sending messages or commands without muscular activity, [3].

Different electrophysiological experiments have shown that neurons in the visual cortex are synchronized with the frequency of a flashing light [4], generating in the EEG responses showing the same frequency as the light stimulus [5].

In this work, a BCI based on Steady-State Visual Evoked Potentials (SSVEP) was used in order to select a word from a predefined set of words, in a fast and robust way. This BCI allows a person to express a word quickly, instead of typing letter by letter as in [7] or [8], which can result in a longer process and tiring for the user. This approach is focus in the first stages on rehabilitation, when an easy, no stressed and fast communication between patient and physician is needed.

2. EEG database

The BCI was implemented to allow simple communication. Hence, three lights were set on the sides of a computer monitor: two flashing bars of 10cm x 2.5cm illuminated with green light-emitting diodes (led) and a red led, placed on above, below and left sides of the monitor,

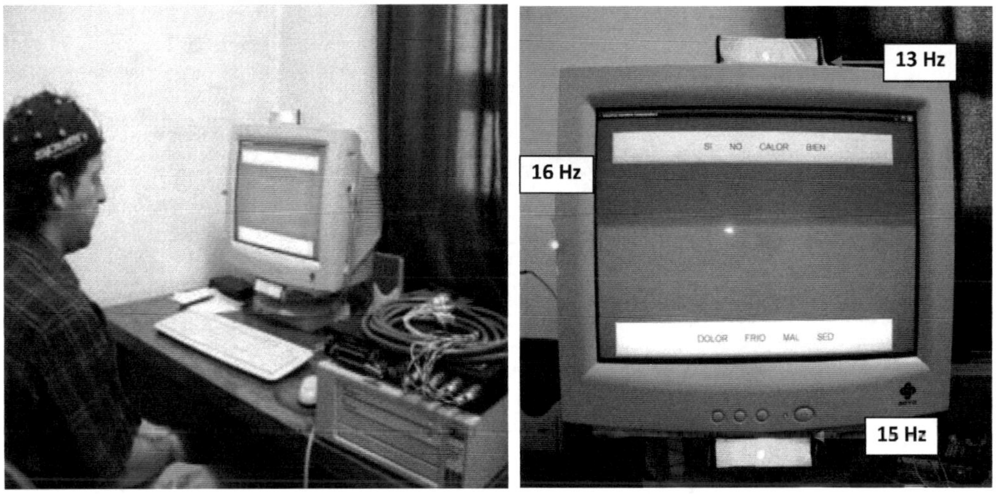

Figure 1: Signal acquisition equipment and subject with EEG cap (left), screen with words and stimuli (right).

respectively. Top bar flicker at 13 Hz, low bar flicker at 15 Hz and red left led flicker at 16 Hz, these frequencies were controlled precisely with an FPGA Xilinx Spartan 2E (Figure 1).

Before using the BCI, it is required to acquire an EEG base signal, where the patient is asked to concentrate on the center of the screen for 30 seconds. This stage is needed in order to normalize the powers of each of the stimuli (see Eq. 2).

EEG signals were obtained with a Grass 15LT amplifier system and digitized with a NI-DAQPad6015 (sampling frequency= 256 Hz). Analogical bandpass filter were set at 1 and 100 Hz and a notch filter at 50 Hz to eliminate powerline interference. Only one EEG channel was measured, with electrodes placed at O_2 and P_4, and grounded at linked A_1-A_2.

3. Methods

The EEG signal processing method was based on [6]. A scheme of the proposed BCI is showed in Figure 2. First, the periodogram with rectangular window of 1024 points over 1 s EEG segments was computed as:

$$Pxx(f) = \frac{T_s}{N}\left|\sum_{n=1}^{N} x[n]e^{-j2\pi f n T_s}\right|^2 \tag{1}$$

where $Pxx(k)$ is the periodogram, Ts is sampling period, N is number of samples on EEG segment and f is frequency. Then, normalized power P_N is computed for each interest frequency as:

$$P_N(f_i) = \frac{P_{f_i}}{Pb_{f_i}} + \frac{P_{2f_i}}{Pb_{2f_i}} + \frac{P_{3f_i}}{Pb_{3f_i}} \tag{2}$$

where f_i is the interest frequency (13, 15 or 16 Hz); P_{fi} is the power on each f_i and their second ($2f_i$) and third harmonic ($3f_i$), P_b is the power on base EEG acquired previously at the experiment. In Figure 3 is presented the spectral content of an EEG segment when the subject observes the light placed above the monitor, hence an SSVEP at 13 Hz is elicited and it is observed a power increase at 13 Hz and its second harmonic at 26 Hz.

Later, the maximum P_N is detected and, if this maximum is maintained for a time interval of at least 1.5 s, then that segment is classified as a one of the stimulus (top, down or left).

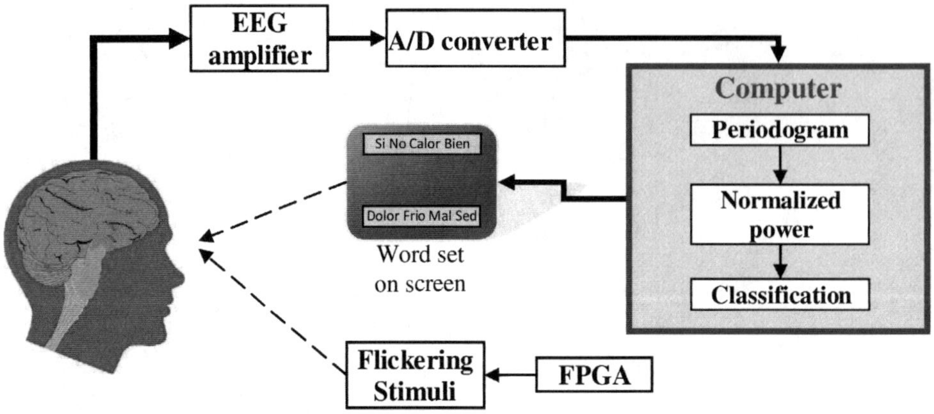

Figure 2: Scheme of proposed BCI

3.1. Word selection

This setup was designed in order to allow an easy communication between the patient and the medics or relatives. In that way, if medic (or relative) takes a question to the patient, he/she can respond it easily.

The patient is presented with a series of words, separated into two sets, one at the top of the screen and one on the bottom (Figure 1). These words have been chosen so that the patient can answer some questions ("yes", "no", "good", "bad") or expressing some situations ("heat", "pain", "cold", "thirst"). Thus, the subject may speak quickly, for example, during a rehabilitation session with his physician.

The patient starts by choosing the word to express, and then through a binary classification process he/she can select the chosen word, as shown in Figure 4. The patient should gaze the stimulus that corresponds to the set where the desired word is. For example, to select the word "no" the subject should gaze at the top light (Figure 4A), consequently the BCI should detect the patient intention and the upper set is subdivided into two new subsets (Figure 4B). Then, the process is repeated until the chosen word is reached (Figures 4C and 4D). At the beginning of each selection step is emitted a beep, indicating the patient can start gazing at the stimulus corresponding to the chosen bar on the screen.

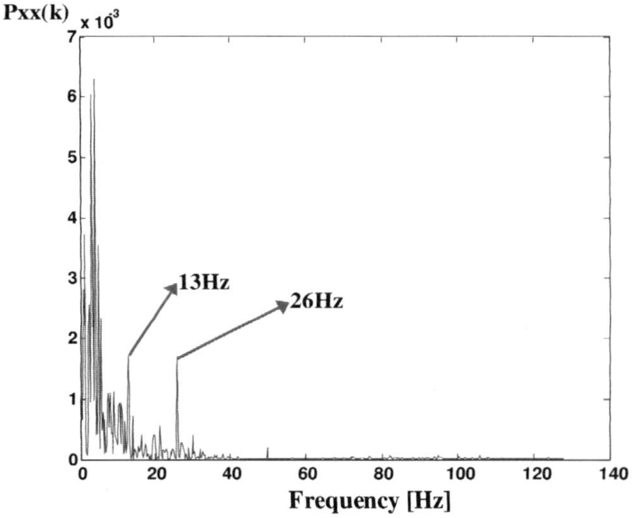

Figure 3. EEG Periodogram of a subject gazing at 13 Hz stimulus, the second harmonic at 26 Hz can be observed as well.

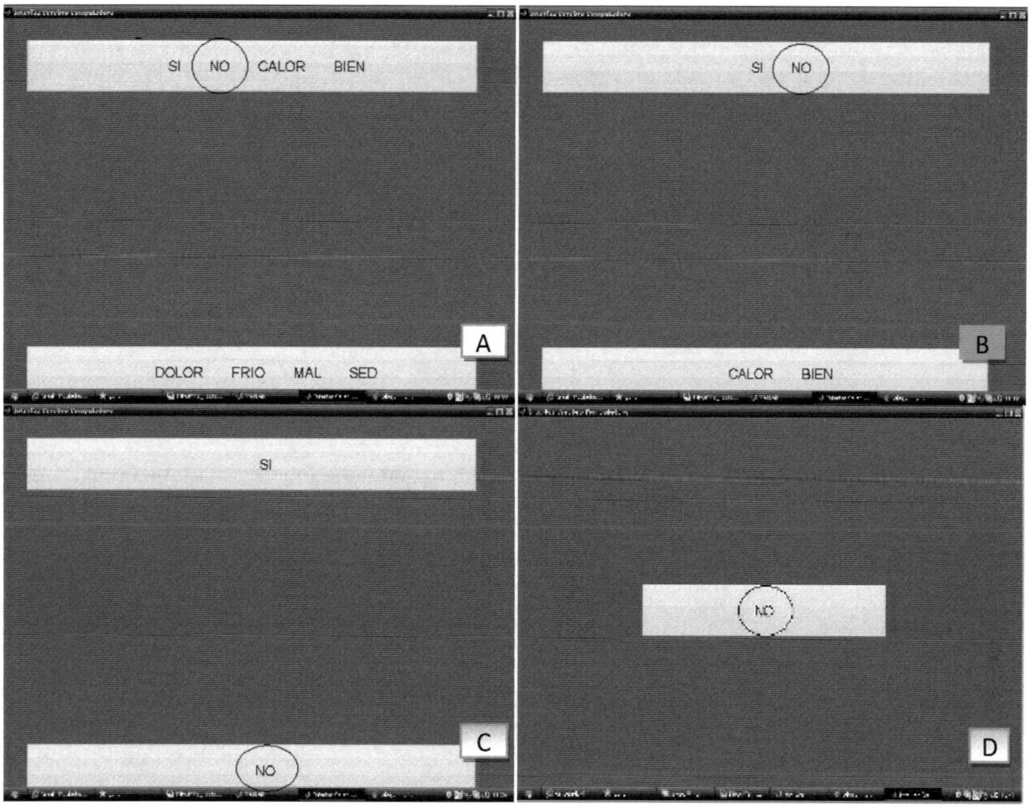

Figure 4: Selection word process for "no". A, entire sets. B, first division of upper set. C, second division of upper set; D, final selection of "no" word. These words are in Spanish, si = yes, no = no, calor = heat, bien = good, dolor = pain, frío = cold, mal = bad, sed = thirst.

Whether an erroneous detection is performed, the patient could correct it, just gazing on the red led at the left of the monitor, returning to the previous step.

Finally, once the patient gets to the desired word, it is presented on the screen and also listening on the computer speaker.

4. Results

To evaluate the proposed BCI performance, four healthy subjects (men, 29 to 33 years) were seated in front of a computer monitor and they were asked to select different words from the word set. All subjects provided written consent to participate in the experiment.

Each subject was evaluated according to accuracy in selecting the desired word, and the time spent in the process, these results are presented in Table 1. Additionally, a measure of the information transferred between the brain and the computer, the Information Transfer Rate (ITR), is showed. The ITR is calculated as [7]:

$$ITR = (1 - Pd)\left(log_2 N + (1 - Pe)\, log_2(1 - Pe) + Pe\, log_2\left(\frac{Pe}{N-1}\right)\right) \qquad (3)$$

where *Pd* is the rate of unknown cases (i.e., the BCI was not able to detect a SSVEP), *Pe* is the error rate (i.e., the BCI detects a SSVEP but it is not the stimulus the subject gazes) and *N* is the number of stimuli. In equation (3), the ITR is expressed in bits/trials; the maximum ITR achievable for the proposed BCI is 1.58 bits/trial. Additionally, the ITR could be expressed in bits/min considering the number of trials per minute.

The minimum number of steps to select a word is 3 steps. However, eventually it could be necessary more than 3 steps. For example, if a stimulus is not detected when subject gaze it, resulting in an undefined output; then the subject must gaze the stimulus once again after start beep signal. Another case, when a stimulus is detected erroneously, the subject can correct the wrong choice gazing at the red light on the left and the word sets is backward to previous state.

Table 1: Results on experiments

Subjects	Average time [s]		ITR		Accuracy
	Total	by trial	bits/trial	bits/min	
1	45.1 ±17.8	9.465	1.185	7.625	100%
2	26.3±9.7	9.346	1.25	9.56	75%
3	17.3±0.5	5.73	1.58	16.616	100%
4	17±0.8	5.667	1.58	16.82	100%
Average	**26.42**	**8.159**	**1.185**	**12.121**	**94**

5. Discussion

In this work was implemented a BCI that allows a disabled person respond quickly and concisely to the questions that might make his caregivers. This is accomplished through a robust system that gets on average 94% accuracy in reaching the desired word.

In this preliminary stage, the BCI was tested in four healthy subjects. All subjects always achieved the desired word, except subject 2. Subjects 3 and 4 had better performances as they reached the desired word within the shortest possible time (17 s) and the maximum possible ITR for the BCI is reached, i.e. 1.58 bits/trial.

In [8] a speller was implemented through a decision tree and an undo command to correct any errors. This BCI, was tested in eight healthy subjects with an accuracy rating of 92.25% and 37.62 bits/min, which translates into an average speed of 5.51 letters per minute. In this work was attained a lower ITR (up to 16 bits/min), but it is possible to select a whole word in 26.42 s.

In [9] a spatial filter was used for the detection of SSVEP with an average ITR of 27 bits/min, however this algorithm is more complex and with higher computational cost than proposed method on this paper.

Generally, in related works the stimulation flickering was performed on the same screen. In [10] was evaluated various parameters using the screen (such as frequency range, resolution, stability, etc.) and it was founded that cannot be render frequencies above 30 Hz (or half the monitor refresh rate). In this work, using stimulation with leds, it is possible to achieve higher frequencies than 30 Hz. These high-frequency SSVEP range has the advantage of a great decrease of visual fatigue caused by flickering, making the SSVEP-based BCI a more comfortable system [11].

Although, the word set is limited, it was designed in order to answer a wide range of questions, where the patient is able to respond fast and accurate to doctor or relative questions.

6. Conclusions

In this work, was implemented a SSVEP-based BCI that could allow a person to communicate fast and without much effort, i.e., it is possible to select a word in an average time of 26 s with accuracy in word selection on average of 92%.

This BCI approach is intended to be used in the first stages on rehabilitation or even in first moments of some diseases (such as stroke), when the person is desirous to communicate and the physician need to talk with the patient. Eventually, depending on the patient recuperation, a more sophisticated BCI could be used or no use a BCI.

As future perspectives, it is planned to extend the set of word and to develop a faster method to select words on the screen and moreover, to test the system in disabled people.

Referencias

[1] http://www.who.int/mediacentre/events/annual/day_disabilities/es/index.html

[2] Wolpaw J. R., N. Birbaumer, McFarland D. J., Pfurtscheller G., and. Vaughan T. M "Brain-Computer Interfaces For Communication And Control" Electroenceph. Clin. Neurophysiol, Vol.113, No. 6, June 2002.

[3] Schalk G., McFarland D.J, Hinterberger T., Birbaumer N., and Wolpaw J., "BCI2000: A General-Purpose Brain-Computer Interface (BCI) System" IEEE Transactions on Biomedical Engineering, Vol. 51, No. 6, June 2004.
[4] Regan D, "Human Brain Electrophysiology: Evoked Potentials and Evoked Magnetic Fields in Science and Medicine". Dalhousie University, Halifax; York University, Toronto; University of Toronto, 1989
[5] Herrmann C.S. "Human EEG responses to 1-100Hz flicker: resonance phenomena in visual cortex and their potential correlation to cognitive phenomena". Max Planck Institute of Cognitive Neuroscience, Leipzig, Germany, March 2001.
[6] Diez, P, Mut, V., Laciar E. Avila E. A Comparison of Monopolar and Bipolar EEG Recordings for SSVEP Detection. 32nd Annual International Conference of the IEEE EMBS Buenos Aires, Argentina, August 31 - September 4, 2010
[7] Millán J. del R, Renkens F., Mouriño J., Gerstner W.: Noninvasive Brain-Actuated Control of a Mobile Robot by Human EEG, IEEE Trans. Biomed. Eng 2004, 51(6):1026-1033
[8] Cecotti H., A self-paced and calibration-less SSVEP based Brain-Computer Interface speller, IEEE Trans, Neural Syst. Rehab Eng., vol. 18, no.2 pp.127-133 April 2010.
[9] Friman O., Lüth, Volosyak I. and Gräser A., Spelling with Steady-State Visual Evoked Potentials, Proc. of the 3rd International IEEE EMBS Conference on Neural Engineering Kohala Coast, Hawaii, USA, May 2-5, 2007
[10] Sugiarto I., Allison B., Gräser A., Optimization Strategy for SSVEP-Based BCI in Spelling Program Application 2009 International Conference on Computer Engineering and Technology pp223-226.
[11] Wang Y., Wang R., Gao X., Gao S.: Brain-computer Interface based on the High-frequency Steady-state Visual Evoked Potential. Proc. 1st International Conference on Neural Interface and Control Proceedings, May 2005, Wuhan, China; 2005, pp.26-28.

Design and implementation of an electrocardiographical signal acquisition and digital processing system orientated to the detection of paroxysmal arrhythmias.

Agustín Iriart Braceli[1], Jorge Exequiel Morani[2]

[1] Universidad de Mendoza, Facultad de Ingeniería. Peatonal Descote 750. (5500) Mendoza, Argentina
E-mail: agusi20@gmail.com

[2] Universidad de Mendoza, Facultad de Ingeniería. Peatonal Descote 750. (5500) Mendoza, Argentina
E-mail: exe.morani@gmail.com

Abstract. This article describes the design, technical aspects and implementation of a device capable of acquiring electrocardiograph signals; visualize them in real time over a graphic liquid crystal display (GLCD), and the storage of these ECG registers on a SD memory card. It also details a noise suppression algorithm using the Wavelet Transform. This system was specially developed to cover some bankruptcy that presents actual Holters or ECG regarding the detection of paroxysmal arrhythmias. The contribution of this work is settled on its portability and low production cost. The filtering method used provides an ECG signal without any significant noise and appropriate to the diagnosis of cardiac pathologies.

1. Introduction

Arrhythmia is any irregularity on the natural rhythm of the heart beat. The presence or absence of symptoms depends on the heart condition, type, seriousness, frequency and length of the arrhythmic event. Paroxysmal arrhythmias appear and revert spontaneously. This characteristic makes its detection and registration very difficult. The most appropriate study in this case is a Holter; but, at the same time, it is very inefficient if the arrhythmia does not appear while the Holter is connected to the patient. This work presents the design and implementation of a device, of reduce dimension, that allows to be used even by the patient oneself, simply connecting three electrodes to the chest and activating the device any time the arrhythmia appears. It makes possible to register the paroxysmal event and store a file with the ECG signal on a memory card for further analysis from qualified personal. This characteristic makes the device more efficient than the Holter for this kind of arrhythmias. Furthermore it is very comfortable to the patient because there is no need to be always connected, just the times there is an arrhythmic event.

2. System Overview

This project consists of two well delimited parts (Figure 1). In first place the device makes possible the acquisition, visualization and storage of the ECG signals. Subsequently there are digital processing algorithms and the conditioning process of the signal.

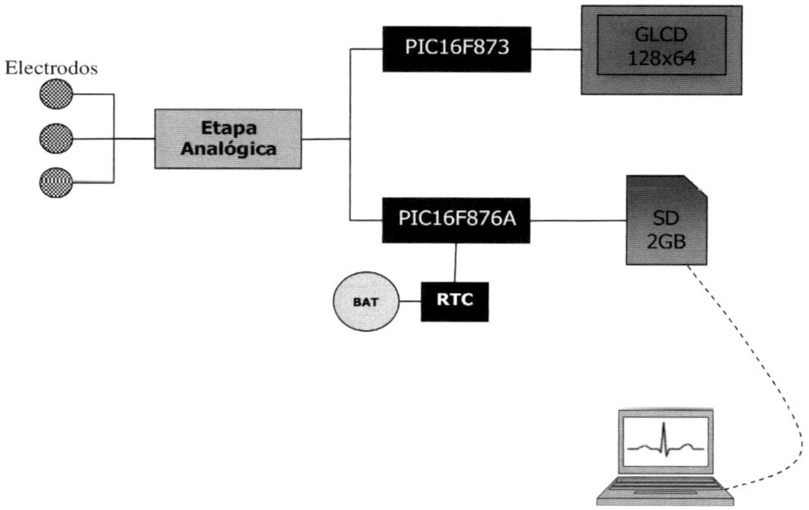

Figure 1. System Block Diagram

2.1. ECG acquisition
The designed device is equipped with a 110 dB CMRR instrumentation amplifier [1] [2]. The analog filter bank leaves a band-pass signal that goes from 0.05 Hz to 105 Hz [1][2][3]. Besides, there is a Notch filter at 50 Hz that reduces the currents induced over the patient body and the device by the electrical distribution line [4] [5]. Two mid-range PIC microcontrollers (16F87X) carry out the storage and visualization processes in an independent way. The sampling frequency is 400 Hz and the resolution is 10 bits [7] in order to satisfy the conditions establish by the American Heart Association (AHA) and the SCP-ECG (Standard Communication Protocol for Computerized ECG). These protocols set up the minimum requirements for the digitalization of ECG signals in a sampling frequency of 250 Hz and a resolution of 8 bits.

The graphic display is a transmissive type and has a blue backlight that allows working in places dimly lit. The speed of the ECG signal over the display is 25 mm/sec which is one of the standards speeds of actual electrocardiographs.

The second microcontroller implements a FAT16 [8][9] file system over the SD memory card[10][11]. It generates text files that contain the samples of the ECG signal. The sizes of these files are 96 KB and contain 30 seconds registers. The name contains information of the date the signal was acquired. While in the files properties, the exact time and date of acquisition can be found. This information is taken from a real time/calendar clock (RTCC) that is read just before each file is generated.

2.2. Noise suppression using the Wavelet Transform
The ECG signal presents different kinds of noise-problems. The most frequents ones are the distribution line interference and the baseline wander. The last one is due to the movement of the electrodes over the patient body, physico-chemical alteration on the electrode-skin interface and the noise produced by the muscles involved in the respiration during the study.

2.2.1 Fourier Transform vs. Wavelet Transform
The Fourier Transform presents certain limitation such as temporal information of the processed signal. It means that FT fails to determine the exact moment at which certain frequency component appears. On the other side, the Wavelet Transform gives specific information in both, time and frequency [12], which makes it a very handy tool in the study of non stationary signals like ECG.

According to some properties of the Wavelet Transform, it could be seen like a band-pass filter [12] [13]. Considering this, the Wavelet Transformation of a signal could be represented as a signal passing through a filter bank, making possible to use it in the noise suppression process.

2.2.2. Wavelet de-noising

The de-noising [13] [14] procedure aims to suppress any noise, including the line frequency interference (50 Hz) added to the acquired ECG signal. There are three processes to be considered:

1. Decomposition. Choose a wavelet, and choose a level N. Compute the wavelet decomposition of the signal s at level N.
2. Detail coefficients thresholding. For each level from 1 to N, select a threshold and apply soft thresholding to the detail coefficients.
3. Reconstruction. Compute wavelet reconstruction based on the original approximation coefficients of level N and the modified detail coefficients of levels from 1 to N.

In order to filter the signal, the MATLAB "wden" function was used. The algorithm that provides the best results consists of a decomposition process of the signal up to the level 4, a soft threshold and a Daubechies 6 wavelet.

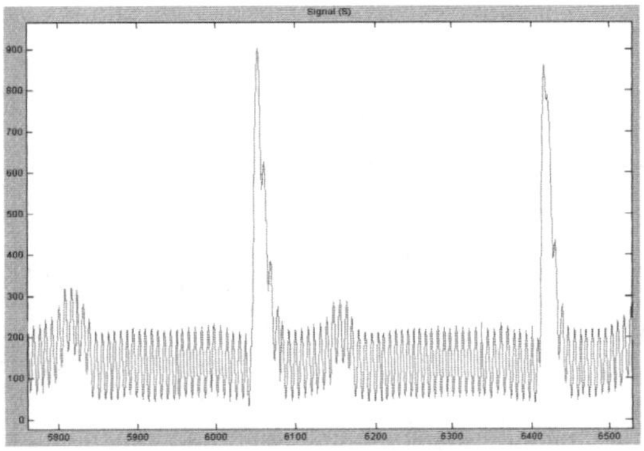

Figure 2. 50 Hz wave added to the ECG signal

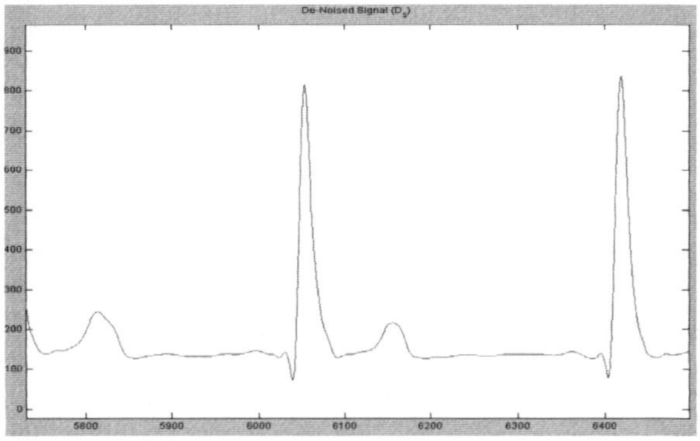

Figure 3. Filtered signal

2.2.3.Baseline wander reduction

The baseline wander is due to low frequency interference. To reduce this interference the "wavedec" and "waverec" [13][14] MATLAB functions were used. The procedure consists on decomposing the signal up to level 9. Considering that the signal sampling frequency in the digitalization process was 400 Hz, this level represents a frequency of 0,78 Hz. The next step is to eliminate the coefficient that contains the frequency that goes from 0 to 0,78 Hz (Figure 4), and then reconstruct the signal using the details coefficients (Figure 5 and 6).

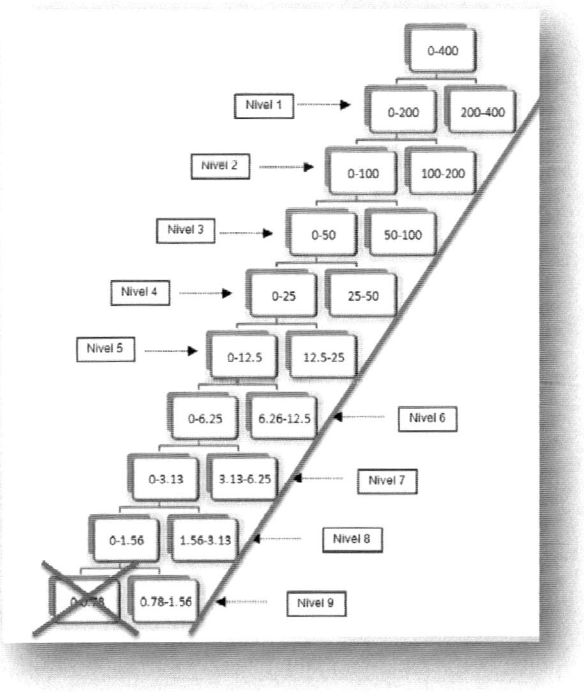

Figura 4. Decomposition Wavelet tree at level 9. Approximation Coefficient suppression and reconstruction using the Detail Coefficients.

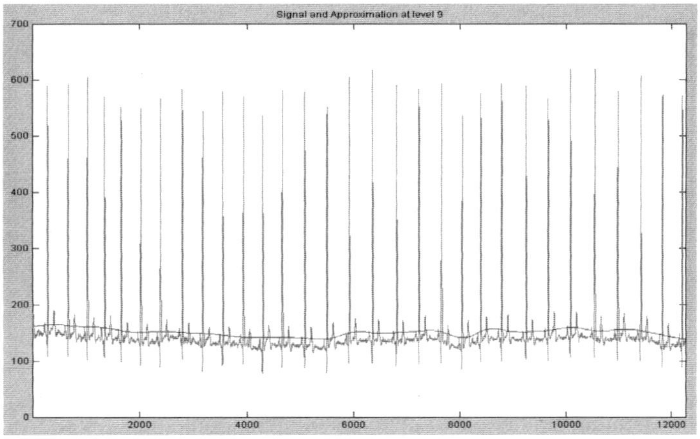

Figure 5. Baseline Wander

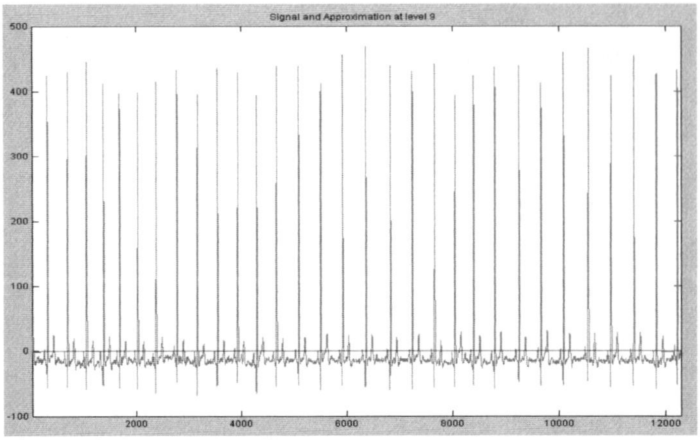

Figure 6. Constant baseline

3. Results

The device dimensions are 13 x 9 x 4 cm and it is powered by two 9 Volt batteries and a 3V coin battery that keeps actualized the time and date registers. Two LEDs indicate the beginning and ending of the storage process. The user is able to adjust the gain, baseline level, and GLCD brightness and contrast using potenciometers.

A successful implementation of a FAT16 file system over the SD card was achieved. The file extensions, characteristics and the contained information were as expected; as well as the used of the real time/calendar clock (RTCC).

Once the device was built, several studies were made using silver/chloride-silver electrodes. Leads I, II and III were successfully measured. Figure 7 shows the ECG registers displayed on the GLCD screen.

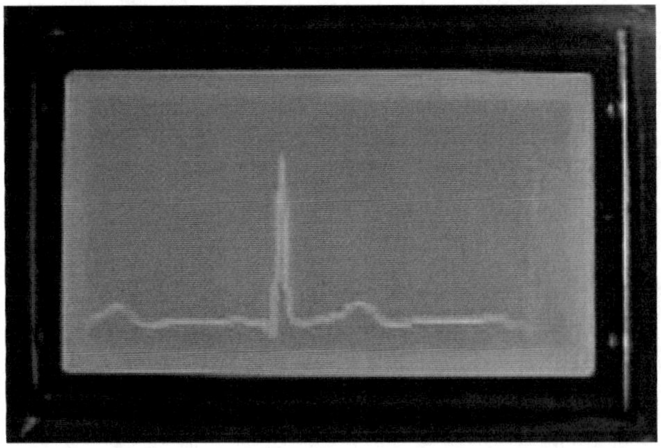

Figure 7. ECG signal on the GLCD

The results on the de-nosing algorithm and the baseline restoration process were a complete success. In the first case, the algorithms show to be extremely effective removing noise of nearly a 1% of the signal amplitude. Regarding to the baseline wander, constant baseline ECG registers were obtained after the process.

This work have already been showed to cardiologists, which were satisfied by the ECG visualization. Also, the device was tested in some patients. In one case, a patient was able to record himself his own ECG signal during an arrhythmic event (Figure 8).

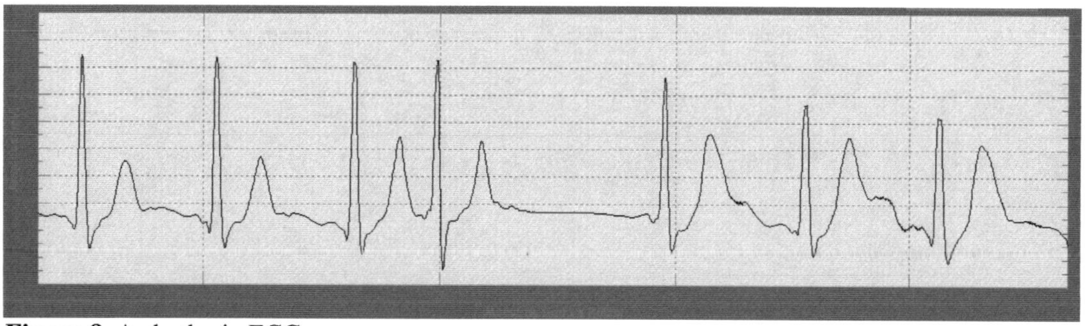

Figura 8. Arrhythmic ECG

The good results obtained in this work encourage continuing the research and developing of this device and the studies of new filtering algorithms using the Wavelet Transform.

References

[1] Webster, John G. [et. al] 1998: *"Medical Instrumentation Application and Design"*. John Wiley & Sons Inc - Estados Unidos
[2] *"Wiley Encyclopedia of Biomedical Engineering"* Volume 1-Metin Akay
[3] Oliveri, Marttin 2004: *"Elementos de diseño de circuitos de Amplificación del ECG"* Uruguay , Montevideo XII Seminario de Ingeniería Biomédica, Facultades Medicina de Ingeniería
[4] *"Análisis de Esquemas de Filtrado Análogo Para Señales ECG"* Scientia et Technica Año XIII, No 37, (2007) Universidad Tecnológica de Pereira
[5] Amor, Mariano; Lasarte, Ignacio; Liria, Luciana; Poblete, Felipe; González, Mariano 2007: *"Monitor de ECG"*. Universidad Tecnológica Nacional. Facultad Regional San Nicolás
[6] *"PICmicro™ Mid-Range MCU Family Reference Manual"*
[7] Sumit Mitra, Stan D'Souza, Russ Cooper *"AN546 Using the Analog-to-Digital (A/D) Converter"*
[8] Ortiz Verdu, Jordi; Gigante Beneito, Luis; Cucart Galbis, Pau; Asmae el Moudden; Abad Linares, Daniel 2008: *"Acceso a Tarjetas de Memoria de Estado Sólido"* Laboratorio de Sistemas Electrónicos Digitales Universidad Politécnica de Valencia
[9] *"Hardware White Paper Designing Hardware for Microsoft Operating Systems FAT: General Overview of On-Disk Format"* Version 1.02 - Microsoft Corporation 1999
[10] *"SD Specifications Part A2 SD Host Controller Simplified Specification"*. Version 2.00 – 2007
[11] *"SanDisk Secure Digital Card-Product Manual"*. Version 1.9 – 2003
[12] Azor Montoya, J. R., 2005. *"La transformada Wavelet (WT) Aplicada a la Ingeniería."*, Universidad de Mendoza - Facultad de Ingeniería
[13] Mathworks, 2009. *"Wavelet Toolbox"* MATLAB Documentation
[14] Semmlow, John L., 2004, *"Biosignal and Biomedical Image Processing. MATLAB-Based Applications"*. Marcel Dekker Inc.

Impact of head models in N170 component source imaging: results in control subjects and ADHD patients

L. Beltrachini[1,†], A. Blenkmann[1,2,†], N. von Ellenrieder[1], A. Petroni[4], H. Urquina[3], F. Manes[3], A. Ibáñez[3,5] and C.H. Muravchik[1]

[1] Laboratory of Industrial Electronics, Control and Instrumentation (LEICI), National University of La Plata.
[2] Epilepsy Center, IBCN - CONICET - University of Buenos Aires.
[3] Institute of Cognitive Neurology (INECO); & Institute of Neuroscience, Favaloro University, Buenos Aires, Argentina.
[4] Integrative Neuroscience Laboratory, Physics Department, University of Buenos Aires, Buenos Aires, Argentina.
[5] Universidad Diego Portales, Santiago, Chile.
[†] Both authors contributed equally to this work

Abstract.
The major goal of evoked related potential studies arise in source localization techniques to identify the loci of neural activity that give rise to a particular voltage distribution measured on the surface of the scalp. In this paper we evaluate the effect of the head model adopted in order to estimate the N170 component source in attention deficit hyperactivity disorder (ADHD) patients and control subjects, considering faces and words stimuli. The standardized low resolution brain electromagnetic tomography algorithm (sLORETA) is used to compare between the three shell spherical head model and a fully realistic model based on the ICBM-152 atlas. We compare their variance on source estimation and analyze the impact on the N170 source localization. Results show that the often used three shell spherical model may lead to erroneous solutions, specially on ADHD patients, so its use is not recommended. Our results also suggest that N170 sources are mainly located in the right occipital fusiform gyrus for faces stimuli and in the left occipital fusiform gyrus for words stimuli, for both control subjects and ADHD patients. We also found a notable decrease on the N170 estimated source amplitude on ADHD patients, resulting in a plausible marker of the disease.

1. Introduction

Researchers use source localization techniques of event-related potential (ERP) to identify the loci of neural activity that give rise to a particular voltage distribution measured on the surface of the scalp. This problem, usually called electroencephalogram (EEG) inverse problem, is affected by the uncertainties of the model used to simulate the situation, such as the head geometry, electrical conductivity or electrode positions. However, it has been shown that if the model used does not differ considerably from the true model, those differences tend to be of the same order of the electronic noise, and then can be individually negligible [1, 2, 3]. To build those realistic models it is necessary to have extra information, usually provided by magnetic resonance images (MRI) and an electrode acquisition system. However, most of ERP studies consist of making repeated measurements over a large number of patients/controls, making individual model acquisition impractical. Therefore, a standard head model is customarily used.

The simplest and historically accepted model is the three layered homogeneous spherical model, where each layer represents the brain, skull and scalp (inside out), and the electrical conductivities are assumed to be homogeneous and isotropic [4]. As is known, this model has plenty inaccuracies that might affect the EEG forward and inverse problems, such as the lack in geometrical adherence of the assumed shape with respect to a real human head, as well as its simplified electrical conductivity model. In order to avoid these problems, realistic average atlas-based head models jump on stage. These models are generally made from an average of one hundred or more healthy subjects and guarantee an improvement over wrong shaped approximation models [5, 6].

To evaluate the effect of head model on source localization we consider the widely accepted standardized low resolution brain electromagnetic tomography (sLORETA) algorithm [7]. This method allows to localize distributed brain sources by performing a location-wise inverse weighting of the results of a minimum norm least squares analysis with their estimated variances, leading to a smooth solution. Based on the sLORETA we evaluate the impact of the head model in the inverse problem solution by comparing the estimated source variance. We show differences between head models and compute solutions for control subjects and patients.

In this work we focus on solving the EEG inverse problem for the ERP N170 component. The N170 is a cortical marker linked to facial, object and word processing [8, 9]. Several studies have looked for the N170 cortical sources, leading to different solutions. Some of them [9, 10] conclude that the brain generators of this ERP are localized in the fusiform gyri (FG), or in the FG and additional structures such as the lingual gyri [11]. Other works found the brain generators of the same component in the lateral occipital-temporal cortex [12] or in the superior temporal sulci [13]. Those differences may be due to the kind of the stimuli used or to the model assumed to solve the problem. As is shown in this work, unrealistic head models lead to an increase in the localization variance, and therefore its usage cannot be recommended.

We distinguish between control subjects and patients with attention deficit hyperactivity disorder (ADHD). ADHD is a neuropsychiatric condition with onset in childhood that extends over adolescent and adult life with a considerable symptomatic burden and functional impairment. It has been shown [14, 15] that the N170 component represents an ideal brain marker to assess possible cortical markers of emotion processing for faces in ADHD. Therefore, accurate source imaging is fundamental in order to report differences in cortical activation areas between control subjects and ADHD patients. In this work we analyze such differences and show that the spherical head model may lead to erroneous results, even worst if a single dipole model is assumed.

2. Material and Methods

This study is part of a broader research project on the association between cortical processing of facial emotional expressions, social cognition skills and executive functioning [16]. Only relevant data is presented in this work.

2.1. Participants

Ten adult participants with ADHD (one male; three left-handed; mean age of 33.1 years old, standard deviation of 3.42 years old) and ten healthy subjects (one male; two left handed; mean age of 33.3 years old, standard deviation of 3.64 years old) were recruited. ADHD was diagnosed according to the American Psychiatric Association criteria [17], and a questionnaire was given to healthy participants to rule out hearing, visual, psychiatric or neurological deficits. All participants gave signed, informed consent in agreement with the Helsinki declaration. All experimental procedures were approved by the Ethics Committee of the Institute of Cognitive Neurology. More details regarding participants selection procedure may be found in [16].

2.2. ERP stimuli
Faces and word images were presented to test the effects of stimulus type (faces, words or face-word stimuli), valence (positive vs. negative) and compatibility (compatible vs. incompatible in word and face valence combinations) in a dual valence task. Participants were asked to categorize words or faces displayed on a computer screen according to their valence, into one of two categories (positive or negative), as quickly as possible. In this study variations due to valence and compatibility were not analyzed, i.e. only the three different stimuli type were studied: faces, words and combined face-word. The task comprised of two blocks of 320 trials each, the simultaneous stimuli and the single stimuli block.

Figure 1 shows a diagram of the experimental design. Trials started with a fixation cross for 1000 ms. Then a stimulus was presented for 100 ms, followed by a fixation cross until participants responded. If the response was incorrect, a red cross was presented as feedback for 100 ms and the trial ended. No feedback was provided in trials with correct responses. After responses or feedback, an inter-stimulus interval (ISI) of 1000 ms was added.

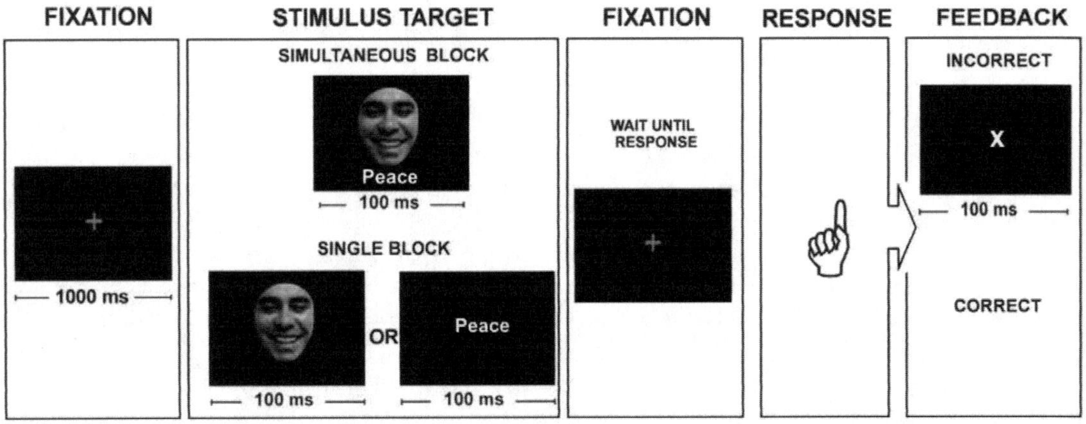

Figure 1. Experimental design. The trial started with a fixation cross, followed by a target stimulus (face or word) or two simultaneous stimuli (face and word) depending on which of the two main blocks was being performed. After the target stimulus, a fixation cross appeared until the response. If the response was incorrect, a red cross appeared and the trial ended. Otherwise, the trial ended without feedback. After responses or feedback, an ISI of 1000 ms was added (not shown).

In the simultaneous stimuli block, participants were exposed in each trial to a face in the center of the screen and a word four degrees below, simultaneously for 100 ms. Participants were asked to indicate the emotion shown by the face and to ignore the word. In the single stimulus block, participants were exposed on each trial to a face or a word in the center of the screen and asked to indicate the emotion of the stimulus. More details regarding experimental conditions can be found in [16].

2.3. Data acquisition
EEG signals were recorded with a BioSemi 128-channel system (BioSemi, Amsterdam, the Netherlands). Sampling rate was set at 500 Hz and signals were band-pass filtered between

0.1 and 100 Hz. Then, data were filtered off-line to remove unwanted frequency components between 0.3 and 30 Hz, down-sampled to 128 Hz and re-referenced to instantaneous average potential. Two bipolar derivations monitored vertical and horizontal ocular movements. EEG data were segmented from 200 ms prior, to 800 ms after the stimulus onset. All segments with eye movement contamination were removed from further analysis using an automatic algorithm [18] and a visual procedure. Only artifact-free segments were averaged to obtain ERPs.

2.4. Head model and forward problem

In order to solve the inverse problem we must first choose a head model. Since individual MRI data was not available, a standard model was considered. The first model proposed was the widely used three layered isotropic spherical model, where each sphere represents the boundaries of different tissue domains (i.e., brain, skull and scalp). The electrical conductivities adopted for each layer are the accepted 0.33 S/m for the brain and scalp, and 0.0096 S/m for the skull [19].

The second model considered in the present study was an average head model built from a sample of 152 MRIs of typical participants provided by the International Consortium of Brain Mapping (ICBM) [20]. The scalp and skull surfaces were extracted from the image using the brain extraction tool [21] and then corrected using MeshLab. To make a fully realistic model we considered the white matter anisotropy using a diffusion tensor atlas based on 81 healthy participants [22] co-registered with the ICBM model. This allows to consider the electrical conductivity as a tensor field through a linear transformation of the diffusion tensor eigenvalues [23]. In Figure 2(a) we show a sagital slice ($x = 9$ in the Montreal Neurological Institute coordinate system) where the boundary surfaces of both models are overlapped. There can be seen the geometric differences between them. In Figure 2(b) we show in the same slice the electrical conductivity map (mean conductivity across each voxel) considered for the realistic (average) model.

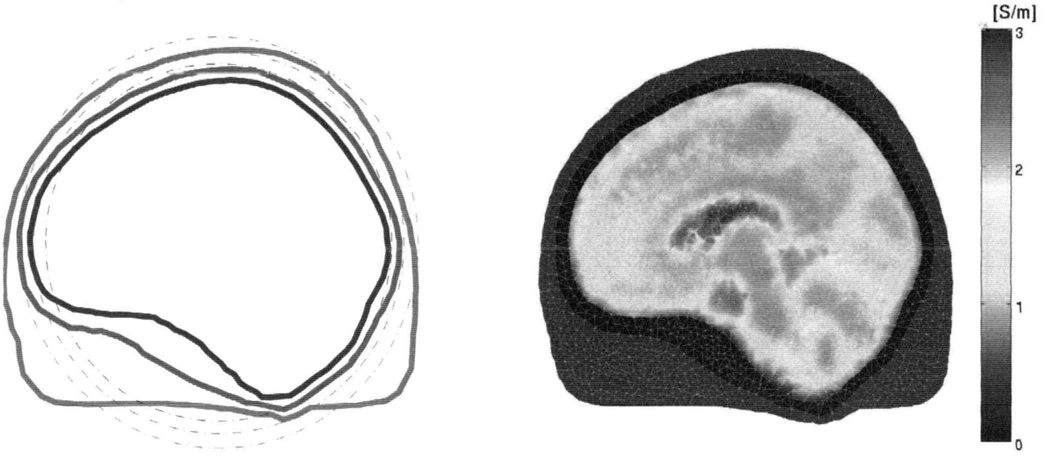

(a) (b)

Figure 2. Head models considered in this work. In Figure 2(a) a sagital slice where both models are overlapped is shown. In Figure 2(b) we show the real-shaped model tessellation colored by the electrical conductivity map (mean conductivity).

In order to solve the EEG forward problem we use the linear finite element method (FEM) [24]. This method, unlike the boundary element method, allows to handle anisotropy of the electrical conductivity in the soft tissues. The original models were tessellated using the ISO2Mesh toolbox [25], resulting a tetrahedral mesh with 2 mm side length average (see

Figure 2(b)). Possible solutions were constrained to the cortical surface but were not constrained for orientation [6].

2.5. Source imaging

To localize the generators of brain activity we considered the widely accepted sLORETA [7]. This is a statistical source imaging method that performs a distributed and smooth source localization. Although its solution is a distributed source activation map, it has been shown that sLORETA method localizes dipolar sources with zero error under ideal conditions [26, 27], allowing its usage also for dipolar source estimation. Therefore, it can be used to compare both dipolar and distributed solutions.

Basically, sLORETA employs a minimum norm current density estimation, and then localization is computed by standardizing the obtained values with respect to the variance of those estimators, leading to the here called source current power density (SCPD) factor. This estimation is done by assuming that the current density variance receives contributions from possible noise in the EEG measurements due to electric neuronal activity. In its simplest form, this noise can be assumed to be independent and identically distributed all over the cortex, allowing equal contribution of all cortical neurons to the biological noise. Under these conditions, it can be shown [7] that the estimated current density covariance matrix $\mathbf{S}_{\hat{\mathbf{j}}}$ depends only on the head model used to compute the forward problem, and then can be used to compare the influence of models in the EEG inverse problem. To quantify this influence we compute the ratio between the mean variance of the estimated current density across the directions for both spherical and realistic head models, i.e.

$$R_l = \frac{tr\left\{\left[\mathbf{S}_{\hat{\mathbf{j}}_S}\right]_l\right\}}{tr\left\{\left[\mathbf{S}_{\hat{\mathbf{j}}_R}\right]_l\right\}}, \tag{1}$$

where $tr\{\cdot\}$ is the trace operator, S and R refers to the spherical and realistic head models, and the l subindex refers to the lth test location over the cortex. Equation (1) allows to compare between the variance of the estimated current density map for both spherical and realistic head models. If the mean value of the estimated current density is of the same order for both models, the variance ratio will account for the differences on the SCDP factor, and then will explain differences on source location. Larger variance ratio values will be associated with cortex regions where the realistic model improves the source localization over the spherical model. We refer to [7] for sLORETA related formulae.

2.6. Data selection

sLORETA images were calculated at N170 component negative peak for each subject and condition using a spherical and a realistic head model. To determine the N170 peak, the average signal of representative electrodes over occipital lobes was obtained (see Figure 3), within a 167-229 ms time window for faces and simultaneous faces-words stimuli, and within a 182-284 ms time window for word stimuli, for each subject. The N170 peak amplitude was found as the local minimum of this average.

SCDP was calculated for each condition and subject using sLORETA method with the presented head models at the N170 peak. Finally, the average of this source images for faces, words and simultaneous faces-words stimuli were obtained for both groups under study (patients and controls).

3. Results

In this Section we show results concerning the N170 component source localization considering both realistic and spherical head models, for control subjects and ADHD patients. First we

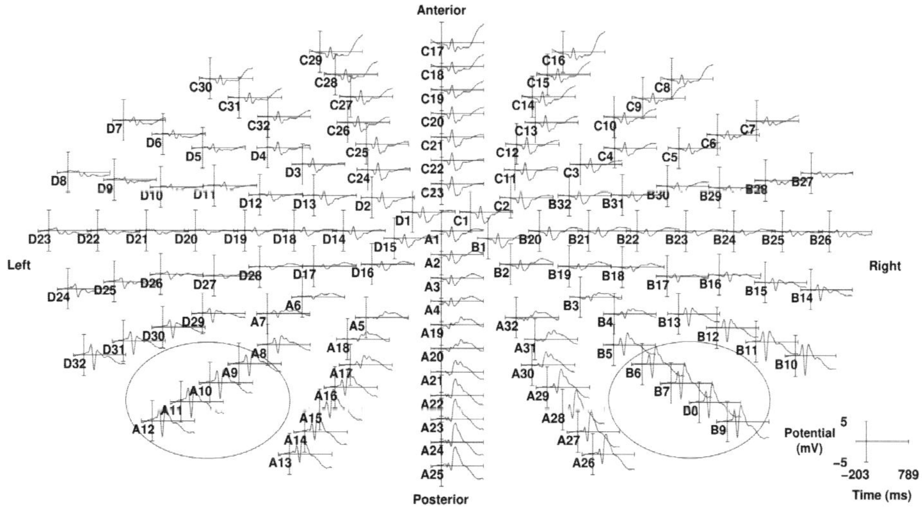

Figure 3. Measured potentials placed in the 128 electrode BioSemi standard position. Representative electrodes are marked (see Section 2.6).

perform the source localization. For each subject and condition two sLORETA source images were obtained at N170 peak, one for each head model. These images were averaged across control subjects (Figure 4) and ADHD patients (Figure 5).

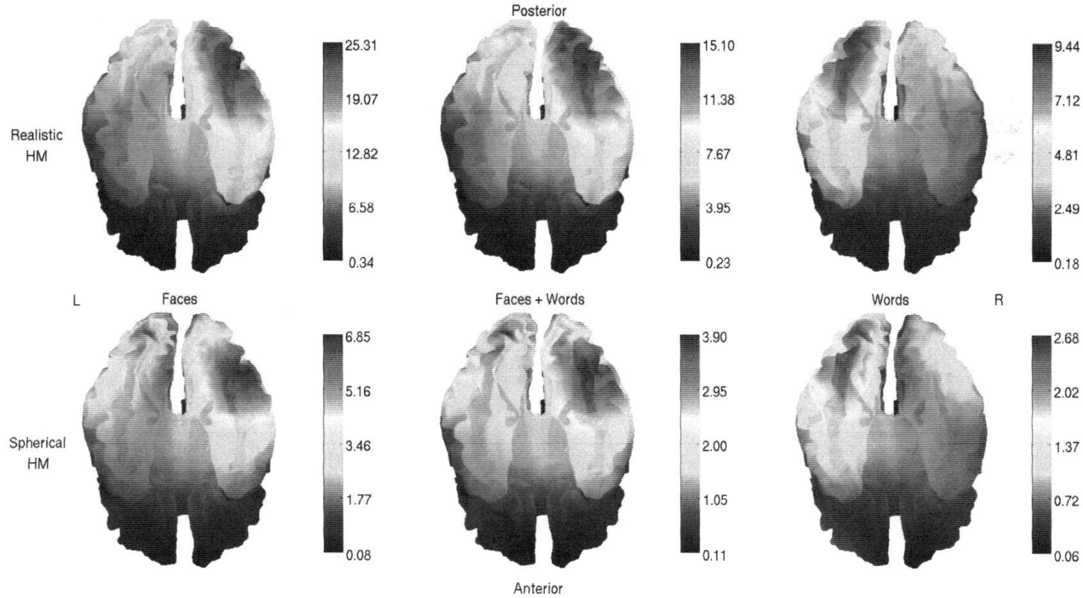

Figure 4. sLORETA source images (SCDP$\times 10^{-6}$) of N170 component peak for control subjects using the realistic head model (upper row) and the spherical head model (lower row) from a basal view. Three different stimuli are depicted: faces (first column), simultaneous faces and words (second column) and words (third column).

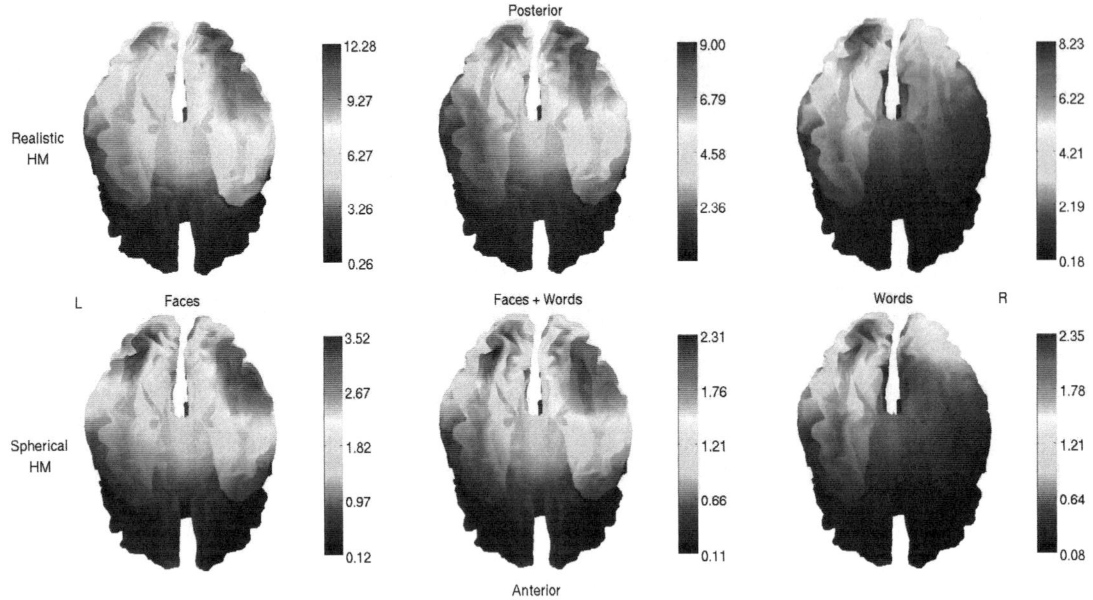

Figure 5. sLORETA source images (SCDP$\times 10^{-6}$) of N170 component peak for ADHD patients using the realistic head model (upper row) and the spherical head model (lower row) from a basal view. Three different stimuli are depicted: faces (first column), simultaneous faces and words (second column) and words (third column).

In a general view, all images obtained using the spherical head model found the activation zones in more anterior areas and more bilateral than using the realistic model. The intensity of SCDP is considerably larger in the realistic model, implying more reliable solutions to the inverse problem. These differences may be decisive if only one dipolar source placed on the test point where the SCDP is maximum is considered. Table 1 shows the different MNI coordinates and the anatomical description obtained from the Harvard-Oxford Cortical Structural Atlas [28] of these maxima. When the realistic head model was used, maximum SCDP values for faces and simultaneous faces-words stimuli in control subjects were found in the right occipital fusiform gyrus (OFG), whereas the right temporal-occipital fusiform cortex (TOFC) was found when the spherical model was used. This difference is even higher when analyzing the results for ADHD patients, since the right OFG is found when the realistic head model is used, whereas the left OFG is found when the spherical head model is considered. Note also that when analyzing SCDP for words stimuli, the left OFG was found in all cases. However, it has to be remarked that the SCDP value is higher when the realistic model is considered, and therefore results will be more accurate.

These differences between models may be explained in terms of the variance ratio introduced in Section 2.5. Figure 6 shows the variance ratio between the spherical and the realistic head models.

It can be seen that this measure is predominantly larger than unity, which means that the realistic head model is more reliable than the spherical model. Note that in the occipital pole (OP) and OFG the relation is up to 5 times, which means that the realistic head model will improve the spherical head model even better in these cortex areas, which are of special interest in the source localization of the N170 component.

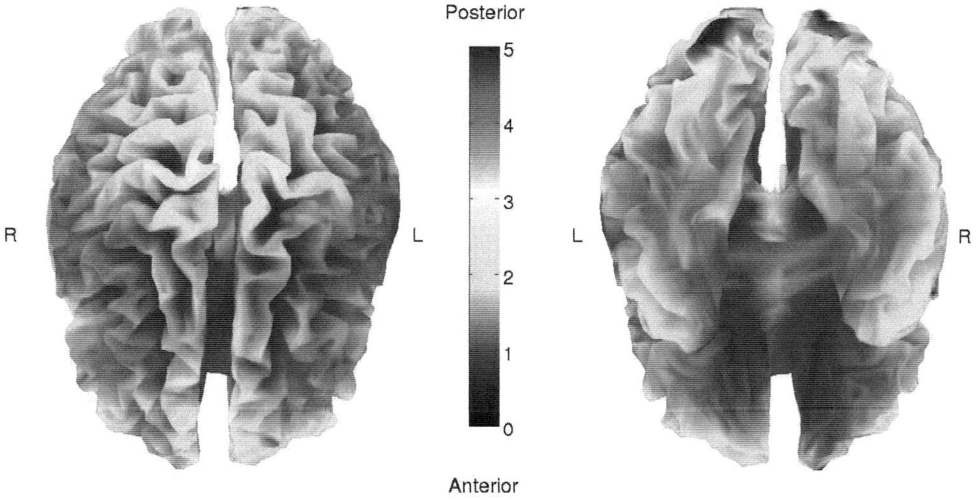

Figure 6. Superior (left) and basal (right) view of the brain showing the bias ratio (equation (1)). Larger values will be associated with cortex regions where the realistic model improves the source localization over the spherical model. Note that in the occipital pole and occipital fusiform gyrus the relation is up to 5 times.

4. Conclusions

In this work we show the influence of the head model considered in the forward problem for the ERP N170 component source localization. We show an accurate way to build an atlas based fully-realistic model, considering both geometry and electrical conductivity. Based on the sLORETA source imaging method we compare the influence of both models for source localization by means of the variance ratio on source reconstruction over the cortex. We find that the realistic head model improves the spherical head model, specially when the dipolar source model is considered. We also find that, using the proposed model, N170 sources are mainly located in the right OFG for both faces and simultaneous faces-word stimuli, and in the left OFG for words stimuli. Although these results are found to be the same for control subjects and ADHD patients, a notorious decrease of the SCPD value is found between them. This decrease in SCPD could be a possible marker of ADHD. In these situations, the realistic head model leads to more reliable results. Further work will be focused to analyze an accurate method to report activation areas considering distributed source localization results.

Acknowledgements

This work was funded by ANPCyT PICT 2007-00535, UNLP 11-I-0127, and ANPCyT PICT 2005-35423. The work of LB, AB, NvE, AP and AI was supported by CONICET. The work of CHM was also supported by CICpBA.

Group	Head Model	Faces			Faces + Words			Words		
		MNI Coordinates (x, y, z)	Hemisphere/ Anatomical Description	SCDP (10^{-6})	MNI Coordinates (x, y, z)	Hemisphere/ Anatomical Description	SCDP (10^{-6})	MNI Coordinates (x, y, z)	Hemisphere/ Anatomical Description	SCDP (10^{-6})
Control	Realistic	(40,-66,-15)	Right 45% OFG 15% LOC(id) 5% TOF	25.30	(26,-75,-15)	Right 74% OFG 7% LG	15.09	(-29,-80,-19)	Left 48% OFG 9% LOC(id)	9.44
	Spherical	(39,-57,-13)	Right 29% TOFC 8% OFG	6.85	(34,-58,-18)	Right 64% TOFC 16% OFG	3.90	(-22,-82,-20)	Left 26% OFG 5% LG	2.68
ADHD	Realistic	(25,-85,-17)	Right 49% OFG 9% LOC(id) 5% OP	12.27	(20,-85,-19)	Right 20% OFG 6% LOC(id)	9.00	(-24,-86,-20)	Left 35% OFG 8% LOC(id)	8.23
	Spherical	(-27,-72,-17)	Left 67% OFG	3.52	(-27,-72,-17)	Left 53% OFG	2.31	(-22,-82,-20)	Left 26% OFG 5% LG	2.35

Table 1. Anatomical locations of maximum estimated current density power (SCDP) for each group and stimuli condition (faces, words and simultaneous faces and words) using spherical and realistic head models. OFG: Occipital Fusiform Gyrus, LOC(id): Lateral Occipital Cortex (inferior division), TOFC: Temporal Occipital Fusiform Cortex, LG: Lingual Gyrus, OP: Occipital Pole. Cortex areas with probability less than 5% were not considered.

References

[1] von Ellenrieder N, Muravchik C and Nehorai A 2006 *IEEE Trans. Biomed. Eng.* **53** 249–257
[2] Güllmar D, Haueisen J and Reichenbach J 2010 *Neuroimage* **51** 145–163
[3] Beltrachini L, von Ellenrieder N and Muravchik C H 2010 *Comput Methods Programs Biomed.* In press
[4] de Munck J and Peters M 1993 *IEEE Trans. Biomed. Eng.* **40** 1166–1174
[5] von Ellenrieder N, Muravchik C, Wagner M and Nehorai A 2009 *IEEE Trans. Biomed. Eng.* **56** 587–597
[6] Valdés-Hernndez P, von Ellenrieder N, Ojeda-Gonzalez A, Kochen S, Alemán-Gómez Y, Muravchik C and Valdés-Sosa P 2009 *Journal of Neuroscience Methods* **185** 125–132
[7] Pascual-Marqui R 2002 *Methods & Findings in Experimental & Clinical Pharmacology* **24** 5–12
[8] Proverbio A, Riva F, Martin E and Zani A 2010 *PLoS.One* **5** e11242
[9] Rossion B, Joyce C, Cottrell G and Tarr M 2003 *Neuroimage* **20** 1609–1624
[10] Deffke I, Sander T, Heidenreich J, Sommer W, Curio G and Trahms L 2007 *Neuroimage* **35** 1495–1501
[11] Mnatsakanian E and Tarkka I 2004 *Clin. Neurophysiol.* **115** 880–886
[12] Schweinberger S, Pickering E, Jentzsch I, Burton A and Kaufmann J 2002 *Cognitive Brain Research* **14** 398–409
[13] Itier R and Taylor M 2004 *Neuroreport* **15** 1261–1265
[14] Marsh P and Williams L 2006 *Neuroscience and Biobehavioral Reviews* **30** 651–665
[15] Ibáñez A, Manes F, Escobar J, Trujillo N, Andreucci P and Hurtado E 2010 *Neuroscience Letters* **471** 48–52
[16] Ibáñez A, Petroni A, Urquina H, Torralva T, Blenkmann A, Beltrachini L, Muravchik C, Baez S, Cetckovih M, Torrente F, Hurtado E, Guex R, Sigman M, Lischinsky A and Manes F 2011 *Social Neuroscience* In press
[17] American Psychiatric Association 2000 *Diagnostic and statistical manual of mental disorders (4th ed., text rev.)* (Amer Psychiatric Pub)
[18] Gratton G, Coles M and Donchin E 1983 *Electroencephalography and Clinical Neurophysiology* **55** 468–484
[19] Dannhauer M, Lanfer B, Wolters C and Knösche T 2010 *Human Brain Mapping* In press
[20] Mazziotta J, Toga A, Evans A, Fox P, Lancaster J, Zilles K, Woods R, Paus T, Simpson G, Pike B, Holmes C, Collins L, Thompson P, MacDonald D, Iacoboni M, Schormann T, Amunts K, Palomero-Gallagher N, Geyer S, Parsons L, Narr K, Kabani N, Goualher G L, Boomsma D, Cannon T, Kawashima R and Mazoyer B 2001 *Philos. Trans. R. Soc. Lond. B Biol. Sci.* **356** 1293–1322
[21] Jenkinson M, Pechaud M and Smith S 2005 *Eleventh Annual Meeting of the Organization for Human Brain Mapping*
[22] Mori S, Oishi K, Jiang H, Jiang L, Li X, Akhter K, Hua K, Faria A, Mahmood A, Woods R, Toga A, Pike G, Neto P R, Evans A, Zhang J, Huang H, Miller M, van Zijl P and Mazziotta J 2008 *Neuroimage* **40** 570–582
[23] Tuch D, Wedeen V, Dale A, George J and Belliveau J 2001 *Proc. Natl. Acad. Sci. USA* **98** 11697–11701
[24] Beltrachini L, von Ellenrieder N and Muravchik C H 2008 *Actas del XVII Congreso sobre Métodos Numéricos y sus Aplicaciones - ENIEF*
[25] Fang Q and Boas D 2009 *Proceedings of IEEE International Symposium on Biomedical Imaging* 1142–1145
[26] Sekihara K, Sahani M and Nagarajan S 2005 *Neuroimage* **25** 1056–1067
[27] Greenblatt R, Ossadtchi A and Pflieger M 2005 *IEEE Trans. Signal Process.* **53** 3403–3412
[28] Desikan R, Segonne F, Fischl B, Quinn B, Dickerson B, Blacker D, Buckner R, Dale A, Maguire R, Hyman B, Albert M and Killiany R 2006 *Neuroimage* **31** 968–980

Study of T-wave morphology parameters based on Principal Components Analysis during acute myocardial ischemia

Fabricio Hugo Baglivo[1], Pedro David Arini[2,3]

[1]Facultad de Ingeniería, Dto. de Electrónica, Universidad de Buenos Aires, Argentina
[2]Facultad de Ingeniería, Instituto de Ingeniería Biomédica, Universidad de Buenos Aires, Argentina
[3]Instituto Argentino de Matemática *'Alberto P. Calderón'*, CONICET, Buenos Aires, Argentina

E-mail: `pedro.arini@conicet.gov.ar`

Abstract. Electrocardiographic repolarization abnormalities can be detected by Principal Components Analysis of the T-wave. In this work we studied the effect of signal averaging on the mean value and reproducibility of the ratio of the 2nd to the 1st eigenvalue of T-wave (T_{W}^{21}) and the absolute and relative T-wave residuum ($T_{\mathrm{WR}}^{\mathrm{rel}}$ and $T_{\mathrm{WR}}^{\mathrm{abs}}$) in the ECG during ischemia induced by Percutaneous Coronary Intervention. Also, the intra-subject and inter-subject variability of T-wave parameters have been analyzed. Results showed that $T_{\mathrm{WR}}^{\mathrm{rel}}$ and $T_{\mathrm{WR}}^{\mathrm{abs}}$ evaluated from the average of 10 complexes had lower values and higher reproducibility than those obtained from 1 complex. On the other hand T_{W}^{21} calculated from 10 complexes did not show statistical differences versus the T_{W}^{21} calculated on single beats. The results of this study corroborate that, with a signal averaging technique, the 2nd and the 1st eigenvalue are not affected by noise while the 4th to 8th eigenvalues are so much affected by this, suggesting the use of the signal averaged technique before calculation of absolute and relative T-wave residuum. Finally, we have shown that T-wave morphology parameters present high intra-subject stability.

1. Introduction

The electrocardiographic ventricular repolarization dispersion (VRD) can be evaluated by parameters based on the Principal Component Analysis (PCA) such as the ratio of the $2nd$ to the $1st$ eigenvalue (T_{W}^{21}) and either relative or absolute T-wave residuum ($T_{\mathrm{WR}}^{\mathrm{rel}}$ and $T_{\mathrm{WR}}^{\mathrm{abs}}$). These parameters, applied to 12-lead standard ECG quantify the spatial complexity of ventricular repolarization and may become a noninvasive diagnostic tool in different cardiac pathologies [1]. The aim of this work is to evaluate the PCA-based parameters derived from signal averages in ischemic patients. We hypothesize, according to the results of Batdorf et al. in healthy subjects [2], that the PCA-based morphology parameters obtained during ischemia in averaged ECG will present lower values and better reproducibility than when computed from a single beat. For that purpose, we evaluated VRD in the course of transmural ischemia induced by Percutaneous Coronary Intervention (PCI) procedure [3]. The balloon-inflation PCI provides an advantageous model of the first minutes of acute ischemia in humans, due to a total interruption of blood flow through the ventricles.

2. Materials and Methods

2.1. Data Set

The study group consisted of 93 ECG records from patients at the Charleston Area Medical Center in West Virginia undergoing elective prolonged balloon occlusion during PCI in one of the major coronary arteries (STAFF-III study) [3]. This group was selected from a total of 108 patients, with the condition that T-wave could be delineated during the complete time course of ischemia. The mean inflation duration was 4' 28" with a standard deviation of 74". Eight independent ECG leads (V_1-V_6, I, II) were recorded using equipment by Siemens-Elena AB (Solna, Sweden) and digitized at sampling rate of 1000 Hz and amplitude resolution of 0.6 μV. Leads III, aVR, aVL and aVF were derived from leads I and II. ECGs were acquired for each patient in supine position. Two control records C_k (k=1,2) were continuously acquired for 5-minute some time before the PCI and one ECG was recorded during PCI.

2.2. ECG processing

For a total of 93 patients, the 8 ECG leads were processed in control situation and PCI procedure respectively. The QRS fiducial points were detected on a multilead basis by an automatic QRS detector [4]. Cubic spline interpolation was used for baseline wander rejection and the ECG delineation system based on the Wavelet Transform has been used for T-wave location and delineation [5].

We analyzed the morphological T-wave parameters for both single complex (1-Cx) and average of 10 complexes (10-Cx) using Singular Value Decomposition (SVD). First of all a "*template beat*" was calculated. For that purpose 10 beats were segmented using a constant window and aligned using a cross-correlation technique. After that the template beat was obtained through direct averaging of 10 beats. Second new jitter-corrected beats (arising from analysis windows) were used into the "*final average beat*" if the cross-correlation coefficient between the new beats and the "*template beat*" in each channel was greater than 98%, otherwise the beat is rejected. Two rejected beats were the accepted limit on the computation of each ith "*final average beat*". The cross-correlation coefficient was calculated from the QRS, which was segmented by taking a 200 msec. interval centered at the QRS-fiducial point previously detected [4] [6] .

The computation of T-wave morphological parameters, calculated from 1-Cx and 10-Cx, were obtained from onto N-samples windows (W_i), which defines ventricular repolarization phase for the ith beat of ECG. The beginning of these windows was calculated as a median value of T-wave onset in the 8 independent ECG leads. In the same way, the end of these windows was obtained with the median value of T-wave offset fiducial points. Afterward SVD was applied obtaining one set of 8 singular values ξ_j $(j = 1, \ldots, 8)$, which are ordered such that $\xi_1 \geq \xi_2 \geq \xi_3 \geq \ldots\ldots \geq \xi_8 \geq 0$. SVD technique finds a system of eight orthogonal directions (S1...S8), in which S1 contains most of the ECG energy, S2 contains most of the remaining ECG energy, and so forth. It has been shown that the first three directions S_1 S_2 S_3, contain 98 % of the whole ECG energy [7] representing the dipolar component, while the last 5 represent the so-called non-dipolar components of the T-wave respectively. In this sense, we have computed and evaluated the T-wave morphology parameters based on PCA; these parameters have been named $\mathcal{T}$ parameters and were calculated as:

- The ratio of second to the first eigenvalue for the ith beat, corresponding with the T-wave, expressed as:

$$T^{21}_{W_i} = 100 * (\xi^2_{i,2}/\xi^2_{i,1}) \tag{1}$$

Journal of Physics: Conference Series **332** (2011) 012020 doi:10.1088/1742-6596/332/1/012020

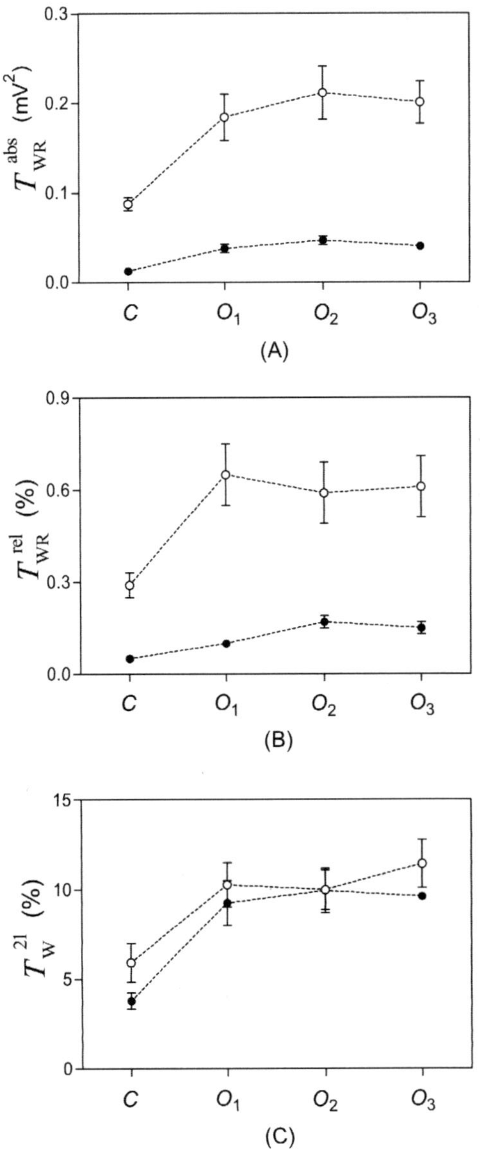

Figure 1. The effect of using a single complex versus 10 complex signal average on the mean±SEM (standard error mean) in control and during PCI procedure. (A) Absolute T-wave residuum, (B) Relative T-wave residuum and (C) the ratio of second to the first eigenvalue. 1-Cx (blank circle) and 10-Cx (filled circles). Control Situation (C), first minute of occlusion (O_1), second minute of occlusion (O_2) and the third minute of occlusion (O_3).

- The absolute T-wave residuum, for the ith beat, to quantify the non-dipolar components as the sum of the fourth to the eight eigenvalues, calculated as:

$$T^{\text{abs}}_{\text{WR}_i} = \sum_{l=4}^{8} \xi^2_{i,l} \tag{2}$$

- The relative T-wave residuum, for the ith beat, to quantify the relative contribution of $T^{\text{abs}}_{\text{WR}_i}$ with respect the total energy, calculated as:

$$T^{rel}_{\text{WR}_i} = 100 * (\sum_{l=4}^{8} \xi^2_{i,l} / \sum_{l=1}^{8} \xi^2_{i,l}) \tag{3}$$

The $\mathcal{T}$ parameters, $T^{\text{abs}}_{\text{WR}}$, $T^{\text{rel}}_{\text{WR}}$ and T^{21}_{W}, were obtained from a single complex and from the 10 complexes in the signal averaging.

2.3. *Statistical analysis*

We analyzed the statistical distribution of $\mathcal{T}$ parameters using D'Agostino-Pearson normality test. The $\mathcal{T}$ parameters, in both control and PCI procedure, did not follow a Gaussian distribution. Therefore, a two-tailed non-parametric Mann-Whitney test and Wilcoxon Signed Rank Test were applied. When p value was <0.05, differences were considered statistically significant.

2.4. *Intra-individual variability of $\mathcal{T}$ parameters*

To assess the intra-individual variability of $\mathcal{T}$ parameters [8], the standard deviation (SD) of $\mathcal{T}$ expressed as $\xi^{\mathcal{T}}_k(j)$ was computed for each patients ($j = 1, ..., J$) in each of the two control ECG recordings C_k, $k = 1,2$. Also, the difference between SDs of the two control recordings was calculated as: $D^{\mathcal{T}}(j) = \xi^{\mathcal{T}}_1(j)$ - $\xi^{\mathcal{T}}_2(j)$. Then, the Wilcoxon Signed Rank Test was applied to the difference $D^{\mathcal{T}}(j)$ from the total of population and $\mathcal{T}$ parameters, with the aim to evaluate the next hypothesis:

- $\mathcal{H}_0$: the intra-individual change is zero
- $\mathcal{H}_1$: the intra-individual change is different to zero

2.5. *Inter-individual variability of $\mathcal{T}$ parameters*

To assess the inter-individual variability of $\mathcal{T}$ parameters [8], the mean value of the two SDs $\xi^{\mathcal{T}}_k(j)$, ($k = 1,2$), was calculated as $\xi^{\mathcal{T}}(j) = \frac{1}{2}$ { $\xi^{\mathcal{T}}_1(j) + \xi^{\mathcal{T}}_2(j)$ }. Then SD of $\xi^{\mathcal{T}}(j)$ over all patients ($j = 1, ..., J$) was denoted by $\chi^{\mathcal{T}\updownarrow}$. Moreover, the SD of the two SDs $\xi^{\mathcal{T}}_k(j)$, $k = 1,2$ for each patient j was computed and denoted by $\chi^{\mathcal{T}\leftrightarrow}_k(j)$.

The Wilcoxon Signed Rank Test was applied to compare intra-individual variability $\chi^{\mathcal{T}\leftrightarrow}_k(j)$, for $j = 1, ..., J$, with the inter-individual SD of the whole population $\chi^{\mathcal{T}\updownarrow}$.

3. Results

The series $T^{\text{abs}}_{\text{WR}}$, $T^{\text{rel}}_{\text{WR}}$ and T^{21}_{W} were calculated during control and balloon inflation in PCI procedure and, for the sake of robustness, it was applied a median filter with a windows size of 7.5 sec on the three series of parameters. We characterized 4 different time instants: $\mathcal{T}$ parameters during Control Situation (C) (mean value of C_k, $k = 1,2$) associated to the median value of 5 minutes just before the start of occlusion, $\mathcal{T}$ parameters during the first minute (O_1) associated to the median of 7.5 sec centered around the first minute, the second (O_2) and the third (O_3), for their respective minutes of the occlusion procedure.

Table 1 shows the results represented by the mean value $\pm$ of the Standard Error of the Mean (SEM) of C, O_1, O_2 and O_3 values for each $\mathcal{T}$ parameter considering the 93 analyzed records. The results showed that $T^{\text{abs}}_{\text{WR}}$, $T^{\text{rel}}_{\text{WR}}$ and T^{21}_{W} values during PCI compared against control situation

Table 1. Mean values ± SEM of $\mathcal{T}$ parameters. ${}^{A}p<0.001$ and ${}^{B}p<0.0001$ against control. ${}^{C}p<0.0001$ represent significant differences between 10-Cx and 1-Cx.

	beats	C	O_1	O_2	O_3
$T_{\mathrm{WR}}^{\mathrm{abs}}$ (mV^2)	1Cx	0.087±0.007	0.18±0.026 ^{A}	0.211±0.029 ^{A}	0.200±0.023 ^{A}
	10Cx	0.012±0.001 ^{C}	0.037±0.004 ^{C,B}	0.046±0.004 ^{C,B}	0.041±0.004 ^{C,B}
$T_{\mathrm{WR}}^{\mathrm{rel}}$ (%)	1Cx	0.29±0.04	0.65±0.10 ^{A}	0.59±0.10 ^{A}	0.61±0.10 ^{A}
	10Cx	0.05±0.01 ^{C}	0.10±0.01 ^{C,B}	0.17±0.02 ^{C,B}	0.15±0.02 ^{C,B}
T_{W}^{21} (%)	1Cx	5.92±1.07	10.27±1.23 ^{A}	9.99±1.10 ^{A}	11.45±1.34 ^{A}
	10Cx	3.80±0.45	9.26±1.25 ^{B}	9.94±1.25 ^{B}	9.62±0.02 ^{B}

Table 2. Reproducibility estimated as SDs of within-subject variance of $\mathcal{T}$ parameters. A ${}^{*}p<0.0001$ represents significant differences between 10-Cx and 1-Cx.

Cx	$T_{\mathrm{WR}}^{\mathrm{abs}}$ (mV^2)	$T_{\mathrm{WR}}^{\mathrm{rel}}$ (%)	T_{W}^{21} (%)
1	0.027	0.09	1.27
10	0.004*	0.01*	0.99

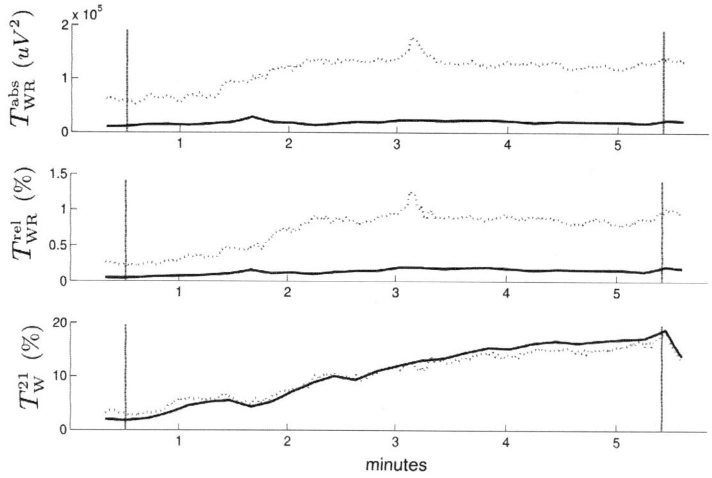

Figure 2. One example (patient number 10) of time evolution of $\mathcal{T}$ parameters for both 1-Cx (dotted line) and 10-Cx (black line) during PCI procedure. The vertical lines represents the start and end of occlusion.

were statistically significant for 1-Cx as well as 10-Cx analysis. Also we have studied the effect of ECG signal averaging *vs.* a single complex. The results shows that $T_{\mathrm{WR}}^{\mathrm{abs}}$ and $T_{\mathrm{WR}}^{\mathrm{rel}}$ values for 1-Cx compared against 10-Cx value are statistically significant, for both control and during PCI procedure. However, non significant differences were found for T_{W}^{21} parameter.

The mean value ± SEM of $\mathcal{T}$ parameteres are plotted in Figure 1, also it can be observed that T-wave morphological parameters show an increased tendency during PCI.

Table 2 shows standard deviations (SDs) of within-subject variance of $T_{\mathrm{WR}}^{\mathrm{abs}}$, $T_{\mathrm{WR}}^{\mathrm{rel}}$ and T_{W}^{21} . For minute-to-minute observation intervals of control situation we have observed significantly less variance for both $T_{\mathrm{WR}}^{\mathrm{abs}}$ and $T_{\mathrm{WR}}^{\mathrm{rel}}$ as the quantities of complexes were increased from 1 to 10. We have not observed $\mathcal{T}_{\mathrm{W}}^{21}$ statistically significant differences between 1-Cx to 10-Cxs respectively. In Figure 2 we show time-course evolution of $T_{\mathrm{WR}}^{\mathrm{abs}}$, $T_{\mathrm{WR}}^{\mathrm{rel}}$ and T_{W}^{21} during PCI procedure (patients #10).

A high intra-individual stability of $\mathcal{T}$ parameters was obtained (Section 2.4). The p value of the Wilcoxon Signed Rank Test was <0.05 and the hypothesis $\mathcal{H}_0$ of the intra-individual variability being negligible was accepted. Moreover, the inter-individual variability of $\mathcal{T}$ parameters was obtained (Section 2.5). The Wilcoxon Signed Rank Test showed that, the differences between intra-individual variations $\chi_k^{\mathcal{T}\leftrightarrow}(j)$ and SD of the whole population $\chi^{\mathcal{T}\updownarrow}$ were highly significant (p<0.05), being $\chi^{\mathcal{T}\updownarrow} > \bar{\chi}_k^{\mathcal{T}\leftrightarrow}$; where $\bar{\chi}_k^{\mathcal{T}\leftrightarrow} = \frac{1}{J}\sum_{j=1}^{J} \chi_k^{\mathcal{T}\leftrightarrow}(j)$.

4. Discussion and conclusions

We have hypothesized that $\mathcal{T}$ parameters obtained from ECG signal averaging with 10 complexes would have lower mean values and more reproducibility than those derived from a single complex analysis during control and PCI procedure. The obtained results indicate that:

1) $T_{\mathrm{WR}}^{\mathrm{abs}}$ and $T_{\mathrm{WR}}^{\mathrm{rel}}$ computed from 10-Cx show lower values, and higher reproducibility than those obtained from 1-Cx. This results suggest that a big amount of absolute and relative T-wave residuum on one beat is due to noise.

However T_{W}^{21} computed on the average cycle did not show statistical significant with respect to T_{W}^{21} on single beats; possibly because the numerator of T_{W}^{21} ($2nd$ eigenvalue) is not so much affected by noise as it is the numerator of relative T-wave residuum; obtained from the smaller 4-to-8 eigenvalues.

2) The results were obtained as from control situation (corresponding to patients with cardiac disease) and PCI (corresponding acute ischemic process). The results observed were concordant with the study from healthy subject done by Batdorf et al. [2].

3) The results showed that signal averaging technique can provide more significant differences than a single complex analysis. The ECG ventricular repolarization dispersion parameters (i.e, both relative or absolute T-wave residuum) can be improved through the use of signal-averaged technique.

4)Intra-individual analysis showed that the $\mathcal{T}$ parameters variability have high stability in control recordings. Moreover, the inter-individual analysis showed that the $\mathcal{T}$ parameters should be normalized in order to diminish such variation.

On the other hand, we have found the following limitations for this study. Batdorf et al. [2] shows that 200 complexes of signal averaging have lower noise and higher reproducibility than 10 complexes of signal averaging. In this work we did not increase the beats quantity into the signal average technique because dynamic ST-changes during PCI could be affected the measure of T-wave morphology. Furthermore, we have observed increases in noise of the ECG during PCI procedure, a phenomenon that should be quantified for different stages of the angioplasty procedure. Then, future studies are needed to investigate the optimal and maximum beats number that it is feasible to include into the average technique for PCI studies; as well as a method of normalization to reduce the variability between subjects. Also, we have analyzed the normal reproducibility or dispersion in $T_{\rm WR}^{\rm abs}$, $T_{\rm WR}^{\rm rel}$ and $T_{\rm W}^{21}$ during five minutes.

Acknowledgments
This work has been supported by *Consejo Nacional de Investigaciones Científicas y Técnicas* (PIP 538) and *Agencia Nacional de Promoción Científica y Tecnológica* (PICT 2108), both institutions from Argentina.

References
[1] Priori S, Mortara D, Napolitano C, Diehl L, Paganini V, Cant F, Cant G, Schwartz P 1997 Evaluation of the spatial aspects of T wave complexity in the long-QT syndrome *Circulation* **96** 3006-3012.
[2] Batdorf BH, Feiveson AH, Schlegel TT 2006 The effect of signal averaging on the reproducibility and reliability of measures of T-wave morphology *J. Cardiovasc. Electrophysiol.* **39** 266-270.
[3] García J, Lander P, Sörnmo L, Olmos S, Wagner G, Laguna P 1998 Comparative study of local and Karhunen-Loève-Based ST-T indexes in recording from human subjects with induced myocardial ischemia *Computer and Biomedical Research* **39** 271-292.
[4] Moody GB, Mark RG 1982 Development and evaluation of a 2 lead ECG analysis program *Computers in cardiology* 39-44.
[5] Martínez JP, Almeida R, Olmos S, Rocha AP, Laguna P 2004 A wavelet-based ECG delineatior: Evaluation on standard databases *IEEE Transactions on Biomedical Engineering* **51(4)** 266-270.
[6] Laciar E, Jané R, Brooks D 2003 Improved alignment method for noisy high-resolution ECG and Holter records using multiscale cross-correlation *IEEE Transactions on Biomedical Engineering* **50** 344-353.
[7] Acar B, Yi G, Hnatkova K, Malik M 1999 Spatial, temporal and wavefront direction characteristics of 12-lead T-wave morphology *Med. Biol. Eng. Comput.* **37** 574-584.
[8] Romero D, Ringborn M, Laguna P, Phalm O, Pueyo E 2011 Depolarization changes during acute myocardial ischemia by evaluation of QRS slopes: standard lead and vectorial approach *IEEE Transactions on Biomedical Engineering* **58** 110-120.

On the use of LDA performance as a metric of feature extraction methods for a P300 BCI classification task

Iván Gareis[1], Yanina Atum[1], Gerardo Gentiletti[1], Rubén Acevedo[1], Verónica Medina Bañuelos[2], Leonardo Rufiner[3, 4]

[1] Laboratorio de Ingeniería en Rehabilitación e Investigaciones Neuromusculares y Sensoriales; Facultad de Ingeniería, Universidad Nacional de Entre Ríos, Argentina
[2] Laboratorio de Investigación en Neuro Imagenología; División Ciencias Básicas e Ingeniería, Universidad Autónoma Metropolitana (Iztapalapa), México.
[3] Centro de I+D en Señales, Sistemas e Inteligencia Computacional; Facultad de Ingeniería y Ciencias Hídricas, Universidad Nacional de Litoral, Argentina.
4 Laboratorio de Cibernética; Facultad de Ingeniería, Universidad Nacional de Entre Ríos, Argentina.

E-mail; ivangareis@bioingenieria.edu.ar

Abstract. Brain computer interfaces (BCIs) translate brain activity into computer commands. To enhance the performance of a BCI, it is necessary to improve the feature extraction techniques being applied to decode the users' intentions. Objective comparison methods are needed to analyze different feature extraction techniques. One possibility is to use the classifier performance as a comparative measure. In this work the effect of several variables that affect the behaviour of linear discriminant analysis (LDA) has been studied when used to distinguish between electroencephalographic signals with and without the presence of event related potentials (ERPs). The error rate (ER) and the area under the receiver operating characteristic curve (AUC) were used as performance estimators of LDA. The results show that the number of characteristics, the degree of balance of the training patterns set and the number of averaged trials affect the classifier's performance and therefore, must be considered in the design of the integrated system.

1. Introduction

Brain computer interfaces (BCI) are devices that provide a direct link between the brain and a computer [1]. Such interfaces can be considered as being the only way of communication for people affected by a number of motor disabilities [2].

Figure 1 shows the general architecture of a BCI proposed by Millán *et al* [4] where the functional blocks are described. In this figure it can be seen how a subject (user) could control a device (e.g. a motorized wheelchair).

Most BCI systems are based on electroencephalography (EEG), but there are several ways (or paradigms) to obtain the desired control signals. One of them is the oddball paradigm, which is based on event related potentials (ERP). The ERP are evoked potentials with latencies higher than 100 ms whose expression depends on psychological and behavioral processes. When infrequent visual or auditory stimuli are mixed with frequent stimuli, the former evoke a potential in the EEG in the parietal cortex with a peak located around the 300 ms called P300. In order to estimate or detect the

ERP, the initial signal to noise ratio (SNR) must be improved because many recorded epochs are immersed in high levels of noise; background brain activity and electromyogram are some examples.

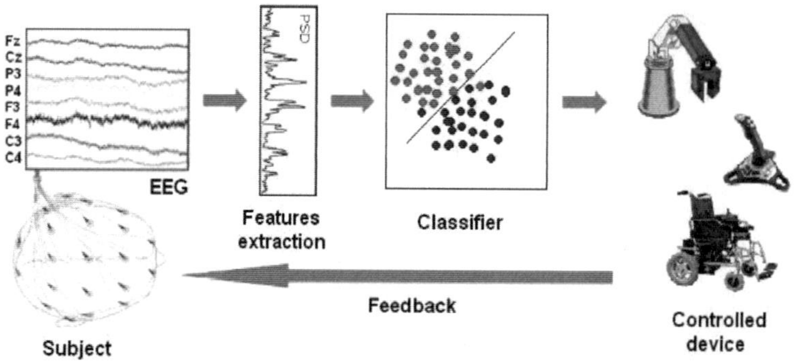

Figure 1. General architecture of BCI for device control.

With the recent advancement of machine learning algorithms and digital processing techniques, a part of the BCI research lies on exploration of feature extraction and classification techniques. The performance of a brain computer interface is highly dependent on these signal processing techniques used to extract the features that encode the BCI user intentions [3]. Therefore, there is a need for objective comparison methods to analyze different feature extraction techniques [4]. One straightforward solution is to feed a classifier with the features that are being compared, and use its performance as a measure of the separation power of such features. In this case, care must be taken in order to consider only the variations in the system's performance caused by the particular properties of the feature extraction techniques that are being evaluated. Any other issues, like those that may result from changes in the parameters of the classifier, must be ignored. In order to do so, it is important to study the behavior of the classifier to be used in conditions and with data similar to the ones that will be presented when using it as a feature extraction techniques comparison method.

One of the most popular classifiers for BCI applications is the Fisher´s linear discriminant analysis (LDA) [5, 6]. Even though the LDA has been extensively studied [7-9], the effect of unbalanced training datasets using electroencephalographic (EEG) data and the number of patterns necessary to reach a performance plateau have not been tested. That is, the point at which no significant performance gain will exist when adding more training patterns has not been determined.

In this paper, the problem of studying the behavior of LDA when used to discriminate between EEG signals containing event related potentials (ERPs) and EEG signals without the presence of ERPs is addressed.

2. Methodology

2.1. P300-Speller

When infrequent or particularly significant auditory, visual, or somatosensory stimuli, are mixed with frequent or routine stimuli, ERPs are typically evoked over the parietal cortex. This phenomenon can be used to implement a BCI commonly called P300 speller, which allows the user to select symbols from a matrix in a computer screen [10].

In the classical P300 speller the user faces a 6 x 6 matrix that contains all letters and characters. During the experiment a single row or column is intensified randomly with a predefined frequency; and, in a complete block of 12 intensifications, each row or column flashes once. To make a selection the user focuses on the character he/she desires to choose. As a result, assuming the intensification of one character of the matrix elicits ERPs, there will be two target trials and ten non target trials in each block. Typically the block of intensifications has to be repeated to effectively determine the character the user is focusing on. Figure 2 shows a matrix during the intensification of the second row.

To determine which intensification elicits an ERP the system has to be able to solve the binary classification problem (two possible classes: recordings with ERP and recordings without ERP).

Figure 2. Classical P300 speller stimulation matrix.

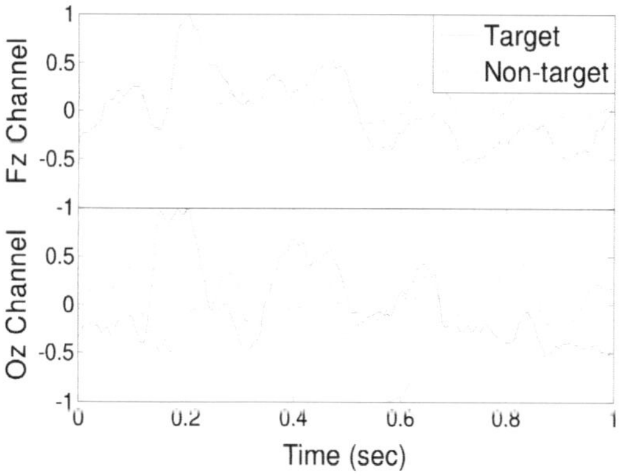

Figure 3. The coherent averages of 600 trials with and without ERP for channels Fz and Oz.

2.2. Recordings

The Neuroimaging Research Laboratory at Universidad Autónoma Metropolitana (UAM) provided a database containing the recordings of 30 healthy subjects using the P300 speller on a BCI2000 platform [11, 12]. Ten channels of ERP (*Nc*) were recorded using a sample frequency of 256 Hz. Channels Fz, C3, Cz, C4, P3, Pz, P4, PO7, PO8, Oz were recorded using a right ear reference and a right mastoid ground. A complete description of the parameters used for the speller and the data are available on the database website: http://akimpech.izt.uam.mx/p300db. In figure 3 the coherent averages of 600 trials with and without ERP can be seen. The signals were taken from channels Fz and Oz from a subject of the database.

Each subject in the database participated in four sessions with fifteen sequences per session. This yields a number *Nt* of labeled target trials equal to 630 and a number of labeled non target trials *Nnt* equal to 3150.

One of the premises of the creation of this database was to provide a realistic sample of the recordings, thus many of them present a significant number of outliers. A selection of ten subjects has been made among the ones without a large number of outliers, in order to prevent these variables to influence the results and to avoid using an artifact rejection block.

2.3. Preprocessing

As a first preprocessing stage the data were filtered and downsampled. Lowpass eighth-order forward-backward Chebyshev filters were used and a downsampling step was performed by selecting each Nth sample from the lowpass filtered data. As it will be explained in section 3 three sets of experiments were performed, each at different cutoff (3.5, 7 and 14 Hz) and decimation (Fsi of 8, 16 and 32 Hz) frequencies.

The signals from each electrode were normalized independently as to have a zero mean and a unitary standard deviation.

Single trials were extracted from the data, having one second duration and starting at the beginning of the intensification of a character. Due to the trial duration and the downsampling rates, the numbers of samples per trial or Ns_i are 8, 16 and 32 for each set of experiments.

The feature vectors (or patterns) were constructed by concatenating the single trials from the ten channels. Therefore the dimension of the feature vectors was Nc x Ns_i, or 80, 160 and 320.

Finally coherent averaging was applied in some experiments as it will be explained in section 3. This technique improves the signal to noise ratio, but may diminish the bit-rate of the BCI and reduces the available number of training patterns.

2.4. Classifier

The objective of LDA is to compute a discriminant vector $w \in \mathbb{R}^D$ that, given a set of training patterns $x_j \in \mathbb{R}^D$, $j \in \{1...N\}$ with their corresponding class labels, separates the classes as well as possible. This is achieved in LDA by maximizing the criterion function represented by

$$J(w) = \frac{(\tilde{m}_1 - \tilde{m}_2)^2}{\tilde{s}_1^2 + \tilde{s}_2^2} \qquad (1)$$

where

$$\tilde{m}_k(w) = \frac{1}{N_k} \sum_{i \in Y_k} w^T x_i, \quad \tilde{s}_k^2 = \sum_{i \in Y_k} (w^T x_i - \tilde{m}_k)^2, \qquad (2)$$

Y_k is the set of indices i corresponding to class k and N_k is the number of training patterns corresponding to class k.

It can be proven that the w that maximizes (1) can be found by [13].

$$w \propto S_w^{-1}(m_1 - m_2), \qquad (3)$$

where

$$m_k = \frac{1}{N_k} \sum_{i \in Y_k} x_i, \quad (4)$$

$$S_w = \sum_{k=1}^{2} \sum_{i \in Y_k} (x_i - m_k)(x_i - m_k)^T. \qquad (5)$$

In the LDA the between-class scatter matrix S_w can become singular, and the inverse of S_w can become ill defined. This happens when the number of features becomes larger than the number of training patterns, and is called the small sample size problem [14]. For these cases several solutions have been proposed, one of which is to carry out the calculation of the inverse by the Moore-Penrose pseudo-inverse [18]. That was the alternative chosen in this work.

2.5. Performance evaluation

The error rate (ER), is the most widely used evaluation metric. However, as it is an average over all the observations that are classified, it favors the majority class, i.e. the class with higher prior probability [15].

For two-class discrimination of unbalanced data, the area under the receiver operating characteristic curve (AUC) is commonly used. The receiver operating characteristic (ROC) curve is a plot of true positive rate vs. false positive rate, and hence a higher AUC generally indicates a better classifier. In contrast to ER, AUC is invariant to the prior probabilities [16, 17].

Considering the different characteristics of ER and AUC, both were used to estimate the performance of the classifiers.

ER and AUC are not useful parameters to estimate the capacity that a system has to accurately recognize one class, independently from its capacity to recognize the other [16]. Therefore the sensitivity and the specificity were also computed. The sensitivity is the fraction of correctly classified objects in the target class (in our case the target class is constituted by the patterns with ERP). The specificity is the fraction of non target objects that are not classified into the target class.

3. Experiments

Three sets of experiments were carried out, each having a different downsampling ratio according to $Fs_i = 2^i \times 4\text{Hz}$, where i defines the set.

For each set, five subsets of experiments were defined using different unbalance ratios to compute the classifiers, ranging from one to five target patterns per non target pattern. The different ratios were generated by random under-sampling of the non target patterns [15]. The first set of experiments represents the balanced situation, while the fifth corresponds to the situation when all the data are included. The other three subsets represent less natural situations when using a 6 x 6 stimulation matrix, but when modifying the matrix size this target vs. non-target ratios can be present. The inclusion of these three subsets also allows us to analyze trends.

In addition, the effect of coherent averaging was studied, by taking the average of the signals to train and test the classifiers. Averages from two to five trials were tested. In each subset of experiments the classifiers were computed by varying the number of training target patterns Nt_{jkl} and the number of training non target patterns Nnt_{jkl} according to

$$Nt_{jkl} = \left\lceil \frac{0.9^k}{l} \times Nt \right\rceil, \quad Nnt_{jkl} = \left\lceil \frac{j \times 0.9^k}{l} \times Nt \right\rceil, \qquad (6)$$

where j takes an integer value between one and five corresponding to the subset of experiments considered, k corresponds to the integers ranging from one to twenty, l varies from one to five and depends on the amount of trials averaged and Nt is the number of target patterns for each subject in the database (630).

The performance was estimated by cross-validation [13]. With each experimental configuration, the classifiers were trained and tested thirty times with different randomly selected training and validation datasets and the results were averaged. It is important to note the difference between this process and the m-fold cross validation, where the training set is randomly divided into m disjoint sets of equal size Nt/m.

4. Results

Figures 4, 5 and 6 show the averaged performance results, where all evaluation metrics were plotted against the total amount of training patterns. Each figure corresponds to a different set of experiments, and each row of graphics corresponds to a different subset.

It must be noted the logarithmic scale in the abscissas axis, and that the number of training patterns shown is different for each subset of experiments; this is due to the inherent unbalance of the problem and to the balancing approach used. These values are the sum of the number of target training patterns Nt_{jkl} and the number of non target training patterns Nnt_{jkl}. It is important to consider that even though the number of patterns used for training is different for the corresponding points in the graphics (*i.e.*, the points corresponding to the same k with different j values), the time the user should spend in the training session is the same for those cases when using a 6 x 6 stimulation matrix. Also the different scale in the ordinates axis in the sensitivity plot of the fifth subset in figure 4 must be noted.

The minimum peaks seen on all metrics when the number of features equals the number of training patterns might appear abnormal, but they are due to the transition between the LDA and the Moore-Penrose pseudo-inverse LDA. This phenomenon is explained on [18].

5. Discussion

In this work the effect of several variables that affect the behavior of LDA has been studied. The variations in the amount of averaged trials have influence on the difficulty of the problem, as does the subsampling rate (or number of features), but this is determined by the amount of discriminating information available on the higher frequencies of the EEG.

The analysis of results shows that when using fewer features the classifiers reach the performance plateau with fewer training patterns, as expected. It is interesting to notice that the number of training patterns at which the performance reach the plateaus is not dependent on the number of training trials corresponding to each class, but rather on the total number of training trials. Another variable that has influence over this is the coherent averaging, since it can be seen that rendering the problem easier, diminishes the amount of training patterns needed to reach the plateau. Although the results for the different subjects are not shown nor analyzed in the paper, there were also differences between them in this regard. Given all these variables affecting the point where the performance reaches a plateau, it is important to be very careful when using or defining a relationship among the necessary amounts of training patterns with the other variables; however the most significant dependence seen was on the number of features. Taking this into account, it can be said as a general rule that, in order to reach the point at which no significant performance gain will exist when adding more training patterns, more than ten training patterns per feature will be needed.

The AUC has not shown any variations between the experimental subsets (i.e. between different unbalance ratios), stabilizing at similar values after reaching the plateau, when averaging the same amount of trials. Regarding the ER, a variation for the different experimental sets could be observed, favoring, as expected, the ones with larger unbalances. On the other hand in some cases it was not possible to reach the plateaus with the balanced datasets, and so for these cases a better performance was obtained using unbalanced datasets. Regarding the other metrics the effect of unbalance over the LDA behavior can be clearly seen when analyzing the variations of specificity and sensitivity from figures 4, 5 and 6. These performance measures have very similar values when the classes are balanced, but as the unbalance grows so does the specificity, while the sensitivity decreases accordingly. This type of behavior is also mentioned in [8].

As expected, the performance was superior when applying coherent averaging if the results having the same amount of training patterns are compared, being the biggest difference the one between the average of two trials and the single trial case. However this is not always true when comparing points that require training sessions of the same duration.

Even though it was not the objective of this work, an overall good performance was obtained with the proposed system, considering its simplicity, since the AUC values were over 0.9 and the ER below 0.1.

6. Conclusions

In this work the effect of several variables that affect the behavior of LDA has been studied. It is important to consider these factors when using the LDA as a tool for evaluating the feature extraction stage, as these strategies tend to influence these variables. These techniques usually change the number of features and in some case the available amount of training patterns, which can affect the system's performance without changing the inherent discrimination power of these features, or in some cases disguising an actual improvement in the quality of the features by a detriment in the classifier performance.

In this study the number of training trials that are necessary to reach a performance plateau using LDA to classify EEG with and without ERP were estimated. However, other variables may influence these results.

The variations seen in specificity and sensitivity provide important information about the response of LDA. In addition of being significant factors when using the classifier to measure the discriminating power of a feature set, it can prove being useful information when designing a BCI based on LDA.

In further work, the problem of evolution of the density distribution of the mappings of the classes as the amount of patterns in each dataset is increased should be studied, as it could give a more complete explanation of the obtained results. Also, an extension to other types of classifiers with different characteristics could be considered.

Acknowledgments

This work was supported by the Universidad Nacional de Entre Ríos (grant PID 6101), the Universidad Nacional de Litoral (grant CAI+D 012-72), STIC-AMSUD program (09STIC01) and the Consejo Nacional de Investigaciones Científicas y Técnicas (CONICET) from Argentina.

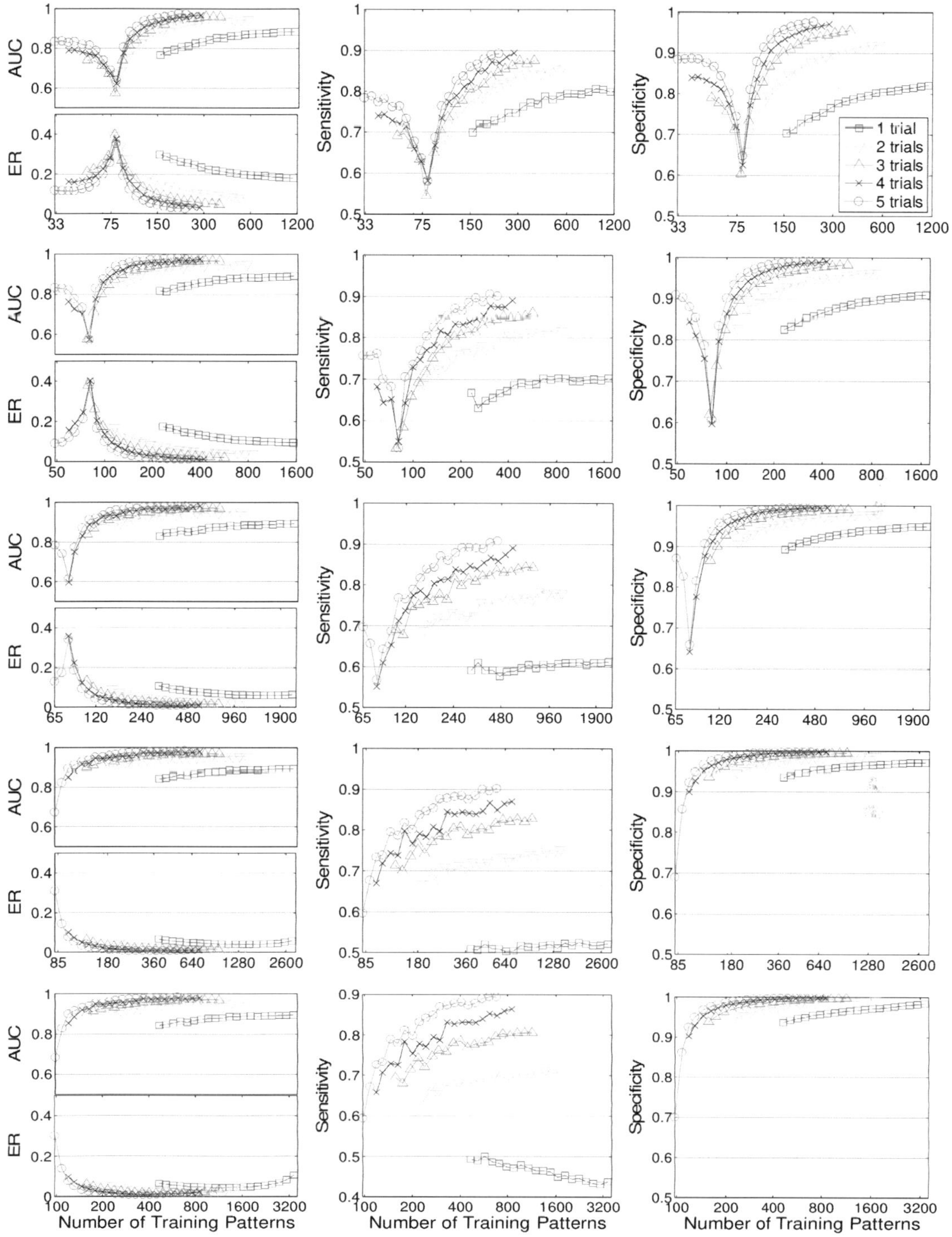

Figure 4. Graphics of performance estimates for the first set of experiments averaged over subjects vs. number of total training patterns obtained from the different subsets of experiments. From top to bottom: using one non target per target pattern, using two non target per target pattern, using three non target per target pattern, using four non target per target pattern, using five non target per target pattern. Legend acronyms: ER (error rate), AUC (area under ROC curve).

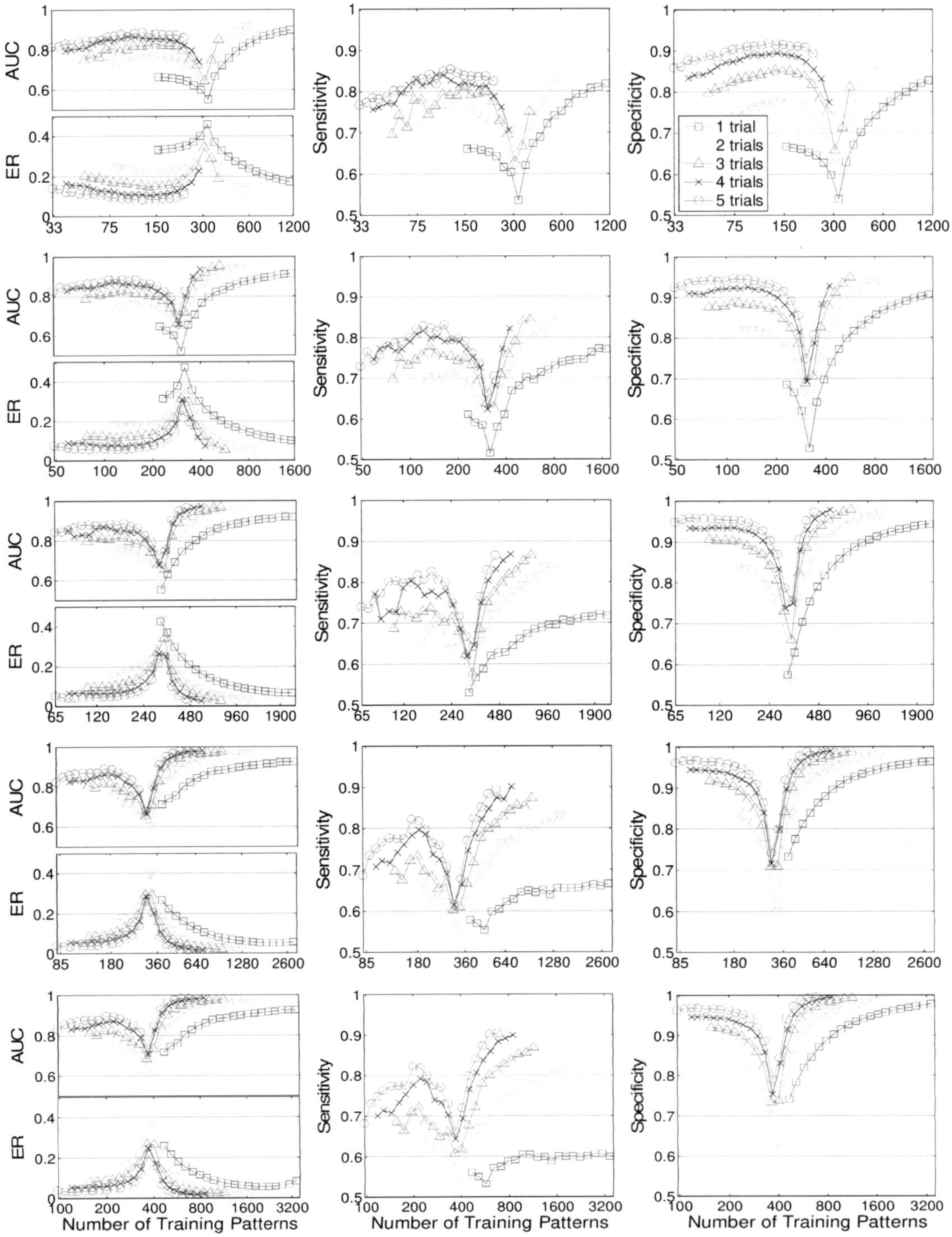

Figure 5. Graphics of performance estimates for the second set of experiments averaged over subjects vs. number of total training patterns obtained from the different subsets of experiments. From top to bottom: using one non target per target pattern, using two non target per target pattern, using three non target per target pattern, using four non target per target pattern, using five non target per target pattern. Legend acronyms: ER (error rate), AUC (area under ROC curve).

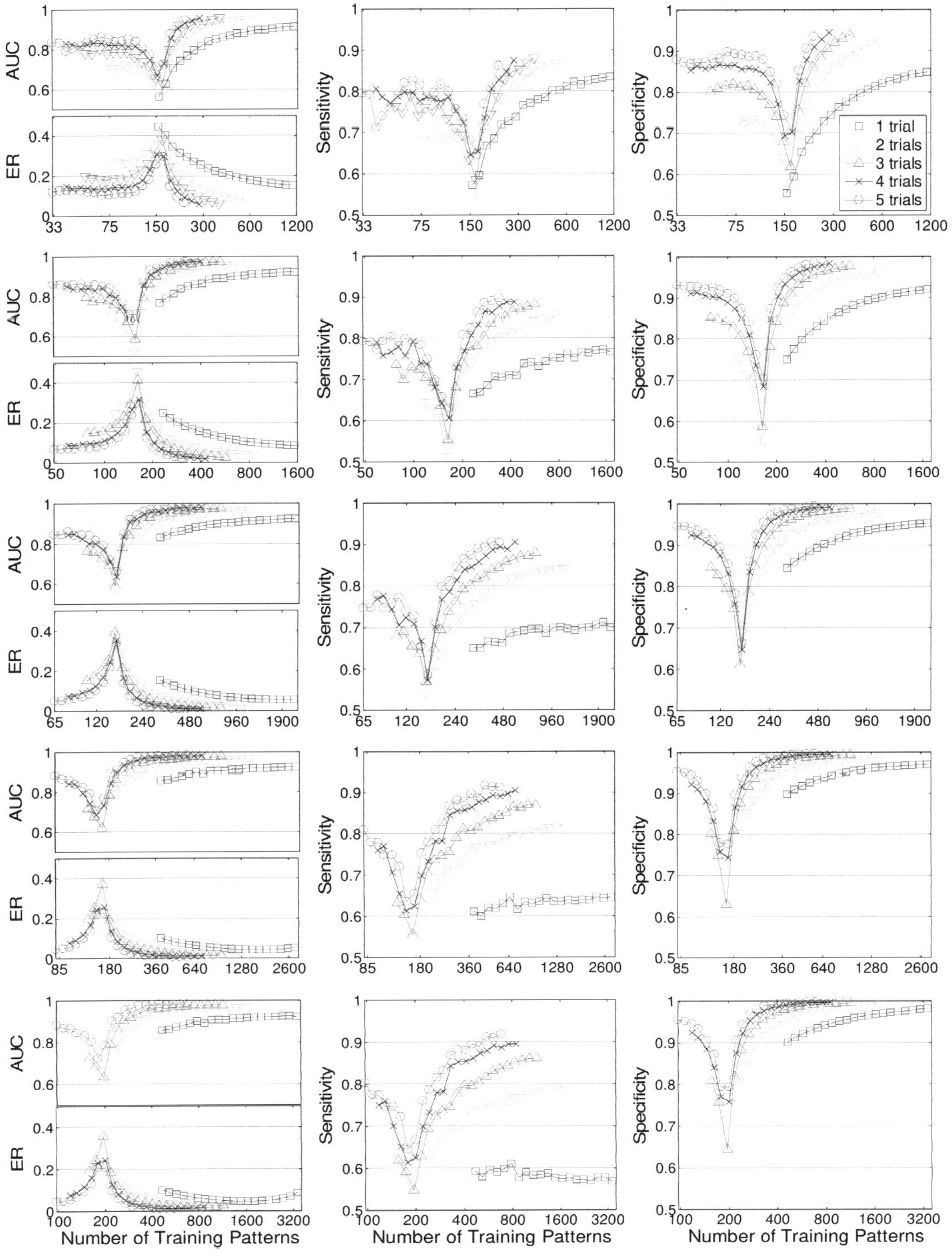

Figure 6. Graphics of performance estimates for the third set of experiments averaged over subjects vs. number of total training patterns obtained from the different subsets of experiments. From top to bottom: using one non target per target pattern, using two non target per target pattern, using three non target per target pattern, using four non target per target pattern, using five non target per target pattern. Legend acronyms: ER (error rate), AUC (area under ROC curve).

References

[1] Wolpaw J. R., Birbaumer N., McFarland D. J., Pfurtscheller G., and Vaughan T. M., 2002 Brain Computer Interfaces for communication and control. Clin. Neurophysiol., **113**(6): 767-791.

[2] Kübler A., Kotchoubey B., Kaiser J., Wolpaw J.R. and Birbaumer N., 2001 Brain–computer communication: unlocking the locked in. *Psychol. Bull.* **127:**358–75.

[3] Bashashati A., Fatourechi M., Ward R. K. and Birch G. E., 2007 A survey of signal processing algorithms in brain-computer interfaces based on electrical brain signals. *Journal of Neural Engineering*, **4**: 32-57.

[4] Atum Y., Gareis I., Gentiletti G., Acevedo R., Rufiner L, 2010 Genetic feature selection to optimally detect P300 in brain computer interfaces, Engineering in Medicine and Biology Society (EMBC), Annual International Conference of the IEEE, 3289 – 3292.

[5] Lotte F., Congedo M., Lecuyer A., Lamarche F. and Arnaldi B., 2007 A review of classification algorithms for EEG-based brain-computer interfaces, *Journal of Neural Engineering*, **4**: R1-R13.

[6] Krusienski DJ, Sellers EW, Cabestaing F, Bayoudh S, McFarland DJ, Vaughan TM, Wolpaw JR, 2006 A Comparison of Classification Techniques for the P300 Speller, *Journal of Neural Engineering*, **3**:299-305.

[7] Fisher R. A. 1936 The use of multiple measurements in taxonomic problems, *Ann. Eugenics* **7** 179–88.

[8] Xue J.H. and Titterington, D.M. 2008 Do unbalanced data have a negative effect on LDA?, *Pattern Recognition*, **41** (5). pp. 1575-1588. ISSN 0031-3203.

[9] Xie J.G., Qiu Z.D., 2007 The effect of imbalanced data sets on LDA: a theoretical and empirical analysis, *Pattern Recognition* **40** (2) 557–562.

[10] Farwell L.A., Donchin E. 1988, Talking off the top of your head: toward a mental prothesis utilizing event-related brain potentials. *Electroenceph. clin. Neurophysiol.* **70**:510–523.

[11] http://www.bci2000.org/BCI2000/Home.html.

[12] Ramírez C. L., Bojorges Valdez E., Yañez Suárez O., Saavedra C., Bougrain L. and Gentiletti G. 2010 An open-access P300 speller database, *Fourth international BCI meeting*, Paper L-12, Monterrey California.

[13] Duda R., Hart P. and Stork D., 2000 *Pattern Classification* (2nd Edition), Wiley-Interscience.

[14] Fukunaga K., 1990 *Introduction to Statistical Pattern Recognition,* Academic Press.

[15] Weiss G.M., 2004 Mining with rarity: a unifying framework, *SIGKDD* Explor. **6** (1), 7–19.

[16] Bradley A.P., 1997 The use of the area under the ROC curve in the evaluation of machine learning algorithms, *Pattern Recognition* **30** (7) 1145–1159.

[17] Zweig MH, Campbell G., 1993 Receiver-Operating Characteristic (ROC) Plots: A Fundamental Evaluation Tool in Clinical Medicine. *Clinical Chemistry.*

[18] Raudys S., Duin R. P., 1998 Expected classification error of the Fisher linear classifier withpseudo-inverse covariance matrix. *Pattern Recognition Letters* archive vol. **19** Issue 5-6.

Estimation of electrical conductivity of a layered spherical head model using electrical impedance tomography

M Fernández-Corazza, N von-Ellenrieder and C H Muravchik
Laboratorio de Electrónica Industrial, Control e Instrumentación (LEICI)
Facultad de Ingeniería, Universidad Nacional de La Plata (UNLP), Buenos Aires, Argentina
E-mail: marianof.corazza@ing.unlp.edu.ar

Abstract. Electrical Impedance Tomography (EIT) is a non-invasive method that aims to create an electrical conductivity map of a volume. In particular, it can be applied to study the human head. The method consists on the injection of an unperceptive and known current through two electrodes attached to the scalp, and the measurement of the resulting electric potential distribution at an array of sensors also placed on the scalp. In this work, we propose a parametric estimation of the brain, scalp and skull conductivities using EIT over an spherical model of the head. The forward problem involves the computation of the electric potential on the surface, for given the conductivities and the injection electrode positions, while the inverse problem consists on estimating the conductivities given the sensor measurements. In this study, the analytical solution to the forward problem based on a three layer spherical model is first described. Then, some measurements are simulated adding white noise to the solutions and the inverse problem is solved in order to estimate the brain, skull and scalp conductivity relations. This is done with a least squares approach and the Nelder-Mead multidimensional unconstrained nonlinear minimization method.

1. Introduction

Electrical impedance tomography (EIT) is a relatively new imaging method that has evolved over the past 20 years. In the medical field it has a potential clinical value in diagnostics and monitoring of a number of disease conditions, from the detection of breast cancer [1], to monitoring brain function [2], and possibly stroke [3]. EIT consists on the injection of small electrical currents in different points of the body and the measurement of the resulting electric potential distribution through a sensor array. It is safe, portable, inexpensive and fast [4]. At present, EIT does not have the spatial resolution of other methods like Magnetic Resonance Imaging (MRI) or Computer Tomography (CT), while its key advantage is the temporal resolution, which is in the order of milliseconds [3].

In neuroscience, EIT can be used for at least two different applications: to image conductivity changes or to measure absolute conductivities. The former is based on measuring the conductivity changes produced by blood flow and volume alterations during evoked activity in the brain [4]. Also, a slight ($\sim$1%) change of the impedance is produced in the cerebral cortex during neuronal depolarization [4]. But since the skull is more resistive than the other tissues of the head, the current that may reach the brain is relatively small [4] (less than 15% of total

Published under licence by IOP Publishing Ltd

injected current in a simulation study [5]). This is why in order to use EIT as a functional brain mapping tool, the shield effect of the skull to the injecting currents should be estimated first [6]. The second application is a technique for measuring the electrical conductivity of the head, useful to create an anatomical conductivity map. In this application, the EIT forward problem consists of computing the electric potential distribution generated by a known injected current in a known head model with its corresponding conductivity map. The EIT inverse problem is the estimation of the electric conductivity map from the electric potential measured at the electrode positions. As mentioned, the spatial resolution that can be achieved from EIT data alone is poor.

We propose to parameterize the inverse problem by a few parameters being the electric conductivity of the tissues, assuming the spatial distribution of the tissues is known. In this study, this spatial distribution consists of three concentric spheres. But next, realistic geometry obtained from some other techniques such as MR or CT scalp will be used. The approximate conductivity map will have then a very good spatial resolution but with conductivity values that may differ a little from the real ones since they are assumed homogeneous in each tissue. The proposed methodology is expected to work very well to estimate the scalp and skull conductivities, and a combination with some other techniques can complete the model. Diffusion Tensor Imaging (DTI) can be used to estimate the brain and cerebrospinal fluid (CFS) conductivities [7]. The knowledge of the conductivity map is important in the study of electroencephalography (EEG) problems, in particular for the estimation of the position of neural sources.

The EEG source localization problem aims to determine the neural sources inside the brain that best explain the electrical potentials measured on the surface of the scalp. The determination of the sources is made through the use of mathematical models which describe the head as an electrical conductor. Then, the knowledge of the electrical conductivity map of the head is important since it is known that the solution to the source localization problem is highly dependent on the values taken by the scalp, skull, and brain conductivities [8].

The scalp impedance plays a major role in head EIT, as it is engaged with currents of the highest density coming from the electrodes, and therefore it is the first stage in which distribution of current into the rest of the layers is determined. Any inaccuracies at this stage will substantially affect the impedance readings of the inner layers [9]. Several studies lead to a wide range of values (see Horesh graphs and tables [9]) showing the importance of *in vivo* and per patient based information. Regarding the skull, there are two kinds of mature bones: compact bone and spongy bone [9]. Many of the skull bones are flat, consisting of two plates of compact bone (thickness of 1.7 to 4.3 mm) enclosing a narrow layer of cancellous bone (thickness of 3.8 - 5.1 mm) containing bone marrow. Data published regarding skull impedance also covers a wide range of values. Dielectric properties of the skull are influenced by several factors: bone composition (skull layers-structure), non-uniform geometry, anisotropy and age (water content is reduced and bone marrow is changing from red to yellow with age, which decreases the conductivity) [9]. *In vivo* and per patient based information regarding the skull impedance is needed for building more accurate models to use in EEG and functional EIT problems.

In the next section we describe the adopted head model in detail. Then, we explain the mathematical background of the forward problem. Next, we state the inverse problem, and finally, the results of several simulations for different number of electrodes are presented and discussed.

2. Methods

2.1. Electrical Model

The head model consists of three concentric spherical shells with the enclosed space between them representing the three main tissues of the head: scalp, skull and brain (figure 1). Each layer is considered as homogeneous and isotropic, i.e. conductivity is constant and with no preferred direction.

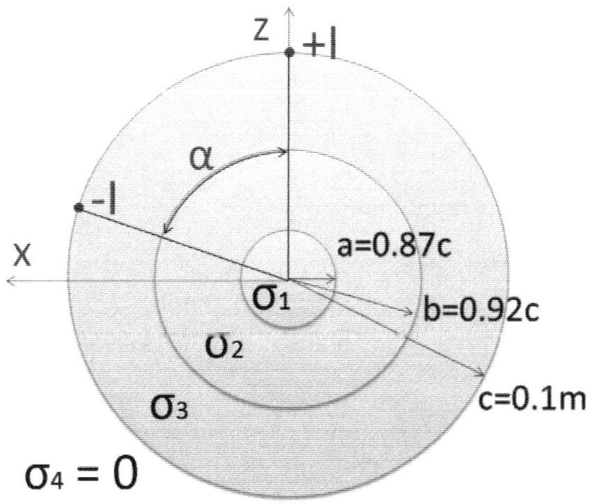

Figure 1. Electrical model. σ_1, σ_2 and σ_3 represent homogeneous and isotropic conductivities of brain, skull and scalp respectively. $+I$ and $-I$ represent the positive and negative current injection points.

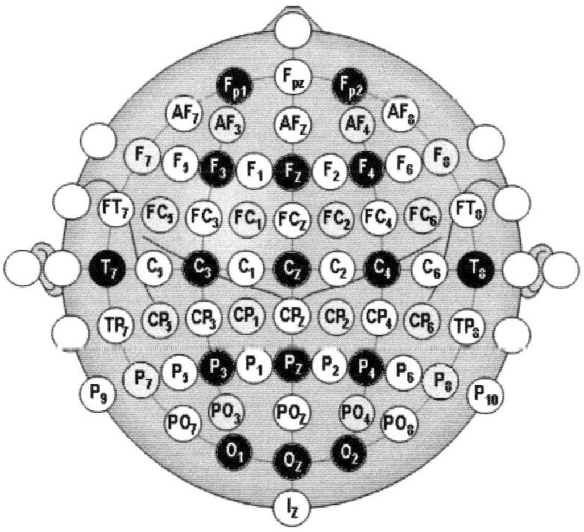

Figure 2. Electrode positions based on the 10-10 system over a spherical head. Black, black and shaded, and labeled circles show the positions used for 16, 32 and 64 configurations respectively.

The electrode placement used is based on the 10-10 system for 64, 32 and 16 sensors (figure 2). A pair of electrodes is selected for the current injection, while the others are dedicated to measure the resultant electric potential difference. One of the current electrodes will supply a current of $+I$ (current source) and the other a current $-I$ (current sink), both with the same absolute value (figure 1).

EIT applications usually comply with the International Electrotechnical Commission (IEC) standard [10] which specify a "patient auxiliary current" limit of $100\mu A$ from 0.1 Hz to 1 kHz. This standard is based on limitation of the current to 10 % of the average threshold of sensation when current is applied to the skin using surface electrodes. This corresponds to a widely cited study in which a mean perception current of 1.1 mA was observed between DC and 200 Hz, for hand holding of a copper wire of d = 3.3 mm in diameter [11]. This gives a threshold current density around 2 to 15 A/m^2. In another study ([12]), with abrasion just under a conventional cup electrode (Ag/AgCl, 11 mm diameter), the threshold of sensation was $100 - 200\mu A$ and it became slightly unpleasant at $400\mu A$. This gives a threshold current density similar to previous study. Then, a current of $100\mu A$ is appropriate for use in real studies with patients. We use this value to produce a realistic simulation.

Also, the model dimensions must be scaled to a human head. So, the outer shell radius is set to $10cm$. The other radii are the ones that are mostly used in spherical models [13]. Table 1 summarizes the interface radii chosen for the model.

The remaining data to complete the model are the isotropic conductivity values. These are found in the literature with a high dispersion (table 2 shows some examples). Some of them are measurements and some others are reference values given in the studies. The values chosen for the present work are shown in the last row. The criteria was to select realistic values, with two exact decimals for brain:skull and brain:scalp conductivity ratios in order to ease the comprehension of the results.

Table 1. Model radii

Interface	Label	Radius [cm]
Brain:Skull	a	10.0
Skull:Scalp	b	9.7
Scalp:Air	c	8.7

Table 2. Conductivity values found in literature [S/m]

Author	Brain	Skull	Scalp
Zheng Xu et al [6]	0.33	0.04	0.33
Gilad et al [12]	~0.30	0.015	~0.30
Abascal et al [5]	0.30	0.039	0.44
Adhazi et al [14]	0.25	0.018	0.44
Gonçalves et al [8]	0.33	0.008	—
Horesh [9]	0.15	0.02	0.25
Liston et al [15]	0.25	0.018	0.44
Zlochiver [16]	0.12 (grey matter)	0.02 (bone)	0.35 (muscle)
Fernández-Corazza	0.30	0.01	0.40

2.2. Forward Problem

Assuming that the scalp, skull, and brain conductivities are known, the problem consists on estimating the electric potential on the outermost surface. This is an electromagnetism problem with surface conditions and spherical geometry. Maxwell equations govern the physics of electromagnetism. In particular, for low frequencies, the quasi-static assumption can be used. This is common in many studies since it reduces the complexity of the problem [3]. The limitation is that the model is valid only for frequencies below ~10 kHz, for which the wave length is much greater than the head dimensions. Then, we can write

$$\vec{\nabla} \times \vec{E} = 0, \tag{1}$$

where $\vec{E}$ is the electric field. From (1), the electric field $\vec{E}$ can be expressed as the gradient of a scalar potential Φ

$$\vec{E} = -\vec{\nabla}\Phi. \tag{2}$$

It is known that the current density $\vec{J}$ is proportional to $\vec{E}$ (3) [17].

$$\vec{J} = \sigma\vec{E}, \tag{3}$$

Where, σ is the electric conductivity of the medium. Applying the divergence to $\vec{J}$ and assuming that σ is constant,

$$\vec{\nabla} \cdot \vec{J} = \vec{\nabla} \cdot (\sigma\vec{E}) = \sigma(\vec{\nabla} \cdot \vec{E}) = \sigma(-\vec{\nabla}^2\Phi). \tag{4}$$

The current conservation law states that the divergence of the current density is zero everywhere in the sphere except in the injection points. With this approximation, the electric potential Φ must satisfy the Laplace equation in each layer:

$$\vec{\nabla}^2\Phi = 0. \tag{5}$$

Taking advantage of the problem symmetries, for each layer i, the solution of the potential Φ in every point of the space (r,θ,ϕ) can be expressed as a linear combination of the Legendre Polynomials as shown in (6) [17] [18].

$$\Phi^{(i)}(r,\beta,\gamma) = \sum_{l=0}^{\infty}(A_l^{(+)(i)}r^l + B_l^{(+)(i)}r^{-l-1}).P_l(cos\beta) - \sum_{l=0}^{\infty}(A_l^{(-)(i)}r^l + B_l^{(-)(i)}r^{-l-1}).P_l(cos\gamma)$$

$$\Phi^{(i)}(r,\beta,\gamma) = \sum_{l=1}^{\infty}(A_l^{(i)}r^l + B_l^{(i)}r^{-l-1}).(P_l(cos\beta) - P_l(cos\gamma)); \qquad i = 1,2,3. \qquad (6)$$

Where the constants $A_l^{(i)}, B_l^{(i)}$ depend on the boundary conditions, $P_l(x)$ is the Legendre Polynomial of order l (where $P_0(x) = 1$), $\beta(\theta,\phi)$ is the angle to the $+I$ current source and $\gamma(\theta,\phi)$ to the $-I$ current sink. In layer 3, the potential is not exactly as (6) because the current injection points are within this layer and their effect must be included. Applying linearity, the potential on this layer can be expressed as the potential in an unbounded medium (Φ_{unb}) plus another with the expression of (6) ($\Phi^{(3)}$) that allows the total potential on this layer ($\Phi_{tot}^{(3)}$) to satisfy the boundary conditions. Then,

$$\Phi_{tot}^{(3)} = \Phi_{unb} + \Phi^{(3)}, \qquad (7)$$

where,

$$\Phi_{unb} = \frac{I}{4\pi\sigma_3}\left(\frac{1}{\sqrt{r^2+c^2-2rc.cos(\beta)}} - \frac{1}{\sqrt{r^2+c^2-2rc.cos(\gamma)}}\right). \qquad (8)$$

Note that the constant $B_l^{(1)}$ must be zero in order to have continuity at the origin. So, for each order l, there remain five constants $A_l^{(1)}, A_l^{(2)}, B_l^{(2)}, A_l^{(3)}, B_l^{(3)}$ to be determined. On the interface between the layers, two conditions are satisfied: the continuity of the normal components of the current density and the continuity of the electric potential. They lead to five equations:

$$\begin{aligned}
\sigma_1\frac{\partial\Phi^{(1)}}{\partial r}\bigg|_{r=a} &= \sigma_2\frac{\partial\Phi^{(2)}}{\partial r}\bigg|_{r=a}\\
\sigma_2\frac{\partial\Phi^{(2)}}{\partial r}\bigg|_{r=b} &= \sigma_3\frac{\partial\Phi_{tot}^{(3)}}{\partial r}\bigg|_{r=b}\\
\sigma_3\frac{\partial\Phi_{tot}^{(3)}}{\partial r}\bigg|_{r=c} &= 0 \qquad (9)\\
\Phi_1\big|_{r=a} &= \Phi^{(2)}\big|_{r=a}\\
\Phi_2\big|_{r=b} &= \Phi_{tot}^{(3)}\big|_{r=b}
\end{aligned}$$

Solving the equation system derived from (9) the constants can be determined. Then, expanding and regrouping terms, the potential over the outermost surface ($r = c$) can be written as:

$$\Phi_{tot}^{(3)}(c,\theta,\beta) = \frac{k}{c}.\left(\Phi_\infty + \sum_{l=1}^{\infty}\left(\frac{K_1.{}^{\sigma_1}/_{\sigma_2} + K_2.{}^{\sigma_1}/_{\sigma_3} + K_3.{}^{\sigma_2}/_{\sigma_3} + K_4}{K_5.{}^{\sigma_1}/_{\sigma_2} + K_6.{}^{\sigma_1}/_{\sigma_3} + K_7.{}^{\sigma_2}/_{\sigma_3} + K_8} - 1\right).(P_l(cos\beta) - P_l(cos\gamma))\right), \qquad (10)$$

where,

$$k = \frac{I}{4\pi\sigma_3}, \tag{11}$$

$$\Phi_\infty(\beta,\gamma) = \frac{1}{\sqrt{|2-2\cos(\beta)|}} - \frac{1}{\sqrt{|2-2\cos(\gamma)|}}, \tag{12}$$

$$K_1 = (l+1+lb_r^{2l+1})l(1-(a_r/b_r)^{2l+1}), \tag{13}$$

$$K_2 = (1-b_r^{2l+1})l(l+(l+1)(a_r/b_r)^{2l+1}), \tag{14}$$

$$K_3 = (1-b_r^{2l+1})l(l+1)(1-(a_r/b_r)^{2l+1}), \tag{15}$$

$$K_4 = (l+1+lb_r^{2l+1})(l+1+l(a_r/b_r)^{2l+1}), \tag{16}$$

$$K_5 = l^2(l+1)(1-(a_r/b_r)^{2l+1})(1-b_r^{2l+1}), \tag{17}$$

$$K_6 = (l+(l+1)(a_r/b_r)^{2l+1})(l^2+(l^2+l)b_r^{2l+1}), \tag{18}$$

$$K_7 = (l+1)l(1-(a_r/b_r)^{2l+1})(l+(l+1)b_r^{2l+1}), \tag{19}$$

$$K_8 = (l+1)l((l+1)+l(a_r/b_r)^{2l+1})(1-b_r^{2l+1}). \tag{20}$$

With a_r and b_r being the a and b radii relative to a sphere of radius 1, (i.e., $a_r = a/c$ and $b_r = b/c$).

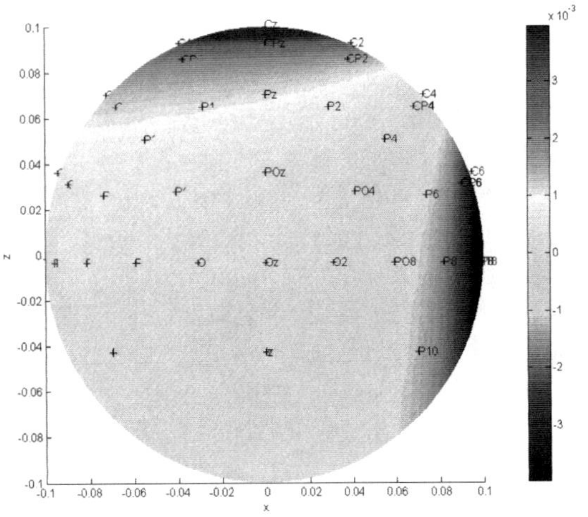

Figure 3. Surface potential for $\sigma_2 = 0.0093 S/m$, as seen from behind. The positions and labels for 64-electrode configuration are also displayed

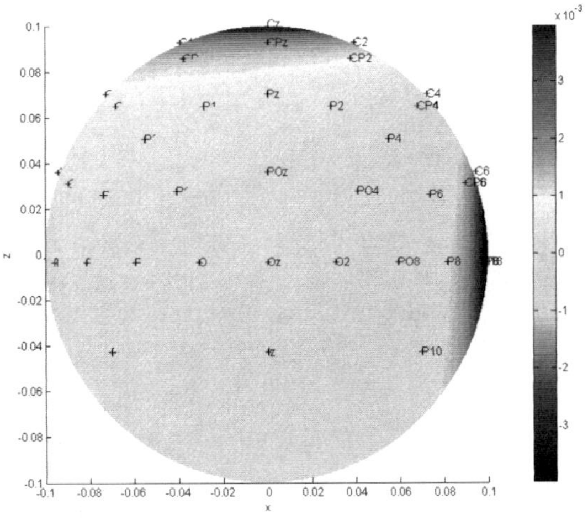

Figure 4. Surface potential for $\sigma_2 = 0.039 S/m$, as seen from behind. The positions and labels for 64-electrode configuration are also displayed

Figures 3 and 4 show the resulting electric potential Φ over the sphere surface for two different skull conductivity values. It can be seen that it changes with that parameter.

2.3. Inverse Problem
For the inverse problem the conductivities are assumed unknown, and the electric potential measured at the electrodes is known. In order to simulate the measurements, white gaussian noise is added to the solution of the forward problem with the previous section formulas (assuming the conductivities of table 2). This noise models the electronic noise of the amplifiers and the skin/electrode contact electrochemical noise. We model it as Gaussian distributed zero mean and $1\mu A$ standard deviation.

We will estimate the conductivity ratios instead of the conductivities themselves. This is convenient, for example, in (10). But only two of them are independent, so $x_1 = {}^{\sigma_1}/_{\sigma_2}$ and $x_2 = {}^{\sigma_1}/_{\sigma_3}$ are chosen as the independent variables implying that ${}^{\sigma_2}/_{\sigma_3} = {}^{x_2}/_{x_1}$.

Then, a least squares approach can be used to estimate the unknowns x_1 and x_2 that minimize the difference between each sensor measurement and its corresponding forward solution:

$$\min_{x_1,x_2} \left\{ \sum_{i=1}^{(n-2)} \left(V_i - \Phi\left(\theta_i, \beta_i, x_1, x_2\right)\right)^2 \right\}, \tag{21}$$

Where, n is the number of electrodes, V_i is the "measured" potential on the electrode i and $\Phi(\beta_i, \gamma_i, x_1, x_2)$ is the surface potential function obtained from the forward problem (10). The upper limit of the summation is $(n-2)$ because the injection nodes are excluded since their potential is not measured.

The optimization algorithm used is the Nelder-Mead multidimensional unconstrained nonlinear minimization method [19]. Note that in the formulation (21) there is a constrain (the unknowns must be positive), but as the starting point can be chosen relatively near to the true value, a faster unconstrained method can be used.

3. Results

The inverse problem (21) was solved for 16, 32 and 64 electrodes with $1\mu A$ standard deviation noise. It can be seen in figure 2 that there is a symmetry on the electrode distribution between left an right sides. This symmetry does not exist between front and back because of the three occipital electrodes (Iz, P9 and P10) in the 64-electrode configuration, and the Oz electrode in the 32 and 16 configurations. So, all different pairs of electrodes without an exact mirrored configuration occurring on the left side were considered. For example, the problem with the injection nodes being C4 and F4 leads to the exact same results of the C3-F3 combination. Then, this last pair was excluded. The number of combinations used are shown in table 3.

Table 3. Injection node combinations

Electrodes	Combinations
64	1551
32	391
16	99

For each combination, 100 different noise realizations were simulated and the mean and variance of the x_1 and x_2 estimates were computed. Figures 5 and 7 show, for the 64 and 32 electrode configurations, the mean of x_1 as a function of the angle α between the injection electrodes. The expected value is also displayed with a horizontal line. Figures 6 and 8 show the same results for x_2.

Figure 9 shows, for the 16-electrode configuration, the mean and mean plus/minus variance of the x_1 estimation for different combinations of injection electrodes. A particular injection pair has a very high variance. This is reasonable since it corresponds to the T7-T8 combination. In this case all measurement electrodes are close to the middle plane between these two positions, where the electric potential is close to zero.

Figure 11 shows the resulting variance for different x_1 simulations. Each cross represents the mean variance for each pair of injection electrodes. The central mark is the median, the edges

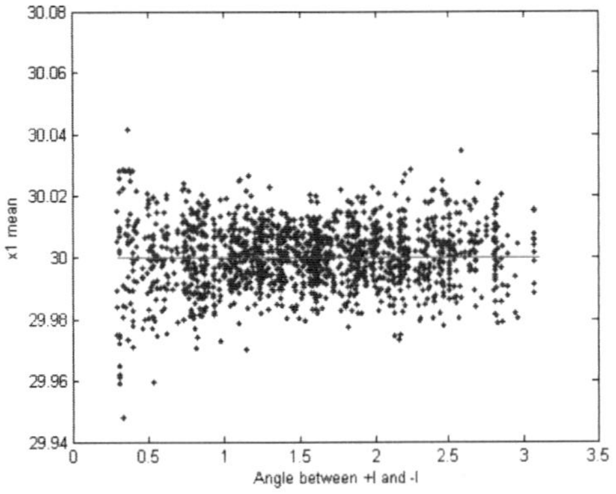

Figure 5. σ_1/σ_2 mean versus α for 64 electrode configuration.

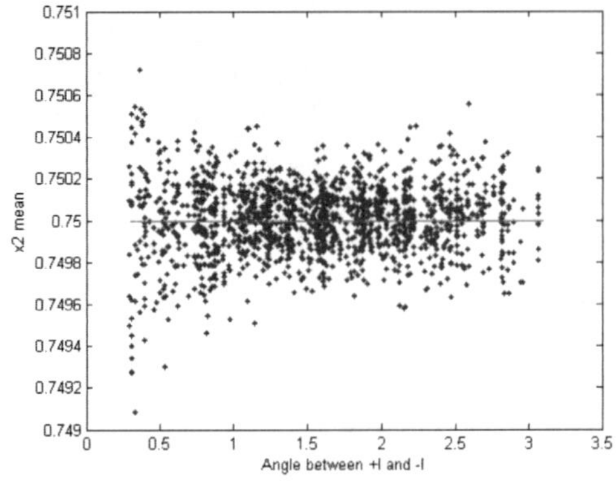

Figure 6. σ_1/σ_3 mean versus α for 64 electrode configuration.

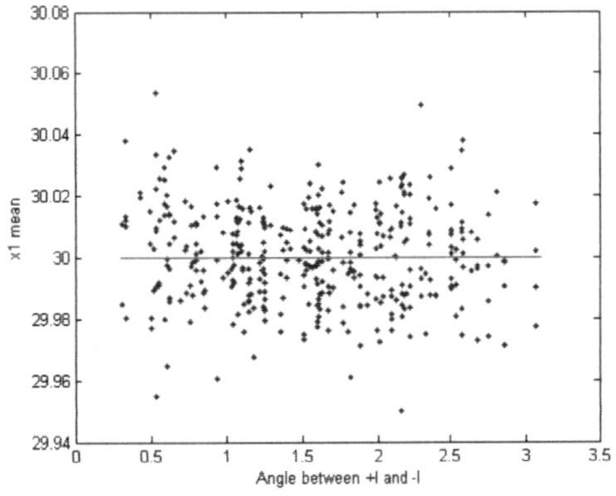

Figure 7. σ_1/σ_2 mean versus α for 32 electrode configuration.

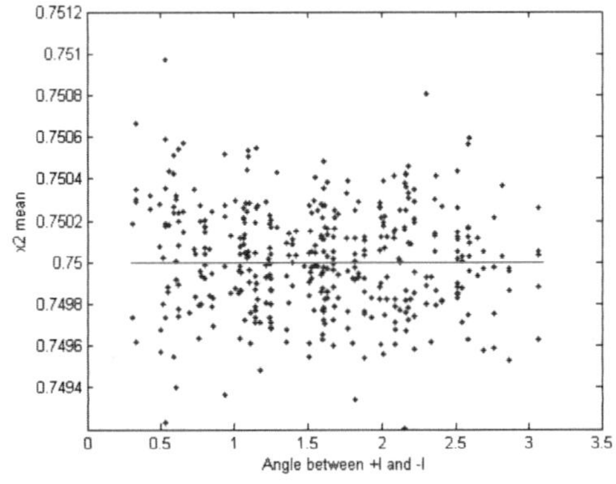

Figure 8. σ_1/σ_3 mean versus α for 32 electrode configuration.

of the box are the 25th and 75th percentiles, and the whiskers extend to the most extreme data points not considered outliers. Figure 12 shows the same result for x_2. The combination T7-T8 was excluded from both plots in the 16-electrode configuration because, as explained before, it is a very atypical case and would impact negatively in the plot readability.

Another simulation was done in order to study the impact of the assumed noise variance. We know that the brain has normal activity that can be detected with the EEG electrodes. This activity means for this application of EIT additional noise with peaks of around $20\mu V$. Although this noise might be disaffected using a proper technique (taking advantage of the fact that the frequency and phase of the injected current is well known), is important to know if such a technique effort is worthwhile. Then, the same simulation for 64 electrodes was ran, but with a noise standard deviation of $10\mu V$. Figure 10 compares the 16 and 64 electrode configurations with $1\mu V$ and $10\mu V$ of noise standard deviation respectively, where the left box of the plot is the same than the right one of figure 11.

4. Conclusions

The forward and inverse problems have been solved successfully for 16, 32 and 64 electrodes over the three shell spherical model and the expected conductivity values were retrieved. It is

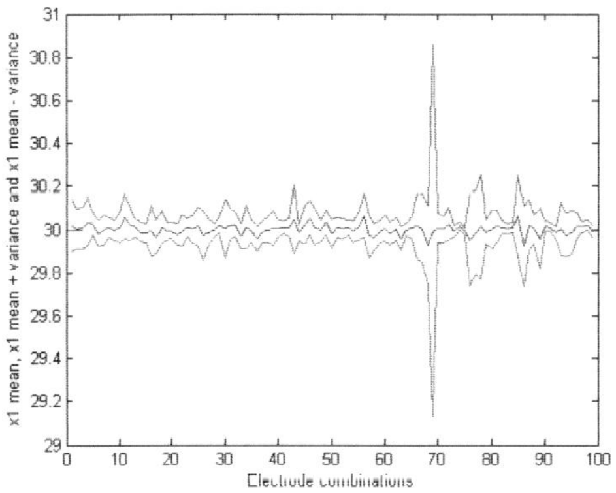

Figure 9. σ_1/σ_2 mean and mean plus/minus variance for 16-electrode configuration versus electrode combinations

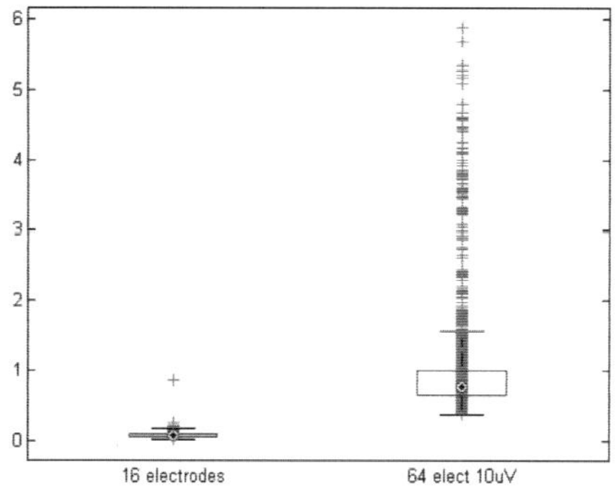

Figure 10. Box plot of the variance for σ_1/σ_2 estimations in 16 and 64 electrode configurations with $1\mu V$ and $10\mu V$ standard deviation noise respectively.

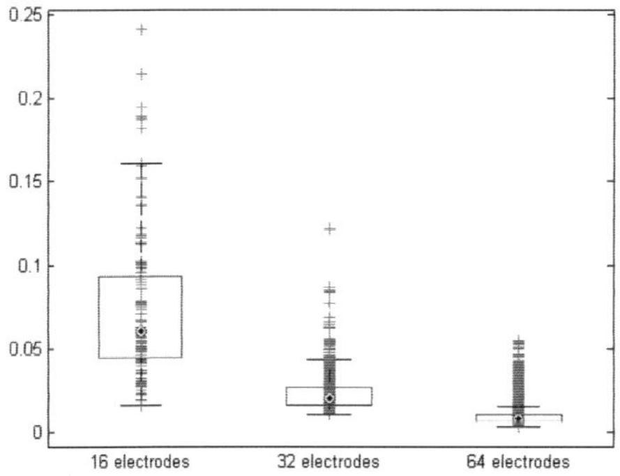

Figure 11. Box plot of the variance for σ_1/σ_2 estimations.

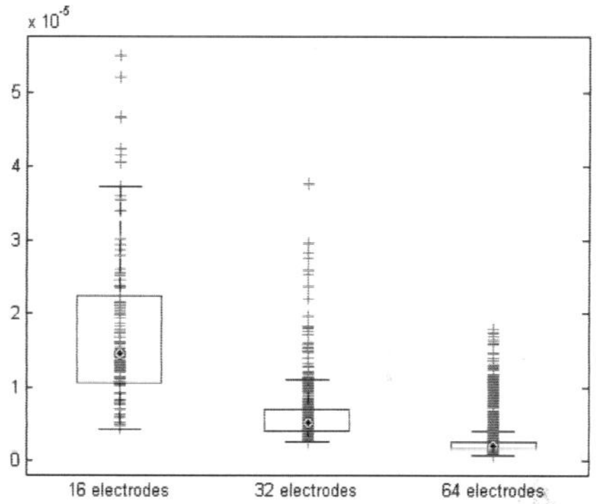

Figure 12. Box plot of the variance for σ_1/σ_3 estimations.

important to highlight that this was achieved with appropriate scales. For instance, the radius of the outer sphere was 10cm, close to a real head size. Also, the injected current value was chosen with the same criterion as if it were a real experiment. Furthermore, the conductivity values were close to real values, and finally, a typical electronic noise standard deviation was chosen.

Injection nodes with greater angle between them generally achieve better results (less variance) but this is not a general rule, since the accuracy not only depends on that angle, but also on the surrounding electrodes for measurement. For example, C3-C4 combination is better than T7-Cz (less variance), although both have very similar angle between the injection nodes. Another example is the atypical T7-T8 pair for the 16 electrode configuration, where the angle α between the electrodes is maximum and the variance of the conductivity estimation is the highest. In conclusion, the combinations with more sensors surrounding the injection electrodes and a large angle between them are preferred.

On the other side, the results show an exponential-like dependence of the variance with the number of electrodes. Of course, more electrodes mean better results, but from this study, and also depending on the measurement precision needed, a good choice on the sensors quantity could be performed. More important than the amount of electrodes is the removal of the brain activity from the recorded signals. A cap with 16 electrodes and a system to remove the background brain activity is much better than a 64 electrodes system without it.

The spherical model mainly serves as a validation tool for numerical algorithms since the forward problem has analytical solution. With numerical methods like Boundary Element Method (BEM) or Finite Element Method (FEM), the algorithms can be independent of the geometry. The solutions for the forward and inverse problems with a real shaped model using FEM are in progress. This model will use a standard shape model of the head and a standard averaged conductivity distribution of the brain and CFS obtained via DTI. Then, it will be used to estimate the conductivities of skull and scalp. An advantage of using FEM instead of BEM is the possibility of include anisotropy in the model. It is known that skull has a tangential to radial anisotropy of around 1:10 [13]. The same seems to happen in the scalp, but in this case the relation is lower (1:1.5) [9] and it is not so relevant. So, the inverse problem will be extended to look for four unknowns: tangential skull, radial skull, tangential scalp and radial scalp conductivities.

The development of an EIT device is also planned. Using the average head model or patient specific models (if possible) jointly with this device, real estimations of scalp and skull conductivities will be performed. Since this information will be used to solve the EEG source localization problems, the real impact of having patient specific conductivity maps will be studied.

Acknowledgments

This work was supported by ANPCyT PICT 2007-00535, UNLP I127, CONICET and CICpBA.

References

[1] Cherepenin V A, Karpov A Y, Korjenevsky A V, Kornienko V N, Kultiasov Y S, Ochapkin M B, Trochanova O V and Meister J D 2002 *IEEE Trans. Med. Imaging* **21**(6) 662–667
[2] Bagshaw A P, Liston A D, Bayford R H, Tizzard A, Gibson A P, Tidswell A T, Sparkes M K, Dehghani H, Binnie C D and Holder D S 2003 *NeuroImage* **20** 752–764
[3] Bayford R 2006 *Annu. Rev. Biomed. Eng.* **8** 63–91
[4] Holder D 2008 *Automation Congress, 2008. WAC 2008. World* pp 1 –6
[5] Abascal J F, Arridge S R, Atkinson D, Shindmes R, Fabrizi L, Lucia M D, Horesh L, Bayford R H and Holder D S 2007 *Neuroimage* **43**(2) 258–68
[6] Xu Z, He W, He C, Zhang Z and Liu Z 2008 *Automation Congress, 2008. WAC 2008. World* pp 1 –5
[7] Tuch D S, Wedeen V J, Dale A M, George J S and Belliveau J W 2001 *Proceedings of the National Academy of Sciences of the United States of America* **98**(20) 11697–11701
[8] Gonçalves S I, de Munck J C, Verbunt J P A, Bijma F, Heethaar R M and da Silva F L 2003 *IEEE Trans. Biomed. Eng.* **50**(6) 754–767
[9] Horesh L 2006 Ph.D. thesis University College London
[10] IEC60601 2005 *Medical electrical equipment - part 1: general requirements for basic safety and essential performance* ed3.0 (Geneva: International Electrotechnical Commission)
[11] Dalziel C F 1972 *IEEE Spectr.* **9**(2) 41–50
[12] Gilad O, Horesh L and Holder D S 2007 *Med. Bio. Eng. Comput.* **45** 621–633
[13] Rush S and Driscoll D 1968 *Anasthesia and analgetica* **47**(6) 717–723
[14] Ahadzi G, Gilad O, Horesh L, Bayford R and Holder D S 2004 *ICEBI'04 - V Electrical Impedance Tomography* 621–624
[15] Liston A D, Bayford R H, Tidswell A T and Holder D S *Physiol. Meas.* **23** 105–119
[16] Zlochiver S 2005 Ph.D. thesis Tel Aviv University
[17] Jackson J D 1975 *Classical Electrodynamics Second Edition* (New York: John Wiley & Sons)
[18] Frank E 1952 *Journal of applied physics* **23**(11) 1225–1228
[19] Lagarias J C, Reeds J A, Wright M H and Wright P E 1998 *SIAM Journal of Optimization* **9**(1) 112–147

Development of an algorithm for heartbeats detection and classification in Holter records based on temporal and morphological features

A García [1], H Romano [1], E Laciar[1] and R Correa[1]

[1] Gabinete de Tecnología Médica, Facultad de Ingeniería, Universidad Nacional de San Juan, San Juan, Argentina

E-mail: alfredorg86@gmail.com, marauder.7@hotmail.com, laciar@gateme.unsj.edu.ar, rcorrea@gateme.unsj.edu.ar

Abstract. In this work a detection and classification algorithm for heartbeats analysis in Holter records was developed. First, a QRS complexes detector was implemented and their temporal and morphological characteristics were extracted. A vector was built with these features; this vector is the input of the classification module, based on discriminant analysis. The beats were classified in three groups: Premature Ventricular Contraction beat (PVC), Atrial Premature Contraction beat (APC) and Normal Beat (NB). These beat categories represent the most important groups of commercial Holter systems. The developed algorithms were evaluated in 76 ECG records of two validated open-access databases "arrhythmias MIT BIH database" and "MIT BIH supraventricular arrhythmias database". A total of 166343 beats were detected and analyzed, where the QRS detection algorithm provides a sensitivity of 99.69 % and a positive predictive value of 99.84 %. The classification stage gives sensitivities of 97.17% for NB, 97.67% for PCV and 92.78% for APC.

1. Introduction

Cardiovascular diseases (CVDs) cause the death of around 17 million people per year, representing almost a third of all deaths globally. The most commonly cardiovascular diseases are heart disease and stroke. It is estimated that in 2015 will die about 20 million people by these diseases and is expected to remain the leading cause of mortality [1].

Holter ECG system is a useful ambulatory diagnostic tool, which has evolved considerably not only in the technological aspect, but in the clinical field as well. It allows a non-invasive diagnosis of heart-rate abnormalities and other cardiovascular diseases for a long period of time (24 – 48 hours). Besides, it is desirable to record for a long period of time, in order to eliminate wrong interpretations from normal spontaneous variabilities.

Through this study it is possible to analyze heart rate variability, detect extra-systoles, tachyarrhythmias and bradyarrhythmias, identify ST disorders or perform more complex analysis like the estimation of sudden death risk using QT interval analysis. Other utilities are the evaluation of the antiarrhythmic drugs efficacy, the performance of pacemakers and defibrillators re-synchronizers, as well as the study of ventricular late potentials [3].

The detection of Premature Ventricular Contraction beats (PVC) and Atrial Premature Contraction beats (APC) is very important for the identification of two kinds of arrhythmias. The APCs are due to auricular arrhythmias, which begin in an atrium or in the AV node [4]. On the contrary, the PVCs are related to ventricular arrhythmias, which begin in a ventricle, being potentially lethal because they can produce ventricular fibrillation and/or sudden death [5]. PVCs have an aberrant morphology, allowing an easy identification. The APCs have a similar morphology as NB, being their detection very difficult. Both PVCs and APCs generate an early contraction of the heart, which produces an increase of instantaneous heart rate.

Diverse techniques were used in different works to detect and classify heartbeats in ECG and/or Holter records, such as classification based on ECG morphology and heartbeat interval features [6] frequency domain analysis [7] and neural networks [8, 9]. The main objective of this work consists of detecting, featuring and classifying the heartbeats in Holter records in the following categories: PVC, APC and Normal Beats (NB), using a modified version of Pan and Tompkins algorithm for the heartbeats detection and the discriminant analysis technique for their classification.

2. Materials

Real Holter records from open-access database such as "*MIT BIH arrhythmia database*" and "*MIT BIH supraventricular arrhythmia database*" were used in this study. They are available free of charge in the website Physionet [10, 11] and they are widely used for the validation of commercial Holter analyzers.

The "*MIT BIH arrhythmia database*" has 48 records of 30 minutes duration. The recordings were digitized at 360 samples per second per channel with 11-bit resolution over a 10 mV range. The subjects were 25 men aged 32 to 89 years, and 22 women aged 23 to 89 years. This database has different morphologies of PVCs and NBs. The "*MIT BIH supraventricular arrhythmia database*" has 78 records of 30 minutes of length. They are sampled at 128 Hz with an amplitude resolution of 11 bits. This database has mainly APCs and NBs.

In most records of both databases, channel one is a modified limb lead II, and channel two is usually a modified lead V1 (occasionally V2 or V5, and in one instance V4). In the development of this work channel one was only used.

3. Methods

The algorithm has 3 stages: the first consist on the QRS complexes detection in the Holter record; this should be done to locate each heartbeat within the recording. In the next stage, the most important characteristics of QRS complexes were extracted. Finally, each heartbeat was classified using discriminant analysis.

Figure 1 shows the block diagram summarizing the different stages of the proposed algorithm. For the development of algorithms and the graphical interface the program MatLab® was used.

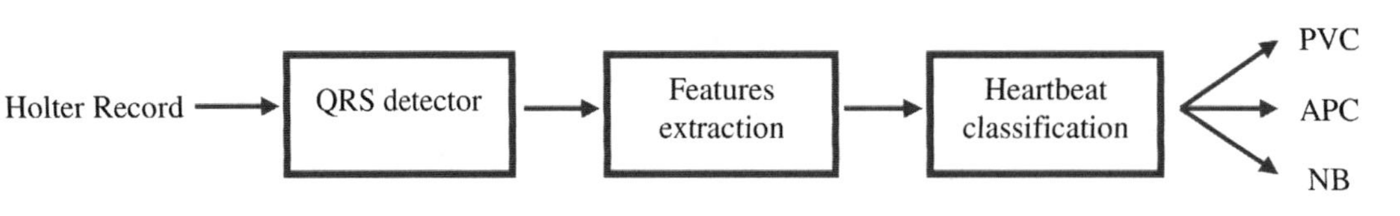

Figure 1. Block diagram of the proposed algorithm for heartbeats detection and classification in Holter records.

3.1. QRS complex detection

The QRS detector implemented is based on the detection method developed by Pan and Tompkins [12] and consists of five stages. Figure 2 shows the block diagram of the implemented QRS detector, showing in each case the result of each stage for an ECG example.

The first stage is a band pass filter, type Butterworth, order 12, with lower cutoff frequency of 5Hz and higher cutoff frequency of 15Hz to reduce the P and T waves, leaving only the QRS complex. After filtering, the signal is differentiated to provide the QRS-complex slope information;

a five-point derivative filter was used. In the third stage, the signal is squared point by point to eliminate negative values, as well as mitigate the points whose value is less than unity. In the fourth stage, a moving average filter of 100 ms was applied to smooth the signal and obtain well-defined maximum. Finally, a robust thresholding technique was proposed to locate the QRS complex in the record. It used an adaptative detection based on the RR intervals.

In order to decrease incorrect detections, the algorithm eliminates QRS marks that are less than 200 ms of an earlier mark (decreasing incorrect detections). If an excessively long RR interval is sensed, the algorithm repeats the analysis in the area with a lower threshold.

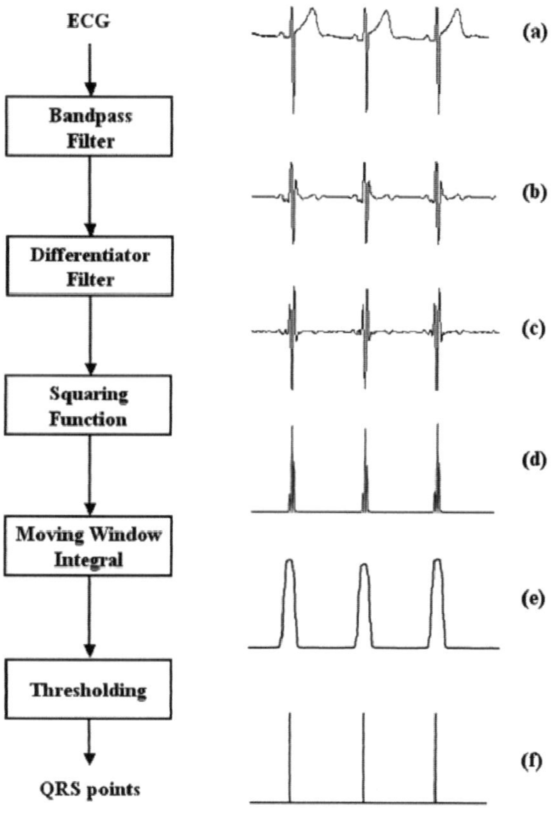

Figure 2. Block diagram of the QRS detector. (*Modified from* [5], [12]).

3.2. Feature Extraction

After detecting the position of each QRS complex in the Holter recording, a number of features were extracted from each heartbeat, which allowed their subsequent classification. This stage is very relevant because the parameters defined here are fundamental for the correct identification of the beating and subsequent clustering.

After several tests and empirical studies, 12 features were selected based on different temporal and morphological parameters, which are very similar to the used for the specialist, and are described below.

3.2.1. Temporal features

Eight temporal features were selected. One of them is the RR interval, defined as the time delay between two QRS peaks. It was found the interval between the current and previous beat, as well as the one between the current and subsequent beat, which are called *RR1* and *RR2* respectively. Another interval used is the distance between the previous beat and its predecessor, called *RR0* (see Figure 3).

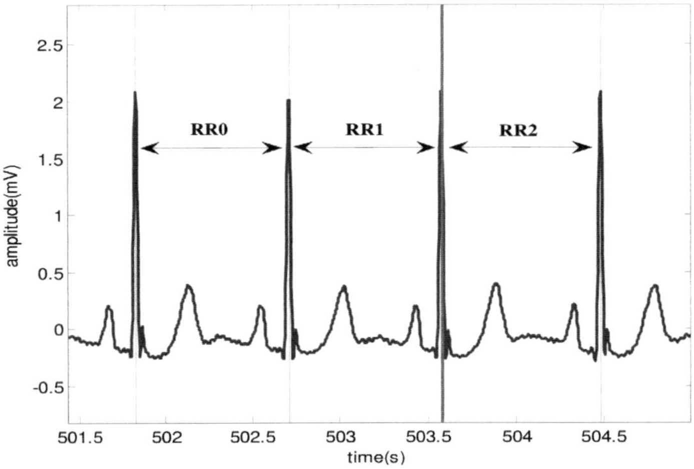

Figure 3. Identification of RR intervals in the Holter record. In red, it is illustrated the current beat. (record # 119 of "*MIT BIH arrhythmia database*")

After that, adimensional time-independent features were looked for, which are adapted to signals coming from different patients. The relations between RR1 and RR2 were defined, as well as the one between RR1 and RR0, called *Ratio1* and *Ratio2* respectively [13]. *Ratio3* was obtained as the relationship between RR1 and RRM, being RRM the average of all RR intervals corresponding to each record. The *MRATIO* was calculated as the mean value of Ratio1, Ratio2 and Ratio3. Finally, the robust median (*MRR*) was computed by applying to a series of RR intervals a median filter, taking a scan window of 10 intervals.

3.2.2. Morphological features

In order to extract 4 characteristics, the QRS complex of each beat was firstly extracted, using the previous 70 ms before the QRS mark up to the 140 ms after the mark. So, a window of 210 ms wide was generated, which contains all QRS complex morphology, without including the P and T waves. Then, each QRS complex was normalized to obtain an independent feature of the range.

$$QRSN_i(n) = \frac{QRS_i(n) - \max(QRS_i(n))}{\max(QRS_i(n)) - \min(QRS_i(n))} \quad (1)$$

i = 1,2,…,N beats

where:

$QRS_i(n)$: is the original *i*-th QRS complex detected in the record.

$QRSN_i(n)$: is the normalized *i*-th QRS complex detected in the record.

Figure 4 shows an example of the normalized QRS complex for: (a) NB, (b) PVC, and (c) APC beats.

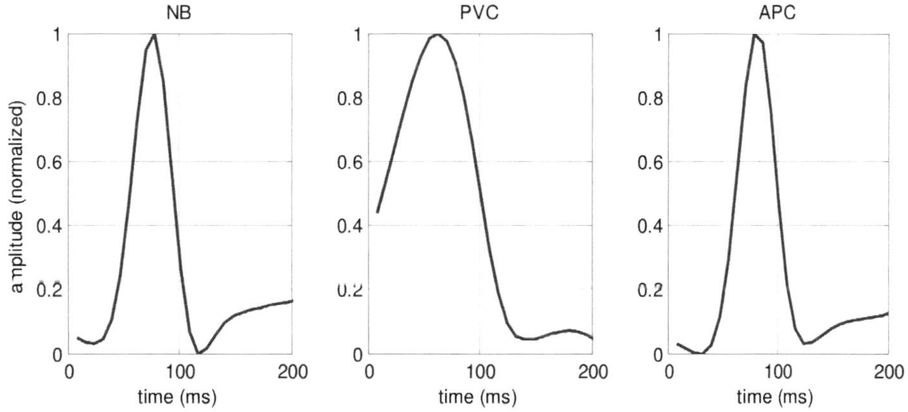

Figure 4. Normalized QRS complex for: (a) NB, (b) PVC, and (c) APC beats.

Using the normalized QRS complexes, the maximum of cross-correlation function between each detected beat and the following beat was calculated, as well as the maximum of cross-correlation between the current beat and the previous beat detected, called respectively *Corr1* and *Corr2* [13].

Another feature was the maximum of cross-correlation between a template of normal beat, with each QRS complex detected, called *Cxy*, was computed. For each Holter record, the template was calculated as the averaged beat of a sequence of 20 normal sinus beats.

Finally, *A* feature was defined as the QRS duration (in ms) when $QRSN_i(n) = 0.5$ in the normalized QRS complex, as shown in Figure 5.

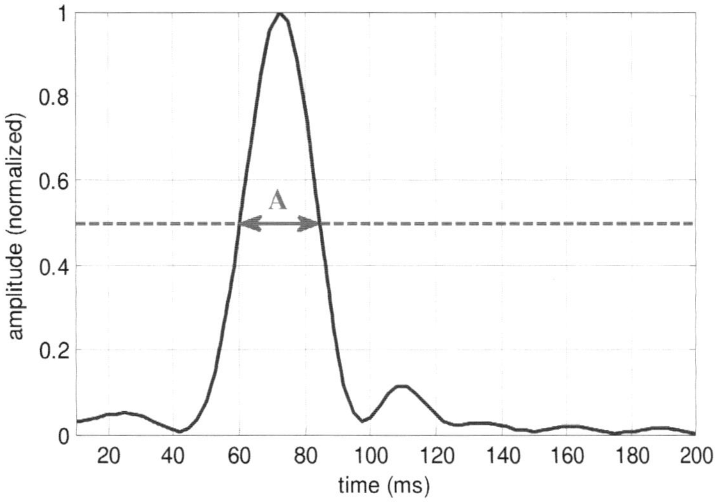

Figure 5. Feature A: QRS duration at 50% of the amplitude of the template ($QRSN_i(n)=0.5$). (record # 119 of "*MIT BIH arrhythmia database*")

3.3. Heartbeat classification

Figure 6 shows the block diagram of the proposed heartbeats classifier.

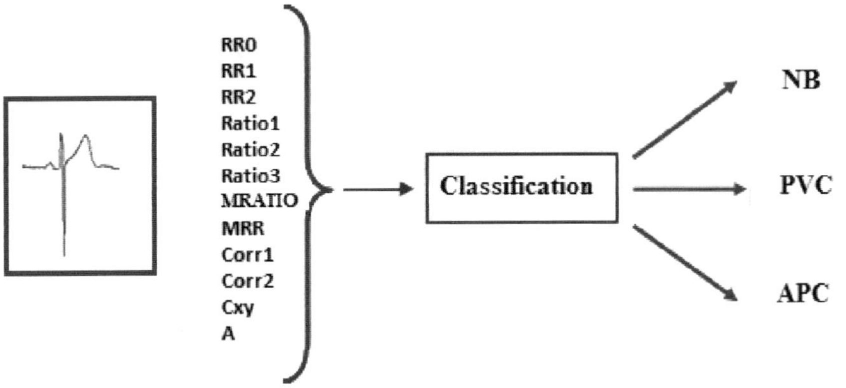

Figure 6. Block diagram of the proposed heartbeats classifier.

The method used is the discriminant analysis method, which is a multivariate statistical technique whose aim is to examine if there are significant differences between groups. Since there are three possible categories for each beat, the classification model has 2 linear discriminant functions. To create that model, 5800 heartbeats were selected from validated open access databases "MIT BIH arrhythmia database" and "MIT BIH supraventricular arrhythmia database", choosing 2000 NBs, 2000 PVCs and 1800 APCs. With these 5800 selected heartbeats, two groups were created:

- A training group, which contains 70% of the heartbeats (4060), the first 1400 corresponding to a NBs, the second 1400 to PVCs and the last 1260 to a APCs.
- A validation group, which contains 30% of the heartbeats (1740), the first 600 corresponding to a NBs, the second 600 to PVCs and the last 540 to APCs.

The heartbeats for both groups were randomly selected. Once materialized this stage, it was proceeded to select the best combination of the 12 features described above, that best discriminated different types, concluding that the best combination was the one that contained the 12 features used in this work. Figure 7 shows a segment of a record, where the different categories of heartbeats detected by the classifier are highlighted with different colours.

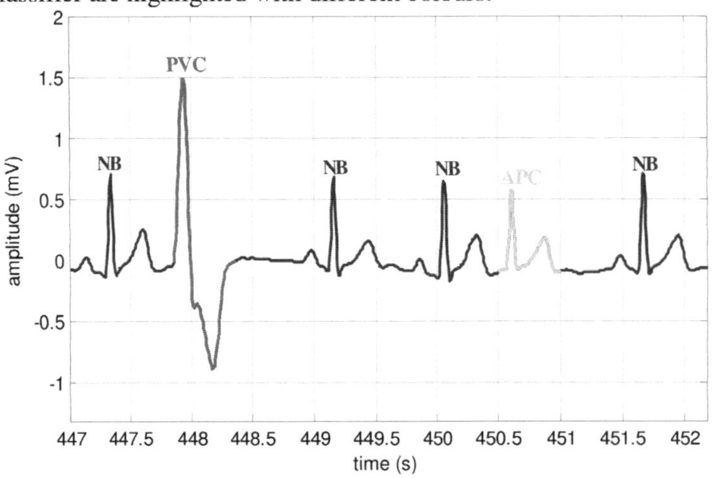

Figure 7. Classifier applied to a real Holter record (record # 872 of "*MIT BIH supraventricular arrhythmia database*")

3.4. Post – Processing

After detecting and classifying PCV, a search of episodes of ventricular bigeminy and trigeminy was made, as well as an identification of ectopic foci. Ventricular bigeminy is the occurrence of a PVC every other beat, where as trigeminy is every third beat. Figure 8 shows an episode of bigeminy and Figure 9 an episode of trigeminy. In Figure 10 are represented PVCs from different foci.

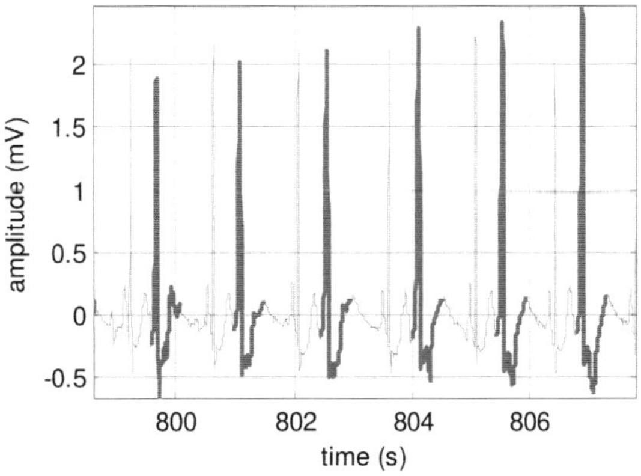

Figure 8. Episode of ventricular bigeminy detected, PVCs in red. (record # 106 of "*MIT BIH arrhythmia database*")

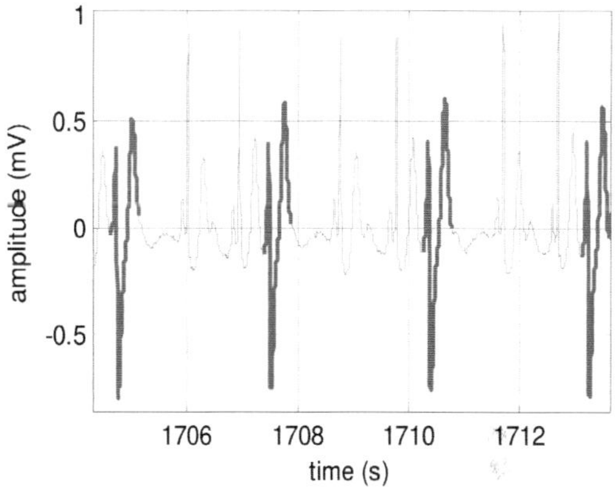

Figure 9. Episode of ventricular trigeminy detected, PVCs in red. (record # 106 of "*MIT BIH arrhythmia database*")

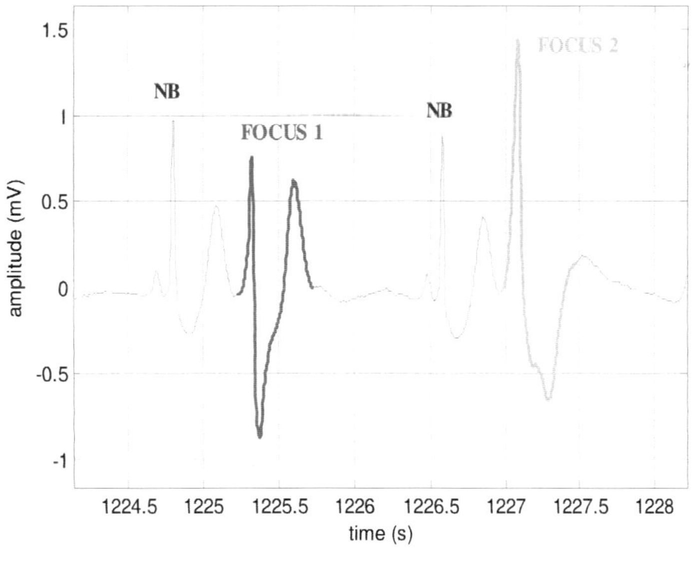

Figure 10. Different PVCs morphologies from two venricular foci. (record # 106 of "*MIT BIH arrhythmia database*")

4. Results

4.1. Heartbeat Detector.

The QRS detection algorithm was validated with the signals of the databases presented above, where 76 records were used. These records contain 166343 heartbeats from different patients, acquired under different conditions (noise, sampling frequency, channel used). The marks obtained by the detector were compared with annotations made by professionals and according to them, the following parameters were calculated.

- True Positives (TP): the number of positive detections that correspond to the annotations of the specialist.
- False Positives (FP): the number of detections that do not correspond with the annotations of the specialist.
- False Negative (FN): the number of heartbeats that were annotated by the specialist that were not detected.
- Sensitivity (Sen): the percentage of heartbeats correctly identified by the algorithm.

$$Sen(\%) = \frac{TP}{TP + FN} * 100$$

- Predictive value (PPV): the detection rate given by the algorithm corresponding to marks made by the specialist.

$$PPV(\%) = \frac{TP}{TP + FP} * 100$$

- Classification error rate (Err): the percentage of false detections over the total number of detected heartbeats.

$$Err(\%) = \frac{FP + FN}{TP + FN} * 100$$

Table 1 shows the results obtained by the QRS detector.

Table 1. QRS detector performance

	Heartbeats	TP	FN	FP	Sen (%)	PPV (%)	Err (%)
MIT BIH arrythmia database	85944	85642	302	83	99,66	99,89	0,45
MIT BIH supraventricular arrythmia database	80399	80214	185	143	99,71	99,80	0,49
Both databases	166343	165856	487	226	99.70	99.86	0.43

In comparison with other works [14] the QRS detector has similar values, but for the validation a 50% more of heartbeats were used, because another database was utilized.

4.2. Heartbeat classification

For the classification stage, the 70/30 method validation was used, the 70% of the heartbeats corresponding to the computation of the discriminating functions and the 30% for their validation. The random selection process of the groups of heartbeats was repeated 100 times. Table 2 shows the average values of sensitivity, specificity, positive predictive value (PPV), TP (true positives), TN (true negatives), FP (false positives) and FN (false negatives) for each of the categories of heartbeats.

Table 2. Results of the heartbeat classification for the validation stage.

	NB	PVC	APC
TP	583	586	501
TN	1119	1119	1172
FP	21	21	28
FN	17	14	39
Sensitivity	97,17%	97,67%	92,78%
Specificity	98,16%	98,16%	97,67%
PPV	97,01%	97,98%	96,15%

Figure 11 shows the scatter diagram obtained, showing that PVCs are notably separated of the NB. On the contrary, the APCs overlap the other two groups.

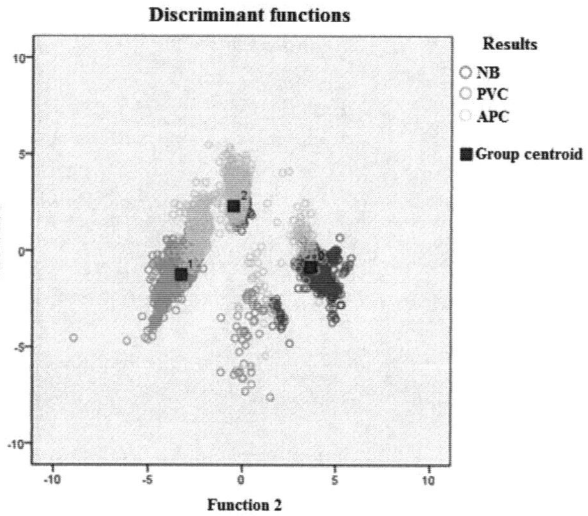

Figure 11. Scatter diagram of the 3 groups.

5. Discussion and conclusions.

In this paper, it was proposed an algorithm for heartbeat detection and classification, which can be used in ambulatory or Holter ECG records.

The results achieved with the QRS detector were similar in comparison to other algorithms, giving a sensitivity of 99.70%, a specificity of 99.86% and an error of 0.43% in contrast with the work of Martinez *et al* [14] which obtains a sensitivity of 99.66%, PPV of 99.56%. The algorithm was adapted to different kinds of signals with different environmental conditions.

The heart beat classification algorithm grouped them in three categories and was compared with the annotations of the specialists, being very efficient in the identification of PVCs (with a 97.67% of sensitivity), since they have very different morphological characteristics from the rest. This performance was similar than others algorithms, such as the proposed for Couceiro *et al* [9] and better than the proposed for de Chazal *et al* [6] in the classification of heartbeats, they obtains sensitivities of 96.35% and 94.4% for the PVCs classification respectively. However, APCs were more difficult to separate, obtaining a sensitivity of 92.78%, because they are morphologically identical to NB, and they also have similar time features as the PVCs. Anyway, better results were achieved compared with other papers, Karraz *et al* obtains a sensitivity of 86% for the APCs classification [15].

In conclusion the classifier proposed in this paper can be applied to real Holter records and can analyze automatically a record beat to beat. It takes approximately 10 seconds to analyze each of the records. This allows an automated analysis without a physician who classifies the different beats. It is also a useful tool for the specialist to make appropriate diagnosis.

In the future, this work will be integrated in a Holter analyzer system, which will have the most important functions of a commercial Holter analyzer.

References

[1] Hoffmann M, Heinke S, Lang K, Leder U, Hoffmann G, Heinke M, Dannberg G, Kiihnert H and Muller S 1996. *Comparative studies of heart rate variability (HRV) using Holter - ECG - systems and high solution ECG - systems in patiens with acute myocardial infarction and dilatative cardiomyopathy and healthy volunteers.* 18th Annual International Conference of the IEEE Engineering in Medicine and Biology Society, Amsterdam. pp. 1256-57.

[2] Rojas Malgarejo E. Monitoría electrocardiográfica ambulatoria de 24 horas (Holter). Historia, indicaciones y elaboración de un informe.

[3] Gamiz Palma JL 1983. Electrocardiografía de Holter. Bases prácticas y aplicaciones clínicas. Ediciones NORMA.

[4] Harrison 2006 *Principios de Medicina Interna* 16a edición http://es.wikipedia.org/wiki/Taquicardia_supraventricular. 21/12/2010 18:58 hs.

[5] Laciar E. and Valentinuzzi M. 2010. Chapter 4. Ventricular Fibrillation Detection, in *Cardiac Fibrillation-Defibrillation. Clinical and Engineering Aspects* by Max E Valentinuzzi, World Scientific Publishing Co, pp. 115-138, 2010.

[6] de Chazal P, Reilly R 2006. A patient-adapting heartbeats classifier using ECG morphology and heartbeat interval features. IEEE transactions on biomedical engineering, vol. BME-53, NO. 12.

[7] Lin C, Du Y, Chen Y and Chen T 2006. Multiple ECG beats recognition in the frequency domain using grey relational analysis. EMBS annual international conference, proceedings of the 28th IEEE.

[8] Palaniappan R, Krishnan S. 2004. Detection of ectopic heartbeats using ECG and blood pressure. International conference on signal processing & communications.

[9] Couceiro R, Carvalho J, Henriques J and Antunes M. 2008. On the detection of premature contractions. 30th annual international IEEE EMBS conference.

[10] MIT-BIH 1990, "MIT-BIH supraventricular arrhythmia database directory", Harvard-MIT Division of Health Sciences and Technology Biomedical Engineering Center.

[11] MIT-BIH 1999, "MIT-BIH arrhythmia database directory", Harvard-MIT Division of Health Sciences and Technology Biomedical Engineering Center.

[12] Pan J and Tompkins W 1985. A real time QRS Detection Algoritm. IEEE transactions on biomedical engineering, vol. BME-32, NO. 3.

[13] Gharavari A, Dehghan F, Teshnelab M and Moghaddam H 2008. Comparison of neural network, ANFIS, and SVM classifiers for PVC arrhythmia detection. Proceedings of the Seventh International Conference on Machine Learning and Cybernetics, Kumming, pp. 12-15.

[14] Martinez JP, Almeida R, Olmos S, Laguna P. 2004. A Wavelet-Based ECG Delineator: evaluation on Standard Databases. IEEE IEEE transactions on biomedical engineering, VOL. 51, NO. 4.

[15] Karraz G, Magenes G 2006. Automatic Classification of Heartbeats using Neural Network Classifier based on a Bayesian Framework. Proceedings of the 28th IEEE EMBS Annual International Conference (New York City, USA).

Stiffness Indices and Fractal Dimension relationship in Arterial Pressure and Diameter Time Series in-Vitro

L. Cymberknop [1], W. Legnani [4,5], F. Pessana [1,2], D. Bia [3] , Y. Zócalo [3] and R. L. Armentano [1,2,3]

[1] Electronics Department, Buenos Aires Regional Faculty, National Technological University, Argentina
[2] Faculty of Engineering and Exact and Natural Sciences, Favaloro University, Buenos Aires, Argentina
[3] Physiology Department, Republic University, Montevideo Uruguay
[4] Centre for Signal & Image Processing, Buenos Aires Regional Faculty, National Technological University, Argentina
[5] Institute of Calculus, Buenos Aires University, Buenos Aires, Argentina

E-mail: ljcymber@electron.frba.utn.edu.ar

Abstract. The advent of vascular diseases, such as hypertension and atherosclerosis, is associated to significant alterations in the physical properties of arterial vessels. Evaluation of arterial biomechanical behaviour is related to the assessment of three representative indices: arterial compliance, arterial distensibility and arterial stiffness index. Elasticity is the most important mechanical property of the arterial wall, whose natures is strictly non-linear. Intervention of elastin and collagen fibres, passive constituent elements of the arterial wall, is related to the applied wall stress level. Concerning this, appropriate tools are required to analyse the temporal dynamics of the signals involved, in order to characterize the whole phenomenon. Fractal geometry can be mentioned as one of those techniques. The aim of this study consisted on arterial pressure and diameter signals processing, by means of nonlinear techniques based on fractal geometry. Time series morphology was related to different arterial stiffness states, generated by means of blood flow variations, during experiences performed *in vitro*.

1. Introduction

During ventricular ejection, the cardiac muscle provides blood flow to the arterial system, exerting hemodynamic forces on the vessel walls. The main function of systemic circulation is to guarantee a continuous blood flow at capillary level. In this sense, mechanic behaviour of large arteries, denoted by its viscoelastic properties, plays a fundamental role [1]. The advent of vascular diseases, such as hypertension and atherosclerosis, is associated to significant alterations in the physical properties of arterial vessels [2].

Evaluation of arterial biomechanical behaviour is related to the assessment of three representative indices: arterial compliance (AC), Peterson's elastic modulus (E_P) and arterial stiffness index (β). The latter are commonly used in clinical practice and can be easily obtained from systolic (maximal) and diastolic (minimal) values of blood pressure and arterial diameter measurements [3, 4]. During the last years, pulse wave analysis (PWA) has regained relevance, due to the development of new time-domain processing methods, which are capable of providing specific information related to the pathological/normal state of the arterial system. It should be emphasized that the study of arterial dynamics necessarily implies the analysis of the stress (applied pressure) – strain (diameter variation) relationship. Although pressure-diameter interactions have been addressed in previous studies [5], the implemented techniques were based on the assumption of a system's linear behaviour. However, elasticity is the most important mechanical property of the arterial wall, whose nature is strictly nonlinear. Intervention of elastin and collagen fibres, passive constituent elements of the arterial wall, is related to the applied wall stress level [6]. Concerning this, appropriate tools are required to analyze the temporal dynamics of the signals involved, in order to characterize the whole phenomenon. Fractal geometry can be mentioned as one of these techniques. A fractal time series is characterized by self-similarity property. Specifically, it behaves similarly at different degrees of magnification (observation scales), which may be deterministic or stochastic. In geometric terms, a fractal signal can not be described or quantified by usual Euclidean measures, owing to its high irregularity [7].

The aim of this study consisted on arterial pressure and diameter signals processing, by means of nonlinear techniques based on fractal geometry [8]. Signals morphology evaluation (by applying a Fractal Dimension estimation algorithm) was related to different arterial stiffness states, during experiences performed *in vitro*. Several experiments were implemented, where the mean blood flow was varied, respecting the pulsatility regimes obtained from *in vivo* measurements. As a result, different arterial stiffness states were considered, as a consequence of the interaction of vessel's elasticity components.

2. Materials and Method

2.1. Self-similarity in arterial pressure and diameter time series

Self-similarity presence in arterial pressure and diameter signals was verified by the application of the wavelet transform (WT) [9]. A wavelet based representation can be explained as the assessment of the resemblance index (wavelet coefficients) between the wavelet mother and the signal involved, at different wavelet dilations, during a specified time interval. A large index indicates high "similarity". Consequently, application of WT generates a coefficient matrix, which provides signal information contained at different scales, or, in other terms, information contained at different "frequency bands". The expression for the WT calculation is defined as follows:

$$W(a,b) = \int_{-\infty}^{+\infty} s(t)\psi^*\left(\frac{t-b}{a}\right)dt \tag{1}$$

where $s(t)$ is the time series and $W(a,b)$ corresponds to its WT coefficient. Parameters a and b indicate the scale and the time displacement of wavelet mother Ψ, respectively.

If a signal manifests similarities at different scales, its wavelet coefficients will indicate such particular behaviour. In a graph where the vertical axis corresponds to the selected scales whereas abscises axis is related to time, the presence of self-similarity will determine a vertical pattern, which is characteristic of sets with equal intensity [10]. In certain cases, the analysis of the properties of the wavelet transform modulus maxima (WTMM) is sufficient, due to its relation with the signal singularities [9-11]. A wavelet coefficient can be considered as a Modulus Maxima, if the following expressions are verified:

$$|W(a,t_0)| \geq |W(a,\tau)|,\ \partial\frac{W(a,\tau_0)}{\partial\tau} = 0, \tag{2}$$

where a is the scale, τ_0 is the specified time and τ belongs either to a left or right neighbourhood of τ_0. Since WTMM positions define geometrical arranges in the time-scale plane, existing self-similarities may be observed throughout a geometric pattern with comparable intensity at all scales [9]. It should be emphasized that self-similarity is a necessary condition for a time series to be considered as a fractal but it is not sufficient. Pressure and diameter signals appear to be "potential fractal signals" due to the presence of self-similarity property. Consequently, fractal dimension estimation is applied in order to perform an irregularity measurement.

2.2. Fractal Dimension. Higuchi's method

Fractal dimension (FD) quantifies how densely a metric space is occupied by the fractal set [12]. Moreover, FD determines the time series complexity measure defined by its geometrical representation [13]. From a theoretical point of view, a fractal can be defined as an affine self-similar set, whose Hausdorff dimension (a measure of the space "filled" by the set at its point's neighbourhood) is strictly larger than its topological dimension [14]. Considering a time series of one time dependent variable, its FD value is included in the interval [1,2]. While the Hausdorff dimension is the most relevant measure, on a practical level, Box counting (BCD) or Correlation dimensions (CD) are implemented more frequently [15]. The former has been selected for FD analysis, during the present study.

Assessment of FD was performed by applying the method proposed by Higuchi [16]. A number of subsets based on the original temporal series (*x(t),* of length *N*) are generated, considering an initial time value (*m*) and a temporal increment (*k*) as parameters, as follows:

$$x_k^m = \left\{ x(m); x(m+k); x(m+2k),.., x\left(m + \left[\frac{N-m}{k} \right] k \right) \right\} \tag{3}$$

The term $[(N-m)/k]$ in (3) denotes the maximal time interval (Gauss notation) that can be considered for a selected *m* value. For each experimental time series, an averaged length is calculated ($L_m(k)$), as can be observed in the following expression:

$$L_m(k) = \frac{\sum_{i=1}^{\left[\frac{N-m}{k}\right]} \left| x(m+ik) - x(m+(i-1)k \right| \frac{N-1}{\left(\frac{N-m}{k}\right)k}}{k} \tag{4}$$

Then, the time series length function for each time increment ($L(k)$) is assessed, according to the expression:

$$L(k) = \frac{\sum_{m=1}^{k} L_m(k)}{m} \tag{4}$$

Finally, if $L(k) \propto k^{-FD}$ is found, the time series morphology may be quantified by its FD value. The latter can be obtained by applying a linear regression method to a doubly logarithmic scale representation of $L(k)$ against $1/k$. In addition, maximal value of time interval k (k_{max}) should be emphasized, especially if adequate accuracy is required in the FD estimation process.

Higuchi proposed a method that may be applied to any kind of time series, stationary or not. However, the obtained value lack of information related to the system involved (deterministic, chaotic or stochastic), which is responsible for the signal being analysed. In consequence, the method should be applied in the evaluation of variations that have occurred in the same signal (before and after significant events or different physiological states) [13]. In the present study, Higuchi's method was applied to time intervals of five cardiac cycles of length, which were acquired at a sampling rate of

400Hz. Established maximal value for k (k_{max}) was of 32. FD error range was estimated by applying linear regression analysis to consecutive groups of 4, 5 and 6 points, belonging to the doubly logarithmic graph. The obtained error values (sensitivity of the method) did not exceed 5%.

2.3. Data analysis

Stiffness (or distensibility) indices assessment was performed based on the following expressions, in accordance with [17]:

$$\beta = \ln\left(\frac{SP}{SD}\right)\frac{DD}{SD-DD}, \quad AC = \frac{SD-DD}{SP-DP}, \quad E_P = \frac{(SP-DP)}{SD-DD}DD \tag{5}$$

where *SD* y *DD* are the arterial systolic and diastolic external diameters and SD y DP are the arterial systolic and diastolic pressures, respectively. Both Arterial compliance and Peterson's modulus evaluate the vessel's buffering capacity [18]. Nevertheless, the difference between AC and E_P values lies in the fact that the latter provides information regardless of the variation in the vessel diameter. Consequently, E_P allows the comparison between arterial segments of different sizes [4]. Stiffness index (β) evaluates elastic arterial condition with no dependence on arterial pressure changes or, similarly to E_P, the geometric properties of the vessel wall.

It may be appreciated then, that E_P and AC considers a linear stress-strain relationship, whereas β suggests a logarithmic adjustment. Considering the above, FD assessment, which is related to the existence of non-linear behaviours, prevents the *a priori* assumption of a linear stress-strain dynamic.

2.4. Surgical Procedure

Six healthy male Corredale sheep were selected for the present study, weighting 25 to 35 kg and aged between 30 to 48 months old. Animals were properly cared, feed and vaccinated and surgically operated under general anaesthesia induced by thiopental sodium (20 mg/kg, intravenous). Subsequently to the intubation, they were maintained with 2.5% of enfluorane in pure oxygen (4 L/min) through a Bain tube connected to a Bird Mark VIII respirator. Each animal were positioned in left lateral decubitus and a sterile thoracotomy was performed at the left third intercostal space. Intra-arterial pressure measurement was performed by placing a high frequency response pressure microtransducer (1,200 Hz; Konigsberg, P7 model) into the aortic arterial lumen, through a collateral branch. Arterial external diameter measurement was obtained by using ultrasound technology. After a minimal dissection, a pair of ultrasound crystals (5 MHz, 4 mm diameter) was saturated at both sides of the aortic trunk's adventitia. A sonometric device (Triton Technology Inc.) was used to convert the ultrasonic signal transit time (1580m/s) into distance. The resultant signal was observed in a digital oscilloscope (Tektronixs, TDS 210 model) in order to ensure optimal diameter signal quality. All experimental procedures ware performed in agreement with ethics norms and international recommendations about research in laboratory animals, ratified in Helsinki and actualized by the Physiology American Society (1981) [19].

After *in vivo* measurements, all aortic segments were extracted in order to perform *in vitro* experiments. Studies were developed on a specially manufactured closed loop device, constituted by a centrifuge pump (fluid drive, electronically controlled), a reservoir (compliant chamber), a perfusion line (polyethylene conducts) and a rectangular chamber with a support system, for the aortic segments placement. The chamber was filled with a Tyrode's solution kept at a temperature of 37ºC, in order to maintain adequate physiological conditions of the specimen under study [20]. Once aortic segments were clamped, anticoagulated blood fluid (2-2.5 mPas viscosity) was putted into circulation, at different mean flow rates (70 to 340 ml/min). Fluid waveforms were pumped through the hydraulic circuit and controlled by the device system, according to *in vivo* measurements. The procedure was implemented in all extracted segments; *in vitro* pressure and diameter signals were digitalized using a data acquisition system (National Instruments CompactDAQ, USA), during at least five cardiac cycles. Sampling frequency (400Hz) was in accordance to the level of detail required for FD estimation [14]. Storage of acquired data was performed by means of specially designed software, developed under MatLab® platform (Mathworks Inc, Massachusetts, USA).

2.5. Signal Processing algorithms development
Signal processing algorithms were developed on MatLab platform (MathWorks INC, Massachusetts, USA) through the design and implementation of a graphic user interface (GUI). Prior to the non-linear processing, existing trends were eliminated (by means of digital filtering) and a WT analysis was applied (denoising method, [21]) with the purpose of eliminating undesired fluctuations originated during the acquisition procedure.

2.6. Statistical Analysis
Data were expressed as mean values plus / minus the standard deviation. Functional relationships between the arterial parameters were calculated by means of linear regression analysis methods. Continuous variables were compared using a Student t test, unpaired, two tails. A Pearson correlation coefficient was applied for all studies, where $p<0.05$ was considered statistically significant.

3. Results
Arterial pressure and diameter measured signals may be observed in Figure I, for a typical case. Similar morphology was found in the rest of the processed cases, for different mean blood flow states.

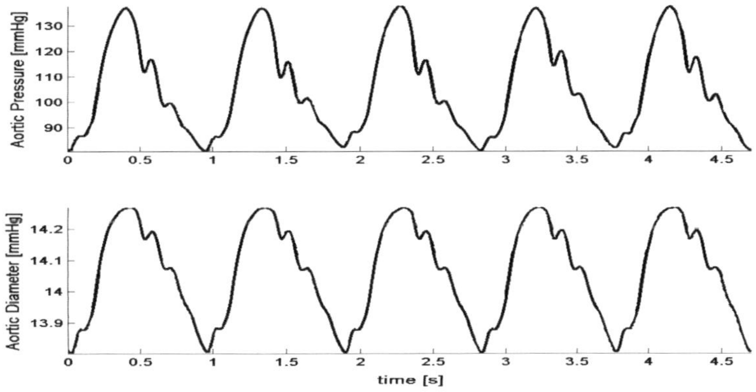

Figure 1. **Higher Panel**: *In-vitro* measured arterial pressure. **Lower Panel:** *In-vitro* measured arterial diameter. Mean blood flow during acquisition process was 140 ml/min

WTMM was applied to arterial pressure and diameter time series. Both of signals revealed the presence of self-similarity, verified in the geometrical pattern described by its coefficients maximal values (Figure 2). Self-similarity condition was also observed for different aortic mean flow states.

FD obtained by applying Higuchi's method, showed differences for different mean aortic values (Figure 3), for all processed time series. It can be observed that, throughout the whole procedure, pressure FD estimation remained higher than diameter FD, for the same aortic flow state.

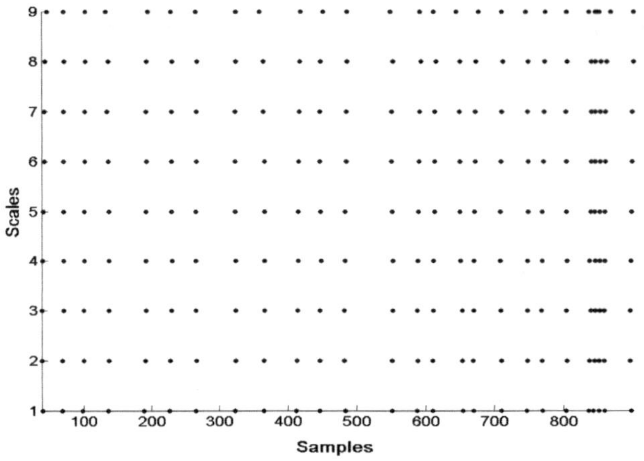

Figure 2. WTMM application for a typical case (mean blood flow 140ml/min). The geometric pattern observed in the time-scale plane indicates the presence of self-similarity.

Table I is constituted by stiffness indices assessments, related to the mean blood flow states performed during the *in vitro* experiments. Mean aortic flow values and hence, the increase of arterial pressure pulsatility, gave rise to a vascular response, expressed by changes in AC, E_P and β indices.

Estimated FD values evolution was in accordance with the variations experimented by the stiffness indices. Obtained calculations showed a decrease until mean aortic flow reached 120ml/min. At the time the mentioned threshold was exceeded, certain "stability" was observed.

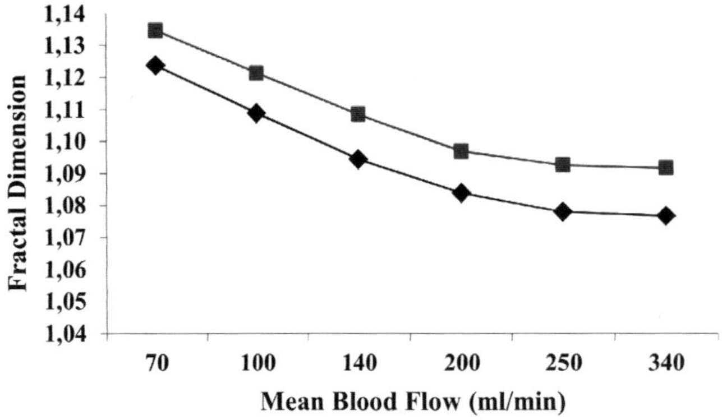

Figure 3. FD variation due to mean blood flow changes. Box line: FD Arterial pressure. Diamond line: FD Arterial diameter.

Table 1. Assessed stiffnes indices for mean blood flow variations. **F** is the mean flow, **β** is the stiffness index, **AC** is the arterial compliance, $\mathbf{E_P}$ is the Peterson's modulus. Values are expressed as mean ± SD

F (ml/min)	β	AC (10^{-3}mm.mmHg^{-1})	E_P (10^2 mmHg)
70	10.52±2.23	16.91±3,88	6.50⊥3.41
100	10.84±2.41	14.48±3,81	7.94±4.23
140	11.17±2.30	13.45±3,53	8.61±4.36
200	12.12±3.45	11.48±3,07	10.48±5.73
250	12.48±3.06	10.59±2,18	11.16±5.16
340	14.06±2.80	9.16±2,00	12.77±4.88

Table 2 specifies the linear regression analysis results, performed between arterial pressure and diameter FD values and arterial stiffness indices, for each blood flow state. Even though all relationships exhibit a high correlation level, a slightly higher adjustment may be appreciated in the FD vs β stiffness index linear relationship.

Table 2. Linear correlation indices (*r*) and statistical significance (*p*), obtained from linear regression analysis between FD and stiffness indices, for different mean blood flow states.

Parameters	Linear correlation index (r)	P
DF_P vs β	0.96	<0.01
DF_P vs CA	0.82	<0.05
DF_P vs E_P	0.83	<0.05
DF_D vs β	0.96	<0.01
DF_D vs CA	0.84	<0.05
DF_D vs E_P	0.81	<0.05

$\mathbf{DF_{P,D}}$ is the arterial pressure and diameter Fractal Dimension, respectively, **AC** is the arterial compliance, $\mathbf{E_P}$ is the Peterson's Modulus, β is the stiffness index.

4. Discussion

The purpose of this study was to process arterial aortic pressure and diameter time series, in order to evaluate the relationship between arterial stiffness and fractal dimension estimation. FD was applied, as a non-linear measure, due to the presence of self-similarity, revealed by the evaluation of the WTMM. Each time series FD assessment was performed by the application of Higuchi's method, which is widely utilized in non-linear signal processing literature. Arterial wall stiffness was evaluated by the calculation of arterial compliance, arterial distensibility and β stiffness indices.

As was previously referred, FD values appeared to be larger in arterial pressure than in diameter time series, for all performed experiences. Accordingly, the phenomenon may be explained based on stress-strain (pressure-diameter) relationship, that implies a filtering effect of high frequency vibrations, generating a loss of detail in morphology waveforms (elimination of an amount of self-similar scales). In addition, both arterial pressure and diameter FD showed a decrease value in accordance to the rise of the mean blood flow. The foregoing might be justified by the fact that the loss of arterial vessel distensibility (or the stiffness increase) precludes fractal manifestations whereas its influence on the content (blood flow) denotes a similar behaviour when arterial pressure signals are analysed. It can be also observed, that the difference between FD values of pressure and diameter time series is higher at low mean blood flow values. An explanation could be that at higher flow rates, dumping function of arterial wall is affected mainly to the activation of smooth muscle cells. The low rate of change of FD values observed after mean blood flow exceeded 120ml/min, could be justified by the same reason.

Another observation to be highlighted is the high correlation index obtained between FD estimation and β index regression analysis. An explanation might be found, based on the assumption of a linear stress-strain relationship, evidenced in the AC and E_P assessments. β index considers a non-linear (exponential) relationship, as the one observed in pressure and diameter curve, more in accordance with the known biological phenomena.

To conclude, numerical/mathematical cost of implementing FD estimation should be taken into account, in comparison with a linear clinical index assessment, widely used and validated. The revenue consists of information provided by the FD estimation, concerning waveform morphology, where significant manifestations associated to the vessel response might be obtained. Regarding the latter, a future study could quantify the degree of linkage between FD and elastic and viscous properties of the arterial wall as well as the individual contribution of each component to the self-similarity property. Such an approach might reduce the existing divergences between experimental data and linear models, essentially due to the non-linear presence in the stress-strain relationship.

5. Conclussion

Fractal characterization of arterial pressure and diameter time series and its linkage to the stress-strain relationship has not been previously documented in literature. As a result, FD references related to arterial stiffness have not been mentioned either. From the expressed results, the existence of a significant trend given by the wall arterial behaviour and morphology acquired by the time series could be inferred. The latter could be quantified by the application of non-linear analysis, as the one provided by fractal geometry. The relationship between estimated FD values of arterial pressure and diameter signals might be able to differentiate local/global arterial stiffness states, improving the information provided by linear-models based estimations.

One of the possible applications of the results of the present work in the near future is to include the calculus of the fractal dimension of the pressure signal acquired with Doppler velocities devices with the aim to estimate a measure of the systemic status of a particular circulatory system.

It is worth noting, that FD estimation requires a specific level of detail of the signal involved, in order to be able to quantify the complete self-similarity degree and to prevent certain fractal manifestations to be ignored. In this sense, sampling rate during data acquisition plays a major role and should not be underestimated. Even more, special consideration should be taken, if a similar analysis is to be implemented in non-invasive measurements obtained *in vivo,* which frequently do not have the appropriate resolution.

6. Acknoledgements

The present study was supported by co-financed research grant provided by National Technological University and National Agency for Scientific and Technological Promotion, FONCyT – IP – PRH – 2007.

7. References

[1] Armentano RL, Meguie JL, Simon A, Belleufant F, Barra JG, Levenson J, "Effects of hypertension on viscoelasticity of carotid and femoral arteries in humans" (1995b), *in Hypertension*, **26**,48-54

[2] Armentano Re, Graf S, Barra JG, Velikovsky G, Baglivo H., Sanchez R, Simon A, Pietrel RH, Lenvenson J, "Carotid wall visosity increase is related to intima media thickening in hypertensive patients" (2004), in *Stroke,* **35**

[3] Bella J., Roman M., Pinni R., Schwartz J., Pickering T,. Devereux R., "Assessment of Arterial Compliance by Carotid Midwall Stress-Strain Relation in Normotensive Adults" (1999) *in Hypertension,* **33**:787-792

[4] Cabrera Fischer E, Bia D, Zócalo Y, Camus J, Forteza E, Armentano RL, "Effects of removing adventitia on the mechanical properties of ovine femoral arteries in vivo and in vitro" (2010), in *J. Circ.*, **74**

[5] Pessana F., Armentano RL, Bía D., Zócalo Y, Cabrera Fischer E, "Endothelium-dependent mechanical effects on the beat to beat energy dissipation in the arterial wall" (2007), in *Rev. Brasileira de Engenharia Biomed.*, v**25**:p. 5-13

[6] Barra JG, Levenson J, Armentano RL, Cabrera Fischer E, Piehel RH, Simón A, "Assessment of smooth muscle contribution to descending thoracic elastic mechanics on concious dogs", (1993), *in Circ. Res,* **73**, 1040-1050,

[7] Falconer K. (1990), *Fractal Geometry: Mathematical foundations and applications*, (Wiley, New York)

[8] Losa G. A., Merlini D., Nonnenmacher T. F., Weibel E. R (2005), *Fractals in Biology and Medicine*, (Birkhäuser Verlag)

[9] Mallat, S. (1999), *A Wavelet Tour to Signal Processing*, (Academic Press)

[10] Meyer Y. (1993), *Wavelets, Algorithms and Applications*, (Society for Industrial and Applied Mathematics, Philadelphia)

[11] Addison P.S. (2002), *The Illustrated Wavelet Transform Handbook*, (Taylor & Francis)

[12] Barnsley M. F. (1983), *Fractals Everywhere, 2ed*, (New York. Academic Press Professional)

[13] Wlodzimierz K., "From conformons to human brains: an informal overview of non linear dynamics and its applications in biomedicine" (2007), in *Nonlin. Biomed. Phys.*, **1:5**,

[14] Mandelbrot B. (1983), *The fractal Geometry of nature*, (Freeman, New York)

[15] Raghavendra B. S., Narayana D., "Computing Fractal Dimension of signals using Multiresolution box counting method" (2010), in *Journal of Math. Sciences*, **6:1**

[16] Higuchi T., "Approach to an irregular time series on the basis of the fractal theory" (1988), in *Physica D.*, **31**:277-283

[17] Bia D, Armentano RL, Zócalo Y, Barmak W, Migliaro E, Cabrera Fischer E, "In vitro model to study wall dynamics through pressure diameter relationship analysis" (2005), in *Lat. Appl. Res.*, **35**:217-224,

[18] Laurent S, Cockcroft J, Van Bortel L, Boutouyrie P, Giannattasio C, Hayoz D, Panner B, Vlachopoulos V, Wilkinson I, Struijker-Boudier H, "Expert consensus on arterial stiffness: methodological issues and clinical applications" (2006), in *Eur. Heart. J*, **27**:2588-2605,

[19] Guide for the Care and Use of Laboratory Animals, National Institutes of Health (NIH Publication N° 86-23, revisited 1985)

[20] E. I. Cabrera Fischer, R. L. Armentano, F. M. Pessana, S. Graf, L. Romero, A. I. Christen, A. Simon, J. Levenson. "Endothelium-dependent arterial wall tone Elasticity modulated by blood viscosity" (2002), in *Am J Physiol Heart Circ Physiol*, **282**:389–394,

[21] Radunovic D. P. (2009), *Wavelets, from Math to Practice*, (Academic Mind, Belgrade, Serbia, Springer-Verlag, Berlin Heidelberg, Germany)

Evaluation of LDA Ensembles Classifiers for Brain Computer Interface

Cristian Arjona[1], José Pentácolo[1], Iván Gareis[1], Yanina Atum[1], Gerardo Gentiletti[1], Rubén Acevedo[1], Leonardo Rufiner[2,3]

[1] Laboratorio de Ingeniería en Rehabilitación e Investigaciones Neuromusculares y Sensoriales; Facultad de Ingeniería, Universidad Nacional de Entre Ríos, Argentina
[2] Centro de I+D en Señales, Sistemas e Inteligencia Computacional; Facultad de Ingeniería y Ciencias Hídricas, Universidad Nacional de Litoral, Argentina.
[3] Laboratorio de Cibernética; Facultad de Ingeniería, Universidad Nacional de Entre Ríos, Argentina.

E-mail; racevedo@bioingenieria.edu.ar

Abstract. — The Brain Computer Interface (BCI) translates brain activity into computer commands. To increase the performance of the BCI, to decode the user intentions it is necessary to get better the feature extraction and classification techniques. In this article the performance of a three linear discriminant analysis (LDA) classifiers ensemble is studied. The system based on ensemble can theoretically achieved better classification results than the individual counterpart, regarding individual classifier generation algorithm and the procedures for combine their outputs. Classic algorithms based on ensembles such as bagging and boosting are discussed here. For the application on BCI, it was concluded that the generated results using ER and AUC as performance index do not give enough information to establish which configuration is better.

1. Introduction

The brain computer interfaces are devices that provide a direct connection between brain and computer [1]. These interfaces could be the only way of communication for people that suffer motor disabilities as the ones described in [2].

The performance of the BCI is highly dependent, on the processing techniques use to extract features from electroencephalographic (EEG) signals that encoded user intentions [3] and on the classification techniques. Between the different evaluation alternatives of classifiers, the most natural one consists on feed a classifier with the features of the signals that are compared and use its performance as a measure of the discrimination power of the classifiers. In this case, you must be very careful for conceptually separate variations in the system performance (feature extractor and classifier) due to modifications in the classifier behaviour from variations due to the feature extraction techniques used.

An alternative technique to improve the performance of the classification process is the use of an ensemble of classifiers in parallel combining their outputs, where each classifier is trained in a different way. Under certain hypothesis systems based on ensemble can achieved better classification results than the individual counterpart, regarding single classifier generation algorithm and the procedures for combine their outputs. With this technique the final classification error is reduced as a result of the contribution of each of the classifiers in the decision making [4].

This study intends to preliminary evaluate and compare the performance of three different configurations of an ensemble composed by three classifiers and a configuration of only one classifier, to distinguish EEG signals with event related evoked potentials (ERP) from signals without ERPs.

In this work we used Fisher linear discriminant analysis (LDA) because it is one of the most used classifiers in BCI applications. [5, 6].

2. Material and methods

2.1. Donchin's P300 speller

When infrequent visual, auditory or somatosensory stimuli are mixed with frequent stimuli, the first ones evoke a potential in the EEG in the parietal cortex of the subject with a peak located around the 300 ms. This phenomenon can be used for BCI implementations, that are known as P300-based paradigm, by the identification of the P300 as the principal ERP. These potentials are used to, for example, select icons from a matrix on the computer screen as is described then [7].

In the classic P300 paradigm, the user is in front of 6 x 6 matrix that has letters and numbers. During the experiment, one row or column is random intensified with a predefined frequency. In a complete block of 12 intensifications, each row and column lights one time. To make a choice of one character, the user watches this character and when it is intensified an ERP is evoked. As a result of the use of the stimulation matrix, two epochs with P300 (corresponding to the infrequent or objective stimuli) and ten epochs without it (frequent or no objective stimuli) for each block are generated. Generally, the intensifications block must be repeat to be sure that the detect character is the one choose by the user.

To detect which intensification evoked an ERP, the system must be capable of solving a binary classification problem (two possible classes: signals with ERP and signals without ERP).

2.2. Database

The Laboratorio de Investigación en Neuroimagenología (LINI) of the Universidad Autónoma Metropolitana (UAM) made available a database that has the registers of thirty healthy subjects, using the BCI2000 P300 speller [8,9]. The registers were sampling at 256 Hz using a number of channels (Nc) equal to ten. The registered channels were Fz, C3, Cz, C4, P3, Pz, P4, PO7, PO8, Oz referred to lobe of the right ear and using as earth the right mastoid.

Each subject was in four sessions of fifteen repetitions of the 12 stimuli of the matrix, with this configuration different sizes of words were spelled. From this procedure a labelled data base of Nt=630 registers with P300 and Nnt = 3150 registers without P300 is obtain for each subject.

One of the premises of this data base is preserving the registers as real as possible. As a result, many of them have significant number of artifacts. A selection of ten subject registers that present the lowest number of artifacts has been made. For this work, a random selection of two subject's registers from the previous ten was made.

2.3. Preprocess

From the original registers, a new data set for each subject was generated by decimation. The decimation frequency was 16 Hz (Fsi).

Before the down sampling, the registers were filter with a low pass eight order Chevysheb filter, with a cutoff frequency of 7 Hz. The filter was passed forward and backward in order to avoid phase distortion. Then the signal from each electrode was normalized independently to obtain zero mean and unit standard deviation.

Individual epochs of one second were extracted from the registers. The first sample corresponds to the instant when the chosen row or column was lighted (stimuli). Due to the duration of the epochs and the values of Fsi, the number of samples per epoch Nei is equal Fsi.

Last, the feature vectors (patterns) were built concatenating the epochs of 4 channels (PO7, PO8, Oz y Cz). These channels were chosen based on a preliminary behaviour analysis done in [10].

Patterns of 64 features were generated. This is the minimum number of features to use in BCI application with LDA according to [10].

2.4. Classifier

The purpose of the LDA is calculate a discriminant vector $w \in \Re^D$ (where D is the number of features), such that, given a set of training patterns, with their $x_j \in \Re^D$ corresponding class label, $j \in \{1..N\}$ separate the classes using a given discriminant function. This is achieved by maximizing the function given by:

$$J(w) = \frac{(\tilde{m}_1 - \tilde{m}_2)^2}{\tilde{s}_1^{\,2} + \tilde{s}_2^{\,2}} \tag{1}$$

where:

$$\tilde{m}_k(w) = \frac{1}{N_k} \sum_{i \in Y_k} w^{\mathrm{T}} x_i, \quad \tilde{s}_k^{\,2} = \sum_{i \in Y_k} (w^{\mathrm{T}} x_i - \tilde{m}_k)^2, \tag{2}$$

E and k represent the set of index i corresponding to the class k and Nk is the number of training patterns of the k class [11].

2.5. Ensemble classifiers

The use of ensemble classifiers is a strategy where more than one classifier are combine taking into account the diversity of their answers to get a better classification performance.

There are many reasons to use ensemble classifiers among which can be mention: statistical reasons, big volumes of data, few data, divide and conquer and data fusion [4].

All of the ensemble classifiers have two key components. The first one is that a strategy to build an ensemble as diverse as possible is required, that is why fusion and selection strategies are used to combine the classifiers [4].

In second place is necessary a strategy to combine each of the classifiers outputs that integrate the ensemble in a way that the right decisions will be amplify and attenuated the wrong ones. Some available options to do this are: simple voting, weighted voting, BKS, addition rule, product rule, mean rule, minimum rule, maximum rule and weighted addition [4].

2.6. Building strategies

Bagging stands for bootstrap aggregating, that is one of the first algorithms based on ensemble. The Bagging method obtains the diversity through bootstrap copies of the training data. For each training subset is used a different classifier. These classifiers are all of the same type. The decisions of individual classifiers are combined through a majority voting. For a particular case, the class chooses by the majority of the classifiers is the decision of the ensemble.

Boosting creates a classifier ensemble by resampling the data, this strategy tends to get a more informative training set to present to the next classifier.

In essence, creates three weak classifiers: the first classifier (C1) is train with a random subset of the available data. The training subset to feed the second classifier (C2) is chosen as the most informative subset, given C1. C2 is train with a database where only half of the patterns were correctly classified by C1. The third classifier (C3) is trained with the patterns that C1 and C2 have classified from different class. The three classifiers are combined through majority voting [4].

K-means is an alternative strategy to select the training group, where the training pattern set is divide into k groups, each group is form with the patterns that are nearest to one of the k centroids, these centroids represent different statistical distributions of the features, this way, there are specific classifiers for different local areas in the feature space [4,11].

2.7. Strategies to combine outputs.

Simple voting: a count of the number of votes that each classifier gives to each class is made and the class select by the ensemble is the one that gets more votes.

Weighted voting: each classifier received a weight in the final voting based on its performance during the training process.

Minimum/maximum/median rule: as the name indicates, they are simple functions to find the minimum, maximum and median of outputs of individual classifiers.

Product rule: multiply the output values of each classifier. This rule is very sensitive to the most pessimistic classifiers.

Mean rule: for each class the average of all the classifiers' output of that class is the parameter that is consider to make the final decision. The mean rule is equivalent to the sum rule (within a normalization factor of *1/T*, where *T* is the number of available patterns).

Behaviour Knowledge Space (BKS): uses a look up table, constructed based on the classification of the training data that keeps track of how often each labelling combination is produced by the classifiers. Then, the true class, for which a particular labelling combination is observed most often during training, is chosen every time that combination of class labels occurs during testing. [4].

2.8. Performance evaluation

The error rate (ER), in the most use metric as a tool to evaluate the performance of a classifier [12].

In the ROC curves the true positive rate versus the false positive rate is represent, and a bigger area under the curve (AUC) generally indicate a better performance of the classifier [13, 14].

Based on the different characteristics of ER and AUC, both performance estimators are use in this work for the ensemble classifiers.

Where the Error Rate is:

$$ER = \frac{Nerror}{Nts}$$

Nerror is number of misclassifications tested registers and *Nts* is the total number of tested registers.

and AUC was calculate:

$$AUC = \sum_{th}^{\max th} S(th) * [Sp(th+1) * Sp(th)]$$

S is sensibility in function of threshold (*th*) and *Sp* specificity in function of threshold too.

3. Experiments

As it was mention before, patterns with a fix number of features were generated, resulting in a 64-dimension space.

The experiment series has 20 iterations, and the results were averaged to valid and evaluated the performance.

An individual classifier (control) and the three groups of classifiers were trained, on iterations. Bagging, Boosting and a variation of Bagging with k-means clustering with three centroids methods were used. The parameters that were modified in the experiments were the threshold; the patterns training sets; the patterns test sets and the method of output combination.

Threshold: the classification threshold is simultaneously modified for each of the LDA classifier of the ensemble to generate a ROC curve for then calculate the AUC.

Pattern training set: the 80% of the available patterns were randomly chosen, this process was repeated for each iteration.

Pattern test set: the 20% of the patterns that not form the training set were used to test the classifiers.

4. Results

In table 1 and 2 are shown the results obtain using the registers of subject 1 and 2 respectively. There you can find the values of AUC and ER of the experiments.

Table 1. Experiments results with subject's 1 registers.

Subject 1	**Simple voting**	**Weighted voting**	**BKS**	**Product rule**	**Addition rule**	**Weighted addition rule**	**Maximum rule**	**Minimum rule**	**Mean rule**
Type	AUC	AUC		AUC	AUC	AUC	AUC	AUC	AUC
Boosting	0,81	0,81	-	0,81	0,81	0,81	0,80	0,80	0,81
Bagging	0,82	0,82	-	0,82	0,82	0,82	0,82	0,82	0,82
Bagging kmeans	0,79	0,79	-	0,80	0,80	0,80	0,80	0,80	0,79
Single LDA	0,82								
Type	ER	ER	ER	ER	ER	ER	ER	ER	ER
Boosting	26,75%	26,75%	26,85%	28,10%	27,70%	26,92%	29,96%	29,96%	26,75%
Bagging	25,32%	25,32%	25,54%	25,30%	25,28%	25,42%	25,04%	25,04%	25,32%
Bagging kmeans	29,09%	29,09%	28,59%	28,57%	28,47%	28,49%	29,27%	29,27%	29,09%
Single LDA	24,82%								

Table 2. Experiments results with subject's 2 registers.

Subject 2	**Simple voting**	**Weighted voting**	**BKS**	**Product rule**	**Addition rule**	**Weighted addition rule**	**Maximum rule**	**Minimu m rule**	**Mean rule**
Type	AUC	AUC		AUC	AUC	AUC	AUC	AUC	AUC
Boosting	0,88	0,88	-	0,88	0,88	0,88	0,87	0,87	0,88
Bagging	0,89	0,89	-	0,89	0,89	0,89	0,89	0,89	0,89
Bagging kmeans	0,87	0,79	-	0,79	0,88	0,79	0,87	0,80	0,87
Single ADL	0,89								
Type	ER	ER	ER	ER	ER	ER	ER	ER	ER
Boosting	20,58%	20,58%	20,60%	21,11%	20,93%	20,48%	21,94%	21,94%	20,58%
Bagging	18,99%	18,99%	18,99%	19,13%	19,13%	19,13%	19,19%	19,19%	18,99%
Bagging kmeans	22,58%	23,81%	23,75%	23,35%	22,44%	23,83%	23,83%	23,83%	22,58%
Single LDA	18,75%								

The figures 1 and 2 belong to subject's 1 registers and figures 3 and 4 belong to subject's 2 registers. In these figures are resume graphically the data from tables 1 and 2. Based on all the obtain information, it can be seed that the performance of the classifiers trained by Bagging method are very near to the performance of the individual classifiers. The results using other training method as Boosting and Bagging k-means show a less performance than the methods mention before.

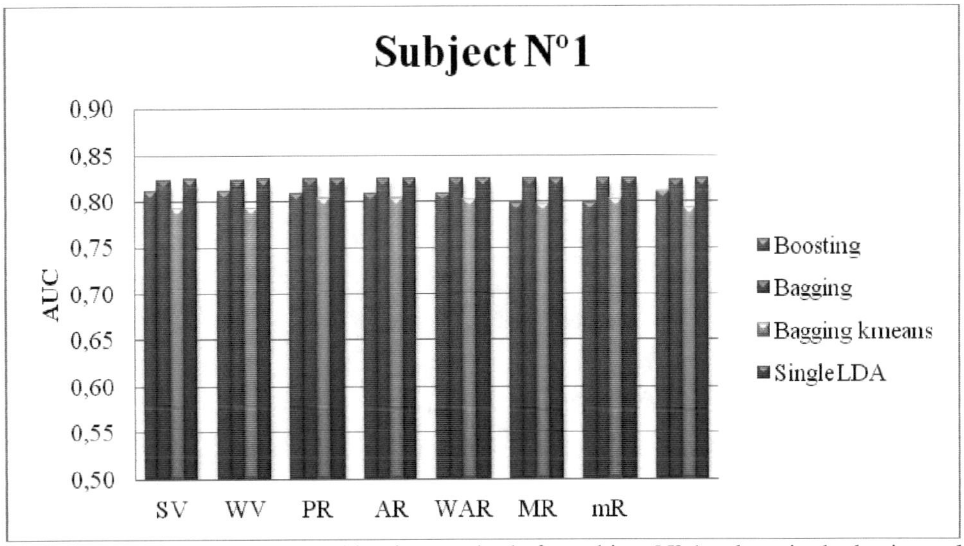

Figure 1. AUC vs. Output combination methods for subject N° 1, where in the horizontal axis are represented the different output combination methods: simple voting (SV), weighted voting (SW), product rule (PR), addition rule (AR), weighted addition rule (WAR), maximum rule (MR), minimum rule (mR), mean rule (meR).

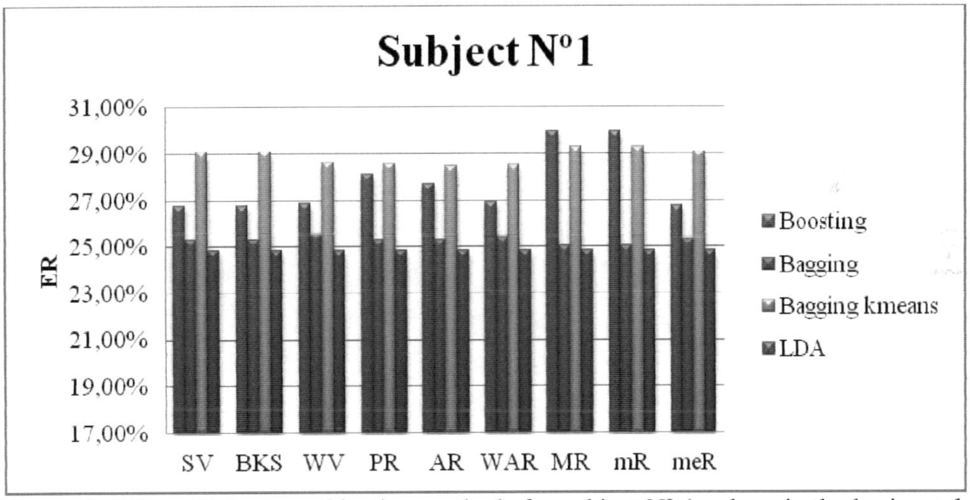

Figure 2. ER vs. Output combination methods for subject N° 1, where in the horizontal axis are represented the different output combination methods: simple voting (SV), weighted voting (WP), product rule (PR), addition rule (AR), weighted addition rule (WAR), maximum rule (MR), minimum rule (mR), mean rule (meR).

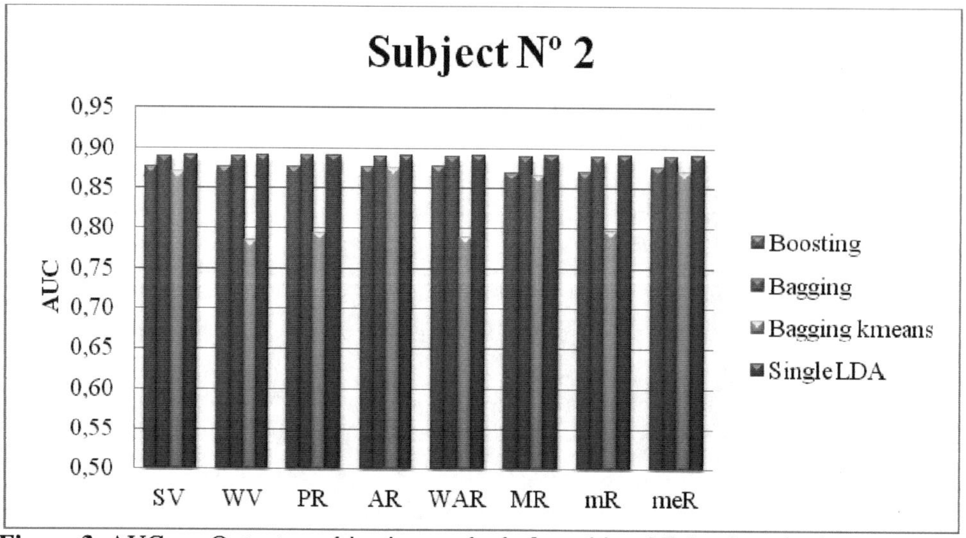

Figure 3. AUC vs. Output combination methods for subject N° 2, where in the horizontal axis are represented the different output combination methods: simple voting (SV), weighted voting (WV), product rule (PR), addition rule (AR), weighted addition rule (WAR), maximum rule (MR), minimum rule (mR), mean rule (meR).

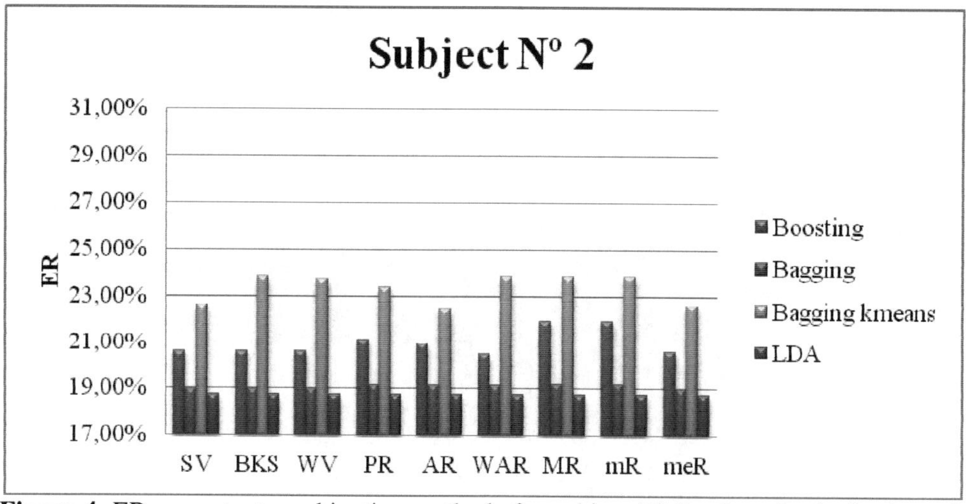

Figure 4. ER vs. output combination methods for subject N° 2, where in the horizontal axis are represented the different output combination methods: simple voting (SV), weighted voting (WV), product rule (PR), addition rule (AR), weighted addition rule (WAR), maximum rule (MR), minimum rule (mR), mean rule (meR).

5. Discusión

We have also evaluated the statistical significance of these results by computing the probability that a given ensemble classifier is better than single LDA classifier. In order to perform this test we assumed the statistical independence of the classification errors for each register and we approached the errors' Binomial distribution by means of a Gaussian distribution. This is possible because we have a sufficiently high number of registers for each subject (3780). In this way, for Subject 1 and Subject 2 we have that the confidence of the relationship obtained between error rates of simple LDA (ref) and each ensemble, ie Pr (Err ensemble > Err ref <), is detailed in Table 3 for each ensemble and each combination of outputs.

Table 3. Percentage of confidence of the results of each ensemble with respect to single LDA.

Confidence Subject 1	Voting	BKS	W. V.	P.R.	S.R.	W. S. R.	Max. R.	Min. R.	Med. R.
Boosting	97,23%	97,23%	97,23%	99,96%	99,85%	97,23%	100,00%	100,00%	97,23%
Bagging	69,92%	69,92%	80,78%	69,92%	69,92%	69,92%	56,91%	56,91%	69,92%
Bagging Kmeans	100,00%	100,00%	99,99%	99,99%	99,99%	99,99%	100,00%	100,00%	100,00%
Confidence Subject 2									
Boosting	97,23%	97,23%	97,23%	99,55%	98,82%	97,23%	99,96%	99,96%	97,23%
Bagging	56,91%	56,91%	56,91%	69,92%	69,92%	69,92%	69,92%	69,92%	56,91%
Bagging Kmeans	100,00%	100,00%	100,00%	100,00%	100,00%	100,00%	100,00%	100,00%	100,00%

From the result analysis, it can be observed that the use of ensemble classifiers instead of individual ones, seems to not generate relevant improvements in the classification performance of ERP using ER and AUC as index of performance. It is estimated that this situation is due to the lack of diversity of the classifiers that form the ensemble, and to the statistical distribution of the features of the used patterns. It should be emphasized that the results of this work were generated with patterns of only one epoch.

Although these results do not lead to a better solution to the classification of ERP problem, they orient the search to the study of other types of classifiers with diversity in the classification method, to be use in ensembles, as are support vector machines (SVM), multilayer perceptrons (MLP) or decision trees (DT). This work could also guide the search to other preprocessing techniques that make a different mapping of the feature space where the separation between classes is maximized. As an extension to this work it is presented as an option to explore the separation of the feature space into subspaces, using the strategy "divide and conquer".

6. Conclusions

In this work probes with ensemble classifiers of LDAs were performed taking basic algorithms as are Bagging and Boosting, also probes using different output combination methods and patterns preselection techniques, could be carried out.

When all the variations, proposed in this work were performed and the AUC and ER were calculated. What could be seen after that was a little different in the performance indexes between ensemble and single classifiers. Although it was done a variation of the training parameters in many ways, the results do not change.

Acknowledgments

This work was supported by the Universidad Nacional de Entre Ríos (grant PID 6101), the Universidad Nacional de Litoral (grant CAI+D 012-72) and the Consejo Nacional de Investigaciones Científicas y Técnicas (CONICET) from Argentina.

References

[1] Wolpaw J R, Birbaumer N, McFarland D J, Pfurtscheller G, and Vaughan T M 2002 Brain Computer Interfaces for Communication and Control. *Clin. Neurophysiol.* **113**, no. 6 767-91.

[2] Kübler A, Kotchoubey B, Kaiser J, Wolpaw J R and Birbaumer N 2001 Brain–computer communication: unlocking the locked in. *Psychol. Bull.* **127** 358–75.

[3] Bashashati A, Fatourechi M, Ward R K and Birch G E 2007 A survey of signal processing

algorithms in brain-computer interfaces based on electrical brain signals. *Journal of Neural Eng.* **4** 32-57.
[4] Polikar R 2006 Ensembe Based Systems in Decision Making. *IEEE Circuits and System Magazine.* **6** 21-40.
[5] Lotte F, Congedo M, Lecuyer A, Lamarche F and Arnaldi B 2007 A review of classification algorithms for EEG-based brain-computer interfaces. *Journal of Neural Engineering.* **4** 1-13.
[6] Krusienski D J, Sellers E W, Cabestaing F, Bayoudh S, McFarland D J, Vaughan T M and Wolpaw J R 2006 A Comparison of Classification Techniques for the P300 Speller. *Journal of Neural Eng.* **3** 299-305.
[7] Farwell L A and Donchin E 1988 Talking off the top of your head: toward a mental prothesis utilizing event-related brain potentials. *Electroenceph. clin. Neurophysiol.* **70** 510.
[8] http://www.bci2000.org/BCI2000/Home.html.
[9] Ledesma Ramírez C, Bojorges Valdez E, Yañez Suárez O, Saavedra C, Bougrain L and Gentiletti G 2010 *An open-access P300 speller database.* Fourth international BCI meeting, Paper L-12, Monterrey, California.
[10] Gareis I E, Acevedo R C, Atum Y V, Medina Bañuelos V, Rufiner H L and Gentiletti G G 2011 Efecto de la cantidad y dimensión de los patrones en una interfaz cerebro computadora basada en LDA, Proceedings of V Latin American Congress on Biomedical Engineering (CLAIB2011), La Habana, May 16 – 21, 2011.
[11] Duda R O, Hart P E and Stork D G 2000 *Pattern Classification (2nd Edition)* Wiley-Interscience (Estados Unidos) pp 117-24 526-28
[12] Weiss G M 2004 Mining with rarity: a unifying framework. *SIGKDD Explor.* **6** 7.
[13] Zweig M H and Campbell G 1993 Receiver-Operating Characteristic (ROC) Plots: A Fundamental Evaluation Tool in Clinical Medicine. *Clinical Chemistry.* **39** 561.
[14] Bradley A P 1997 The use of the area under the ROC curve in the evaluation of machine learning algorithms. *Pattern Recognition* **30** 1145.

Vocal Dynamic Visual Pattern for voice characterization

M E Dajer[1], F A S Andrade [2], A N Montagnoli [3], J C Pereira [2], D H Tsuji [1]
[1]Department of Otorrinolaryngology, School of Medicine, University of São Paulo-SP, Brazil
[2]Department of Electric Engineering, School of Engineering of São Carlos, University of São Paulo, São Carlos-SP, Brazil
[3]Department of Bioengineering, School of Engineering of São Carlos, University of São Paulo, São Carlos-SP, Brazil

E-mail: eugedajer@yahoo.com.ar

Abstract. Voice assessment requires simple and painless exams. Modern technologies provide the necessary resources for voice signal processing. Techniques based on nonlinear dynamics seem to asses the complexity of voice more accurately than other methods. Vocal dynamic visual pattern (VDVP) is based on nonlinear methods and provides qualitative and quantitative information. Here we characterize healthy and Reinke´s edema voices by means of perturbation measures and VDVP analysis. VDPD and jitter show different results for both groups, while amplitude perturbation has no difference. We suggest that VDPD analysis improve and complement the evaluation methods available for clinicians.

1. Introduction

The effect of voice pathologies causes many physiological modifications that result in an unhealthy pattern of vocal folds vibration. Consequently pathological voice assessment should search the different aspects of voice production.

Reinke's edema is a chronic inflammatory disease of the vocal folds and is considered a benign lesion. It is usually bilateral but asymmetric in volume and manifests different voice characteristics depending on the degree of inflammation and the severity of voice impairment. In general, patients with Reinke´s edema are women, heavy smokers and report vocal abuse [1].

Voice disorders as Reinke´s edema are usually evaluated by a perceptual analysis and perturbation measures such as jitter and shimmer. However, Titze [2] suggested that perturbation analysis might not be applicable to aperiodic signals, as patients' voices with benign mass lesions of the vocal fold.

In the last years, techniques based on nonlinear dynamic models have been applied to investigate the dynamic behaviour of biomedical systems, including voice signals [3], [4], [5].

Here we characterize normal and Reinke´s edema voices using qualitative and quantitative analysis of Vocal Dynamic Visual Pattern (VDVP) and acoustic perturbation measures.

[1] E-mail eugedajer@yahoo.com.ar

2. Methods

2.1. Data base
Forty eight female voices of sustained vowel /a/ were analyzed by means of perturbation measures, fundamental frequency and VDVP analysis. Voices were divided in two groups; one group includes 23 signals of female healthy volunteers ranging from 25 to 45 years old (mean age 33), with no voice complaints and no laryngeal pathology. The other group includes 25 voice signals from female patients with Reinke´s edema in different stages of disease evolution, ranging from 28 to 45 years old (mean age 37). All signals were from Signal Processing Laboratory database of the School of Engineering of São Carlos, University of São Paulo.

Procedures:

For recordings, subjects were asked to produce a sustained vowel /a/ at a comfortable pitch and loudness level for about 5 seconds. The microphone was placed at 45 degrees and 5 cm from the person's mouth [2]. Consecutive trials were performed, selecting the signal with less voice variability.

For data analysis we selected 500 milliseconds and the amplitude of the signal was normalized according to its absolute maximum value. Signal samples were quantized in amplitude with 16 bits and recorded in mono-channel .WAV format to preserve the fidelity of the signal. The sampling frequency was 44,100 Hz.

For statistical analysis we applied T-test on perturbation measures, fundamental frequency (F_0) and on quantitative measures of VDVP. Statistical significance level was set at p = 0,05.

2.2. Acoustic analysis
Perturbation measures jitter and shimmer, and fundamental frequency were extracted from healthy and Reinke´s edema voices. Voice analysis was performed using "*Analise de Voz 4.0*" [6]. Frequency perturbation or Jitter and amplitude perturbation or shimmer were estimated using correlation technique.

2.3. Vocal Dynamic Visual Patterns
Vocal Dynamic Visual Pattern (VDVP) analysis was developed from phase space reconstruction technique with time delay technique [7] and Poincaré Section [8]. The analysis is run in Matlab 7.0 and provides qualitative and quantitative measures for healthy and pathological signals.

According to Packard et al the time delay technique can be used to reconstruct the phase spaces of systems. A time series with length *N* is measured and recorded as $x(t_1), x(t_2), x(t_3), \ldots,$ where $x(t_i) \in$ R, $t_i = t_0 + i\tau$ $(i=1,2,\ldots,N)$ at the discrete time interval τ [9].

Then, the time delay vector creates the reconstructed phase space as

$$X(t) = \{x(t), x(t-\tau), \ldots, x(t-(m-1)\tau)\} \qquad (1)$$

The reconstructed phase space preserves the topological properties and geometrical invariance of the original attractor, so the system dynamics can be studied in the reconstructed phase space [10]. According to Titze it is important to choose a proper time delay for phase space reconstruction because the time series have a finite length and finite precision. If τ is too small, the time delay vector will be strongly correlated. But if τ is too long, the reconstructed attractor will produce a self-intersection [11]. Figure 1 shows a single cycle wave and VDPV, and 10 cycle waves and VDVP of the same sustained vowel /a/.

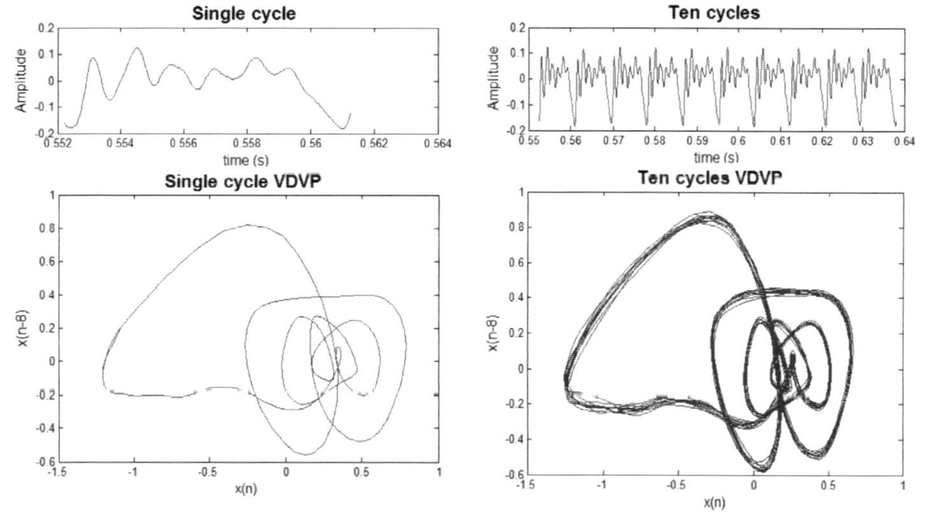

Figure 1. (Right) Single cycle of sustained vowel /a/ signal and the resultant VDVP. (Left) Ten cycles of the same sustained vowel /a/ signal and the resultant VDVP.

2.3.1. Qualitative measures
VDVP analysis describes the nonlinear dynamic characteristics of voice signal in two and three dimensional plots. Therefore qualitative and quantitative measures can be extracted from these VDVP. For qualitative analysis we observed three characteristics of the visual pattern: a) configuration or number of loops (L); b) the trajectories regularity degree (R) – how stable and smooth or rough and irregular are the trajectories – and c) convergence degree (C) – how close or disperse are the trajectories. These characteristics were scored on a five-point scale.

Number of loops: a decreasing scale form 4 to 0 describes the VDVP configurations. Patterns with more than three loops are scored with degree 4; three loops= degree 3; two loops= 2; one loop= 1, and 0= non defined configuration. For vowel /a/ degree 4 and 3 are the expected degrees [5].

Trajectory Regularity: was classified from 4 to 0. Where 4 is the highest degree and corresponds to completely regular and smooth trajectories and the lowest degree is 0 that describes completely irregular or rough trajectories. Middle scores – 3, 2 and 1 – are different degrees for combine behaviour of regularity.

Convergence: For trajectory convergence scores are: 4= trajectories with high degree of trajectories convergence and 0 for disperse or no-convergence trajectories; scores 3, 2 and 1 are different degrees of convergence.

2.3.2. Quantitative measures
We performed an algorithm based on Pointcaré Section to calculate the dispersion of points in one or more sections in the VDVP. Each voice signal was cut in ten different Poincaré sections, and then a mean value was calculated for each of the forty eight voices, as shown in Figure 2.

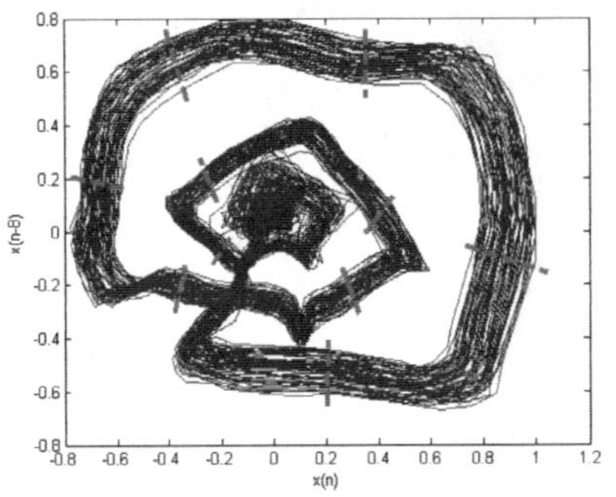

Figure 2. Vocal dynamic visual pattern with ten Poincaré sections.

To determine the Pointcaré Section we first select a tangent line in a trajectory segment. Then, we select three points that determine the cut region and two other points for establishing the perpendicular line to the Poincaré plan (Figure 3).

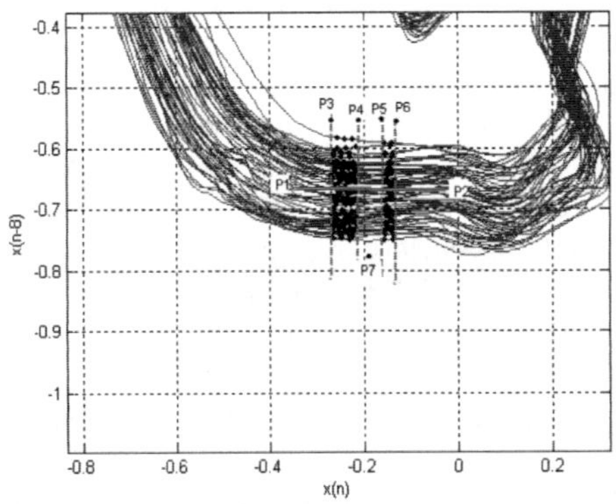

Figure 3. Poincaré section: points P1 and P2 compose the tangent line; P3, P4 and P7 are limit points; and P5, P6 and P7 are the orientation points for determine the perpendicular line to Poincaré section.

The next step is to calculate the dot product between the perpendicular line to Poincaré plan and the lines formed by the midpoint of the cut region with the other points of the cut region. Dot product near zero guarantees that points belong to Poincaré Section.

Then, a 3 dimensional rotation of the points is performed using the perpendicular line to the plane. Next, all points will be parallel to the x axis, and all points will be projected in the x=0 plane. When the points are in the x=0 plane we extract the principal components, in order to avoid the correlation of the points and finally, the Poincaré section is achieved (Figure 4).

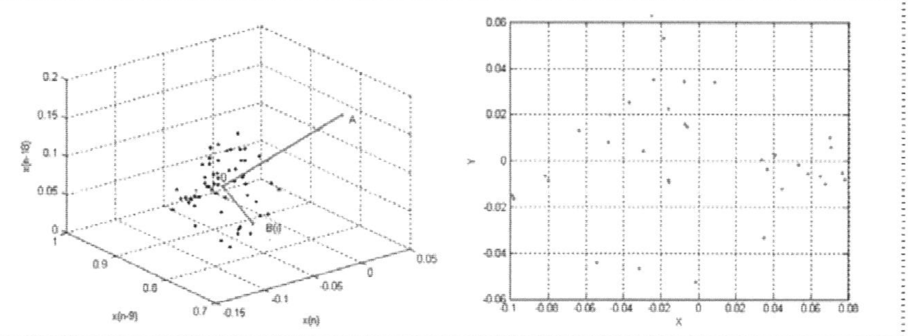

Figure 4. Determination of dot product and the points on Poincaré section.

In order to quantify the dispersion of the points we applied traditional statistical methods:

•DSD: Dispersion Standard Deviation - deviation from the midpoint of the section.
•ASD: Axis standard deviation – related the coordinate axes x and y.

3. Results

3.1. Perturbation Measures

Results for perturbation measures Jitter and Shimmer, and fundamental frequency of healthy and Reinke´s edema female voices are on Table 1.
Statistical analysis with T-Test for jitter showed a significant difference between the groups (P = 0,005). Power of performed test with alpha = 0,050: 0,785.
For shimmer T-test showed no significant difference between the (P = 0,075). Power of performed test with alpha = 0,050: 0,303.
And for fundamental frequency there was a significant difference between the groups (P = <0,001). Power of performed test with alpha = 0,050: 1,000.

Table 1. Values of maximum, minimum, mean and standard deviation for jitter and shimmer percentages and fundamental frequency (F_0) in healthy and Reinke´s edema voices.

	Healthy			Reinke´s edema		
	Jitter (%)	Shimmer (%)	Fo (Hz)	Jitter (%)	Shimmer (%)	Fo (Hz)
Mean	0,18	1,01	218,12	0,35	1,24	149,88
Max	0,34	1,55	249,33	1,55	2,26	227,64
Min	0,12	0,51	191,34	0,11	0,46	105,18
SD	0,05	0,24	18,87	0,55	0,55	37,20

3.2. Vocal Dynamic Visual Patterns

3.2.1. Qualitative measures

Generally female healthy voices showed various loops configuration (degrees 4 and 3), regular line trajectories with 4 and 3 degrees of regularity and a strong convergence (degree 3) while Reinke´s edema voices showed minor loops configuration (degrees 1 and 2), irregular trajectory lines (degrees 1 and 0) and weaker convergence (degrees 3 and 1).

Figures 5A and 5B show two examples of VDVP configuration. Figure 5A presents loops 3, regularity 4 and convergence 3 (L3R4C3) while Figure 5B shows loops 1, regularity 1 and convergence 1 (L1R1C1).

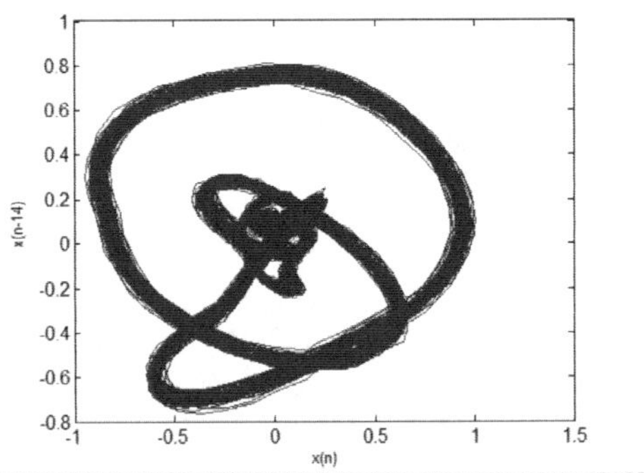

Figure 5A. VDVP of a healthy woman voice. Classification L3R4C3

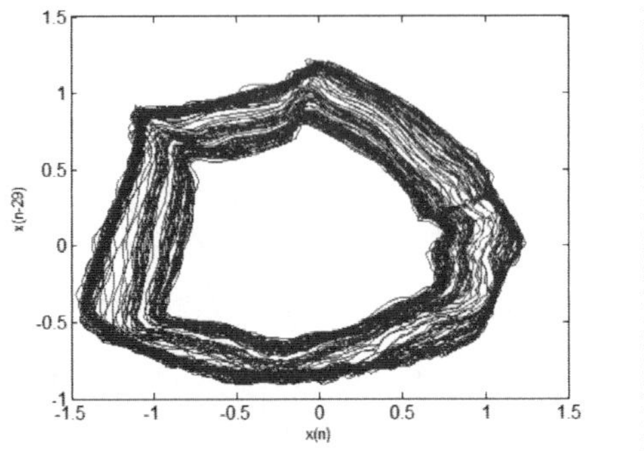

Figure 5B. VDVP of a woman voice with Reinke´s edema. Classification L1R1C1

Results of qualitative analysis for loops (L), Regularity (R) and Convergence degree (C) for both groups are presented in table 2. The number in each parameter indicates de five different degrees presented in section 2.3.1.

Table 2. Results for the 23 healthy voices and 25 Reinke´s edema (RE) voices

Loops	Healthy N=23	RE N=25	Regularity	Healthy N=23	RE N=25	Convergence	Healthy N=23	RE N=25
L4	10	0	**R4**	7	-	**C4**	2	1
L3	9	1	**R3**	4	2	**C3**	14	9
L2	3	9	**R2**	6	4	**C2**	5	6
L1	1	15	**R1**	6	14	**C1**	2	9
L0	-	-	**R0**	-	5	**C0**	-	-

3.2.2. Quantitative measures

VDVP of healthy and Reinke´s edema were analyzed by means of ten Poincaré Section.

Ten Poincaré sections were performed in each signal. Values of Axis Standard Deviation ASD for healthy voices were from 0,1859 to 0,3046 (mean 0,2435) and for Reinke´s edema voices were from 0,2383 to 0,4082 (mean 0,3109).

Dispersion Standard Deviation DSD values for healthy voices were from 0,0127 to 0,0293 (mean 0,0208) and for Reinke´s edema were from 0,0199 to 0,0661 (mean 0,0365).

For ASD and for SDS the T- Test showed statistical difference between the groups (P = <0,001). Power of performed test with alpha = 0,050: 1,000.

4. Discussion

The methods for assessing healthy and pathological voices are not new, but acoustic techniques used previously were base on linear models, and according Titze, this approach might not be indicate for aperiodic voices, as Reinke´edema voices [2]. Other procedures, like video-laryngoscopy, analyzes the vocal folds movements, but are considered an invasive method. The authors used an acoustic nonlinear approach in order to asses the vocal dynamic in a no invasive way.

Results of perturbation measures showed statistical difference for jitter, this may be correlated with the characteristic functionality of chronic inflammatory disease, as Reinke´s edema. Shimmer showed no statistical significance, suggesting that traditional perturbation measures, may not adequately characterize these kind of pathology. Fundamental frequency was different for each groups, and according to Tsuji et al, females with Reinke´s edema usually present lower fundamental frequency due to the increasing volume of mass of the lesion [12].
Qualitative analysis of VDVP showed a differential dynamic for both groups. Due to fluid collection the vocal folds vibrate slower and in an irregular pattern producing a rough voice with perceptive noise. These characteristics appear in the VDVP as irregular trajectories and weak convergence. In healthy voices the vocal folds vibration is quasi periodic and produce regular trajectories with strong convergence.
We observed that voice signals with higher jitter and shimmer are correlated with VDVP configuration of irregular and disperse trajectories and with higher values of Standard Deviation ASD and Dispersion Standard Deviation in the Poincaré Section.

5. Conclusion

This paper shows that tools based on nonlinear dynamical methods, as phase space reconstruction and Poincaré section seem to be suitable techniques for voice signals characterization and analysis, because of the chaotic component of the human voice. The VDVP analysis was able to discriminate between different stages of Reinke´s edema evolution and healthy voices.
Vocal Dynamic Visual Pattern qualitative analysis has shown a potential value to describe normal and Reinke´s edema voices. The standard healthy VDVP of vowel /a/ showed complex configuration and nearly periodic behavior, while Reinke´s edema voices presented simpler loops configuration, irregular trajectory lines and lower convergence degree. Values of Standard Deviation ASD and Dispersion Standard Deviation in the Poincaré Section were different for each group.
We consider that perturbation analysis associated with quantitative and qualitative analysis of the VDVP provide more information and represent non-invasive and helpful tools to describe objectively the vocal folds dynamic.

6. Acknowledgements

The authors acknowledge the Research Foundation of the State of São Paulo (FAPESP) for the financial support. We also acknowledge the School of Engineering of São Carlos (EESC/USP) for the research facilities and the use of its voice database.

7. References

[1] Marcotullio D, Magliulo G, Pezone T. (2002). Reinke's and risk factors: clinical and histopathologic aspects. *Am J Otolaryngol.* 23:81-84.
[2] Titze IR (1995) Workshop on Acoustic Voice Analysis: Summary Statement. Denver, National Center for Voice and Speech.
[3] Pitsikalis B; Kokkinos I; Maragos P. (2003). Nonlinear Analysis of Speech Signals: Generalized Dimensions and Lyapunov Exponents. Proc. European Conf. on Speech Communication and Technology - Eurospeech Geneva, Switzerland.
[4] Jiang JJ, Zhang Y; MacCallum J; Sprecher A; Zhou L. (2009). Objective Acoustic Analysis of Pathological Voices from Patients with Vocal Nodules and Polyps. *Folia Phoniatr Logop*; 61:342–349.

[5] Scalassara PR; Dajer ME; Maciel CD; Guido CR; Pereira JC. (2009). Relative entropy measures applied to healthy and pathological voice characterization. Applied Mathematics and Computation 207 95–108 (2009)

[6] Montagnoli A.N., Pereira J.C., [Análise de Voz, Versão 4.10] Sistema de Auxílio à Análise Acústica da Voz. 2010.

[7] Fraser, AM; Swinney, HL. Independent coordinates for strange attractors from mutual information. Phys. Rev. Lett. 1986, vol. 33, pp. 1

[8] Kantz H; Schreiber, T. Nonlinear Time Series Analysis. Cambridge University Press, Cambridge (1997).

[9] Packard NH, Crutchfield JP, Farmer JD, Shaw RS. Geometry from a time series. Phys Rev Lett. 1980;45:712.

[10] Takens F. Detecting strange attractors in turbulence. In: Rand DA, Young LS, editors. Lecture Notes in Mathematics. Berlin: Springer-Verlag; 1981:366–381.

[11] Titze IR, Baken R, Herzel H. Evidence of chaos in vocal fold vibration. In: Titze IR, editor. Vocal Fold Physiology: New Frontiers in Basic Science. San Diego, CA: Singular Publishing Group; 1993:143–188.

[12] Tsuji DH; Inamura R; Ubirajara LS. Fisiologia Da Laringe. In: Pinho, S; Tsuji, Dh; Bohadana, Sc. Fundamentos em laringologia e Voz. Ed. Revinter, Rio de Janeiro, 2006

Characterization and quantification of oxides generated by anodization on titanium for implantation purposes

L Aloia Games[1], J Pastore[2,3], A Bouchet[2,3], J Ballarre[1,3]
[1] División Corrosión. INTEMA. UNMDP.
[2] Laboratorio de Procesos y Medición de Señales. Dpto de Electrónica. UNMDP.
[3] CONICET.

E-mail: jballarre@fi.mdp.edu.ar

Abstract. The use of titanium as implant material is widely known in the surgery field. The formation of natural or artificial compact and protective oxide is a convenient tool for metal protection and a good way to generate phosphate deposits to enhance biocompatibility and bone fixation with the existing tissue. The present work has the aim of superficially modify commercially pure titanium sheets used in orthopedics and odontology, with a potencistatic anodization process with an ammonium phosphate and ammonium fluoride solution as electrolyte. The objective is to generate titanium oxides doped with phosphorous on the surface, to promote bioactivity. The characterization and quantification of the generated deposits is presented as a starting point for the future application of these materials. The applied characterization methods are X ray diffraction, micro-Raman spectroscopy analysis for evaluating the chemical and phase composition on the modified surface and PDI image analysis techniques that allow the segmentation of SEM images and the measurement and quantification of the oxides generated by the anodization process. The samples with polished treated surface at 30V have the deposit of a phosphate rich thick layer covering almost all the surface and spherical-shaped titanium oxide crystals randomly placed (covering more than 20% of the surface area).

1. Introduction

The metallic materials are widely used as orthopedic and odontologic surgical implants due to their excellent mechanical properties and controlled corrosion resistance for these applications [1]. A disadvantage of the prosthetic metallic materials is that most of them cannot make a natural bonding with the existing tissue, so it is necessary a chemical or mechanic fixation. Also the material that is going to be implanted should have good wear resistance and other superficial characteristics [2].

The use of titanium as implant material is widely known in the field. The formation of natural or artificial compact and protective oxide is a convenient tool for metal protection and a good way to generate phosphate deposits to enhance biocompatibility and bone fixation with the existing tissue. The adhesion of these coatings to the metallic substrate and its mechanical properties are of great importance since they are going to be submitted to several mechanical efforts during and after implantation. Also the characterization and quantification of the oxides present in the surface are a key issue for the future calcium phosphate deposition (apatite, first insight of bone formation), to provide bioactivity and mechanical fixation.

Automatic image segmentation is a significant area in Digital Image Processing. One of the most important elements of any automated system of vision is the task of understanding the image, that severely affects the subsequent process of interpretation of the image, providing useful structures such as regions and edges [3].
Mathematical Morphology (MM) has been a successful tool in image segmentation. The MM is a theory based on concepts of geometry, algebra, topology and set theory, created to characterize physical and structural properties of different materials. The main idea of this theory, is to examine if the geometric structures of an image overlap with an small localized pattern (called "structuring element") [4].
In this paper we propose the surface modification of titanium implants of commercial purity, used in orthopedic surgery and dentistry, through potentiostatic anodizing surface treatment, carried out in a solution of ammonium phosphate and ammonium fluoride in order to obtain titanium oxide doped with phosphorus in the surface, to promote bioactivity and generate better interaction with the bone. The characterization and quantification of the oxides generated is presented as an important starting point for further implementation of these materials. For this we use three methods: x-ray diffraction (XRD) and Raman spectroscopy to assess the chemical composition of the surface, and image processing techniques that will enable the segmentation of the image through the use of MM operators for subsequent measurement and quantification of generated oxides on anodized titanium.

2. Materials and Methods

2.1. Sample preparation

Samples of 28 x 13 x 0.3 mm^3 grade-2 commercially pure Titanium sheets (CP-Ti, "Unalloyed Commercially Pure Titanium") were employed. Its chemical composition can be seen in Table 1. Some of its properties (mainly mechanical) are shown in Table 2 comparing with the cortex bone properties.

Table 1. Chemical composition of CP-Ti (% in weight).

	Ti	C	Fe	H	N	O
CP-Ti (grade 2)	99.2	Max 0.1	Max 0.3	Max 0.01	Max 0.03	Max 0.25

Table 2. Some properties of CP-Ti comparing with cortical bone.

	CP-Ti	Bone
Tensile Strength (MPa)	345	85 – 150
Yield Limit, σ f (MPa)	275	-
Elastic Modulus, E (GPa)	105 - 140	7 – 30
Hardness (GPa)	Aprox. 2 GPa	-
Density (g/cm^3)	4.51	18 – 22
Crystalline Structure	Hexagonal	Crosslinked macromolecule

Two surface conditions of the substrates were analyzed: polished to 600 grit paper with water as lubricant (***AP*** samples) and "as-received" (the same condition as the material was received after the fabrication process, ***AR*** samples).
Before the anodizing treatment the samples were cleaned in ultrasonic bath with acetone during 10 minutes and after that with isoprophylic alcohol for another 10 minutes.

The electrolitic solution was perpared with bi-destilated water, 1 mol/L of NH_4 H_2 PO_4 and 0.5 % weigth of NH_4 F.
The potenciostatic anodic treatments were carried out at two different voltages: 20 y 30 Volts, for 1 hour at room temperature with a power source Consort EV231. So the samples were: with rough surface (***AR20*** and ***AR30***) and with polished substrate (***AP20*** and ***AP30***).

2.2. Surface Characterization
For the morphologic evaluation of the surface deposits several images were taken at different magnifications. A scanning electronic microscope was used to obtain the simple images (SEM, Phillips XL 30).
X-ray diffraction (XRD) and Raman spectroscopy essays were done with the aim of identify the different compounds present in the surface deposits. The XRD experiments were done with a diffractometer PANalytical X'Pert Pro, with a Cu-Kα radiation of 40 kV and 40 mA. The spectra were taken at 0.02 degree per second and with a small angle of 1° from 10 to 70 degrees. The Raman equipment used was an Invia Reflex confocal with a 514 nm wave length laser, at 100 % intensity in a window from 100 to 1500 cm^{-1}, with one accumulation and 50 seconds exposition per spectrum.

2.3. Digital Image Processing
In order to determine the amount of oxide present on the surface of the ***AP30*** samples, MM operators were applied to the images obtained through a scanning electron microscope. The main objective of the MM is to extract information of the geometry and topology of an unknown set present in an image. The key of this methodology is the "structuring element", a small set completely defined and of a known geometry, which is compared, based on a transformation, with the whole image. The shape and size of the structuring element allow quantify if the structuring element is contained in the image or not [4].
A gray level image is a function defined on a subset of $\mathbb{Z}^2$ into the natural range $[\![0, L-1]\!]$, where L is a number within the interval $[\![1,256]\!]$. For this work f denotes a gray level image.
The basic morphological operators of the MM are erosion and dilation. Morphological erosion and dilation of the image f by the structuring element b are defined, respectively, as [5, 6]:

$$\varepsilon(f,b)(s,t) = \min\left\{f(s+x,t+y) - b(x,y) \,/\, (s+x,t+y)\in D_f;\,(x,y)\in D_b\right\} \quad (1)$$

$$\delta(f,b)(s,t) = \max\left\{f(s-x,t-y) + b(x,y) \,/\, (s-x,t-y)\in D_f;\,(x,y)\in D_b\right\} \quad (2)$$

where D_f and D_b are the domains of the functions f and b respectively.
Based on the combination of the previous operators, opening and closing can be defined as follows [5, 6]:

$$\gamma_b(f) = \delta_b(\varepsilon_b(f)) \quad (3)$$

$$\Phi_b(f) = \varepsilon_b(\delta_b(f)) \quad (4)$$

The opening is useful to remove small bright details smaller compared to the size and shape of the structuring element, the rest of the image is not modified. When filtering, in the step of erosion, a bright area surrounded by dark one is reduced. Particularily, if the size of the structuring element is greater than the bright area, this bright area totally disappears. Then, in the step of dilation, an bright area surrounded by dark one is expanded and therefore the image recover its current shapes, except of the details removed by erosion in the previous step.
Similarly, the closure is useful to remove small dark details compared to the size and shape of structuring element, the rest of the image is not modified.

The resulting image depends on the gray level, shape and size of the structuring element. Top-hat transform is a well-known and commonly used morphological technique for extracting locally bright (or dark) objects from a grayscale image. Figure 1 shows an opening Top-Hat transform, which emphasizes locally bright objects that have been eliminated from the image by an opening filter. A structuring element larger than the structures to be filtered was used.

This transform is defined as the residual between the identity and a morphological opening [5, 7, 8]:

$$\rho_b(f) = f - \gamma_b(f) \tag{5}$$

Since this transform emphasizes bright objects on a non-uniform background, it is suitable for segment the oxide deposited on the surface of the samples ***AP30***.

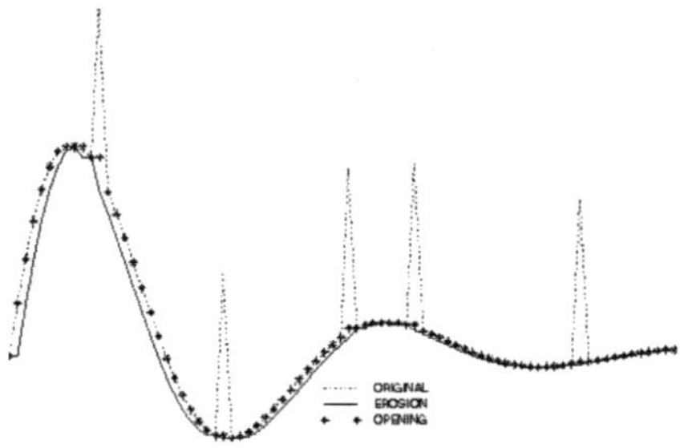

Figure 1. Top-Hat transform on a one-dimensional signal.

The proposed algorithm was implemented in Matlab, using a graphical interface. It selects and loads images, computes the opening Top-Hat transform, binarizes the image and generates a list that provides: name of the selected image, structuring element dimension, threshold used in the binarization, number of black and white pixels presented in the resulting image and the rate of black and white dots (See Fig. 2-b).

Additional buttons allow: to save the results, to modify the size of the structuring element and the threshold, and to reset the default settings. Figure 2 shows the interface of the program, which can display the original image (top left), the resulting image of Top-Hat transform (top right), the binarization of the image above (bottom left), and overlay of the result on the original image (bottom right).

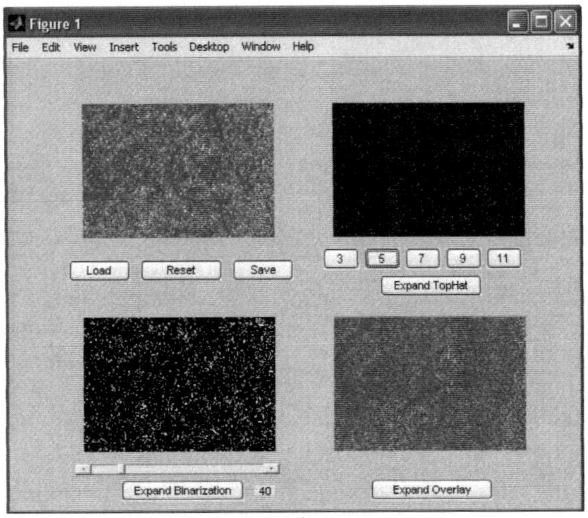

Quantification of oxides generated by anodization on titanium	
Image name:	32_M14.tif
Size of the structuring element:	7
Threshold used in the binarization:	62
Number of white pixels :	103868
Number of black pixels :	982852
Rate of black and white dots:	9,50%

(a) (b)

Figure 2. (a) Graphical user interface. (b) Parameter list.

Journal of Physics: Conference Series **332** (2011) 012027 doi:10.1088/1742-6596/332/1/012027

3. Results

In the polished substrate samples (***AP*'s**) and the ***AR20*** samples can be observed a high amount of phosphates over all the surface. This statement is demonstrated with the XRD spectra showed in Figure 4, where is very clear the presence of phosphorous compounds coming form the electrolytic solution.

From comparing the morphology of the different deposits on the titanium substrates, can be denoted that the final surface roughness highly influences the shape, quantification and Quality of the deposits. The Figure 3 (c) shows the ***AR20*** sample surface by SEM. It shows mostly a non homogeneous deposit, leaving some sites with a clear view of the rough titanium substrate below it. The ***AR30*** sample does not show the presence of cloudy-like deposits by SEM, just a rough surface (Figure 3 (d)).

The polished samples, in other side, have different and maybe better superficial characteristics than the ***AR*** treated sheets. Both the ***AP20*** and the ***AP30*** (Figure 3 (e) and (f)) samples have an homogeneous deposited layer in all the surface extension. The ***AP20*** and ***AR20*** treatment show a similar final surface.

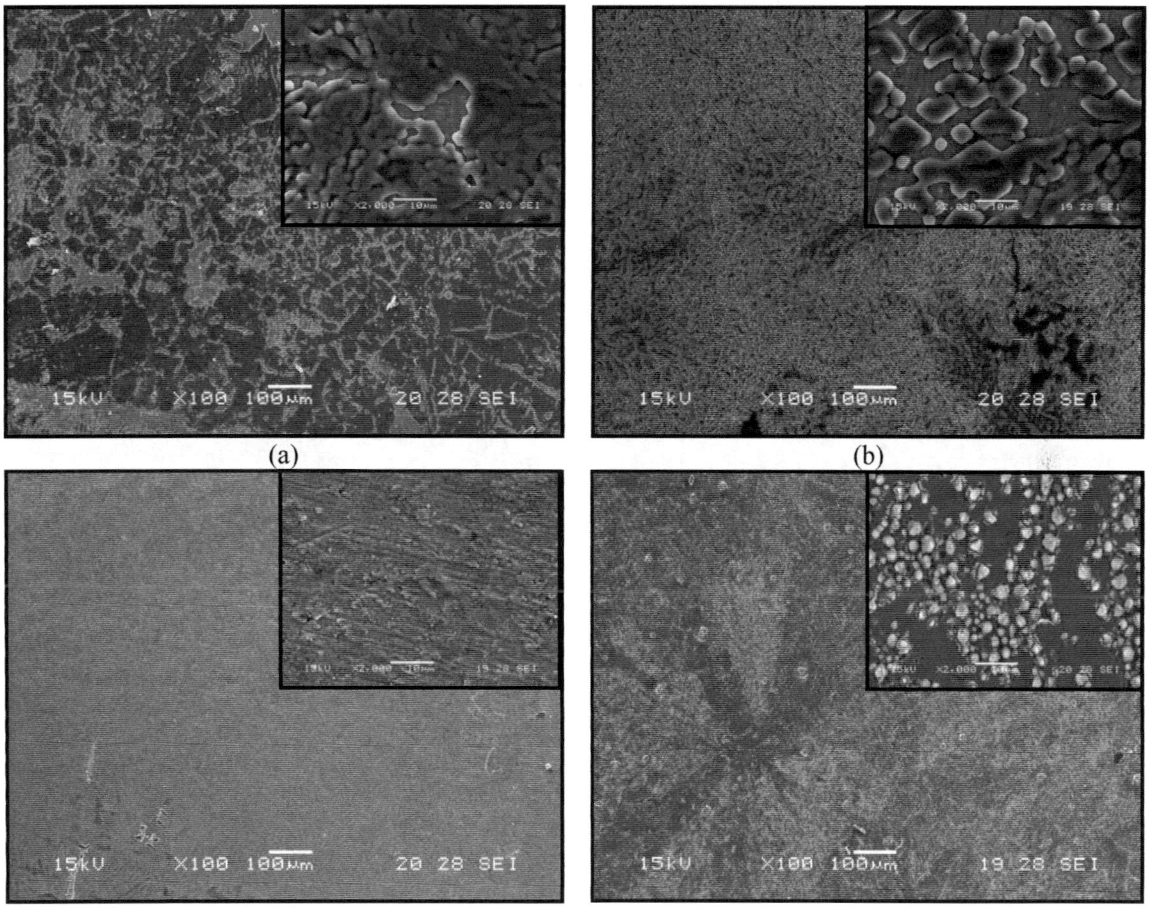

(a) (b) (c) (d)

Figure 3. SEM images obtained for the different substrate roughness and anodization conditions. (a) ***AR20***, (b) ***AP20***, (c) ***AR30***, (d) ***AP30***

The ***AP30*** sample (Figure 3 (d)) presents a distinctive characteristic from the other samples: some spherical-shape white deposits. The identification of the compound was done by X-ray diffraction (Figure 4) showing the presence of two of the characteristics peaks of anatase, a crystallographic phase of titanium oxide. The spectra of the ***AP20*** and ***AR20*** samples also showed the presence high amount of ammonium phosphate at the surface but from the ***AR30*** sample couldn´t be obtained the same results. This could be due to a very thin layer of the deposit present in the surface.

The qualitative analysis of the components of the anodic deposit could also be done by backscattered electron images (Figure 5). Back-scattered electrons (BSE) are beam electrons that are reflected from the sample by elastic scattering. BSE are often used in analytical SEM along with the spectra made from the characteristic X-rays. Because the intensity of the BSE signal is strongly related to the atomic number (Z) of the specimen, BSE images can provide information about the distribution of different elements in the sample. Since heavy elements (high atomic number) backscatter electrons more strongly than light elements (low atomic number), and thus appear brighter in the image, BSE are used to detect contrast between areas with different chemical compositions [9]. In this case the phosphorous-rich compounds are lighter so appear as dark clouds on the surfaces and the titanium substrate and the titanium oxide deposits are lighter points in the image, as it can be seen in ***AP20*** and ***AP30*** samples (Figure 5)

The presence of titanium oxide with anatase structural phase deposits is also denoted by microRaman analysis. The ***AP30*** sample apectrum shows the presence of some characteristic peaks of the anatase phase in 144-148, 400, 516 y 640 cm^{-1} [10, 11]. In Figure 6 there are also shown some Raman spectra of other zones of the sample, where also can be denoted the presence of a characteristic phosphate peak at 925 cm^{-1} [12].

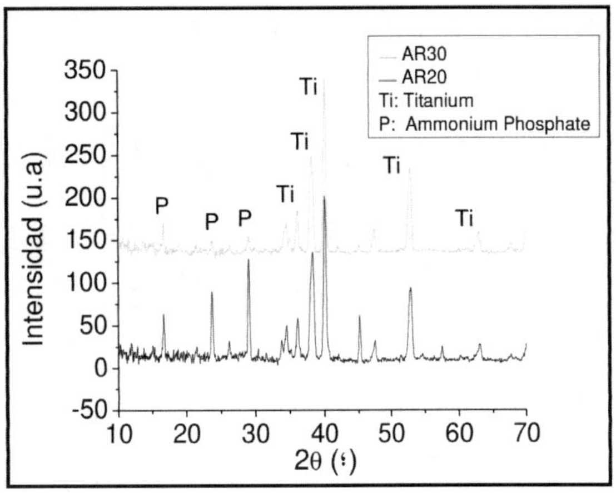

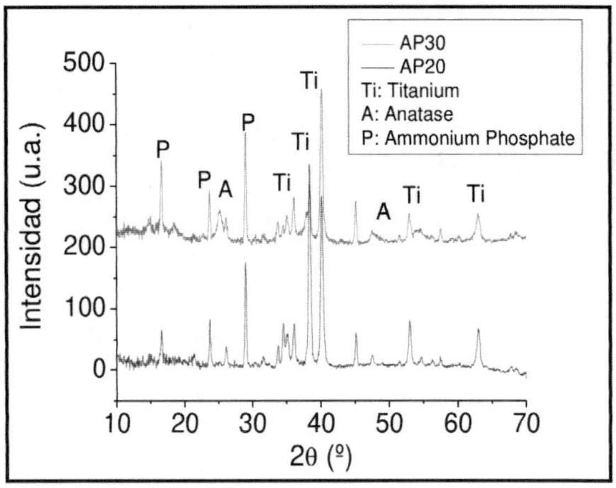

Figure 4. XRD spectra of all the analyzed samples.

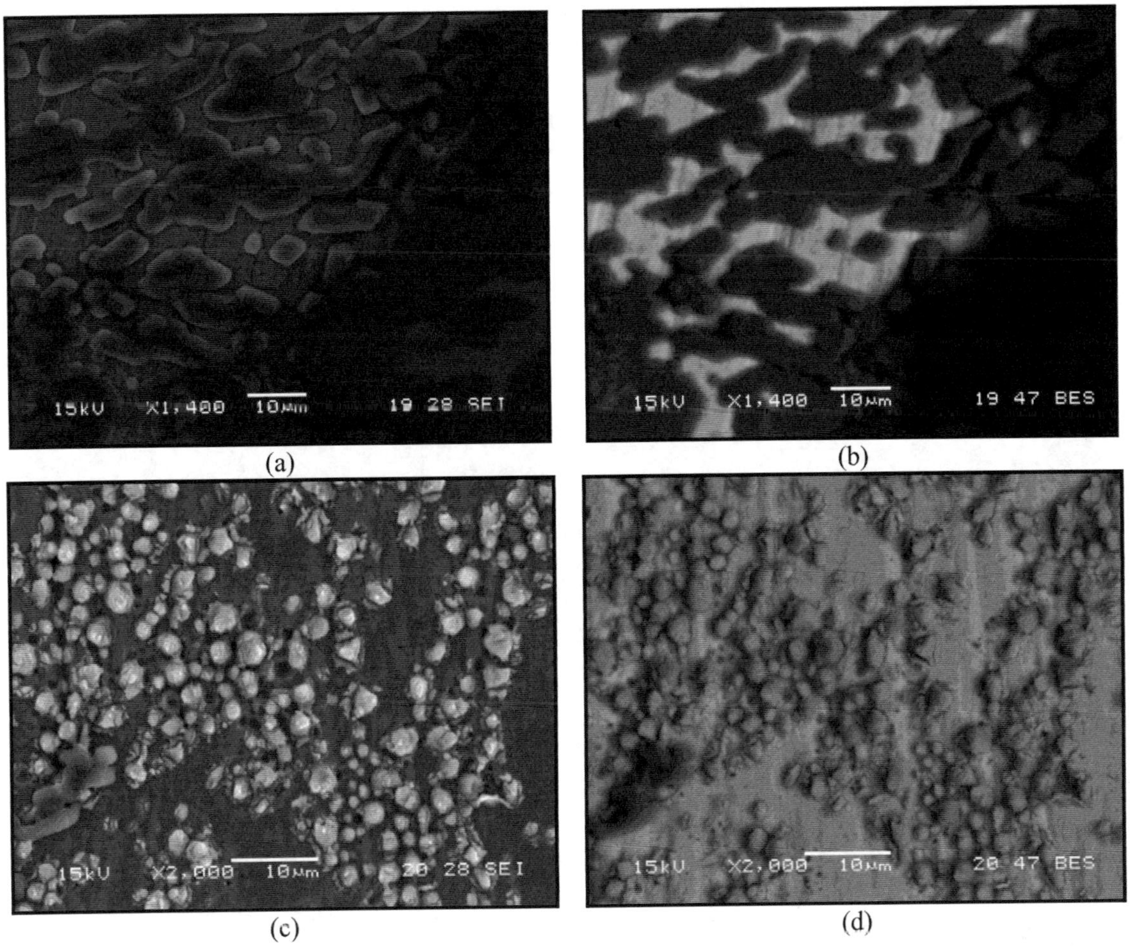

Figure 5. Backscattering images on the AP30 and AP20 samples.
(a) ***AP20***, (b) ***AP20 BSE***, (c) ***AP30***, (d) ***AP30 BSE***

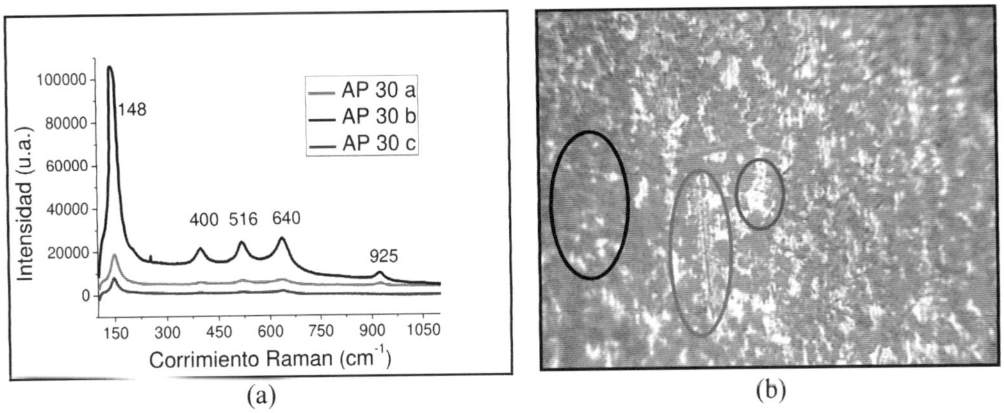

Figure 6. Raman spectra of the ***AP30*** sample showing different areas (a) and the associated optical image (b).

With the aim of quantify the amount of spherical deposits on the samples, the SEM images were processed by a costume-made Mathlab program.

Two sample magnifications were used for the mathematical image analysis. Figures 7 and 8 show the analyzed images of the ***AP30*** sample at 100 and 500X of magnification of SEM microscopy (a), the binary images (b) and both images superposition. The results give that the percentage of oxides at low magnification was 24% and at high, 42%. The calculated proportion shows a high variation since the Figure 8 was taken in a denser particle zone of the ***AP30*** sample. The Figure 7 can be taken as the more representative one since it takes a larger zone of the sample. Even though the mathematical image analysis is a powerful tool that can be applied to different images with similar contrast and color, it has to be taken into account that in this case the real oxides proportion could be more since they grow under and between the phosphate layer.

(a) (b) (c)

Figure 7. (a) SEM image of the ***AP30*** sample (100X). (b) Binary image. (c) Superposed image.

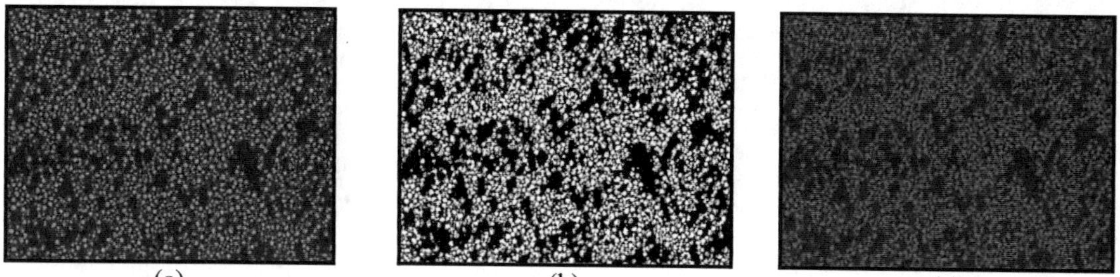

(a) (b) (c)

Figure 8. (a) SEM image of the ***AP30*** sample (500X). (b) Binary image. (c) Superposed image.

4. Conclusions

This work presented the surface modification of commercially pure titanium sheets used in orthopedics and odontology, with a potencistatic anodization process with an ammonium phosphate and ammonium fluoride as the electrolytic solution. The process allowed to generate titanium oxides doped with phosphorous on the surface, thought as a way promote bioactivity and implant fixation as the deposits can bond directly with the existing bone. The characterization and quantification of the generated deposits is presented as a staring point for the future application of these materials. The applied characterization methods (X ray diffraction, micro-Raman spectroscopy) allowed to identify phosphate and titanium oxides phases present after the anodic treatment. The PDI analysis with segmentation of SEM images technique permitted the measurement and quantification of the oxides generated by the anodization process. The samples with polished initial surface and with an anodic treatment of 30V presented a phosphate/titanium oxide layer that could enhance bioactivity and bone fixation.

5. References

1. Jacobs J J, Jeremy, M D, Gilbert, J L, and Urban, R M 1998 Current concept review - Corrosion of metal orthopaedic implants *The journal of bone & joint surgery, Incorporated* **80** 268 -82.
2. Bowditch M, and Villar, R 2001 Is titanium so bad? Medium - term outcome of cemented titanium stems *The journal of bone & joint surgery (Br)* **83 - B** 680 - 85.
3. González R, and Woods, R. 1996. Tratamiento Digital de Imágenes, Addison Wesley.
4. Facon J. 1996. Morfología Matemática. Teoría e ejemplos., Editora Universitária Champagnat da Pontífica Universidade Católica do Paraná, Curitiba, Brasil.
5. Serra J. 1982. Image analysis and mathematical morphology, Academic Press, London.
6. Serra J. 1988. Image analysis and mathematical morphology, Academic Press, London.
7. Bouchet A, Pastore, J, and Ballarin, V 2007 Segmentation of Medical Images using Fuzzy Mathematical Morphology *Journal of Computer Science and Technology* **7** 256 62.
8. Bouchet A, Brun, M, and Ballarin, V 2010 Morfología Matemática Difusa aplicada a la segmentación de angiografías retinales *Revista Argentina de Bioingeniería* **16** 7-10.
9. Goldstein G I, Newbury, D E, Echlin, P, Joy, D C, Fiori, C, and Lifshin, E 1981 Scanning electron microscopy and x-ray microanalysis *New York: Plenum Press.*
10. Sul Y-T, johansson, C B, Petronis, S, Krozer, A, Jeong, Y, Wennerberg, A, and Albrektsson, T 2002 Characteristics of the surface oxides on turned and electrochemically oxidized pure titanium implants up to dielectric breakdown: the oxide thickness, micropore configuration, surface roughness, crystal structure and chemical composition *Biomaterials* **23** 491 - 501.
11. Shibata T, and Zhu, Y-C 1995 The effect of temperature on the growth of anodic oxide film on titanium *Corrosion Science* **37** 133 - 44.
12. Ciceo Lucacel R, Maier, M, and Simon, V 2010 Structural and in vitro characterization of TiO_2-CaO-P_2O_5 bioglasses *Journal of Non-Crystalline Solids* **356** 2869–74.

Fabrication and characterization of porous PHBV scaffolds for tissue engineering

I Ruiz[1], É B Hermida[1,2] and A Baldessari[2,3]

1: Instituto de Tecnología "Prof. Jorge A. Sabato" UNSAM-CNEA Av. Gral. Paz 1499, B1650KNA San Martín, Buenos Aires, Argentina

2: Consejo Nacional de Investigaciones Científicas y Técnicas (CONICET), Av. Rivadavia 1719, C1033AAJ CABA, Argentina

3: Laboratorio de Biocatálisis. Departamento de Química Orgánica, Facultad de Ciencias Exactas y Naturales, UBA, Intendente Güiraldes 2160, Pabellón 2. Piso 3. Ciudad Universitaria, C1428EGA CABA, Argentina

E-mail: iruiz@cnea.gov.ar

Abstract. Porous scaffolds of poly(3-hydroxybutyrate-co-3-hydroxyvalerate) were elaborated by three different techniques: salt leaching (SL), emulsion solvent evaporation (ESE) and temperature induced phase separation (TIPS). For SL partially fused sieved grains of sodium chloride (106-355 μm) were used as porogen. Emulsions, prepared from a solution of PHBV in chloroform allow getting flexible films; the content of surfactant may be used to control the pore size. The pore size of the TIPS scaffolds decreased on increasing the cooling rate and the morphology of the interconnected structure could be controlled by changing the temperature gradient. Finally, chemical changes associated to the enhancement of hydrophilic behaviour of the scaffolds after alkaline and enzymatic hydrolysis as well as after sterilization by γ irradiation are presented.

1. Introduction

Tissue engineering is a fast-moving interdisciplinary pursuit that combines life sciences, medicine, materials science and bioengineering. It intends to overcome the lack of available organs and donor shortages for transplantation as well as to enhance self- regeneration of damaged tissue. Novel tissue engineering technique uses three dimensional (3D) scaffolds as structural templates for cell adhesion and subsequent tissue formation. They must be biodegradable, biocompatible, non-toxic and with high interconnected porosity to allow ingrowths of a significant number of cells, maintained in a viable state by proper diffusion of nutrients and passage of waste. Polyhydroxyalkanoates (PHAs) — polyesters produced by bacterial fermentation— satisfy those requirements and are being tested in different biomedical applications: controlled drug delivery membranes, sutures, wound dressing, bone scaffolds, artificial heart valves, etc [1,2].

PHAs scaffolds with a great surface area for cellular adhesion can be achieved by different techniques; salt leaching [3,4], thermal induced phase separation [5], electrospinning [6] and emulsion solvent evaporation [7-9] are some of them. A great deal of research was devoted to control the parameters of each of these processes in order to get the porous size and interconnected structure suitable for each cell. Furthermore, the process employed to get the scaffolds must be chosen according to the physico-chemical characteristics of the surface, since the biochemical reactions and physiological processes, that enable a proper proliferation of cells, depend on the cell-substrate interaction [10]. Finally, γ or UV irradiation, plasma treatment, chemical functionalization [11], chemical or biochemical hydrolysis [12,13] and coatings are some of the several physical, chemical and biochemical techniques that could be applied to enhance the cell-substrate interaction.

Therefore, the aim of this work is to apply three different methods to produce PHBV scaffolds: salt leaching (SL), emulsion solvent evaporation (ESE) and thermal induced phase separation (TIPS), remarking the parameters to be controlled in order to get different porous structure. Furthermore, the effect of chemical and enzymatic treatments to enhance the hydrophilic behavior of the scaffolds and the chemical changes due to the γ-ray sterilization will be discussed in detail.

2. Materials and Methods

2.1. Materials

Poly(3-hydroxybutyrate-co-3 hydroxyvalerate) (PHBV) 12% wt HV M_w= 185397 Da, M_w/M_n= 1.47, was purchased from Sigma Aldrich and was used as received. All the chemicals used in this work such as chloroform, absolute ethanol, dioxane, Tween 80, sodium chloride and sodium hydroxide were analytical grade. Lipase enzymatic preparation Stick Away®, with an activity of 15 KLU[1]/g was provided by Novozymes.

2.2. Methods

2.2.1. Salt leaching (SL)

Porous scaffolds were prepared by solvent casting/particulate leaching process combined with salt fusion technique using NaCl as particulate porogen. Sieved NaCl crystals (106-355 μm) were dispersed in a Petri dish (100 mm diameter) at 90% relative humidity for partial fusion of the salt crystals during 18 h. Then the fused salt crystals were dried in a vacuum desiccator for 24 h.

Chloroform solution of PHBV was poured over the bed of the porogen crystals and let the solvent evaporate at room temperature. The NaCl was leached out three times with aqueous solution of ethanol 10% vol. and then with distilled water. The scaffolds were dried over a filter paper and then under vacuum to eliminate water completely.

2.2.2. Emulsion solvent evaporation (ESE)

Emulsions have been used as templates to make not only scaffolds for tissue engineering but also devices for controlled drug delivery [14,15]; it is produced by a homogenized mixture of a non-polar solvent solution of the polymer and water, stabilized with a surfactant.

In this work the emulsion was generated by a homogenized mixture of a 5% w/v chloroform solution of PHBV mixed with absolute ethanol. Polysorbate 80 (or Tween 80), a synthetic non-ionic surfactant approved for paranteral use [16], was chosen as a suitable surfactant. Because of the spontaneous diffusion of the ethanol, an interfacial turbulence was promoted between the two phases by sonication, leading to the formation of smaller particles. Once sonicated, the emulsion was extended onto a PTFE plate (using a knife separated 600 μm from the plate) and left at room temperature to evaporate the solvents [17]. Then, the membranes were immersed in an aqueous ethanol 10% v/v, renewing the solution three times in 24 h in order to leach the surfactant, Finally, they were rinsed with a large amount of distilled water, dried first over a filter paper and then under vacuum to eliminate water completely.

---

[1] KLU: kilo lipase unit correspond to 1μm butyric acid produced from tributyrin per minute

2.2.3. Thermal induced phase separation (TIPS)

PHBV was dissolved in dioxane at 70 °C with agitation during 2 h to give a 6%w/v solution. 5 ml alicuots of the solution were poured into borosilicate flasks; these flasks were divided in four groups, each suddenly cooled down in air to the following quenching temperatures: 8°C (fridge), 0 °C (water-ice bath), -5°C (freezer) and -196°C (liquid nitrogen) during 30 minutes. Each group of samples was freezed-dried at -86°C during 25 hs in a Labconco Free Zone Plus lyophilizer.

2.2.4. Enzymatic and alkaline hydrolysis

The enzymatic hydrolysis was performed using a lipase (Stick Away®, Novozymes) solution in phosphate buffered saline characterized by pH=7.3 and an activity of 33 KLU/l. Salt leached scaffolds samples of 1x1 cm^2 were immersed into the enzymatic solution for 20 h at 37°C. Afterward, the samples were washed with aqueous solution of ethanol 70 %v/v and then with a large amount of distilled water. Finally they were dried over a filter paper and then under vacuum to eliminate water completely.

For the alkaline hydrolysis the samples (1x1 cm^2) were sunk in a 0.5 N NaOH aqueous solution for 45 minutes at 55°C. The samples were washed and dried as explained above.

2.2.5. Sterilization

Samples were sterilized with γ irradiation in air at 10 kGy/h, up to a dose of 25 kGy, at the semi-industrial ^{60}Co pilot plant of the Ezeiza Atomic Centre (Buenos Aires, Argentina).

2.3. Characterization

2.3.1. Scanning electron microscopy

In order to preserve the inner porous structure of the scaffolds, samples were prepared via freeze-cutting. Surfaces were coated with gold and observed under high vacuum in a FEI, Quanta 200 SEM. The pore size was measured from the micrographs using the image analyzer software ImageJ®.

2.3.2. Contact angle measurement

Static water contact angles of the modified samples were measured by an *ad-hoc* device to get digital images of 5 μl drops onto the surface. ImageJ® was used to process the images.

2.3.3. GPC analysis

Molecular weight distribution was determined by gel permeation chromatography (GPC) in a Waters system equipped with a 5μm Phenogel column. Tests were performed at 30°C and chloroform was the eluent and sample solvent. The set-up was calibrated at 1ml/min with narrow molecular weight polystyrene standards.

2.3.4. Spectroscopic analysis

Fourier transform infrared (FTIR) spectra were collected in the range of 400 to 4000 cm^{-1}. The analyses were performed over a solution casted onto a KBr window. ^1H Nuclear Magnetic Resonance (NMR) spectra were performed for solutions of the sample in deuterated chloroform.

2.3.5. Dyeing of the treated surface

The surface density of carboxyl groups can be determined by technique based on dyeing the sample with 0.01 g/ml of Toluidine Blue O (C.I. Basic Blue) at pH 10 and 30°C for 5 h [17]. After dyeing, the films were rinsed with distilled water, followed by soaking in 0.1 mM NaOH to remove adsorbed dye molecules. Finally, the associated dye molecules were desorbed in 50 %v/v acetic acid.

The dye concentration was determined at 633 nm using a spectrophotometer; then the carboxyl group density was calculated from the calibration curve.

3. Results and Discussion

3.1. Morphological characterization

3.1.1. SL scaffolds

The main advantage of this processing technique is the ease of fabrication without the need of specialized equipment. Porous substrates 80 mm in diameter were obtained. The mean size of the pores on the surface was (110±20) μm; pores observed in the transversal cuts showed the same size. The thickness of the substrates could be tailored by changing the thickness of the salt bed inside the Petri dish; it ranged from (0.50 ± 0.02) to (1.50 ± 0.02) mm. These samples could be compressed easily, however, the manual bending without breaking decreased as the thickness increases.

The surface of the scaffold, showed in figure 1, shows pores with regular shape, resembling a replica of the salt grains. On the other hand, the transverse cross-section, pictured in figure 2, reveals the interconnected pores that extend uniformly along the whole thickness of the scaffolds. Furthermore, the partial fusion of the salt grains leads to the interconnection of the inner pores through small voids, as illustrated in figure 3. Figure 4 (enlargement of figure 1) remarks the rounded and rather thick walls of the open pores (on the surface). However, inside the SL scaffolds the interconnected pores have thinner walls, as follows from figure 5.

Figure 1 SL scaffold surface; the square porous structure resembles a replica of the salt grains.

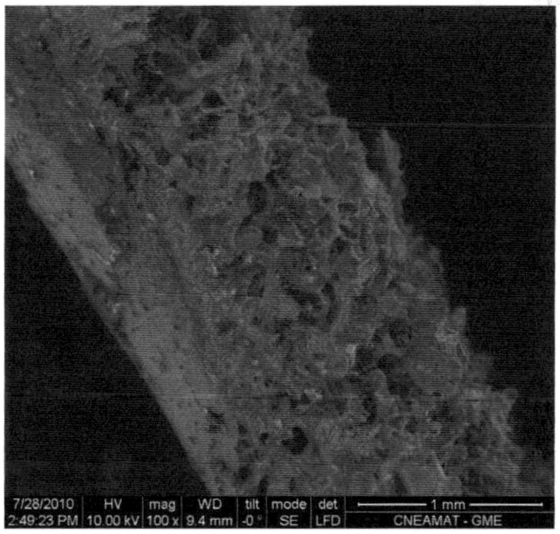

Figure 2 Interconnected pores observed in a cross-section of a SL scaffold.

Journal of Physics: Conference Series **332** (2011) 012028 doi:10.1088/1742-6596/332/1/012028

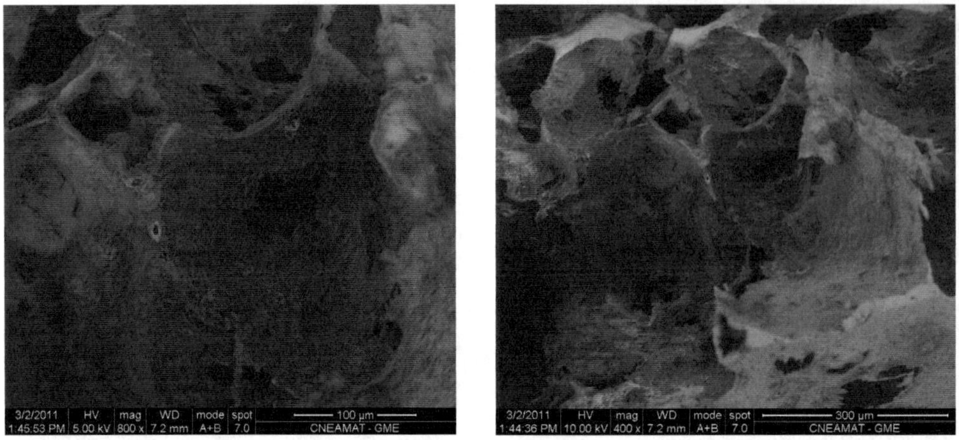

Figure 3 Voids in the walls inside a SL scaffold enhance pore interconnection.

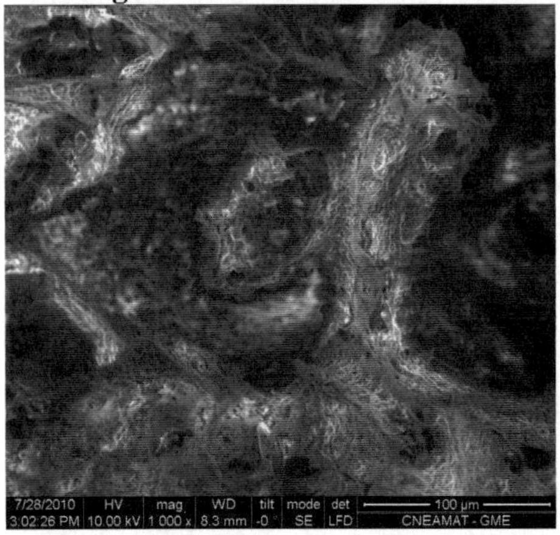

Figure 4 Wide walls of pores at the surface of a SL scaffold (enlargement of figure 1).

Figure 5 Interconnected pores inside a SL scaffold; they can be compared with the superficial pores of figure 1.

3.1.2. ESE scaffolds

It is well known that emulsion solvent evaporation is useful for microparticle preparation, where changes in the emulsion stirring speed and the polymer or surfactant concentration may enable to control the particle size [16]. Even when in this work the stirring speed and the polymer concentration were kept constant, changes in the surfactant concentration led to different porous structures as shown in figure 6.

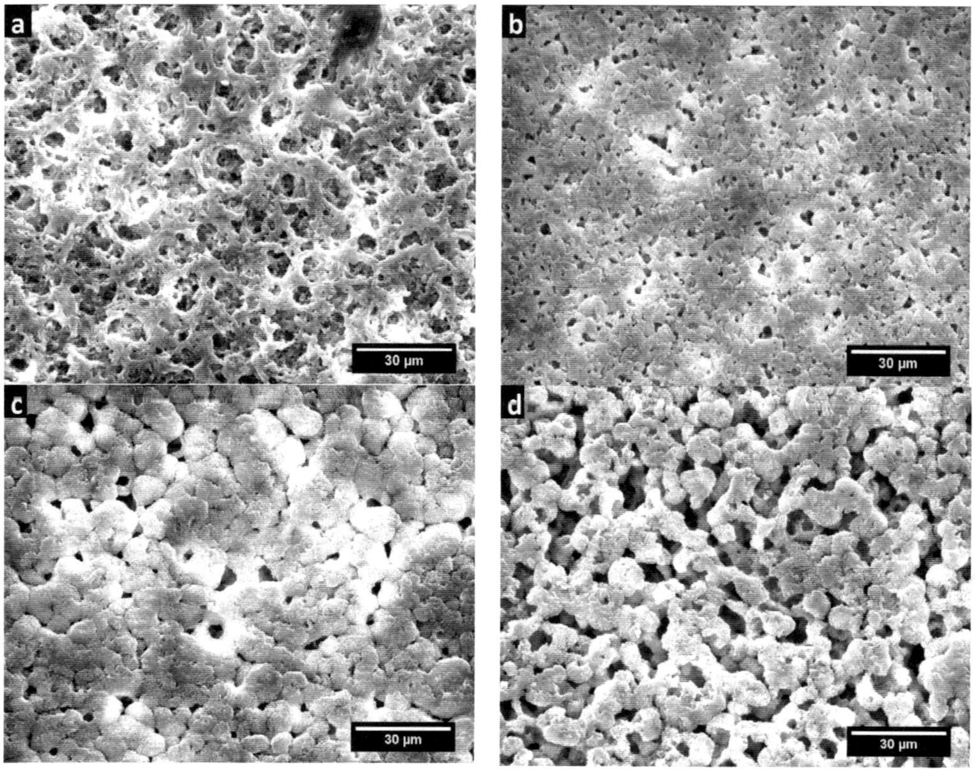

Figure 6 Morphology of ESE scaffolds: 0 (a), 2.8 (b), 4.0 (c) and 7.5 (d) %v/v of surfactant.

The first interesting feature is that even a small amount of surfactant (2.8%) stabilizes the emulsion, leading to the formation of microparticles, which size increases with the volume fraction of the surfactant, as observed in figures 6 b to d. The spreading of the emulsion onto the PTFE surface distorts the shape of the microspheres and sticks them in a fairly closed packed structure. Consequently, the pores formed between the close packed structure follows a fairly linear increase with the volume fraction of surfactant, as shown in table 1. However, the emulsion without surfactant has an irregular structure as shown in figure 6a with bigger pores.

Table 1 Dependence of the pore size of ESE scaffolds on the surfactant content.

Surfactant (%v/v)	**Pore size (µm)**
0	7 ±2
2.8	3 ± 1
4.0	4 ± 1
7.5	7 ±2

3.1.3. TIPS scaffolds

Porous cylindrical samples ca. 10 mm in diameter and 15 mm in height were obtained. Different orientation cuts were made in order to determine the pore sizes and orientations. The morphology of the samples were analysed by SEM and pore size were measured from the micrographs using ImageJ®. Results, summarized in table 2, showed that the pore size tend to decrease as the quenching temperature decreases.

Table 2 Pore size of TIPS scaffolds

Quenching temperature (°C)	Pore size (µm)		
	Transversal cut	**Longitudinal cut**	**On the surface**
8	52 ± 8	50 ± 17	44 ± 9
0	47 ± 13	58 ± 20	17 ± 3
-5	41 ± 13	46 ± 20	19 ± 4
-196	9 ± 2	8 ± 3	4 ± 1

At each quenching temperature the pore structure was homogeneous in size and no preferred orientation was observed, as illustrated in figure 7.

In figure 7a, the structure obtained at 8°C has a thin wall and a highly fractured morphology. On the other side the sample quenched in a water-ice bath (0°C) shows a similar morphology with thin walls (figure 7b) and the same magnitude in the pore size. The spongy structure obtained by freezer quenching has small pore up to 50 µm and open spaces up to 150 µm, as it can be seen in figure 7c. Detailed micrographs of the sample walls show a rounded platelet structure that forms the smaller pores. The sample quenched in liquid nitrogen (figure 7d) allowed getting the smaller pore sizes. The dendritic structure of pores oriented toward the center of the sample, could be attributed to the fast cooling rate.

3.2. Hydrophilic enhancement and sterilization

The hydrophilic behavior of the surfaces was quantified by measuring the contact angle of a sessile drop of distilled water. The first row of table 3 points out that both the porous scaffolds fabricated by SL as those produced by solvent casting were hydrophobic before any surface treatment.

Both enzymatic and alkaline hydrolysis were effective to enhance the hydrophilic behavior of the scaffolds. The effect is more noticeable in the porous surface because it has a higher effective area for hydrolysis. However, neither FTIR nor ^{1}H NMR spectral analyses were suitable to determine any change in the chemical structure of PHBV due to the alkaline or enzymatic hydrolysis.

Table 3 Contact angle of a sessile drop of water on the original and treated scaffolds.

	Porous Scaffolds	Solvent casting
Original	103±5°	103±2°
Alkaline Hydrolysis	0°	13±4 °
Enzymatic Hydrolysis	0°	58±2°

Neither GPC showed major differences between the molecular weight of the untreated SL scaffolds nor the enzymatic hydrolyzed substrates (table 4). However, the GPC analysis revealed a decrease in the molecular weight after the alkaline hydrolysis and even a greater reduction in the after the gamma ray sterilization, as summarized in table 4.

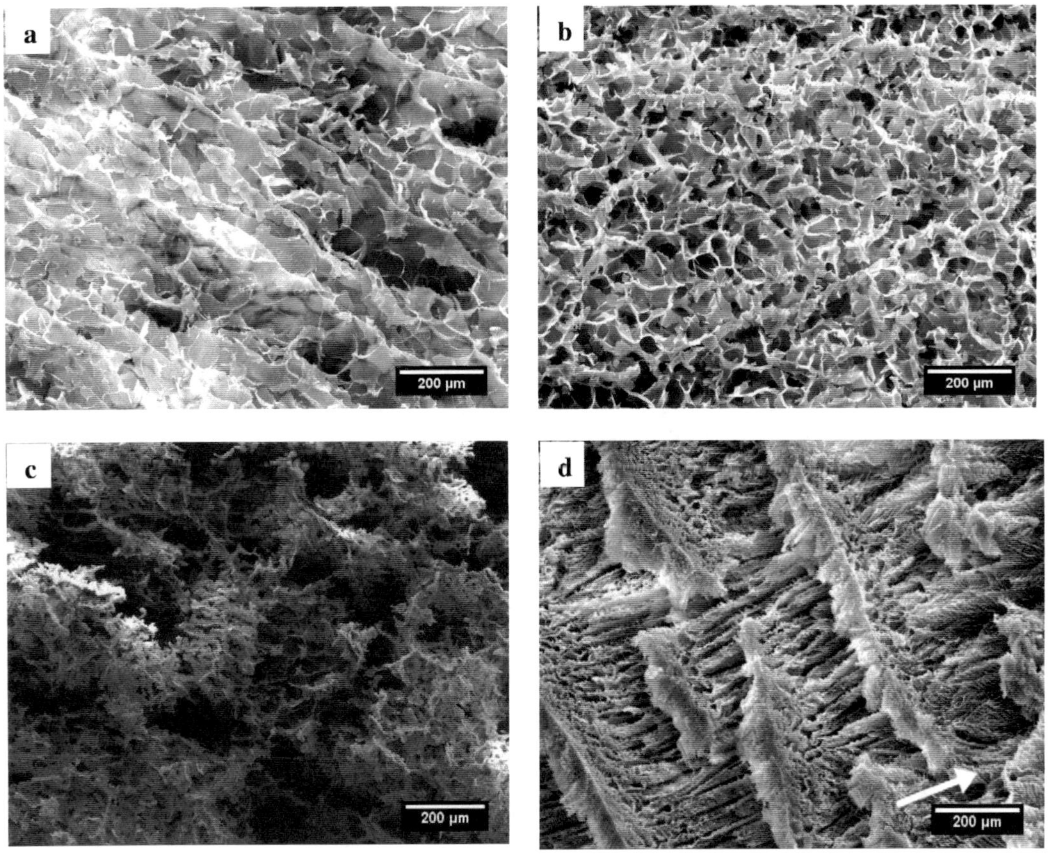

Figure 7 SEM images of scaffolds produced by TIPS under different quenching temperatures: a) 8°C, b) 0°C, c) -5°C and d)-196°C; white arrow indicates the cooling direction towards the centre of the sample.

Table 4 Changes in the molecular weight distribution of a PHBV SL scaffold due to hydrophilic enhancement and γ irradiation (25 kGy @ 10 kGy/h)

	Mn (Da)	Mw (Da)	Mw/Mn
Original	130709	187990	1.44
Alkaline Hydrolysis	94794	168176	1.76
Enzymatic Hydrolysis	130953	187228	1.43
Original + γ	69439	116964	1.68
Alkaline Hydrolysis + γ	74427	117085	1.57
Enzymatic Hydrolysis + γ	67877	111849	1.64

Finally, in order to establish whether the alkaline hydrolysis produces carboxyl groups mainly on the surface of the scaffold, they were dyed with Toluidine Blue O; this technique allows quantifying carboxyl groups in a certain surface area. Since it could have been very difficult to determine the effective area of a porous scaffold, flat scaffolds were prepared by solvent casting (PHBV dissolved in chloroform, poured onto a Petri dish followed by evaporation of the solvent under fume hood and finally under vacuum). The dyeing procedure was applied to cast films subjected to alkaline hydrolysis during different periods of time. The results, summarized in figure 8, showed an increment of carboxyl surface density with the hydrolysis time. These results demonstrated that the hydrophilic enhancement is due to hydrolysis just on the polymer surface.

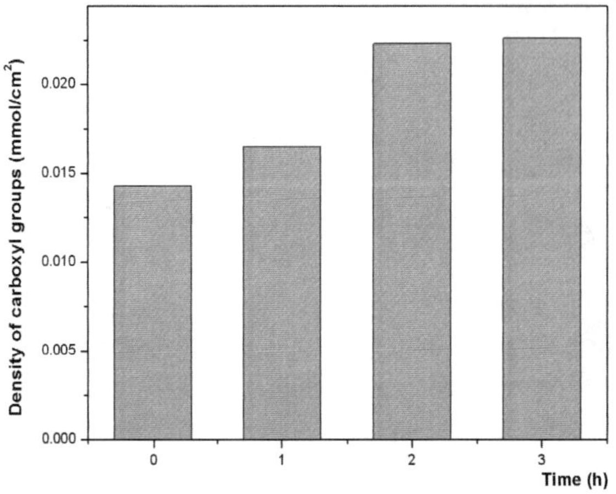

Figure 8 Alkaline hydrolysis: carboxyl group density vs. hydrolysis time.

4. Conclusions

The three methods presented in this work are suitable to produce 3D porous scaffolds. Nevertheless each has advantages and drawbacks, to be considered in order to get the proper morphology according to the cells to be harvested.

Salt leaching is a simple technique with little equipment requirements, where the pore size is easily controlled by the grain size of the porogen. However, this technique is not adequate for large scale scaffold production because of the great number of steps. Furthermore, SL substrates more than 1 mm thick are rather brittle so they can be applied only onto flat surfaces.

Emulsion solvent evaporation is more suitable for applications in which flexibility is a requirement, for instance, to adapt the scaffold to anatomic curvatures. Tween 80 proved to be a proper surfactant to stabilize the emulsions. Furthermore, a linear correlation between the surfactant content and the pore size allows producing an interconnected network of pores according to the cells to be harvested.

Thermal induced phase separation is a versatile technique to get different pore sizes and change its orientation by controlling the quenching temperature and the thermal gradient. The ability to tailor the scaffold morphology, make TIPS suitable for different tissue engineering application with wide different pore sizes and structural requirements.

Both the enzymatic and alkaline treatments lead to the hydrophilic enhancement of the surface. This is mainly due to the increase in the surface density of carboxylic groups, which could be quantified by the Toluidine Blue O dyeing. GPC, FTIR or RMN are not suitable techniques to evaluate the hydrolysis because it only affects the molecules on the surface while the spectral analysis considers changes in the bulk. It is expected that a highly hydrolyzed surface would enhance the biodegradation rate; in vitro biodegradation experiments are being run to evaluate this statement.

Finally, it was shown that alkaline treatment and sterilization by gamma irradiation partially reduces the molecular weight.

Acknowledgements

This work has been partially supported by the National Agency for the Promotion of Science and Technology (Project: PICT2006 Start up 00935).

References

[1] Chen G Q and Wu Q 2005 The application of polyhydroxyalkanoates as tissue engineering materials *Biomat.* **26** 6565-78

[2] R. Rai et al. 2011 Medium chain length polyhydroxyalkanoates, promising new biomaterials, *Mat. Science and Eng.* R **72** 29–47

[3] Murphy W L, Dennos R G and Kileny J L 2002 Salt Fussion: An Approach to Improve Pore Interconnectivity within Tissue Engineering Scaffolds, *Tissue Eng.* **8**, 43-52

[4] Kai Z, Ying D, Jin Chun C and Guo-Qiang C 2003 Polyhydroxyalkanoate (PHA) scaffolds with good mechanical properties and biocompatibility *Biomat.* **24** 1041-45

[5] Jack K S *et al.* 2009 The fabrication and characterization of biodegradable HA/PHBV nanoparticule polymer composite scaffolds *Acta Biomat.* **5** 2657-67

[6] Ying T H *et al.* 2008 Scaffolds from electrospun of polyhydroxyalkanoate copolymers: Fabrication, characterization, bioabsorption and tissue response *Biomat.* **29** 1307-17

[7] Zhijiang C et al. 2004 Layered-lamination Preparing of Tissue Engineering Scaffolds via Emulsion Templates Method *J. Polym. Res.* **11** 225–30

[8] Whang K, Thomas C H , Healy K E and Nuber G 1995 A novel method to fabricate bioabsorbable scaffolds, *Polym.* **36** (4) 837-842

[9] Yan J, Xiao-Tao L, Guo-Qiang C, 2008 Interactions between a poly(3-hydroxybutyrate-co-3-hydroxyvalerate-co-3-hydroxyhexanoate) terpolyester and human keratinocytes *Biomat.* **29** 3807-14

[10] Reis R L 2008 Natural-Based polymers for biomedical applications, ed Neves N M, Mano J F, Gomes M E, Marques A P and Azevedo H S (Cambridge: Woodhead Publishing Limited) pp 165-92

[11] Lao H-K, Renard E, Fagui A E, Langlois V, Vallée-Rehel K and Linossier I 2011 Functionalization of poly(3-hydroxybutyrate-*co*-3-hydroxyvalerate) films via surface-initiated atom transfer radical polymerization: Comparison with the conventional free-radical grafting procedure *J. App. Polym. Sci.* **120** 184–94

[12] Kai Z, Ying D and Guo-Qiang C 2003 Effects of surface morphology on the biocompatibility of polyhydroxyalkanoates, *Biochem. Eng. J.* **16** 115-23

[13] Ya-Wu W, Qiong W and Guo-Qiang C 2003 Reduced mouse fibroblast cell growth by increased hydrophilicity of microbial polyhydroxyalkanoates via hyaluronan coating, *Biomat* **24** 4621-29

[14] Ito Y, Ochii Y, Fukushima K, Sugioka N, Takada K 2010 Three-layered microcapsules as a long-term sustained release injection preparation, *Int. J. of Pharm* **384** 53-9

[15] Nilkumhang S and Basit AW 2009 The robustness and flexibility of an emulsion solvent evaporation method to prepare pH-responsive microparticles *Int. J. of Pharm.* **377** 135–41

[16] Aulton M E 2002 Pharmaceutics. The science of dosage from design. (Edinburgh: Churchill Livingstone)

[17] Hu S-G Jou C-H and Yang M-C 2004 Biocompatibility and antibacterial activity of chitosan and collagen immobilized poly(3-hydroxybutyric acid-co-3-hydroxyvaleric acid) *Carbohyd. Polym.* **58** 173-79

Construction of a 3D coronary map to assess geometrical information *in-vivo* from coronary patients

M E Casciaro[1], D Craiem[1,2], S Graf[1,2], E P Gurfinkel[2,3], R L Armentano[1]
[1]Favaloro University, FICEN, Av Belgrano 1723, C.P. C1093AAS, Ciudad de Buenos Aires, Argentina
[2]CONICET, Argentina
[3]MSCT Unit of the Favaloro Foundation, Argentina

E-mail: mecasciaro@gmail.com

Abstract. Traditional risk factors are involved in the development of coronary artery disease (CAD), but geometrical risk factors have also proved to be determinant. In this work we present a new method to construct a 3D map of the left coronary artery tree using CT image processing and skeletonization techniques. We computed cumulative length-volume functions through the coronary tree bifurcations and also the relationships between total tree volumes, total length, number of segments and bifurcations with respect to the presence and severity of atherosclerotic plaques. A total of 65 patients, 40 with and 20 without CAD were recruited. We found more vascular segments and bifurcations per patient in the CAD group. Accordingly, total cumulative length was longer in CAD patients ($p<0.01$) whereas total cumulative volume was similar between groups. Cumulative length-volume (L-V) relationships conformed to an allometric function $L= K_V.V^{\beta}$. The allometric slope $\beta \approx 0.75$ did not change with disease, whereas the allometric constant K_V was lower in the healthy group ($p<0.05$) consistent with the literature. These results suggest that CAD patients follow a compensatory mechanism, dilating the vessels to maintain a normal coronary flow. The 3D coronary map offers useful quantitative information of the coronary morphometry.

1. Introduction

Traditional risk factors (age, gender, smoking, sedentary habits, etc) are involved in the development of coronary artery disease (CAD). Recently, geometrical risk factors proved to be determinant to explaining the lesions location and inherent risk of atherosclerotic plaques [1]-[4]. It is well known that bifurcations and curvatures condition blood flow and might create turbulences or low shear stress regions that might promote plaque proliferation [2]-[4]. At the same time, coronary geometry reconstruction has been improved by new imaging techniques such as planar reconstruction using angiography (AngioCT) and intravascular ultrasound (IVUS). Multislice computed tomography (MSCT) is being also widely employed [5]-[9]. In some of those works, allometric functions relating arterial diameter, length, volume and bifurcation angles were proposed to detect pathological patterns in the coronary tree [10], [11]. The logic behind this idea is that some geometric rules that determine the coronary architecture to ensure a reliable blood transport to the myocardium might have been reached through evolution of blood transport systems. The best known example is Murray´s Law describing the optimal relationship between parent and daughter vessel diameters and bifurcation angles to minimize metabolic and functional cost in the organism [12], [13]. This Law, local in nature, was generalized to determine the relationship between cumulative lengths and volumes of the coronary tree, resulting in an allometric potential relationship to predict coronary geometry, relying on the fractal nature of the circulatory tree [15]. To evaluate this potential function the coronary tree must be divided in a stem-crown way [19] to define cumulative magnitudes (figure 1). Stems are defined as the vessel segments between two bifurcation points (nodes) or a node and a distal point of the coronary tree. Crowns are defined as the group of vessels distal to a certain stem.

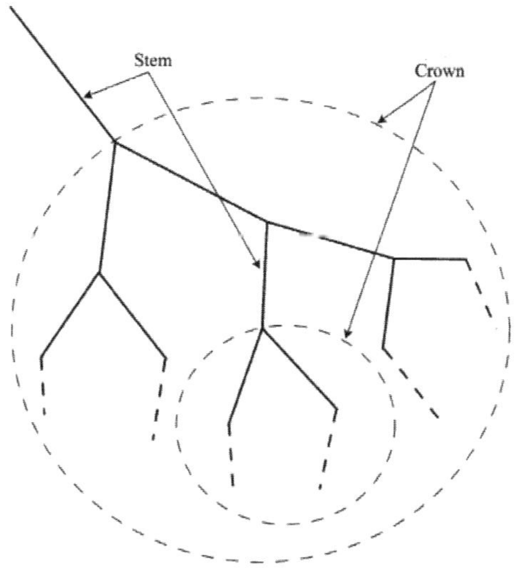

Figure 1. Division of a tree segment in a stem-crown way. Stems are the vessels between two successive bifurcation points, or between a bifurcation point and the end of the tree. Crowns are defined as the group of vessels distal to a certain stem (image reproduced from [19]).

Allometric functions were verified in the past on porcine heart casts and human autopsies [5]. Precise information of the porcine coronary tree in vivo was also obtained through MSCT [11]. Recently we published a work where this same technology was used to verify the same allometric relationships for human patients *in vivo* [16] verifying that the left coronary tree morphology adjusts to the potential function:

$$L = K_V V^{\beta} \tag{1}$$

where V [mm3] and L [mm] stand for cummulative length and volume in each node, respectively. Allometric constants K_V and β determine the geometry of the coronary tree [19] and can be obtained for each patient though least squares fitting of equation (1) in log scale.

In this work we present the methods to construct a 3D map of the left coronary artery tree using tomographic reconstructions of patients with and without CAD. We use this map to obtain morphometrical parameters of the left coronary circulation: vessel lengths and volumes, total amount of reconstructed segments and bifurcations, among others. This quantitative anatomical information is used to verify allometric length-volume functions and to study the relationships between total tree volumes, total length, number of segments and bifurcations with the presence and severity of CAD.

2. Methods

2.1. Population

Studies used in this work belong to consecutive patients of the MSCT Department of the Favaloro Foundation arriving in one month of 2011. Selected subjects were referred due to suspected coronary artery disease. Patients with previous coronary surgeries were excluded. Signed inform consents were obtained from each patient according to the good clinical practices of the Institution. Patients were separated in 2 groups according to the presence or absence of CAD.

2.2. Computed tomography scanning

All scans were performed using a 64-row MSCT scanner (Aquilion 64, Toshiba, Japan). Details of the scanning process can be found in a recent publication of the group [14]. Briefly, each subject received a bolus of ionated contrast material injected through an arm vein, followed by 50 mL of saline infusion at the same injection speed. Heart rate was controlled to remain below 60 bpm. The scan parameters were: a collimator detector of 64x0.5 mm, a rotation time of 0.4 s, tube voltage 120-135 KV and mAs of 320-440 depending on patient weight. Images were acquired during diastole using ECG-gating in order to reduce motion facts. Transaxial images were reconstructed with 0.5 mm slice thickness and 0.3 mm increments. An experienced technician measured Calcium Score Coefficient (CAC) using Agatston method [17] in studies where calcified plaques were found. For each CAD patient, the amount, type, degree of stenosis and localization of every plaque was registered.

2.3. 3D reconstruction

All digital imaging processes were done using a custom software designed in the Favaloro University. DICOM sequences of 350 to 400 images (512 by 512 pixels) were imported from each scan. First, the user selects a volume of interest (VOI) where the whole heart remains inside. Next, the user initiates an automatic process to create a coronary map. This process can be separated in 3 stages: 1) Heart isolation, 2) artery segmentation and 3) skeletonization for the estimation of the length-volume relation (L-V).

2.3.1. Heart Isolation

Heart isolation begins with a k-means clustering process inside the VOI [18] where 3 groups are discriminated: lungs, myocardial mass and blood pools (including coronary arteries, ventricles and atriums). The voxel representing the center of gravity of the whole volume is determined and is considered as the initial seed point for a slice-by-slice region growing segmentation process:

a) Region growing from i-seed point in the ith-slice, grouping all pixels belonging to myocardial mass or blood pool, and excluding all points belonging to lungs.
b) From the region segmented in a), a new center of gravity is calculated. This point is the i+1 seeding point for the (ith+1)-slice segmentation process.
c) Steps a) and b) are subsequently repeated until all slices in the VOI are segmented. This requires the procedure to be split in two parts: the first one is the segmentation of all slices below the initial seed point i=0, and the second one the segmentation of all the slices below the initial seed point.
d) All regions segmented in a) for each slice are merged as Heart-Region (HR). Every voxel not belonging to HR is considered as Not Heart Region (NHR), and its value is set to the minimum pixel value of the whole volume in Hounsfield Units (HU). The result is a fast partial isolation of the heart (processing time is below 10 seconds) where all lung tissue and lung vessels are completely removed (figure 2).

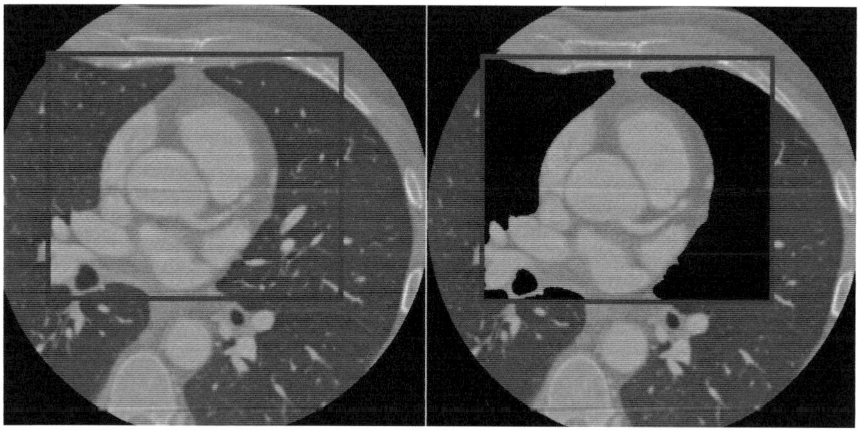

Figure 2. Representative case of the heart isolation process in an arbitrary slice. Note that all lung tissue and pulmonary vessels were automatically removed.

2.3.2. Coronary artery segmentation
The procedure to segment coronary arteries inside the heart-isolated VOI is an adaptation for TC from other reports [19], [20] and is described hereafter:

- Three-dimensional median filtering using a 5x5x5 mask to suppress noise and improve tissue extraction (figure 3a).
- Three-dimensional morphological opening filtering, where a spherical structuring element with a radius slightly greater than the radius of the biggest coronary segment (usually belonging to the left main trunk) is used. This radius is manually selected by the user, and is often between 6 and 8 pixels. The result of this step is a volume in which every spherical/cylindrical structure with a lesser radius than that of the structuring element is removed. Consequently, large blood pools and myocardial mass are enhanced (figura 3b).
- The volume obtained in the previous step is subtracted from the heart-isolated volume. Thus, all structures removed by the morphological filter are enhanced, including coronary arteries, coronary veins and cone-shaped blood pools near to papillary muscles of the ventricles (figure 3c).
- Contrast is doubled in order to enlarge the difference between coronary veins and arteries in HU.
- A windowing process follows in order to visualize coronary arteries and to neglect coronary veins. Threshold values were typically C:-700HU and W: 700HU (C=Center and W=Width). This empirically selected window performs a double function: in one hand it neglects all the voxels corresponding to coronary veins (lower HU values), and in the other hand it neglects all calcified plaques around coronary vessels (higher HU values).
- To select left coronary arteries and neglect all spurious structures inside the VOI that may have survived the windowing process, the user is required to manually select a voxel belonging to the left coronary tree. This point is the initial point for a volumetric region growing segmentation, where each voxel 26-connected to the seed-point and with HU values between the interval [C-W; C+W] is iteratively merged (figure 3d).
- Manual correction of the result is performed using a custom 3D erasing tool, where traces of coronary veins, ventricle blood pools regions near papillary muscles or spurious noise connected to coronary arteries are removed.

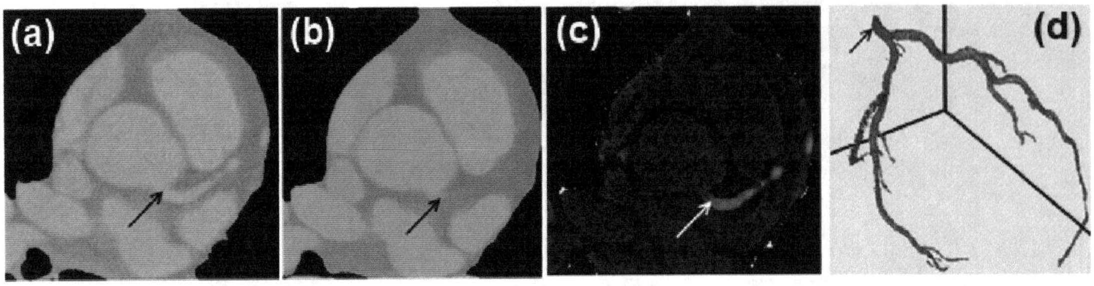

Figure 3. Left coronary artery segmentation: (a) Representative case of the heart isolation process in an arbitrary slice. (b) Morphological opening filter. (c) Images a) and b) are subtracted for vessel enhancement. (d) Three-dimensional result after volumetric region-growing segmentation. In every case, the arrow points to the seed-point selected by the user over the main trunk of the left coronary artery.

2.3.3. Skeletonization and length-volume Measurements

Medial axis skeleton of the segmented coronary artery tree is obtained by a simple-point identification approach using a template matching 12-Steps subiteration algorithm [21]. Skeleton clustered regions are eliminated and approximated by straight lines connecting exit-points to the center of mass of the clustered regions (generally belonging to stenotic portions of arteries with plaque deposition) [22] (figure 4). The coronary skeleton obtained is manually pruned and labeled, where all spurious branches not representing real bifurcations are deleted. All real branches resulting from pruning are smoothed with B-Splines for branch length calculations (L). Each voxel of the coronary tree is associated to the nearest branch by a least squares distance approach (figure 5: left). Branch volume (V) is calculated as the amount of voxels belonging to the branch times the voxel volume, typically $5.4x10^2$ mm^3. Every bifurcation point is labeled as a Node. Data from each stem and crown is assigned to its corresponding node and a coronary map is constructed (figure 5, right). This map is used to calculate L-V measurements [11], [19], [20].

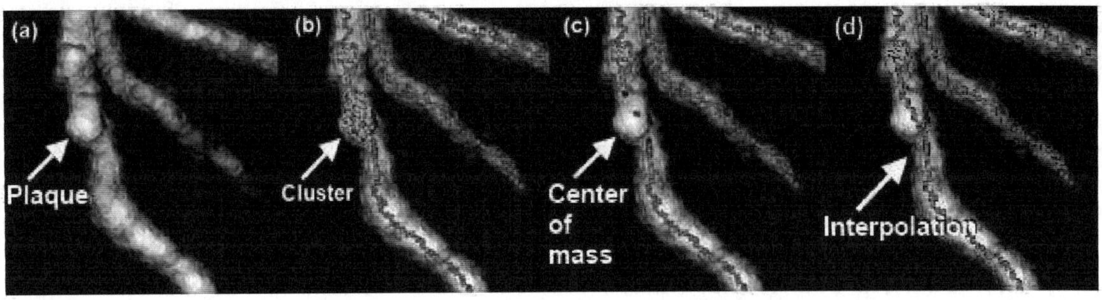

Figure 4. Coronary skeleton cluster elimination. a) 3D reconstruction of the coronary tree. b) Skeletonization of (a) where clusters appear around atheromatous plaque formations. c) Cluster are eliminated and replaced by the center of mass. d) Linear interpolation of clusters exit points and center of mass.

2.4. Statistical analysis

Values were expressed as mean±SD. Cumulative Log(Length)-Log(volume) correlations were plotted for each patient and the allometric constants, β and K_V, were calculated from linear regression models (equation 1). Comparisons between groups were made using a t-test with significant levels when $p<0.05$.

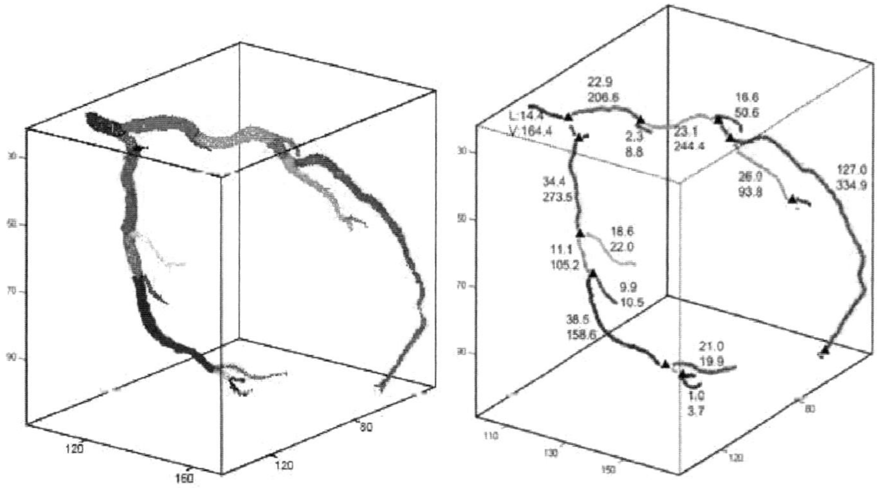

Figure 5. (a) *Left:* Voxel-to-Branch association through minimum square distance. (b) *Right:* 3D coronary map. Triangles represent bifurcation point. V and L represent the vessel lumen volume [mm3] and length [mm] estimated with the proposed algorithm.

3. Results

The total number of subjects was 65, divided into 2 groups. As expected in this kind of study, we found more atherosclerotic than healthy patients. Group 1 (healthy): subjects without plaques (n=25, age 54±8 y.o., male gender 48%). Group 2 (CAD): subjects with at least one plaque in the left coronary tree (n=40, age 59±10 y.o., male gender 57%). We found an average of 4±2 plaques in the second group with a median value of coronary calcium score (CAC) of 233 [quartiles 63-690]. Plaques were 85% calcified. A high correlation was found between Log(CAC+1) and the number of plaques as shown in figure 6.

The morphometric results are presented in Table 1. We detected more vascular segments and bifurcations in CAD group. Accordingly, total cumulative length was longer in CAD patients ($p<0.01$) whereas total cumulative volume was similar between groups. The allometric slopes did not change with disease, whereas the allometric constant Kv was lower in the healthy group ($p<0.05$).

Table 1. Morphometric results.

	1. Healthy subjects (n=25)	2. CAD subjects (n=40)
Number of segments	16±6	22±8††
Number of bifurcations	7±3	10±4††
Total cumulative length, cm	38.5±11.8	47.3±10.8††
Total cumulative volume, ml	2.4±1.1	2.7±1.0
Allometric slope β	0.77±0.08	0.74±0.07
Allometric constant K_V	1.11±0.53	1.48±0.74†

Values are expressed as mean±SD. †$p<0.05$, ††$p<0.01$. CAD = coronary artery disease.

4. Discussion

In this work we present an original method to reconstruct in 3D the left coronary tree of healthy and CAD patients using MSCT images *in vivo*. Previous works, where the coronary tree geometry was assessed with tomography to study allometric relations, were reported in autopsies from pigs [23] and humans [5,24]. We found also in vivo studies but only in pigs [11]. When human patients were involved, invasive methods were employed [6], [7]. As far as we know, this is the first time human patients are tested in vivo with this kind of methodology.

We divided each coronary segment in the left tree and estimated its length and volume. Our main findings were that CAD subjects had a similar cumulative volume and a longer cumulative length than healthy patients. In other words, the presence of atheromatous plaques did not change the volume of the coronary arteries but increased the detected length. Accordingly, the numbers of segments (and bifurcations) were significantly more in the CAD group. We can explain these results with 2 hypotheses: i) Volume did not change due to a compensatory mechanism that self-regulates blood flow. Accordingly, coronary flow reserve is defined as the ability to reduce resistance in the coronary vascular bed to recover a normal blood circulation. When a lesion obstructs a branch, a distal dilating effect is expected to increases the branch perfusion within the coronary flow reserve limits. This dilation might explain the similar cumulative volumes found in both group. ii) The number of detected segments increased in CAD subjects because these dilated vessels became visible to the tomography study. When distal arteries dilate, they enhance blood circulation and might turn visible to the CT detectors. Usually, CT images can capture only arteries with a diameter >1mm. In fact, this first limitation has to be taken into account when this in vivo study is compared with other reports [5,23,24]. We cannot disregard the idea that CAD patients may also promote collateral circulation and the formation of new vessels to recover a proper coronary blood flow. These two mechanisms could also contribute to increase the number of detected segments. Further studies with more patients and detailed hemodynamic information should be addressed to corroborate these hypotheses.

Recently, our group analyzed the allometric relations between cumulative length and volume in the coronary arteries [16]. The idea behind this work was to identify if the coronary geometrical configuration of certain patients could induce the formation of plaques. We found that the allometric slope in patients with focal lesions in the left coronary tree did not differed from healthy patients. In other words, the cumulative length and volume in each bifurcation thought the coronary tree kept a constant allometric relation. In this work, we decided to include patients with a more advanced diffuse CAD. Again, the allometric slope did not change between groups (Table 1). Values of β=0,75 are similar to results from pigs and humans [10,11,20,25]. However, the allometric constant was significantly higher in CAD patients. If the allometric slopes are considered to be invariable, the cumulative Log(L)-Log(V) lines would represent parallel lines [16], [25]. In this scheme, a higher allometric constant would imply that for the same cumulative length, CAD patients have lower cumulative volume due to the presence of lesions. Moreover, this idea is coherent with the lesions simulations introduced in other allometric models [26].

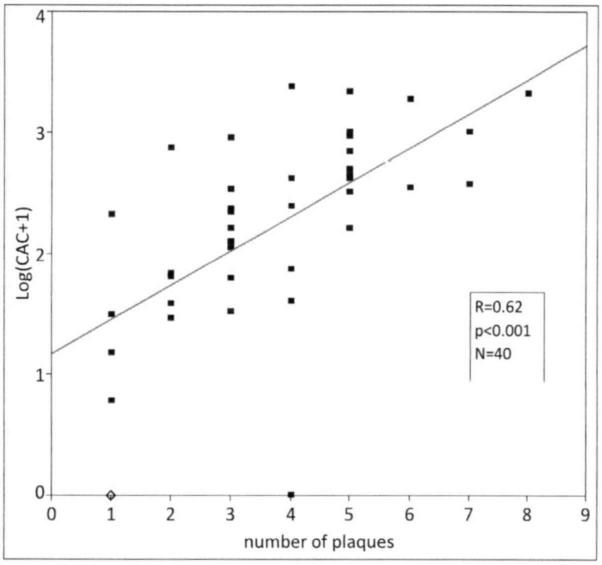

Figure 6. Correlation between coronary calcium score (CAC) in logarithmic scale and the number of plaques in the left coronary tree.

The skeletonization method proved to be efficient even with patients presenting diffuse coronary disease. In spite of the reduced number of patients, our results show that the allometric relation did not change between subjects with and without CAD. This result confirms the existence of an allometric length-volume relationship in the coronary tree. More importantly, this relationship was verified in human patients and with CT techniques *in vivo*. As mentioned before, MSCT technique has an intrinsic limitation to reconstruct vessels with diameters below $1mm^2$ due to cardiac movement artifacts and resolution, not present in ex vivo studies that can go far below this limit [11]. Results obtained in this work could be improved with additional morphometrical information of the vascular tree (tortuosity, bifurcation angles, etc) that could be used to identify geometrical risk factors explaining the location and proliferation of atheromatous plaque [14].

Although the algorithm presented in this work could effectively reproduce the coronary artery tree with minimum intervention, the user still required to select the VOI, to select a seed-point for the region growing segmentation, and to perform a manual pruning of spurious skeleton branches. All these interventions require 3 to 6 minutes per study. A higher amount of spurious branches that need to be manually removed increment this time. The presence of diffuse coronary arteries might also introduce errors in vessel length measurements because of the topology preserving algorithms used to skeletonize the arterial tree. This algorithms cannot fill the holes that surround a plaque, thus introduces clustered regions in the artery skeleton which must be manually erased and linearly interpolated. In this process, the real curvature of vessels may also be altered. Once the arterial map is complete, allometric measuring is automatic. Further efforts are made to reduce user interventions and improve repeatability. In spite of these limitations, the combination of CT technology and image processing techniques are reliable tools for assessing quantitative information of the human coronary tree. They state the basis to develop new diagnostic tools to be applied in clinical practice.

5. Acknowledgments
We want to specially thank Dr. Gurfinkel for his persistent motivation that encouraged us to initiate this work. We are sure his ideas will rest with us beyond his physical presence to continue and improve it in the near future.

6. References

[1] Ikeda U, Kuroki M, Ejiri T, Hosoda S and Yaginuma T 1991 Stenotic lesions and the bifurcation angle of coronary arteries in the young *Jpn. Heart J.* **32** 627-33.

[2] Friedman M H, Deters O J, Mark F F, Bargeron C B and Hutchins G M 1983 Arterial geometry affects hemodynamics. A potential risk factor for atherosclerosis *Atherosclerosis* **46** 225-31.

[3] Gibson C M, Diaz L, Kandarpa K, Sacks F M, Pasternak R C and Sandor T 1993 Relation of vessel wall shear stress to atherosclerosis progression in human coronary arteries *Arterioscler. Thromb. Vasc. Biol* **13** 310-15.

[4] Seiler C, Kirkeeide R L and Gould K L 1992 Basic structure function relations of the coronary vascular tree. The basis of quantitative coronary arteriography for diffuse coronary artery disease *Circulation* **85** 1987-2003.

[5] Brinkman A M, Baker P B, Newman W P, Vigorito R and Friedman M H 1994 Variability of human coronary artery geometry: an angiographic study of the left anterior descending arteries of 30 autopsy hearts *Ann. Biomed. Eng* **22** 34-44.

[6] Kassab G S, Choy J S, Svendsen M, Sinha A K, Alloosh M, Sturek M, Huo Y, Sandusky G E and Hermiller J 2009 A novel system for the reconstruction of a coronary artery lumen profile in real time: a preclinical validation *Am. J. Physiol. Heart. Circ. Physiol.* **297** 485-92.

[7] Zeina A R, Rosenschein U and Barmeir E 2007 Dimensions and anatomic variations of left main coronary artery in normal population: multidetector computed tomography assessment *Coron. Artery Dis.***18** 477-82.

[8] Sarry L and Boire J Y 2001 Three-dimensional tracking of coronary arteries from biplane angiographic sequences using parametrically deformable models *IEEE Trans. Med. Imaging* **20** 1341-51.

[9] Pflederer T, Ludwig J, Ropers D, Daniel W G and Achenbach S 2006 Measurement of coronary artery bifurcation angles by multidetector computed tomography *Invest. Radiol.* **41** 793-8.

[10] Le H Q, Wong J T and Molloi S 2008 Allometric scaling in the coronary arterial system *Int. J. Cardiovasc. Imaging* **24** 771-81.

[11] Wischgoll T, Choy J S and Kassab G S 2009 Extraction of morphometry and branching angles of porcine coronary arterial tree from CT images *Am. J. Physiol. Heart Circ. Physiol.* **297** 1949-55.

[12] Hutchins G M, Miner M M and Boitnott J K 1976 Vessel caliber and branch-angle of human coronary artery branch-points *Circ. Res.* **38** 572-6.

[13] Schreiner W, Neumann M, Neumann F, Roedler S M, End A, Buxbaum P, Müller M R and Spieckermann P 1994 The branching angles in computer-generated optimized models of arterial trees *J. Gen. Physiol.* **103** 975-89.

[14] Craiem D, Casciaro M E, Graf S, Glaser C E, Gurfinkel E P and Armentano R L 2009 Coronary arteries simplified with 3D cylinders to assess true bifurcation angles in atherosclerotic patients *Cardiovasc. Eng.* **9** 127-33.

[15] Zhou Y, Kassab G, Molloi S 1999 On the design of the coronary arterial tree: a generalization of Murray's law *Phys. Med. Biol.* **44** 2929-45.

[16] Craiem D, Casciaro M E, Graf S, Gurfinkel E P, Armentano R L 2010 Non-invasive assessment of allometric scaling laws in the human coronary tree *Artery Research* **5** 15-23.

[17] Agatston A S, Janowitz W R, Hildner F J, Zusmer N R, Viamonte M, Detrano R 1990 Quantification of coronary artery calcium using ultrafast computed tomography *J Am Coll Cardiol*, **15** 827-32.

[18] Kanungo T, Mount D M, Netanyahu N, Piatko C, Silverman R and Wu A Y 2002 An efficient k-means clustering algorithm: Analysis and implementation *Proc. IEEE Conf. Computer Vision and Pattern Recognition* **24** 881-92.
[19] Le H Q, Wong J T and Molloi S 2008 Allometric scaling in the coronary arterial system *Int. J. Cardiovasc. Imaging* **24** 771-81.
[20] Le H, Wong J T and Molloi S 2008 Estimation of regional myocardial mass at risk based on distal arterial lumen volume and length using 3D micro-CT images *Comput. Med. Imaging and Graph.* **32** 488-501.
[21] Palágyi K and Kuba A 1999 A Parallel 3D 12-Subiteration Thinning Algorithm *Graphical Models and Image Processing* **61** 199-221.
[22] Wang T and Cheng I 2008 Generation of Unit-Width Curve Skeletons Based on Valence Driven Spatial Median (VDSM) *Lecture Notes in Computer Science* **5358** 1051-60.
[23] Wischgoll T, Choy J S, Ritman E L and Kassab G S 2008 Validation of image-based method for extraction of coronary morphometry *Ann. Biomed. Eng.* **36** 356-68.
[24] Reig J and Petit M 2004 Main trunk of the left coronary artery: anatomic study of the parameters of clinical interest *Clin. Anat.* **17** 6-13.
[25] Mittal N, Zhou Y, Ung S, Linares C, Molloi S and Kassab G S 2005 A computer reconstruction of the entire coronary arterial tree based on detailed morphometric data *Ann. Biomed. Eng.* **33** 1015-26.
[26] Zhou Y, Kassab GS, Molloi S 2002 In vivo validation of the design rules of the coronary arteries and their application in the assessment of diffuse disease *Phys. Med. Biol.* **47** 977-93.

A survey of medical images and signal processing problems solved successfully by the application of Type-2 Fuzzy Logic

D S Comas[1, 3], G J Meschino[2], J I Pastore[1, 3] and V L Ballarin[1]
[1] Laboratorio de Procesos y Medición de Señales, Dpto. de Electrónica, UNMDP.
[2] Laboratorio de Bioingeniería, Dpto. de Electrónica, UNMDP.
[3] Consejo Nacional de Investigaciones Científicas y Técnicas (CONICET).

E-mail: diego.comas@fi.mdp.edu.ar

Abstract. Typical problems concerning to Digital Image Processing (DIP) and to Digital Signals Processing require specific models for each particular problem and the characteristics of the data involved. However, usually these data show a high degree of uncertainty due to the acquisition system itself, noise or uncertainties related to the nature of the problem. They often require considering different points of view of experts in a single model to determine a set of rules or predicates that would achieve the desired solution. Type-2 fuzzy sets can adequately model such uncertainties. This paper presents a study on different applications of type-2 fuzzy sets in image and signal processing, analyzing the main advantages of this type of fuzzy sets in modeling uncertainties. We also review the definitions of type-2 fuzzy sets, their main properties and operations between them.

1. Introduction

Typical problems concerning to Digital Image Processing (DIP) and Digital Signals Processing require specific models for each particular problem and the characteristics of the data involved. These models make a manipulation of the data in order to obtain useful information from them to achieve an appropriate solution [1, 2].

The segmentation of the images is a common requirement in the medical image processing. This consists in detecting structural components. This process allows characterizing the components or regions of the image to make further analysis for diagnosis tasks [3]. Since medical images often present textures, acquisition noise and inaccuracies in the definition of edges, traditional techniques sometimes are inadequate for processing, because they do not allow modeling satisfactorily such uncertainties. Type-1 Fuzzy Logic (T1FL) [4, 5] is able to make models taking account of these uncertainties. It allows an element to have a fuzzy membership degree to a set. It has been successfully used to model the uncertainties in the images gray levels. Some applications of T1FL involves image processing from different approaches, like: an extension of Mathematical Morphology (MM) to be applied on gray images, giving Fuzzy Mathematical Morphology (FMM) [6, 7]; fuzzy inference systems for edge detection [8]; tissue detection based on their features; image enhancement [9], among others.

Another field related to DIP is the signals processing, which requires consideration in the modeling uncertainties from the acquisition system, from the specific signal and from the processing algorithm. A major application of a Fuzzy System in signal processing is the control systems [10]. In addition, several models base on T1FL have been developed to solve specific problems in signals [11, 12].

Type-1 Fuzzy Sets (T1FS's) present some limitations in uncertainties modeling and minimization, mainly in those what are based on predicates logic or based on rules generated from expert's opinions. These sets only have one membership degree for each element of the sets [13]. The main sources of uncertainties are [14]:

- the meaning of the words used in the evaluation of the predicates may be uncertain and, therefore, they cannot model differences between points of view of different experts on the same word;
- measurements that activate the inputs of a Fuzzy Logic System (FLS) and the data used for parameterization can be noisy and this noise cannot be adequately represented with a single degree of membership.

Type 2 Fuzzy Sets (T2FS's) are an extension of T1FS's in which the degree of membership of each element of the set is fuzzy itself, i.e., it is a T1FS. Due to this characteristic, these sets allow modeling and identifying uncertainties and they are appropriate in circumstances in which it is difficult to determine exactly a membership function to be used in the FLS [10, 14]. For example:

- data generation systems where the description of its variation over time is unknown;
- non-stationary noise measurements whose model is unknown;
- statistical features of unknown nature in the input of pattern recognition systems;
- knowledge taken from an expert group involving words whose interpretation is uncertain.

T2FS's have been applied in the generation of decision-making processes, learning and optimization of fuzzy sets, images pre-processing, function approximation and others [14].
In this work we present a survey of T2FS's applications that have been developed in recent years for solving problems related to signal processing and medical imaging and the advantages gained in solving such problems using Type-2 Fuzzy Logic (T2FL). We define T2FS's and their main properties. We analyze the fundamental operations between T2FS's, seen as an extension of T1FS's. Finally we discuss the main techniques and methods developed in T2FS's.

2. Type-2 Fuzzy Sets

In this section we introduce the definitions related to T2FS's and operations between them.

2.1. Definitions and fundamental properties

Definition 1: Let $A=(X,\mu)$ a T1FS and $\mu: X \to [0,1]$ the membership function of the set A. The *FOU* (*Footprint of Uncertainty*) is defined as follows:

$$FOU_{(X,\mu)} = \bigcup_{x\in X}\left\{[\underbrace{\mu(x)-\varepsilon_x}_{\varphi_x^-},\underbrace{\mu(x)+\delta_x}_{\varphi_x^+}]: 0\le\mu(x)-\varepsilon_x\le\mu(x)+\delta_x\le 1\right\} \quad (1)$$

$$FOU_{(X,\mu)} = \bigcup_{x\in X} J_x = \{[\varphi_x^-,\varphi_x^+]\} \quad (2)$$

$$FOU_{(X,\mu)} = \bigcup_{x\in X} \{(x,\varphi): \varphi\in J_x\} \quad (3)$$

$FOU_{(X,\mu)}$ defines a bounded region of uncertainty on the membership function of variable x. J_x is called *primary membership function* of $x \in X$ [14].

The *Upper Membership Function* (UMF) is defined as:

$$\overline{\mu}(x) = \varphi_x^+ \quad \forall x \in X \tag{4}$$

The *Lower Membership Function* (*LMF*) is defined as:

$$\underline{\mu}(x) = \varphi_x^- \quad \forall x \in X \tag{5}$$

Definition 2: A T2FS, denoted $\tilde{A}$, is defined by [14-16]:

$$\tilde{A} = \int_{x\in X} \int_{u \in J_x \subseteq [0,1]} \mu_A(x,u)/(x,u) = \int_{x\in X} \left[\int_{u \in J_x \subseteq [0,1]} \mu_A(x,u)/u \right] / x \tag{6}$$

or

$$\tilde{A} = \left\{ \left((x,u), \mu_A(x,u) \right) / x \in X, u \in J_x \subseteq [0,1] \right\} \tag{7}$$

where x is the primary variable with domain X, $u \in U$ is the secondary variable with domain J_x in each $x \in X$ and $\mu_A : X \times U \to [0,1]$ is the *secondary membership grade* of $\tilde{A}$ [14]. The operator $\int$ means the union of all elements of the set for a T2FS whose variable x is continuous. The Figure 1 shows the *FOU* for a T2FS and his *UMF* and *LMF* .

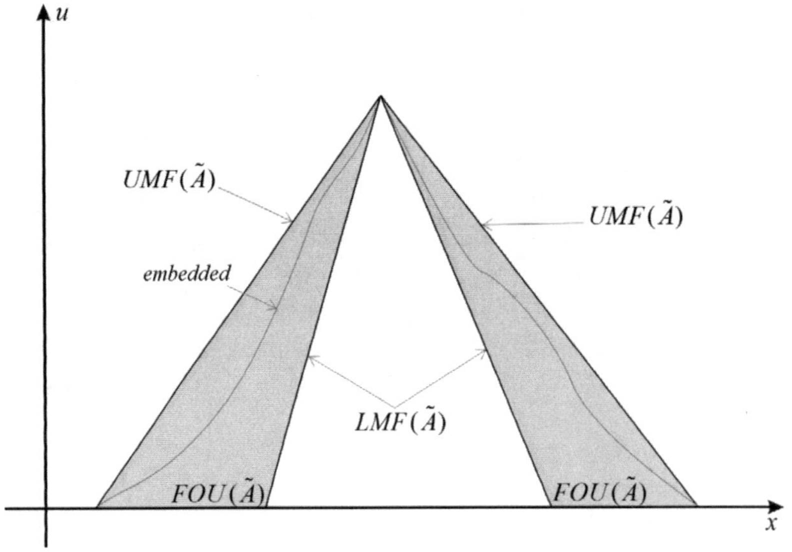

Figure 1. *FOU* for a T2FS.

An *Interval Type-2 Fuzzy Set* (IT2FS) $\tilde{A}$ is a particular case of a T2FS in which the secondary membership grade equals 1 ($\mu_A \equiv 1$) [14-17].
Figure 2 shows an example of a type-2 membership function and his *FOU* for the discrete domains X and U .

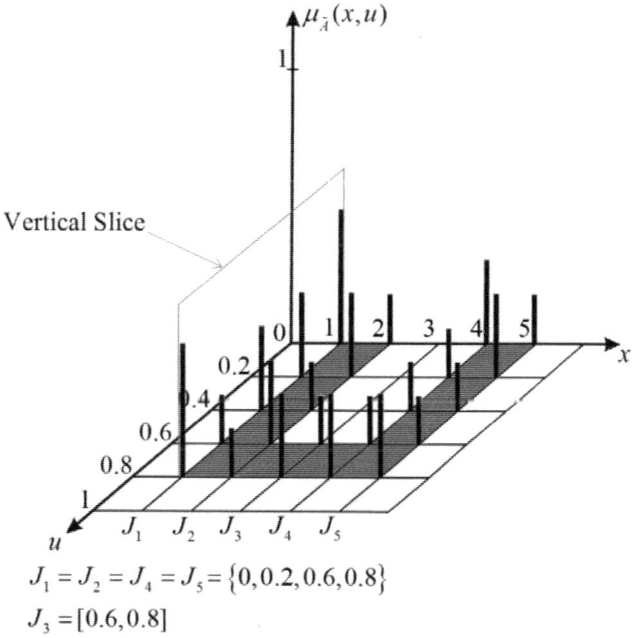

Figure 2. Type-2 membership function for the discrete domains X and U.

Definition 3: For each value of $x \in X, x' = x$ the 2-D plane whose axes are u and $\mu_A(x',u)$ is called *vertical slice* of μ_A [14]. A *secondary membership function* $\mu_A(x',\bullet): U \rightarrow [0,1]$ is a vertical slice of μ_A, and this is defined as:

$$\mu_A(x',\bullet) = \int_{u \in J_{x'}} \mu_A(x',u)/u \, , \, J_{x'} \subseteq [0,1] \tag{8}$$

with $0 \leq \mu_A(x',u) \leq 1$. As this is true $\forall x' \in X$, we denoted $\mu_A(x')$ as the *secondary membership function*, this is a T1FS referred to a secondary set. The type-2 membership function showed in Figure 2 has five vertical slices associated, for $x = 1$ the secondary membership function is:

$$\mu_A(1) = 0.5/0 + 0.35/0.2 + 0.35/0.4 + 0.2/0.6 + 0.5/0.8 \tag{9}$$

Using the concept of secondary sets, we can write a T2FS as the union of all secondary sets, i.e.:

$$\tilde{A} = \{(x, \mu_A(x))/x \in X\} \tag{10}$$

or

$$\tilde{A} = \int_{x \in X} \mu_A(x)/x = \int_{x \in X} \left[\int_{u \in J_x} \mu_A(x,u)/u \right] /x \text{ with: } J_x \subseteq [0,1] \tag{11}$$

Definition 4: The domain of a secondary membership function is called *primary membership function* of x, in the above definition J_x is a primary membership of x.

Definition 5: The amplitude of a primary membership function is called *secondary grade*. $\mu(x',\bullet)$ is a secondary grade.

Definition 6: If X and U are discrete sets, a *embedded type-2 set* $\tilde{A}_e$ has N elements, where $\tilde{A}_e$ contains exactly one element of $J_{x_1},\dots,J_{x_N}$, appointed $u_1,\dots,u_N$ each with a secondary grade $\mu_A(x_1,u_1),\dots,\mu_A(x_N,u_N)$:

$$\tilde{A}_e = \sum_{i=1}^{N}\left(\mu_A(x_i,u_i),u_i\right)/x_i \;,u_i \in J_{x_i} \leq U \leq [0,1] \tag{12}$$

The set $\tilde{A}_e$ is embedded in $\tilde{A}$, and there are a total of $\prod_{i=1}^{N} c(J_{x_i})$ embedded sets, where $c(\bullet)$ denotes the cardinal of the set.

Definition 7: For discrete X and U, a *type-1 embedded set* A_e with N elements, one in each $J_{x_1},\dots,J_{x_N}$ denoted by $u_1,\dots,u_N$:

$$A_e = \sum_{i=1}^{N} u_i / x_i \;,u_i \in J_{x_i} \subseteq [0,1] \tag{13}$$

In the given definitions 6 and 7, the operator $\sum$ means the union of all elements of a set.

Definition 8: A T1FS too can be expressed like a T2FS. This type-2 representation is $(1/\mu_F(x))/x$ or $1/\mu_F(x)$, $\forall x \in X$.

2.2. *Operations between Type-2 Fuzzy Sets*

Let $\tilde{A}$ and $\tilde{B}$ two type-2 fuzzy sets:

$$\tilde{A} = \int_{x\in X}\int_{u \in J_x^u \subseteq [0,1]} \mu_A(x,u)/(x,u) = \int_{x\in X}\left[\int_{u \in J_x^u \subseteq [0,1]} \mu_A(x,u)/u\right]/x \tag{14}$$

$$\tilde{B} = \int_{x\in X}\int_{w \in J_x^w \subseteq [0,1]} \mu_B(x,w)/(x,w) = \int_{x\in X}\left[\int_{w \in J_x^w \subseteq [0,1]} \mu_B(x,w)/w\right]/x \tag{15}$$

In (14) and (15), u and w are dummy variables to differentiate between various secondary membership functions of x in $\tilde{A}$ and $\tilde{B}$ respectively. J_x^u and J_x^w are the *primary membership functions* of the sets $\tilde{A}$ and $\tilde{B}$ respectively. The following defines the union, intersection and complement for the general case of T2FS's and later for the case of IT2FS. The derivation of these operations for T2FS's are performed from T1FS's through the *Zadeh's extension principle* [5] and its analysis can be found in [14, 15].

2.2.1. Union of Type-2 Fuzzy Sets

Just as the union of two T1FS's is another T1FS, the union of T2FS's is another T2FS, defined by [14, 18]:

$$\tilde{A}\cup\tilde{B} \Leftrightarrow \mu_{\tilde{A}\cup\tilde{B}}(x,v) = \int_{x\in X}\int_{v\in J_x^v\subseteq[0,1]} \mu_{\tilde{A}\cup\tilde{B}}(x,v)/(x,v) = \int_{x\in X}\left[\int_{v\in J_x^v\subseteq[0,1]} \mu_{\tilde{A}\cup\tilde{B}}(x,v)/v\right]/x \qquad (16)$$

where

$$\int_{v\in J_x^v\subseteq[0,1]} \mu_{\tilde{A}\cup\tilde{B}}(x,v)/v = \underset{v\in J_x\subseteq[0,1]}{\varphi}\left(\mu_A(x),\mu_B(x)\right) \qquad (17)$$

and φ indicates a t-norm between the secondary membership functions, $\mu_A(x)$ and $\mu_B(x)$, they are T1FSs, and v is defined for a t-conorm, denoted by $\blacklozenge$, between the primary membership grades of $\mu_A(x)$ and $\mu_B(x)$, i.e. $\left\{v/v=u\blacklozenge w,\ u\in J_x^u, w\in J_x^w\right\}$.

When the applied t-conorm is the maximum operator ($\vee$), we have:

$$\mu_{\tilde{A}\cup\tilde{B}}(x)\Big|_{x\in X} = \int_{v\in J_x^v\subseteq[0,1]} \mu_{\tilde{A}\cup\tilde{B}}(x,v)/v = \int_{u\in J_x^u}\int_{w\in J_x^w} \mu_A(x,u)*\mu_B(x,w)/(u\vee w) \qquad (18)$$

where $*$ indicates any t-norm, for example minimum or product, between the secondary grades of the sets $\tilde{A}$ and $\tilde{B}$ for each $x\in X, u\in J_x^u, w\in J_x^w$ [15].

2.2.2. Intersection of Type-2 Fuzzy Sets

Like the union, the intersection of two T2FS's, $\tilde{A}$ and $\tilde{B}$, is another T2FS defined by [14, 15, 18]:

$$\tilde{A}\cap\tilde{B} \Leftrightarrow \mu_{\tilde{A}\cap\tilde{B}}(x,v) = \int_{x\in X}\int_{v\in J_x^v\subseteq[0,1]} \mu_{\tilde{A}\cap\tilde{B}}(x,v)/(x,v) = \int_{x\in X}\left[\int_{v\in J_x^v\subseteq[0,1]} \mu_{\tilde{A}\cap\tilde{B}}(x,v)/v\right]/x \qquad (19)$$

where

$$\int_{v\in J_x^v\subseteq[0,1]} \mu_{\tilde{A}\cap\tilde{B}}(x,v)/v = \underset{v\in J_x\subseteq[0,1]}{\varphi}\left(\mu_A(x),\mu_B(x)\right) \qquad (20)$$

and φ indicates a t-norm between the secondary membership functions, $\mu_A(x)$ and $\mu_B(x)$, they are T1FS, and v is defined for a t-norm, denoted by $*$, between the primary membership grades of $\mu_A(x)$ and $\mu_B(x)$, i.e. $\left\{v/v=u*w,\ u\in J_x^u, w\in J_x^w\right\}$.

For the particular case in which the applied t-norm is the minimum operator ($\wedge$) we have:

$$\mu_{\tilde{A}\cap\tilde{B}}(x)\Big|_{x\in X} = \int_{v\in J_x^v\subseteq[0,1]} \mu_{\tilde{A}\cap\tilde{B}}(x,v)/v = \int_{u\in J_x^u}\int_{w\in J_x^w} \mu_A(x,u)*\mu_B(x,w)/(u\wedge w) \qquad (21)$$

where * indicates any t-norm between the secondary grades.

2.2.3. Complement of Type-2 Fuzzy Sets:

The *complement* of $\tilde{A}$ is another T2FS, such as [14, 15, 18]:

$$\overline{\tilde{A}} \Leftrightarrow \mu_{\overline{\tilde{A}}}(x,v) = \int_{x\in X} \mu_{\overline{\tilde{A}}}(x)/x = \int_{x\in X}\left[\int_{u\in J_x^u\subseteq[0,1]} \mu_{\tilde{A}}(x,u)/(1-u)\right]/x = \int_{x\in X} \neg\mu_{\tilde{A}}(x)/x \quad (22)$$

where $\neg$ denotes the negation operator and $\neg\mu_{\tilde{A}}(x)$ means the negation of the secondary membership function.

The operators *union, intersection* and *complement* for the particular case of IT2FS are obtained of definitions of *union, intersection* and *complement* of T2FS's taking the secondary membership grade $\mu \equiv 1$. A full analysis of the operations on this particular type of T2FS is available in [17].

3. Type-2 Fuzzy Sets Applications

In this section we review applications of methods based T2FL for the resolution of problems involving image and signal processing and we summarize the results of each one.

3.1. Support in the diagnosis of brain tumors

In [19] Zarandi *et al* developed an Expert System for processing Magnetic Resonance images (MR) with T2FL for support in the diagnosis of brain tumors. Some features of brain tissues, gray intensity of each tissue (white matter, gray matter, cerebrospinal fluid, cord and abnormalities) and fuzzy edge definition for the location of tumors make the task of tissue recognition a very difficult one. The T2FS's are used to model such uncertainties.

The authors propose a classification method that consists of 4 stages: pre-processing, segmentation, feature extraction and approximate reasoning. In the first stage the image noise is reduced by applying different filters, which are determined by a set of rules. In the segmentation stage they use a rule-based T2FS, which has gray level of pre-processed images as input variable. Rules are determined with the use of expert's knowledge and parameters of membership functions are adjusted to minimize the classification error by presenting to the system the pre-classified images as examples. Once the image is segmented, features are extracted from each of the classes identified and based on these rules the diagnosis and treatment are suggested for the patient by means of another set of rules.

To evaluate the developed system, the authors compare the outcome of information processing of 95 patients with a T1FL based Expert System. From this comparison it is determined that the proposed type-2 Expert System is more accurate in determining the diagnosis than Type-1, as T2FS's provide better modeling of uncertainties in this type of images and experts' opinion.

3.2. Classification of tibia radiographic images

John *et al* present in [20] a T2FS based system for processing radiographic images of the tibia followed by a classification with Neural Networks. T2FS's are used to model the uncertainties in the data and the different opinions of experts and the system is able to provide assistance in the diagnosis of bone lesions, determining the location and length of the lines in the pre-processed images.

T2FS's are determined by surveying experts in diagnosis. They define two variables of analysis of the segmented lines: length and radius of them. In order to describe these variables linguistically with adjectives, fuzzy sets are defined. The values of the variables are the inputs of a Neural Network [21] for classifying injuries in the images into four classes: MTS, Patchy, Stress Fractures and Healing Stress Fractures.

The authors show the results of processing a huge number of images, whose segmentation was known, in order to analyze the performance of the system. The proposed system has higher accuracy in classification of images in cases of MTS, Patchy, Stress Fractures, compared with type-1 FLS, offering a better ability to model the uncertainties inherent to the descriptions given by experts on the images.

3.3. Edges Detection

In [22], Mendoza *et al* propose a new method for edge detection in images. Images are pre-processed with three types of filters: gradient based, high-pass and low-pass [1, 2]. Input vectors for an inference system with T2FL are generated with the intensity values of the resulting images. There are three fuzzy sets for each input variable, called "Low", "Medium" and "High". This system of rules and the T2FS's defined allow modeling intensity levels required in the filtered images to determine when a pixel is considered an edge. The processing results are compared with T1FL system. The T2FS's give better results than T1FS's in the edges detection.

3.4. Images thresholding

Tizhoosh proposes in [23] a method for determining an appropriate threshold value for binarization of images [1, 2]. The author defines a measure of fuzziness weighing the histogram of the image by a membership function of an Interval Type-2 Fuzzy Logic System (IT2FLS) that moves itself along the gray scale using an iterative algorithm. The threshold is determined by the position of the membership function that minimizes the fuzziness value. The results shows better results for determining the threshold value compared to systems using T1FL.

3.5. Fuzzy reasoning modeling for generation of decision support experts systems

En [24] Garibaldi *et al* present a study to research about the introduction of the vagueness or uncertainty in the membership functions of a fuzzy system to model the variation shown by experts in the context of medical decision making. A Type-1 Expert System was previously developed to assess the health of newborns immediately after birth, through biochemical analysis of the blood taken from the umbilical cords. Variety in decision-making was introduced in the Expert System through small changes in their membership functions by default over time. Three types of variation in the membership functions were considered: the variation of the central points, the variation in the width and the addition of white noise. Different levels of uniformly distributed variation were investigated for each of them. Monte Carlo simulations were carried out to propagate the variations through the inference process, in order to determine the distribution of the conclusions reached. IT2FLSs were applied to find the limits of variability in decisions. Results were compared with the expert's decisions to determine what type and size of the variability of the membership functions represent better the variability given by the experts. The new technique of reasoning introduced in this study is called a non-stationary fuzzy reasoning.

3.6. Classification of arrhythmias by EKG analysis

In [25], Tan *et al* develop a Type-2 Fuzzy Logic System (T2FLS) [15, 16] for the classification of arrhythmias in electrocardiograms (ECG). They analyze three types of ECG signals: normal sinus rhythm (NSR), ventricular fibrillation (VF) and ventricular tachycardia (VT).
The authors use as inputs for the classifier the average period and the pulse width, features that are extracted from the pre-processed ECG records.
Three fuzzy sets ("Small", "Med" and "Large") are defined for each input feature and the system rule set is determined, based on that fuzzy sets. The classification of each ECG is determined by a threshold comparator that uses the output of the defuzzification process of T2FLS [16].
FOUs of the membership functions involved in the antecedents of the rules is determined from training data. *UMF* is determined considering the centroids extracted from the application of a Fuzzy C-Means algorithm (FCM) [26] to the training data. *LMF* is defined by measuring the dispersion of the training data. Thus the *FOUs* of the fuzzy sets in the rules antecedents are obtained.

In order to evaluate the performance of the proposed system, the authors compare the results with three methods: a method based on rules, a classifier based on T1FS and a Self-Organizing Map [21]. Experimental results show that the system using T2FS present greater accuracy in the classification of ECG arrhythmias used for evaluation.

3.7. Speech recognition

Zeng and Liu in [27] presents an extension of Hidden Markov Models (HMM's) based on T2FS's, called Type-2 Fuzzy HMMs (T2 FHMM's) for voice recognition. T2FS's are introduced into the HMM's to consider uncertainties in the mean and variance of the probability distributions for state transitions that HMM's model. This allows model adequately the uncertainties of the voice data (phonemes have different values in different contexts, the same phoneme can have different lengths, the beginning and end of a phoneme is uncertain, etc.) affecting the generalization ability of the HMMs after training. The authors derive the type-2 fuzzy forward-backward algorithm and Viterbi algorithm, using operations involving type 2 fuzzy sets.
Algorithm is applied to the classification and recognition of phonemes in the TIMIT speech database. Experimental results show that T2 FHMM's can properly address the uncertainties of the noise and dialect in voice signals and have a better classification performance than traditional HMM's.

4. Conclusions

In this paper we defined the Type-2 Fuzzy Sets, their main properties and operations between them.
We surveyed different applications of these sets both in image processing and signal processing.
We found that the models developed in the papers can see greater capabilities in uncertainties modeling when they used Type-2 Fuzzy Sets than similar systems using Type-1, mainly in the development of systems based on expert opinion, because they provide models that consider that words mean different things to different people. Also we highlighted the large ability of these models on modeling uncertainties in images and signals whose statistical distribution is unknown.
This review allows us to conclude that the applications of Type-2 Fuzzy Logic is a current area of interest, and it suggests that these research directions should be continued to produce new applications and to achieve generalizations of those applications based on Type-1 Fuzzy Logic.

5. References

[1] Gonzalez R C and Woods R E 2002 *Digital image processing* (Upper Saddle River, N. J.) vol 1 (Prentice Hall)
[2] Baxes G A 1994 *Digital image processing: principles and applications* (New York) vol 1 (Wiley)
[3] Pastore J I Moler E and Meschino G 2005 Segmentación de biopsias de médula ósea mediante filtros morfológicos y rotulación de regiones homogéneas *Revista Brasileira de Engenharia Biomédica* **21** 37-44
[4] Zadeh L A 1965 Fuzzy sets *Information and Control* **8** 338-53
[5] Dubois H and Prade D 1980 *Fuzzy Sets and Systems: Theory and Applications* (New York) vol 1 (Academic Press Inc)
[6] Bouchet A Pastore J and Ballarin V 2007 Segmentation of Medical Images using Fuzzy Mathematical Morphology *Journal of Computer Science and Technology* 256-62
[7] Bouchet A Brun M and Ballarin V 2010 Morfología Matemática Difusa aplicada a la segmentación de angiografías retinales *Revista Argentina de Bioingeniería* **16** 7-10
[8] Aborisade D O 2010 Fuzzy Logic Based Digital Image Edge Detection *Global Journal of Computer Science and Technology* **10** 78-83
[9] Tizhoosh H R and Michaelis B 1999 *Image Enhancement Based on Fuzzy Aggregation techniques* 16th IEEE IMTC'99 (Venice, Italy) pp. 1813-7
[10] Wagner C and Hagras H 2010 *Uncertainty and Type-2 Fuzzy Sets and Systems* UK Workshop on Computational Intelligence (UKCI) (Colchester, UK) pp. 1-5

[11] Pérez-Neira A Lagunas M A Morell A and Bas J 2005 *Neuro-fuzzy Logic in Signal Processing for Communications: From Bits to Protocols* Non-Linear Speech Processing (Barcelona) pp. 10-36

[12] Ghosh S Razouqi Q Schumacher H J and Celmins A 1998 A Survey of Recent Advances in Fuzzy Logic in Telecommunications Networks and New Challenges *IEEE Transactions on Fuzzy Systems* **6** 443-7

[13] Mendel J 2003 *Fuzzy sets for words: a new beginning* 12th IEEE International Conference on Fuzzy Systems (Saint Louis, MO) pp. 37-42

[14] Mendel J and John R I B 2002 Type-2 Fuzzy Sets Made Simple *IEEE Transactions on Fuzzy Systems* **10** 117-27

[15] Mendel J M 2001 *Uncertain Rule-Based Fuzzy Logic Systems: Introduction and New Directions* (Upper-Saddle River, NJ) vol 1 (Prentice-Hall)

[16] Mendel J M 2007 Type-2 fuzzy sets and systems: an overview *IEEE Computational Intelligence Magazine* **2** 20-9

[17] Liang Q and Mendel J M 2000 Interval Type-2 Fuzzy Logic Systems: Theory and Design *IEEE Transaction on Fuzzy Systems* **8** 535-50

[18] Karnik N N and Mendel J M 2001 Operations on Type-2 Fuzzy Sets *Fuzzy Sets and Systems* **122** 327-48

[19] Zarandi M H Zarinbal M and Izadi M 2011 Systematic image processing for diagnosing brain tumors: A Type-II fuzzy expert system approach *Applied Soft Computing* **11** 285-94

[20] John R I Innocent P R and Barnes M R 1998 *Type 2 fuzzy sets and neuro-fuzzy clustering of radiographic tibia images* Proceedings of the Sixth IEEE International Conference on Computational Intelligence (Anchorage, AK , USA) pp. 1373-6

[21] Haykin S 1999 *Neural Networks: A Comprehensive Foundation* 2nd Edition (Nueva Jersey, EE.UU.) vol Prentice Hall)

[22] Mendoza O Melin P and Licea G 2007 *A New Method for Edge Detection in Image Processing Using Interval Type-2 Fuzzy Logic* IEEE International Conference on Granular Computing (San Jose, California) pp. 151-6

[23] Tizhoosh H R 2005 Image thresholding using type II fuzzy sets *Pattern Recognition* **38** 2363-72

[24] Garibaldi J M and Ozen T 2007 Uncertain Fuzzy Reasoning: A Case Study in Modelling Expert Decision Making *IEEE Transactions on Fuzzy Systems* **15** 16-30

[25] Tan W W Foo C L and Chua T W 2007 *Type-2 Fuzzy System for ECG Arrhythmic Classification* Fuzzy Systems Conference (London) pp. 1-6

[26] Jain A K Murty M N and Flynn P J 1999 Data Clustering: A Review *ACM Computing Surveys* **31** 264-323

[27] Zeng J and Liu Z-Q 2006 Type-2 fuzzy hidden Markov models and their application to speech recognition *IEEE Transaction on Fuzzy Systems* **14** 454-67

Development of software for digital image processing for analysis of neuroangiogenesis

M A Gonzalez[1,2], V L Ballarin[1], A R Celín [2,3], M Rapacioli[2,3], J J López-Costa[2,4] and V Flores[2,3,4]
[1] Laboratorio de Procesos y Medición de Señales, Facultad de Ingeniería, UNMdP , ARGENTINA
[2] CONICET: Comisión Nacional de Investigaciones Científicas y Tecnológicas.
[3] Grupo Interdisciplinario de Biología teórica. Departamento de Ciencias Bioestructurales, Universidad Favaloro, ARGENTINA.
[4] Instituto de Biología Celular y Neurociencias. "Prof. E. De Robertis", U.B.A. ARGENTINA.

E-mail: mazulgonzalez@fi.mdp.edu.ar

Abstract. The process of formation, growth and distribution of vessels within the developing central nervous system is difficult to analyze due to the complexity of the paths and branches within the system. The study of images of this area poses particular problems because the high levels of noise, blurring and poor contrast often prevent the objects of interest detected correctly. The design of algorithms for digital image processing suitable for this type of imagery remains a constant challenge. The aim of this work is to develop a computer tool to assist the specialist in processing these images. This paper proposes the use of morphological grayscale reconstruction and other morphological operators in order to segment the images properly. The results show that the algorithms allow a suitable segmentation of the objects of interest. Moreover, the interface developed for processing enables easy and simple analysis of them by the specialists.

1. Introduction

Each organ has a typical spatial distribution of blood vessels depending on the requirements of nutrients and gases and structural and functional characteristics [1.2]. This spatial distribution is made during embryonic development. However, the needs of nutrients and oxygen to developing tissues are not dependent on the functions of terminally differentiated organ but of the requirements related to compliance own development cell behavior [3,4]. Nervous and circulatory systems develop early, and they share typical peripheral distribution. Vessels and peripheral nerves in the body usually travel together forming "neurovascular" bundle [5]. The CNS (Central Nervous System) also has a great influence on the organization of the perineural vascular plexus that provide irrigation [6], the number of vessels entering to it [7] and the distribution of the plexus that irrigate it [8]. There is a great difficulty in evaluating the wealth of intrinsic blood irrigation of the CNS, the mass of tissue that each entrant vessel supplies (unit neura-vessel: uN-V) because the vessels that enter it branch profusely and generate a network of varying complexity.

This paper proposes a strategy that requires computer assistance to determine precisely the site of entry for each vessel to the CNS, log and quantification of their relative positions and demarcation of areas of influence (uN-V) of each the pial surface. [9,12].

The aim of this work is to develop a computer tool to assist the specialist in processing and analyzing these images. This paper proposes the use of morphological grayscale reconstruction and other morphological operators to segment images of entrant vessels correctly to the CNS [13-17].

2. Materials and Methods

We analyzed images obtained from the pial surface of the developing CNS (Tectum optical chick embryo 6 days) that were previously processed by the diaphorase technique [18] allowing selectively stain the endothelial cells that line the light developing blood vessels. This technique allows to observe: (a) the surface segments of the vessels that enter the CNS, (b) entry sites and (c) the pial surface areas surrounding each cell (Fig. 1).

All algorithms were implemented using the libraries of Mathematical Morphology and Image Processing Matlab R14. Moreover, the software was developed using the library of Matlab Graphical Interface (GUI).Figure 1 allows us to observe one of these test images.

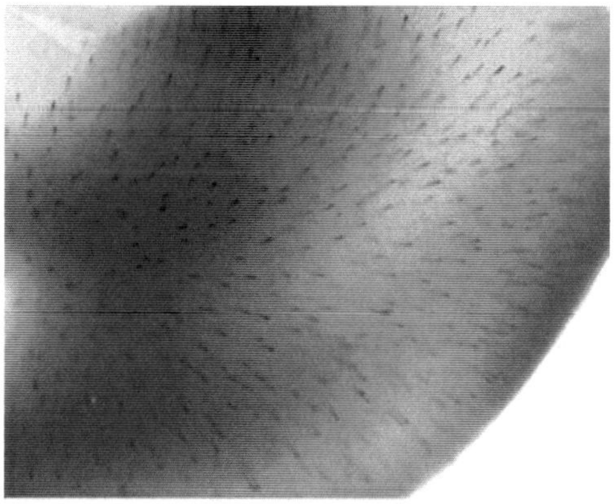

Figure 1. Pial surface image of the developing CNS stained with the diaphorase.

Figure 1 shows small truncated cones that correspond to the segment superficial radial vessels entering the CNS. The thickest point of these "cones" corresponds to the site of entry of each vessel. As can be seen each entrant vessel is surrounded by a "sphere of influence" or uN-V.

2.1.Mathematical Morphology

This paper proposes the application of morphological operators to process test images, images of the CNS entrant vessels. For several years the mathematical morphology, theory based on the algebra of sets, has been applied in biomedical images with very good results and minimal Computational cost making it a helpful tool [9-12]. Morphological operations can be defined through the algebra of sets [13-15]. There is a rigorous representation of the theorems where you can get a morphological expression in terms of more primitive operations. The choice of a morphological operators and a way of structuring element depends on the particular application [16-17]. These techniques are widely used in digital image processing, for example in edge detection, filtering, enhancement and restoration, segmentation and noise suppression. Morphological filters are more suitable for shape analysis than conventional linear filters. Binary morphology is defined from two basic operations called erosion and dilation. These operations compare the subsets within the binary image with a pattern called a structuring element. The structuring element is moved through the entire image.

The erosion of the set A by the structuring element $B(x) \subset \Re_n$ is defined as:

$$A \Theta B = \{x, B(X) \subset A\}$$
$$being\ B(x) = \{b + x, b \subset B\} \tag{1}$$

As it is widely known, the result of the erosion is one if the structuring element is included in the subset and zero if it not included.

The dilation of A by the structuring element $B(x) \subset \Re^n$ is defined as:

$$A \oplus B = \{x | B(x) \cap A \neq \varnothing\} \tag{2}$$

The result of the dilation is one if there is at least one pixel of intersection between the subset of the original image and the shifted structuring element and it is zero when the intersection is empty.
Binary morphological reconstruction dilates small sets included in the original image called markers.
Binary reconstruction can select the original image marked objects (called *mask*) and display them in a new binary image with its size and shape.
Giving a set A called mask, a structuring element B and a set called mask M, binary reconstruction (Eq. 4) is defined by the conditional $A \oplus B_c(M)$ dilation, (Eq. 3) as:

$$A \oplus B_c(M) = (M \oplus B) \cap A \tag{3}$$

$$\rho_A(M) = \lim_{n \to \infty} (\underbrace{A \oplus B_c(A \oplus B_c(A \oplus B_c \ldots (M))))}_{n\ times} \tag{4}$$

Conditional dilation is defined as the dilation of the subset of the image marker, M, as long as the result of this dilation is contained in the mask, A. When conditional dilation is applied successively, it meets idempotency property when no changes occur in the image. The reconstruction is $\rho_A(M)$ met when the resulting image does not changed more.

As mathematical morphology for binary images, morphological grayscale operations are defined by the erosion and dilation. Given an image $f(x, y)$ within domain Df and a structuring element $B(x, y)$ within a Df domain the erosion of f by B is defined as:

$$f \Theta B(s,t) = Min\{f(s+x, t+y) - B(x,y) | (s+x), (t+y) \in D_f; (x,y) \in D_B\} \tag{5}$$

For each pixel of the image, the erosion is defined as the minimum difference between the intensities of the shifted structuring element and the corresponding intensities of the original image Given an image $f(x, y)$ within domain Df and a structuring element $B(x, y)$ within a DB domain is defined by the dilation of f by B is defined as:

$$f \oplus B(s,t) = Max\{f(s-x, t-y) + B(x,y) | (s-x) \in D_f; (x,y) \in D_B\} \tag{6}$$

For each pixel of the image, the dilation is defined as the maximum of the sum of the intensities of the subset of the original image and the intensities of the corresponding structuring element.
Given a gray level image, S, called mask and another gray level image called marker, M. Conditional dilation is defined as the successive dilation by the structuring element B, as follows:

$$d_{B,S}(M) = d_B(M) \cap S \tag{7}$$

$$\rho_S(M) = \lim_{n \to \infty} (\underbrace{d_{B,S}(d_{B,S}(d_{B,S}(d_{B,S}(M) \ldots))))}_{n\ veces} \tag{8}$$

The morphological grayscale reconstruction can be defined the same way that the binary reconstruction based on conditional grayscale dilation. When the conditional dilation is applied successively, ideally it meets idempotency property (no change occurs in the image). The resulting image is called reconstruction.

In Figure 2 we can see the result of the reconstruction of a row of an marker image *g*, using an image *f* as a mask. The g image is dilated until the intersection of the result with the mask f are not exceeded

.

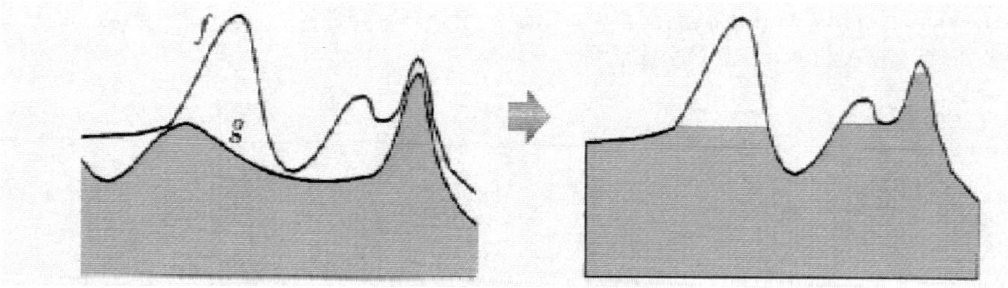

Figure 2. Morphological grayscale reconstruction.

2.3. Proposed Algorithm

First, a morphological gray scale reconstruction from the edges of the image was applied. Then, the reconstructed image was subtracted from the original image. This operation was used to remove noise and background. Then the gradient was obtained by subtracting the reconstructed image form de original. This gradient was equalized to increase the contrast and, thus, highlight objects of interest. Finally, the threshold of the resulting image was obtained using the algorithm proposed by Otsu. This algorithm calculates the optimum threshold separating two classes (e.g. foreground and background) so that their intra-class variance is minimal [19]. Figure 3 shows a block diagram with the main procedures of the proposed algorithm. Figure 4 shows the segmentation obtained for one of these images.

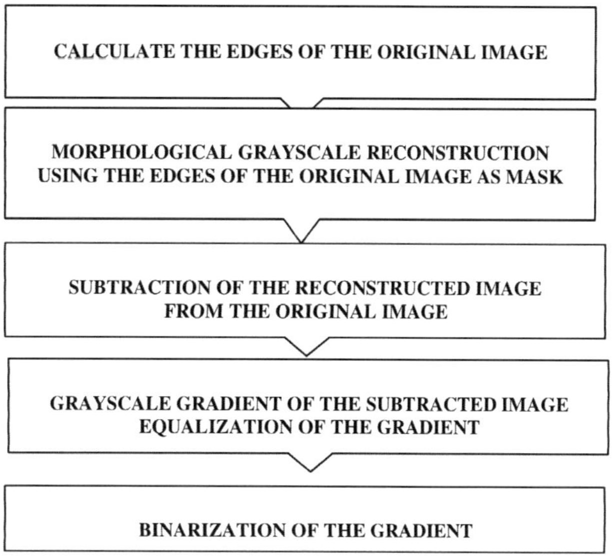

Figure 3. The block diagram illustrates the stages of the segmentation algorithm used.

2.4. Graphical interface
In order to assist the specialist in processing these images we developed a graphical interface that allows the automatic segmentation of images and also the improvement of them by adding missing items and removing irrelevant objects. This tool also allows determining the Watershed Transform of segmented elements. This operation is useful in analyzing the layout of objects of interest in the acquired sample. Figure 4 shows that interface.

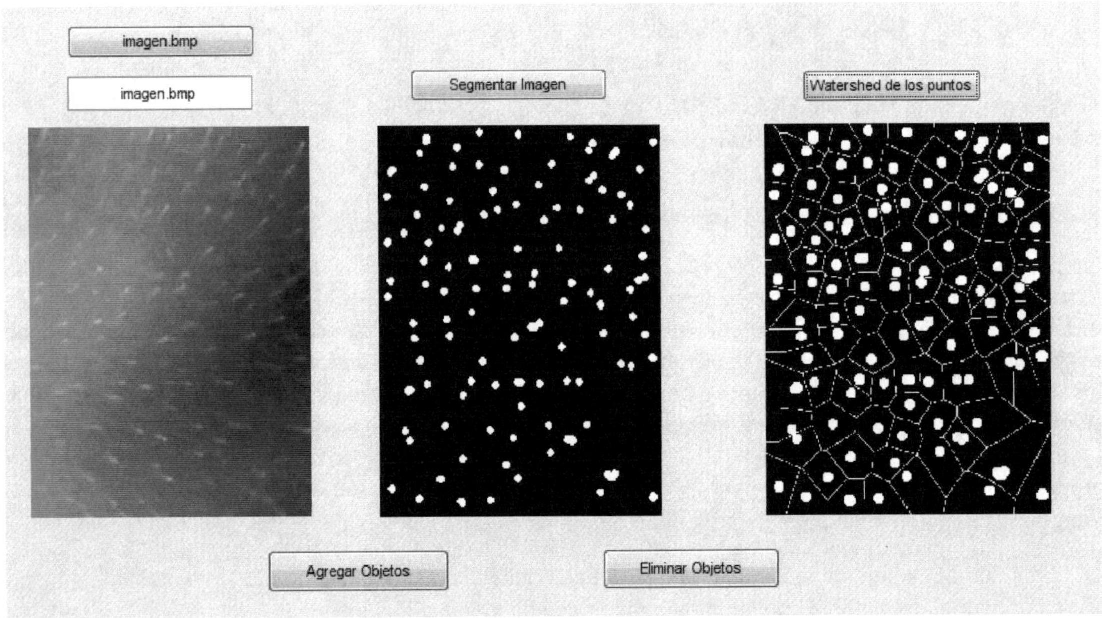

Figure 4: Graphical interface

4. Results
The preliminary results show that the developed algorithms allow a suitable segmentation of objects of interest. The developed algorithms were incorporated into a graphical interface that allows the specialist to automatically segment the images and improve the segmentation result of adding or removing objects from the image obtained. Figure 5 shows some of these results in conjunction with the developed interface.

While the results are still preliminary and the tool evaluation itself is qualitative. All the algorithms that are part of the tool have been validated in previous articles. Despite this, it is expected to perform a full validation in the near future upon completion of the analysis for the normal variations in the Uv-n that occur during experimental conditions in the regulatory processes interference that control the neuroangiogenesis.

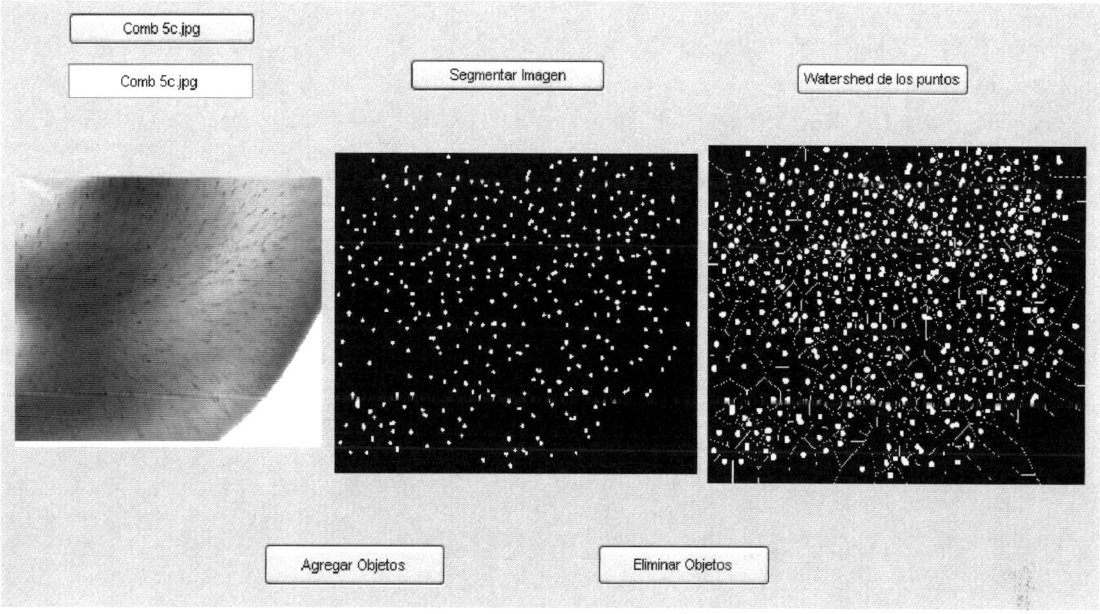

Figure 5: Some results obtained by applying the proposed algorithm

4. Conclusions

Neuroangiogenesis concept refers to all processes to ensure integrated development of vessels and nerve tissue [20]. Such integration is regulated by endothelial cell interactions in development on the one hand, and neuroprogenitor stem cells, neurons or glial cells in development, on the other [21]. These interactions are mediated by signal molecules that spread from cell to cell and exert their effects in the corresponding soft cells [22]. These are networks of messages whose effects fluctuating subsume some randomness difficult to grasp and measure by conventional methods of analysis.

This work shows that computer assistance can detect, record, quantify and highlight properties because of its variability are difficult to describe. The algorithms developed allow detecting and accurately recording the sites of the CNS entry of vessels. The developed software also allows delimiting their areas of influence (Uv-n) in the pial surface and based on this data to estimate the volume of neural tissue supplied by each vessel that enters the system.

Using a graphical interface facilitates the analysis of the images analyzed, allowing the expert to improve the result of automatic segmentation. The interface provides a simple procedure that allows the specialist to process and quantify properties in a large number of images with low computational cost. The software developed will be useful for analyzing the normal variations that occur in the Uv-n during normal development as well as their changes under experimental conditions that interfere in the regulatory processes that control the neuroangiogenesis. Alterations in these regulatory processes are the cause of many abnormalities present at birth or appear throughout postnatal life.

5. References

[1] Risau W. In: Development of the Vascular System of Organs and Tissues. Schaper W, Schaper J, editors. Kluwer Academic Publishers; Boston, pp. 17–28 (1993).

[2] Risau W. Mechanisms of angiogenesis. Nature.386, pp.671–674 (1997).

[3] Crossley DA 2nd, Burggren WW, Altimiras J. Cardiovascular regulation during hypoxia in embryos of the domestic chicken Gallus gallus. Am J Physiol Regul Integr Comp Physiol. 284(1), pp.219-226 (2003).

[4] Lee YM, Jeong CH, Koo SY, Son MJ, Song HS, Bae SK, Raleigh JA, Chung HY, Yoo MA, Kim KW. Determination of hypoxic region by hypoxia marker in developing mouse embryos in vivo: a possible signal for vessel development. Dev Dyn. 220(2), pp.175-86 (2001).

[5] Mukouyama, Y. S., Shin, D., Britsch, S., Taniguchi, M. and Anderson, D. J. Sensory nerves determine the pattern of arterial differentiation and blood vessel branching in the skin. Cell 109, pp.693-705 (2002).

[6] Hogan KA, Ambler CA, Chapman DL, Bautch VL. The neural tube patterns vessels developmentally using the VEGF signaling pathway. Development.131(7), pp.1503-13 (2004).

[7] Herken R, Gotz W, Wattjes KH. Initial development of capillaries in the neuroepithelium of the mouse. J Anat.164, pp.85–92 (1989).

[8] Michael R. Mancuso, B.S., Frank Kuhnert, Ph.D., and Calvin J. Kuo. Developmental Angiogenesis of the Central Nervous System. Lymphat Res Biol. 6(3-4) pp: 173–180 (2008).

[9] Castleman K. R., Digital *Image Processing*, Prentice Hall (1979).

[10] Serra, J. *Image Analysis and Mathematical Morphology,* Academic Press, London (1982).

[11] Glasbey, C. A. y Horgan G. W., *Image analysis for the biological science, Statistics in Practice*, Series Editor Vic Barnett., Jonh Wiley and Sons (1994).

[12] González, R y Woods, R., *Tratamiento Digital de imágenes*, Addison Wesley (1996).

[13] Facon, J., Morfología *Matemática. Teoría y ejemplos*, Curitiba Brasil, CITS (1996).

[14] Serra, J., Image Analysis and Mathematical Morphology- Part II: Theoretical Advances,, Academic Press, London (1988).

[15] Serra J. Image analysis and Mathematical Morphology, Academic Press (1992).

[16] Ballarin, V. y Valentinuzzi, M.: "Segmentación en imágenes de Resonancia Magnética de Cerebro utilizando Morfología Matemática," Actas del Congreso Argentino de Bioingeniería. Tafí del Valle. Septiembre 2001. (Publicadas en CD) (2001).

[17] Heijmans, H., "'Theoretical aspects of gray-scale morphology," *IEEE Trans. Pattern Anal. Mach. Intell.,* 13, pp.568-582 (1991).

[18] Scicolone G, Ortalli AL, Alvarez G, López-Costa JJ, Rapacioli M, Ferrán JL, Sanchez V, Flores V. Developmental pattern of NADPH-diaphorase positive neurons in chick optic tectum is sensitive to changes in visual stimulation. J Comp Neurol. 494(6), pp.1007-1030 (2006).

[19] N. Otsu, "A threshold selection method from gray-level histograms", IEEE Trans. on SMC, 9, pp. 62-66 (1979).

[20] Louissaint A, Jr, Rao S, Leventhal C, Goldman SA. Coordinated interaction of neurogenesis and angiogenesis in the adult brain. Neuron. 34, pp.945–960 (2002).

[21] Shen Q, et al. Endothelial cells stimulate self-renewal and expand neurogenesis of neural stem cells. Science.304, pp.1338–1340.(2004)

[22] Ogunshola OO, et al. Neuronal VEGF expression correlates with angiogenesis in postnatal developing rat brain. Brain Res Dev Brain Res., 119, pp.139–153 (2000).

Mathematical morphology: detection and characterization of directed axonal growth in vitro

M A Gonzalez [1,2], V L Ballarin [1], M Rapacioli [2,3], A Rodríguez Celín[2,3], V Sánchez[2,4] and V Flores [2,3,4]

[1] Measurement and Signal Processing Laboratory, School of Engineering, UNMdP, Juan B. Justo 4302, Mar del Plata, B7608FQD, ARGENTINA

[2] CONICET (Comisión Nacional de Investigaciones Científicas y Tecnológicas), Rivadavia 1917, Ciudad Autónoma de Buenos Aires, C1033AAJ, ARGENTINA.

[3] Interdisciplinary Group in Theoretical Biology, Department de Biostructural Sciences, UF. ARGENTINA.

[4] Institute of Cell Biology and Neurosciences "Prof. E. De Robertis", U.B.A. ARGENTINA.

E-mail: mazulgonzalez@fi.mdp.edu.ar

Abstract. Neurite growth in vitro, neuritogenesis, is a widespread methodology in the field of developmental neurobiology. Morphological analysis of growing neurites is usually a difficult task because of their thinness and their low contrast that makes it difficult to observe clearly their shape, number, length and spatial orientation. This paper presents the use of the granulometric size distribution function to obtain, automatically, information about the shape, size and spatial orientation of growing axons in tissue cultures. The results presented show the efficiency of the granulometric size distribution for this application. The automatic detection of growing axons and the precise characterization of a relevant parameter, indicative of the axonal growth spatial orientation, was obtained by this morphological tool. The developed algorithms facilitates the analysis of these images by automatically quantifying the angle of deviation of the direction of growth, which is important given the large number of images that need to be processed for this type of study.

1. Introduction

In vitro culture of developing neurons is a common strategy to analyze dendro- and axonogenesis regulation. It is known that the specificity of the peripheral-central connections and the establishment of local circuits within the central nervous system depends in part on the directed growth cone migration [1][2]. Developing neurites are highly dynamic and subtle structures. Given the difficulty to measure and quantify their relevant features, i.e. shape, thickness, length, spatial orientation etc, they are usually described qualitatively. A relevant aspect in developmental neurobiology is the statistical analysis of neuritogenesis and this requires precise records of the above mentioned characteristics [3]. This situation reveals the need for designing and implementing algorithms with the ability to precisely reveal some of these neurite features.

The main object of this work was the application of digital image processing algorithms to images obtained from cultures of explants obtained from a specific superior area of the chick central nervous system in order to automatically characterize the shape, size and spatial orientation of the growing axons.

Images obtained from this kind of tissue cultures are usually characterized by noisy background, non-homogeneous gray level (illumination), groups of neuronal bodies, isolated migrating neurons and a large number of neurites with different characteristics depending on the embryonic stage and on the area under analysis. Axons usually differ in shape, gray level, texture, size, orientation etc. All these features greatly difficult the acquisition of a binary image [4].

Conventional image processing methods for analysis of shapes of objects requires a binary image and a subsequent calculation of factors roundness, shape, area, etc. [4][5]. Unlike standard techniques, morphological techniques are based on concepts of geometry, algebra, topology and set theory, to characterize structural properties in images [6]-[10]. The central idea of these techniques is to examine the geometric structures in an image overlaying them with small patterns called structuring elements. Of all mathematical morphology analysis techniques, the most appropriate tool for discriminating shapes is the granulometric size distribution. [11][13].

This paper proposes the computation of the granulometric size distribution with structuring elements of various shapes and sizes to quantify the number and orientation of the growing axons and the subsequent statistical analysis.

2. Materials

We used microscope images of different patterns of growing axons distribution. We analysed 16 images of growing axons. The images were converted to grayscale (255 levels). The color was not providing information to assist the study. Some of those images show clear preferential spatial orientations; others images show irregularly distributed spatial orientations. Images without axons were also analyzed. Figure 1 shows three cases representative of the growth of axons.

All algorithms were implemented using Matlab R14, especially image processing libraries.

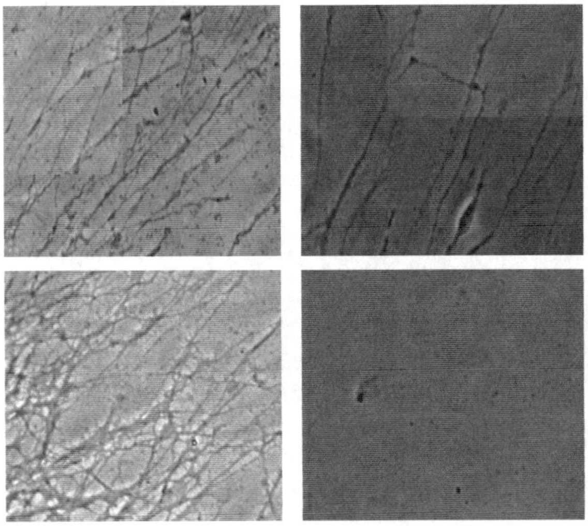

Figure 1. Growing axons Images

3. Methods

3.1. Mathematical Morphology applied to binary images

The basic operations of binary morphology are erosion and dilation [6]. These operations compare subsets of the binary image with a pattern element called structuring element. The structuring element is moved over the whole image [8]-[10].

Formally, the erosion of a set $A \subset \Re^2$ by a structuring element $E \subset \Re^2$ is defined as:

$$A \Theta E(x) = \{x \in A : E_x \subset A\} \tag{1}$$

which represents the translation of the set E at x. The result of erosion is a binary image where pixels are white if the structuring element is included within the subset of the analyzed image and black otherwise.

Similarly, dilation of a ser $A \subset \Re^2$ by a structuring element $E \subset \Re^2$ is defined as:

$$A \oplus E(x) = \{x \in A : E_x \cap A \neq \varnothing\} \tag{2}$$

The result of dilation is a binary image where pixels are white if at least exists one pixel of intersection between the subset of the original image and structuring element and black when the intersection is empty. Dilation is an operation of expansion while erosion is a contraction operation.

These basic operations define new morphological operators. For example, opening is defined as erosion followed by dilation:

$$A \circ E(x) = (A \Theta E) \oplus E \tag{3}$$

Closing is defined as dilation followed by erosion, that is:

$$A \bullet E(x) = (A \oplus E) \Theta E \tag{4}$$

From these operators, the granulometric size distribution is defined. This mathematical morphologic function is a tool for characterization of shapes and sizes of objects in an image. Given the family $\{A \circ E_\lambda\}_{\lambda \in I}$ obtain by the application of successive openings with structuring elements of increasing size is defined by the name of granulometric size distribution function is defined by:

$$G(\lambda) = 1 - \frac{\Omega(A \circ E_\lambda)}{\Omega(A)} \tag{5}$$

where Ω is a measure of the resulting image. For the case of binary images, Ω results to be the area.

3.2. Mathematical Morphology applied to gray level images

As in mathematical morphology for binary images, the basic morphological operations for gray level images are erosion and dilation [8]-[10].

Given two images f and g, with domains $D_f, D_g \subset \Re^2$, erosion of image f by structuring element g is defined as:

$$f \Theta g(s,t) = \min_{\substack{(x,y)\in Dg \\ (s+x,t+y)\in Df}} \{f(s+x,t+y) - g(x,y)\} \tag{5}$$

For each pixel of the image, the erosion operation is defined as the minimum difference between the intensities of the shifted structuring element and the corresponding intensities of the original image.

Similarly, the dilation of image f by structuring element g is defined as:

$$f \oplus g(s,t) = \max_{\substack{(x,y)\in Dg \\ (s-x,t-y)\in Df}} \{f(s-x,t-y) + g(x,y)\} \tag{6}$$

Dilation of each pixel of the image is defined as the maximum value of the sum between the intensities of the subset of the original image and the intensities corresponding structuring element.

The opening operator and the granulometric size distribution are defined similarly to the binary operators [14]-[15].

$$G(\lambda) = 1 - \frac{\Omega(A \circ E_\lambda)}{\Omega(A)} \tag{7}$$

Successive openings are made to the original image with increasing structuring elements. Then a measure Ω is calculated for each opening. In the case of gray level images, Ω is the volume. This result is normalized to the original image.

After calculating the granulometric size distribution function, which is actually a probability distribution function, mean, standard deviation and energy are calculated.

4. Results

We evaluated the statistical on images of growing axons in tissue cultures. The minimum mean value calculated from the granulometric function using lineal structuring elements increased with the number and the thickness of the growing axons. The plot of the percentage of area occupied by the axons versus the minimum mean value obtained illustrates this relationship (Figure 2). The area occupied by neurons was determined by counting the number of pixels that are included in them and then dividing this value by the total number of pixels. Although this analysis cannot determine the number of axons, it showed a linear relationship and a good approximation of the area occupied by axons. The mean value obtained with other structuring elements did not show linear relations.

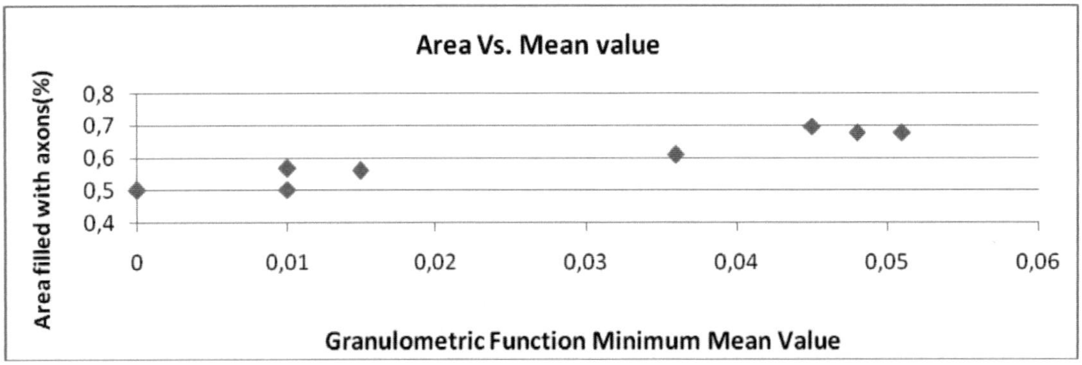

Figure 2: Area Vs, Minimun Mean Value obtained with oriented linear structruring element

The calculation of the granulometric size distribution function with linear structuring elements permitted to know the orientation of the objects in the images. The mean value determined from this function is lower when the axon has the same orientation as the structuring element used in the calculation of the granulometric function. The minimum mean value obtained by means of this function allows a precise quantification of the angle of deviation of the axons longitudinal axes which constitute a reliable parameter indicating the preferential spatial orientation or the preferential growth direction (see figure 3).

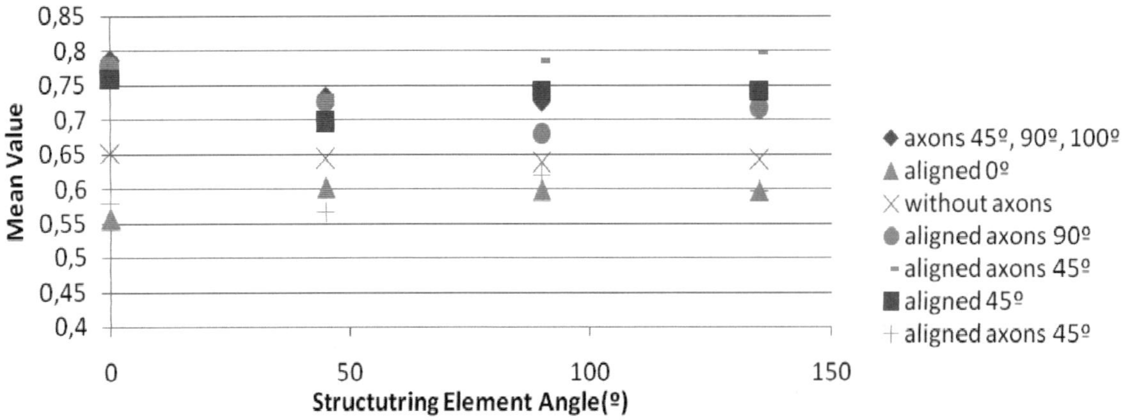

Figure 3: Axon Orientation Vs, Minimun Mean Value obtained with oriented linear structruring element

The granulometric function was also calculated by using circular or square structuring elements. In the absence of a preferential axon orientation, the average values calculated from the granulometric function with these non-linear elements were lower than those obtained with linear elements. This feature allows the characterization of axonal networks where axons are irregularly distributed without any preferential orientation, since this structuring element analyzes all directions simultaneously.

5. Conclusions

The present paper is a preliminary approach to design a reliable method for digital image analysis with the ability to identify and to quantify some relevant parameters of axonal growth indicative of their plasticity, dynamism and sensitivity to different factors during neuritogenesis (growth factors, extracellular matrix components, axon-neuron interactions, axon-glial cell interactions etc).

Our results show that the granulometric size distribution function is a useful tool for the characterization of the spatial orientation of objects both in synthetic images and in real images of growing axons. This utility is not relevant when the growing axons fasciculate (close contact between parallel adjacent growing axon) displaying a defined preferential growth direction.

The granulometric size distribution function, however, can be succesfuly applied in those cases when growing axons do not display a preferential growth direction but spatially arrange as irregular complex networks. Even in these cases the granulometric function allows the detection and quantification of axons displaying different growth direction. This is a remarkable utility of the granulometric function since the most frequent images of axons growing in tissue culture correspond to complex networks with none preferential spatial orientation.

As future work, a similar algorithm able to analyze in vivo the degree of order of growing axons will be implemented. Reliable data to analyze in vivo the establishment of peripheral and central neural circuits requires a precise recording about how growing axons orientate and direct their growth with respect to intrinsic spatial reference axis (cephalic-caudal, dorsal-ventral and radial axes) as well as the time or stage-dependent changes in these parameters during development. These data could help to better understand the neuritogenic process and the development and stabilization of the neuronal circuits.

6. References

[1] S. Fiszer de Plazas, M. Rapacioli, D. Rodriguez, M. Vacotto, V. Flores, "Acute hypoxia differentially affects the gamma-aminobutyric acid type A receptor alpha(1), alpha(2), beta(2), and gamma(2) subunit mRNA levels in the developing chick optic tectum: Stage-dependent sensitivity". *Journal of Neuroscience Research,* vol. 95 No. 5 pp.280-286, 2008.

[2] R. Farías, C. Vidal, M. Rapacioli, V. Flores, "Basics elements for modelling the dynamics of cell migration in cell culture", *Journal of Physics: conference series*. vol. 90 pp.1-7, 2007.

[3] J. Mazzeo, M. Rapacioli, V. Sanchez, G. Scicolone, C. D'Attellis, V. Flores, "Characterizing cell proliferation process in the developing central nervous system", *InterJournal Complex Systems*, No. 1686 pp.1-7, 2006.

[4] Glasbey, C. A. y Horgan G. W., *Image analysis for the biological science, Statistics in Practice*, Series Editor Vic Barnett., Jonh Wiley and Sons, 1994.

[5] González, R y Woods, R., *Tratamiento Digital de imágenes*, Addison Wesley, 1996.

[6] Facon, J., Morfología *Matemática. Teoría y ejemplos*, Curitiba Brasil, CITS, 1996.

[7] Castleman K. R., Digital *Image Processing*, Prentice Hall, 1979.

[8] Serra, J. , *Image Analysis and Mathematical Morphology,* Academic Press, London, 1982.

[9] Serra, J., *Image Analysis and Mathematical Morphology- Part II: Theoretical Advances,*, Academic Press, London, 1988.

[10]Serra J. *Image analysis and Mathematical Morphology*, Academic Press, 1992.

[11]Ballarin, V. y Valentinuzzi, M.: "Segmentación en imágenes de Resonancia Magnética de Cerebro utilizando Morfología Matemática," Actas del Congreso Argentino de Bioingeniería. Tafí del Valle. Septiembre 2001. (Publicadas en CD), 2001.

[12]Heijmans, H., "'Theoretical aspects of gray-scale morphology," *IEEE Trans. Pattern Anal. Mach. Intell.,* vol. 13, pp.568-582, 1991.

[13]Vincent L., Dougherty E., *Morfological Segmentation for Textures and Particles*, Digital Image Processing Methods, Rochester, New York, 1994.

[14] Dougherty, E. R. y Astola J., *An introduction to nonlinear image prossesing*, Tutorial Texts in optical engineering, vol. TT16, 1994.

[15] Marshall, S. y Sicuranza G., *Advances in nonlinear signal and image prcessing*, Eurasip, 2006.

SABI 2011 IOP Publishing
Journal of Physics: Conference Series **332** (2011) 012033 doi:10.1088/1742-6596/332/1/012033

Medical Image Segmentation using the HSI color space and Fuzzy Mathematical Morphology

J P Gasparri[1], A Bouchet[1,2], G Abras[1], V Ballarin[1] and J I Pastore[1,2]
[1] Laboratorio de Procesos y Medición de Señales. Dpto de Electrónica. UNMDP.
[2] Consejo Nacional de Investigaciones Científicas y Técnicas (CONICET).

E-mail: jpastore@fi.mdp.edu.ar

Abstract. Diabetic retinopathy is the most common cause of blindness among the active population in developed countries. An early ophthalmologic examination followed by proper treatment can prevent blindness. The purpose of this work is develop an automated method for segmentation the vasculature in retinal images in order to assist the expert in the evolution of a specific treatment or in the diagnosis of a potential pathology. Since the HSI space has the ability to separate the intensity of the intrinsic color information, its use is recommended for the digital processing images when they are affected by lighting changes, characteristic of the images under study. By the application of color filters, is achieved artificially change the tone of blood vessels, to better distinguish them from the bottom. This technique, combined with the application of fuzzy mathematical morphology tools as the Top-Hat transformation, creates images of the retina, where vascular branches are markedly enhanced over the original. These images provide the visualization of blood vessels by the specialist.

1. Introduction

Diabetic retinopathy is the most common cause of blindness among the active population in developed countries. An early ophthalmologic examination followed by proper treatment can prevent up to 95% of blindness [1-3]. The study of the characteristics of retinal blood vessels, such as width and branching pattern, provides crucial information to the ophthalmologist.

The retinal angiography is a diagnostic test that uses special cameras to evaluate the structure of the ocular background. For this test eye drops are given to dilate the pupil and then pictures are taken of the inside of the eye. After taking the first set of images, a special dye, called fluorescein, is injects into a vein. A special camera takes pictures of the eye as the dye passes through blood vessels in the back of the eye. This test is used to detect leaks or damage in the blood vessels in the retina. It can also be used to diagnose eye problems or to evaluate the success of a treatment.

The images obtained after this test are color images. The use of color images for processing and analysis is increasing relative to gray-level images, because of the additional information that they can provide. Traditionally, color images were not widely used because the capacity of the acquisition and processing equipment were limited and the cost and time of processing were prohibitive. For that reason, very few subjects took into account the potential that they offer. The growth in computing power, storage capacity and the upgrading of low-cost systems for capturing and printing images have prompted a growing interest in the use of color images in the development of many applications.

Image segmentation is a fundamental step for a correct diagnosis. Segmenting an image is a process of decomposing an image into regions of interest, to partition them based on its major structural components according to some criterion. This criterion depends on the outcome expected for the image [4]. The methods for carrying out the segmentation task vary widely depending on the specific application, image type and other factors. Actually there is no a segmentation method to reach acceptable results for all types of medical images and there are no methods that are general and can be applied to any variety of data.

Due to the low contrast, in the images under study, between the background and blood vessels, the segmentation of these vessels is not an easy task. This paper proposes to combine Fuzzy Mathematical Morphology and process of the HSI (Hue, Saturation, Intensity) color space in order to enhance them for better viewing, and thus assist the experts in their diagnosis. The processing is based on changing only the tone, in the hue matrix, without modifying the saturation and intensity matrices, in order to maintain the appearance and effects of shading of the objects in the image.

2. Theoretical Concepts

2.1. HSI color space

Color spaces provide a way to specify order and manipulate colors. While there are several color spaces, the choice of a suitable space for color representation remains a challenge for scientists researching processing and analysis of color images [5, 6]. The most of the color spaces have been developed for specific applications, but all come from the same concept: the trichromatic theory of primary colors red, green and blue.

The HSI color space (Hue, Saturation and Intensity) define a model in terms of its components. This space has the ability to separate the intensity of the intrinsic color information, which would refer to the hue and saturation. For this reason its use is recommended for processing images when they are affected by lighting changes, characteristic of the images under study in this paper [7].

HSI space representation is through a double cone, as shown in figure 1. The center of this double cone is a circumference divided into angles of equal magnitude. For this reason the value of H, which describes the color by its wavelength, covering angles whose amplitude varies between 0° and 360°. The letters R, Y, G, C, B and M refer to Red (0°), Yellow (60°), Green (120°), Cyan (180°), Blue (240°) and Magenta (300°) respectively. The distance from the center of the exterior circumference represents the saturation found in every color and takes values from 0 to 1, indicating how the color is diluted with white light. Finally, the axis through the two cones corresponds to the intensity component. This has a normalized value from 0 (black) to 1 (white) and indicates the amount of light in a color. Removing a small circumference of the figure formed by two cones, colors close to an intensity of 1 are lighter than those close to zero. When the saturation component is close to 0, colors only reflect a change between black and white. When this component is close to 1, the color will reflect the true value represented by the hue [8, 9].

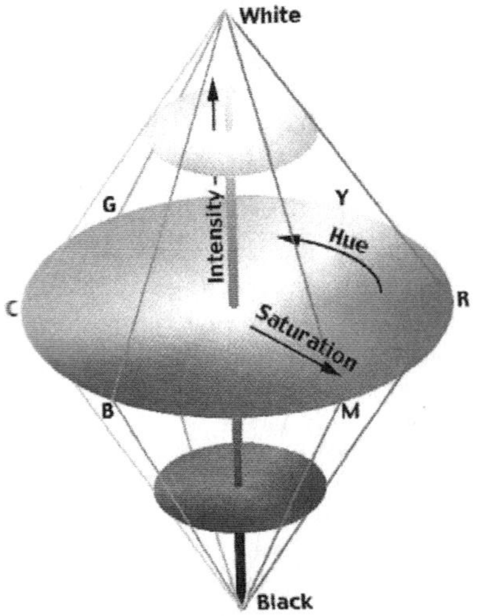

Figure 1: HSI space representation.

2.2. *Fuzzy Mathematical Morphology*

The Mathematical Morphology (MM) is a theory for image processing based on concepts of geometry, algebra, topology and set theory, created originally with the goal to characterize physical and structural properties of different materials [10, 11]. Currently the MM has become a solid mathematical theory based on the powerful tools for Digital Image Processing. The central idea of this theory is to analyze the geometric structures in an image to overlap with small patterns located in different parts of it, called structuring elements (SE). The MM allows enhancing areas, edge detection, analyzing structure and segment regions, among others. Based on solid theoretical foundations, the MM has achieved excellent results in the segmentation of structures and deployment fast and simple algorithms. Anyhow, in images of high texture and vague edges, a new approach is indispensable.

The MM has been applied successfully to a large number of image processing problems. However, MM does not allow a complete representation of the uncertainties in images with high texture or a high degree of uncertainty in structural components. From the binary MM different extensions have been made for grayscale images. One of these extensions is the Fuzzy Mathematical Morphology (FMM). The fuzzy sets have several advantages to represent the uncertainties in the images and turn out to be a useful tool for their segmentation. The FMM has been applied successfully in medical image segmentation [12, 13]. The following operators are defined to be used in this work.

In what follows μ and ν denote two fuzzy sets, where the first corresponds to a grayscale image and the second is the structuring element. Importantly, for most cases the images in gray levels are defined so that the gray level intensity at each pixel is an integer value belonging to the natural range $[\![0,255]\!]$. Therefore, to be able to apply the FMM operators need to generate a function that changes the scale of these images, leading them to the range $[0,1]$. This process of scaling it is called *"fuzzification"*, while the reverse process is called *"desfuzzification"*. For the development of this work fuzzification function $g:\{0,1,2,\ldots,255\}\rightarrow[0,1]$ used is:

$$g(x)=\frac{x}{255} \tag{1}$$

The reverse process by which the intensities of the gray levels of an image, belonging to the interval $[0,1]$, are brought to the set $\{0,1,2,\ldots,255\}$ is defined from the function $h:[0,1]\rightarrow\{0,1,2,\ldots,255\}$ given by:

$$h(x)=[255.x] \tag{2}$$

where $[.]:\mathbb{R}\rightarrow\mathbb{Z}$ represents the integer part function, ie, $[a]$ is the nearest integer to a with $a\in\mathbb{R}$.

Importantly, this process not converts the image into a fuzzy representation of an object, but that models a grayscale image as a fuzzy set in order to apply the theory of fuzzy sets.

The definition of the basic operations of the FMM is shown below [16, 17].

Fuzzy dilation of the image μ by the SE ν [10]:

$$\mu\oplus\nu=\sup_{y\in U}\left[t\left(\mu(y),\nu(y-x)\right)\right] \tag{3}$$

where $t[a,b]$ is a t-norm [16].

Fuzzy erosion of the image μ by the SE ν [15]:

$$\mu\Theta\nu=\inf_{y\in U}\left[s\left(\mu(y),c\left(\nu(y-x)\right)\right)\right] \tag{4}$$

where $s[a,b]$ is a s-norm and $c(a)$ is the fuzzy complement [17].

Fuzzy morphological closing of the image μ by the SE ν is given by:

$$\mu\bullet\nu=(\mu\oplus\nu)\Theta\nu \tag{5}$$

Within the FMM an image segmentation technique very useful is the Fuzzy Top-Hat transform by closing, which highlights locally dark objects that have been deleted in a grayscale image by the closing filtering. For this use a structuring element larger than the structures to detect. The choice of form, size and orientation of this element makes it possible to filter the image greatly increasing the contrast of the areas eliminated, ignoring the regions which are not relevant in the analysis to be performed. This operator is defined, then, as the residue between the fuzzy morphological closing and the image:

$$\rho(\mu,\nu) = (\mu \bullet \nu) - \mu \tag{6}$$

3. Materials and methods

Were used 40 digital retinal images belonging to the public database DRIVE (Digital Retinal Images for Vessel Extraction) [18]. The images were acquired using a Canon CR5 non-mydriatic 3CCD camera with a 45 degree field of view (FOV). Each image was captured using 8 bits per color plane at 768 by 584 pixels.

The purpose of this work is to develop a processing on HSI color space in order to enhance the blood vessels in these images. Color space is divided into subspaces, each of which characterizes an object in the image. A false color in the hue component is assigned to the subspace that characterizes the blood vessels, while saturation and intensity components are not modified so that the resulting image has the same shading effects.

The subspace representing the blood vessels is obtained from the fuzzy segmentation of grayscale image that emerges from the combination of RGB components of the original image. The steps to obtain this subspace are the following:

Step 1: *Chromatic Filter:* A color image consists of three components: Red, Green and Blue. Each of these components is stored in a matrix of $M \times N$ and takes values between 0 and 255. By combining them a RGB color image is generated [19]. The chromatic filter allows modifying the intensity of each color, i.e., it modifies the values in each of the matrices according to a criterion. This criterion depends on the objects available on the image. That allowed enhancing the areas of interest. For this purpose we attenuated the red component, since this component is homogeneous present in the ocular background and in the blood vessels. The green and blue components were set at a level slightly higher than it was originally to be those that contain high frequency information corresponding to the vessels under study. Therefore, the red component is multiplied by the scalar 0.4, while the green component is multiplied by 1.8 and the blue by 2.

Step 2: *Fuzzy Top-Hat transform by closing:* From the image obtained in the previous step, a grayscale image was obtained doing a weighted average of the three components that define it. The way to do it is apply the following formula to the value of each pixel:

$$I_{Gray} = 0.2989 \times R + 0.5870 \times G + 0.1140 \times B \tag{7}$$

where R, G and B are the value of the pixel in the red, green and blue components, respectively. Then it applies a fuzzy Top-Hat transform to the grayscale image, since this transform enhancing dark objects against from a not uniform background.

Step 3: *Binarization*: The result of applying the fuzzy Top-Hat transform is also a grayscale image where enhanced regions represent the blood vessels in the image. To remove noise in the filtered image, binarization is applied, where 0 represents the background of the image and 1 represents the objects of interest.

Step 4: *Enhancement and visualization*: The image obtained in the previous step is the subspace that characterizes the blood vessels. Assigning tonality 0° in the hue component, an enhancement of objects characterized by this subspace is obtained, maintaining the original value of tonality in the rest of the image.

The results obtained from the proposed method are compared in the next section with the gold standard, which is provided by the database.

The algorithm was developed in Matlab. Standard functions of this language and a specific library with functions of Fuzzy Mathematical Morphology [20] were employed.

4. Results

This section presents the results obtained with the proposed method. In figures 1 to 3 show some resulting images. In each case, first the original image is shown, then the segmented image using the operators of the FMM and finally the original image enhanced. To validate the obtained enhanced image, we used the gold standard image obtained from manual blood vessels segmentation by an expert. The computational cost of execution the algorithm is negligible given the low computational cost of the morphological operators used.

Tables 1 to 3 show the classification error as an error matrix or confusion matrix [21]. The boxes on the main diagonal of each matrix show the percentage of pixels correctly classified. The values for the blood vessels and the ocular background are very high. It also follows from the results that the greatest degree of confusion arises in the corresponding pixels in the ocular background that are classified as ocular vasculature. Table 4 shows the average percentages of correct classification the 40 images processed. The percentages of correct classification for blood vessels and ocular background are 98.4737% and 91.7861%, respectively.

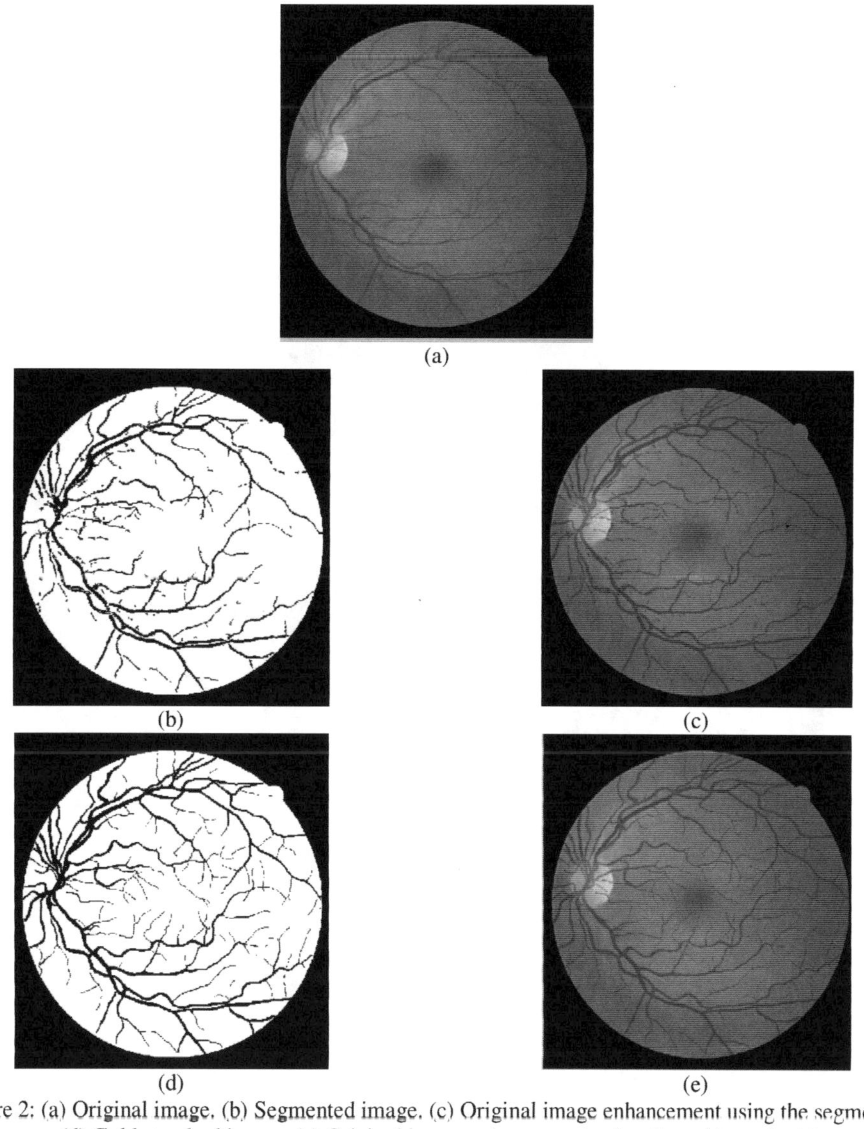

(a)
(b) (c)
(d) (e)

Figure 2: (a) Original image. (b) Segmented image. (c) Original image enhancement using the segmented image. (d) Gold standard image. (e) Original image enhancement using the gold standard image.

Table 1: Average percentage of pixels by subspace for image of figure 2.

	% of pixels to be classified as blood vessels...	% of pixels to be classified as ocular background...
... that would have been classified as blood vessels	97.9955	6.4165
...that would have been classified as ocular background	2.0045	93.5835

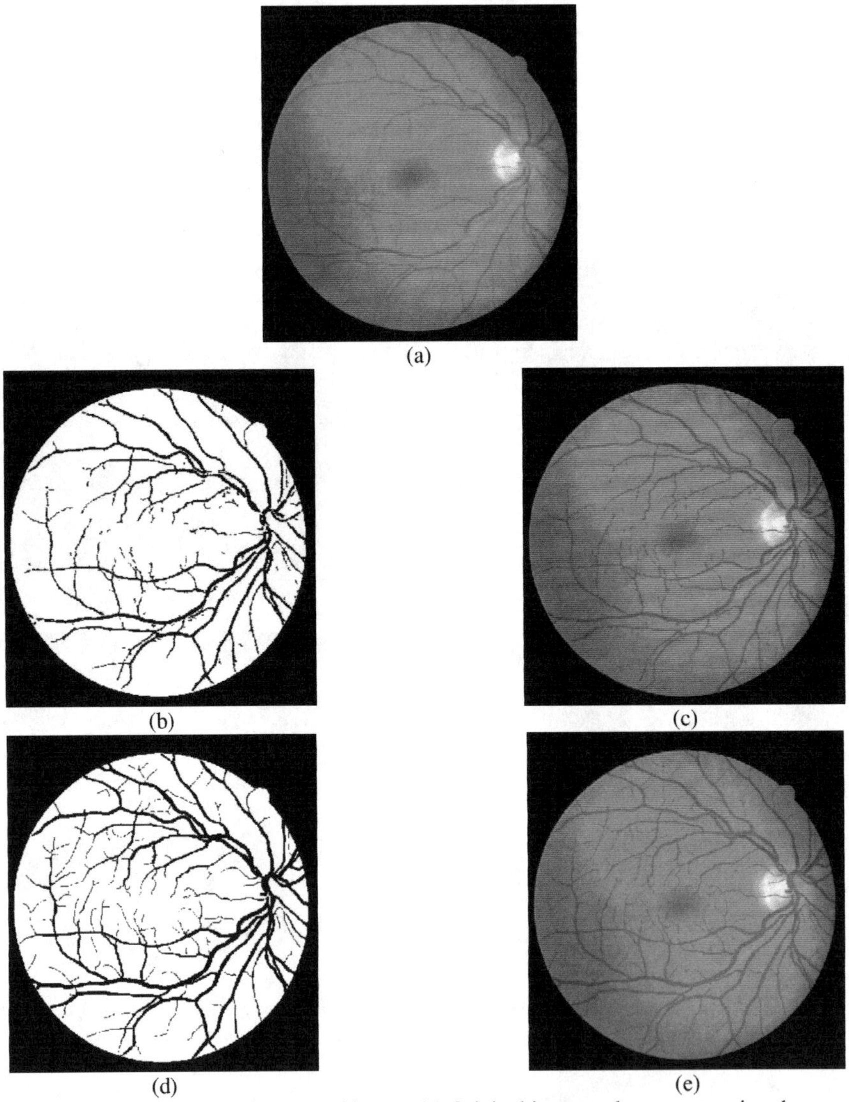

Figure 3: (a) Original image. (b) Segmented image. (c) Original image enhancement using the segmented image. (d) Gold standard image. (e) Original image enhancement using the gold standard image.

Table 2: Average percentage of pixels by subspace for image of figure 3.

	% of pixels to be classified as blood vessels…	% of pixels to be classified as ocular background…
…that would have been classified as blood vessels	98.6596	7.8025
…that would have been classified as ocular background	1.3404	92.1975

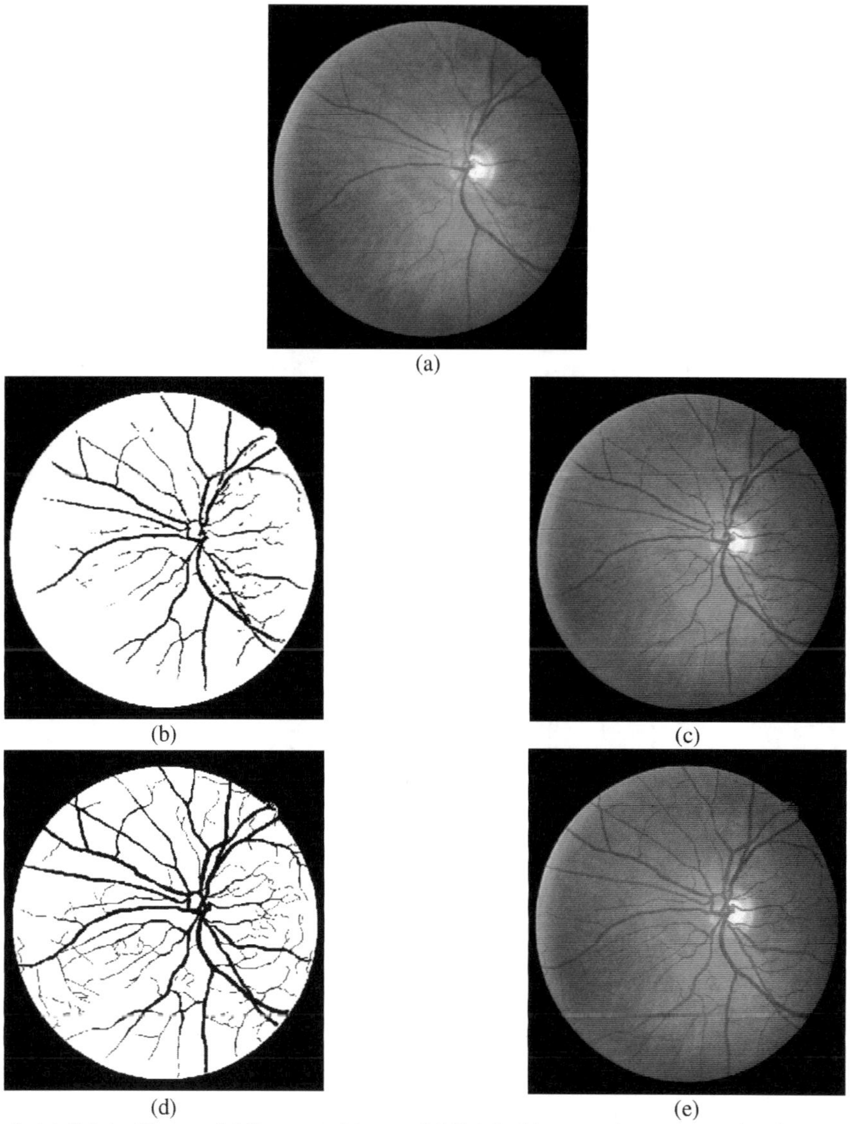

Figure 4: (a) Original image. (b) Segmented image. (c) Original image enhancement using the segmented image. (d) Gold standard image. (e) Original image enhancement using the gold standard image.

Table 3: Average percentage of pixels by subspace for image of figure 4.

	% of pixels to be classified as blood vessels...	% of pixels to be classified as ocular background...
...that would have been classified as blood vessels	99.1898	9.7479
...that would have been classified as ocular background	0.8102	90.2521

Table 4: Average percentage of pixels by subspace for the 40 images analyzed.

	% of pixels to be classified as blood vessels...	% of pixels to be classified as ocular background...
...that would have been classified as blood vessels	98.4737	8.2139
...that would have been classified as ocular background	1.5263	91.7861

5. Conclusions

This paper proposed a method for enhancement of blood vessels in retinal angiography in the HSI color space in order to assist experts in the formulation of diagnoses. The color space was segmented into subspaces using Fuzzy Mathematical Morphology, allowing assign a false color on the hue component of the subspace that representing blood vessels and thereby improve visualization of these ramifications.

To validate the proposed method we used images provided from the public database DRIVE, which includes the ideal segmentation, a tool necessary to carry out the validation of any method of segmentation. As a result of the processing of the 40 images, were obtained that on average 98.4737% of the pixels corresponding to blood vessels have been correctly segmented.

After doing the necessary tests and receiving the views of specialists, as future work, a graphical user interface will be proposed to facilitate the task of the experts.

6. References

[1] Wilson C, Horton M, Cavallerano J and Aiello LM 2005 Addition of primary care-based retinal imaging technology to an existing eye care professional referral program increased the rate of surveillance and treatment of diabetic retinopathy *Diabetes Care* **28** 318-322.

[2] Jiménez S, Alemany P, Fondón I, Foncubierta A, Ancha B and Serrano C 2010 Detección automática de vasos en retinografías *Archivos de la Sociedad Española de Oftalmología (Elsevier)* **85** 103-109.

[3] Fong DS, Aiello L, Gardner TW, King GL, Blankenship G, Cavallerano JD, Ferris FL and Klein R 2003 Diabetic retinopathy *Diabetes Care* **26** 226-229.

[4] Facon J, Morfología 1996 *Matemática. Teoría e ejemplos*. Editora Universitária Champagnat da Pontífica Universidade Católica do Paraná.

[5] Ledley, S., Buas, M., Golab, T. (1990). Fundamentals of true-color image processing. *Proceedings of 10th International conference on Pattern Recognition*, Atlantic City. pp 791-795. 1990.

[6] Plataniotis, K., Venetsanopoulos, A. (2000). *Color Image Processing and Applications.* Springer-Verlag. Plataniotis.

[7] Ortiz, F., Torres, F., Puentes, S., Candelas, F., Gil, P., (2000). *Use of the Hue/Saturation/Intensity Colour Spaces to the Morphological Processing of Colour Images.* Proceedings of First International Conference in Graphics and Image Processing. Saint-Etienne, France. Pp 219-224.

[8] Mary L. Comer and Edward J. Delp (1999). *Morphological operations for color images processing.* Journal of Electronics Imaging; pp: 279-289.

[9] Ortiz, F., Torres, F., Angulo, J., Puente, S., (2001). *Comparative study of vectorial Morphological operations in different colour space.* Proceedings of SPIE, Intelligent Robots and Computer Vision XX, 4572, pp 259-268.

[10] Serra J 1982 *Image analysis and mathematical morphology vol. 1.* Academic Press, London.

[11] Serra J 1988 *Image analysis and mathematical morphology vol. 2*. Academic Press, London.
[12] Bouchet A, Brun M and Ballarin V 2010 Morfología Matemática Difusa aplicada a la segmentación de angiografías retinales *Revista Argentina de Bioingeniería* **16** 7-10.
[13] Bouchet A, Pastore J and Ballarin V 2007 Segmentation of Medical Images using Fuzzy Mathematical Morphology *Journal of Computer Science and Technology* **7** 256-262.
[14] Bouchet A, Pastore J and Ballarin V 2005 Segmentación de Imágenes a través de Morfología Matemática Difusa *In Reunión de Procesamiento de la Información y Control (RPIC 2005)* Universidad Nacional de Río Cuarto, Córdoba.
[15] Bloch I and Maitre H 1995 Fuzzy mathematical morphologies: A comparative study *Pattern Recognition* **28** 1341-1387.
[16] Dubois D and Prade H 1980 *Fuzzy Sets and Systems: Theory and Applications* Academic Press Inc, New York.
[17] Klement EP, Mesiar R and Pap E 2004 Triangular norms. position paper I: basic analytical and algebraic properties *Fuzzy Sets and Systems* **143** 5-26.
[18] Staal JJ, Abramoff MD, Niemeijer M, Viergever MA and Van Ginneken B 2004 Ridge based vessel segmentation in color images of the retina *IEEE Transactions on Medical Imaging* **23** 501-509. http://www.isi.uu.nl/Research/Databases/DRIVE/.
[19] R. Gonzalez and R. Woods 1992 *Digital Image Processing*. Adison–Wesley.
[20] D. S. Comas, A. Bouchet and J. I. Pastore 2011 Interfaz gráfica para desarrollo de prototipos de algoritmos de procesamiento de imágenes con Morfología Matemática Difusa *In 40º Jornadas Argentinas de Informática (JAIIO 2011)* Córdoba, Argentina. 55-66.
[21] F. Provost and T. Fawcett 1997 Analysis and Visualization of Classifier Performance: Comparison Ander Imprecise Class and Cost Distributions *Proc. Third International Conference on Knowledge Discovery and Data Mining* 43–48.

Measuring Leaf Area in Soy Plants by HSI Color Model Filtering and Mathematical Morphology

M Benalcázar [1,3], J Padín [1,3], M Brun [1], J Pastore [1,3], V Ballarin [1], L Peirone [2,3] and G Pereyra [2,3]

[1] Laboratorio de Procesamiento Digital de Imágenes, Departamento de Electrónica, Universidad Nacional de Mar del Plata, Buenos Aires, Argentina.
[2] Laboratorio de Fisiología Vegetal, Unidad Integrada Balcarce (Facultad de Ciencias Agrarias, Universidad Nacional de Mar del Plata - Instituto Nacional de Tecnología Agropecuaria).
[3] Consejo Nacional de Investigaciones Científicas y Técnicas (CONICET).

Mail: marco_benalcazar@hotmail.com

Abstract. There has been lately a significant progress in automating tasks for the agricultural sector. One of the advances is the development of robots, based on computer vision, applied to care and management of soy crops. In this task, digital image processing plays an important role, but must solve some important problems, like the ones associated to the variations in lighting conditions during image acquisition. Such variations influence directly on the brightness level of the images to be processed. In this paper we propose an algorithm to segment and measure automatically the leaf area of soy plants. This information is used by the specialists to evaluate and compare the growth of different soy genotypes. This algorithm, based on color filtering using the HSI model, detects green objects from the image background. The segmentation of leaves (foliage) was made applying Mathematical Morphology. The foliage area was estimated counting the pixels that belong to the segmented leaves. From several experiments, consisting in applying the algorithm to measure the foliage of about fifty plants of various genotypes of soy, at different growth stages, we obtained successful results, despite the high brightness variations and shadows in the processed images.

1. Introduction

An important aspect that emerges in agriculture is the identification of new genes to improve soy crops. This research should allow for new varieties of soy plants with a strong immunity to diseases, requiring fewer quantities of chemical products and water. This kind of research requires constant monitoring and evaluation of plants growth over the time. For this reason, it is very important the accurately measuring of physical characteristics of plants, called the *phenotype*, such as area and leaf angle distribution, which allows the researchers to match it with the genetic structure of the individuals, called the *genotype*. Artificial vision, or computer vision, is a technique from artificial intelligence, which allows for non-invasive automation of these tasks, with a significant increase in the capacity of information processing [1].

One of the most significant problems at applying machine vision for automatic phenotyping of plants is to find an automated and efficient method for segmentation of plants. The main issue is present during the image acquisition, which is usually conducted in complex environments, exposed to continuous lighting changes. Different intensities and shadows in the images make too difficult the segmentation process because it causes deterioration of color and contrast, which are features used to separate the plant from the background of the image.

For the above reasons, it is fundamental to use the right digital image processing techniques, applying a profitable segmentation algorithm with parameters independent from brightness variations, which should reduce significantly the effects caused by lighting changes. In this way, it is possible to achieve optimal results on a large quantity and variety of images. In this paper we propose an automatic method for measure the foliage area after the leaves segmentation of soy plants, which showed a good performance on several light conditions.

2. Materials and Methods

2.1. *Images*

The images used for this work were obtained from an automated platform, which enables to photograph soy plants automatically. This platform is operating at the Unidad Integrada Balcarce that depends on the Facultad de Ciencias Agrarias, from the Universidad Nacional de Mar del Plata, and the Instituto Nacional de Tecnología Agropecuaria (INTA). The digital images have top and side views of soy plants with a size of 640x480 pixels, in the RGB color model and JPG format.

2.2. Automatic measurement method of leaf area

The proposed algorithm for automatic measurement of leaf area from digital images includes the following steps:

1. Color filtering
2. Binarization
3. Noise filtering
4. Foliage segmentation
5. Leaf area estimation

2.2.1. Color filtering

The color filtering is a fundamental and essential step previous the segmentation, because color is a powerful descriptor that simplifies object identification and extraction from a scene [2,3]. However, this process has different issues associated to the image acquisition, such as those caused by the changes in lighting conditions and when the background of the scene contains surfaces and objects with similar colors to the target [4]. Therefore, for color filtering it is important to select an appropriate color model that separates the brightness from chromaticity information. Chromaticity is used to distinguish the interest-object or target from the image background [5].

The color model selected to perform the filtering was the HSI model (hue, saturation and intensity). This model, different from others such as RGB (red, green and blue) that is oriented toward hardware [6], is very attractive for digital image processing because it represents a color using three components represented by the two solid cones shown in Figure 1. In this representation the hue (H) and saturation (S) are intimately related with the chromaticity information and are separated from the intensity (I). Hue is an attribute that describes a pure color and it is associated with the dominant wavelength in a mixture of light waves, their values are normalized in the range from 0 to 1. In this work, in order to simplify the image analysis and the filter definition, it will be ranged from 0 ° to 360 °. The saturation gives the degree to which a pure color is diluted by the white light, and their values are in the range from 0 to 1. The intensity represents the brightness information and their values are in the range from 0 to 1 [2].

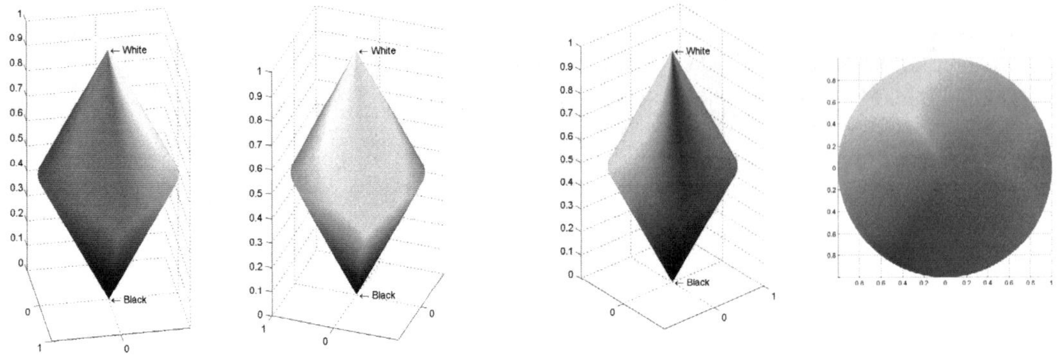

Figure 1. Two cones HSI solid: view with **(a)** H = 0º (red), **(b)** H = 120º (green), **(c)** H = 240º (blue) and **(d)** cross-section of the solid with I = 0.5

The color filter proposed in this paper is intended to distinguish the soy plant from the background of the image, by defining a region in the HSI color space which represents the green color of plants. For the thresholds selection of the filter, the images were transformed from the RGB to the HSI model, using the conversion equations defined in [2]. Then, it was analyzed the color distribution of leaves at the HS plane (Figure 2). From this analysis it was determined that the hue for soy plants varies approximately in the range from 20 ° to 180 °, whereas the saturation varies in the whole range of S (0 to 1). This region belongs to the range of yellow and green colors of the HS plane in the HSI model (Figure 3). The pixels intensity was approximately in the range from 0.2 to 1, whit the highest frequency in the range from 0.3 to 0.8. This variability in the intensity is due to images were taken in a greenhouse environment at different times of a day, with non-controlled daylight.

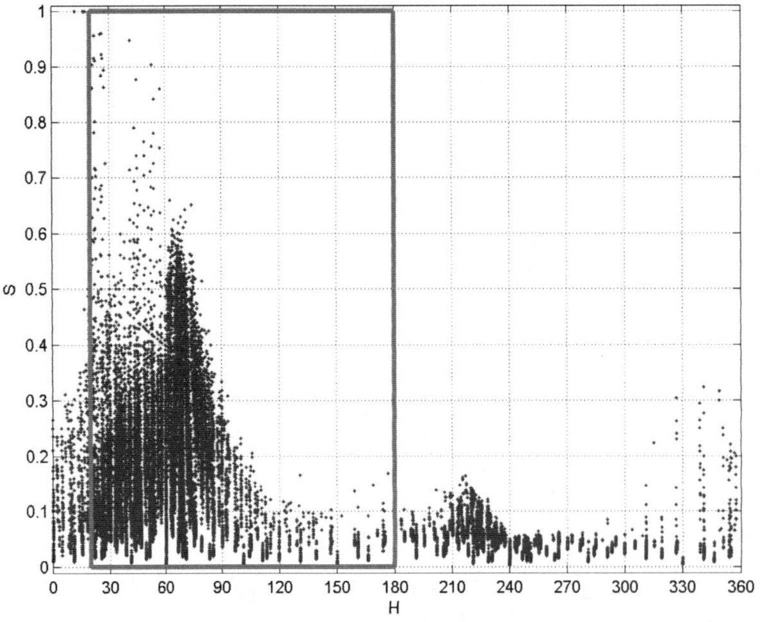

Figure 2. Color distribution of a soy plant image in the HS plane with an intensity variation from 0.2 to 0.8 and a maximum frequency at 0.7

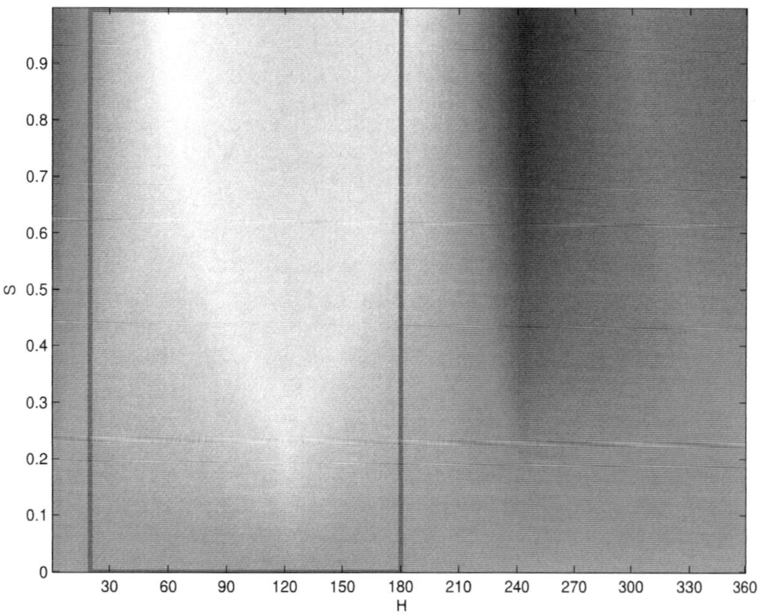

Figure 3. Color distribution in the HS plane of the HSI model with an intensity of 0.7

Depending on the analysis above, the filter between the plant and the rest of the scene is defined by the following equation:

$$H_f = \begin{cases} H & if \quad 20^o \le H \le 180^o \\ 0^o & otherwise \end{cases}$$

$$S_f = \begin{cases} S & if \quad 20^o \le H \le 180^o \\ 0 & otherwise \end{cases} \qquad (1)$$

$$I_f = \{ I_{frec} \in [0.2 \quad 0.9]$$

Where H_f is the filter output for hue and H is the original hue. S_f is the saturation of colors that passed the filter and S is the original saturation. I_f is the output for intensity and I_{frec} is the most frequent intensity in the range from 0.2 to 0.9 in the original image. The most frequent value, of the initial intensities in the mentioned range, was assigned to the filter output intensity to uniform the brightness across the whole image, and thus minimize the possible effects of shadows or excessive lighting at certain regions of the processed image.

After the image color filtering, it was converted back to the RGB model. The conversion was performed using the equations for this purpose defined in [2].

2.2.2. Binarization

Before foliage segmentation, it was performed a binarization of the resulting color filtered image, expressed in the RGB model. For this task the G (green) channel was used, because it contains the most useful information about the target. The image corresponding to this channel was treated as a grayscale image, to which was applied a contrast filter, via a linear transformation in the spatial domain, defined by the following equation:

$$G_c = \frac{G}{G_{\max}} \qquad (2)$$

Where G_c is the gray level of a pixel after the linear transformation, G is its initial gray level and $G_{\max}$ is the highest gray level among all the pixels in the G channel as it is shown in Figure 4.

For the selection of the binarization threshold the Otsu method was used [4]. This method selects a threshold that minimizes the intra-class variance (or the measure of separability of the resulting classes in gray levels) of black and white pixels. The resulting image from the binarization contains the interest-object represented by a 1 (white), and the background represented by 0 (black). This figure contains a certain amount of noise, both in the object of interest as in the background.

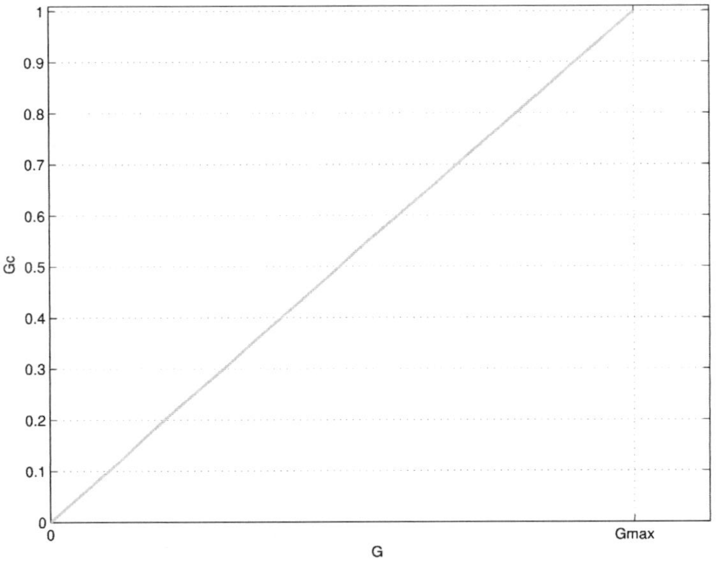

Figure 4. Contrast linear transformation function

2.2.3. Noise filtering

This step involved the removal of noise from the objects of interest, through the removal of objects smaller than 500 pixels. This value was selected heuristically based on the fact that all the size of branches and foliage of the plant, for all images analyzed, were above this value. The identification of objects was done by a connectivity 8 criterion. In a binary image, connectivity 8 defines a neighborhood between pixels in the horizontal, vertical and diagonals axis, as shown in Figure 5.

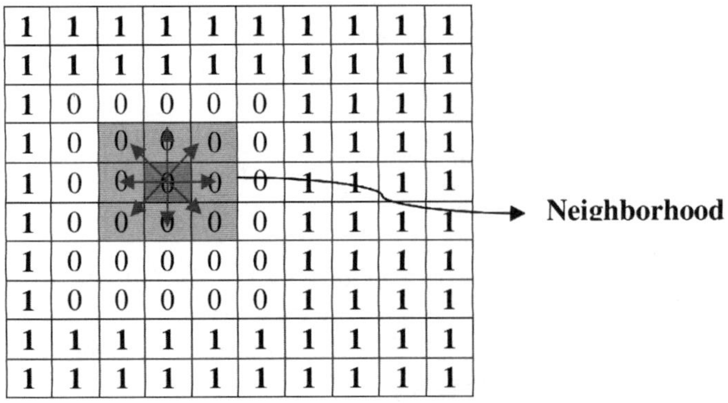

Figure 5. Graphical representation of connectivity 8 in a binary image

Journal of Physics: Conference Series **332** (2011) 012034 doi:10.1088/1742-6596/332/1/012034

Having identified the objects in the image, we proceeded to count the pixels present in each of these, and then eliminate those with smaller size than the threshold, and so we obtained a new noise-free binary image.

2.2.4. Foliage segmentation

The next step was separate the stems from the leaves, to measure the area of the leaves. The foliage segmentation was done by Mathematical Morphology. The operation used was the opening with a disk-type structural element of radius 10 (in pixels). This size corresponds roughly to the radius of the stems to be deleted. Opening is an operation consisting of an erosion, which removes objects smaller than the structuring element, in this case stems, followed by a dilation with the same structuring element used for erosion, which recovers the approximate shape of the eroded objects, in this case the leaves. The mathematical detail of the operations mentioned above can be found in [8,9].

As the final stage of segmentation, it was applied a process of elimination of holes inside the leaves. A hole was defined as a small set of background pixels that is not connected with the edges of the image [10]. The holes appear in the images because in some cases plants have an identification tag on its leaves, which is erased during the color filtering. Holes also originate due to a non-uniform and intense illumination of some leaves.

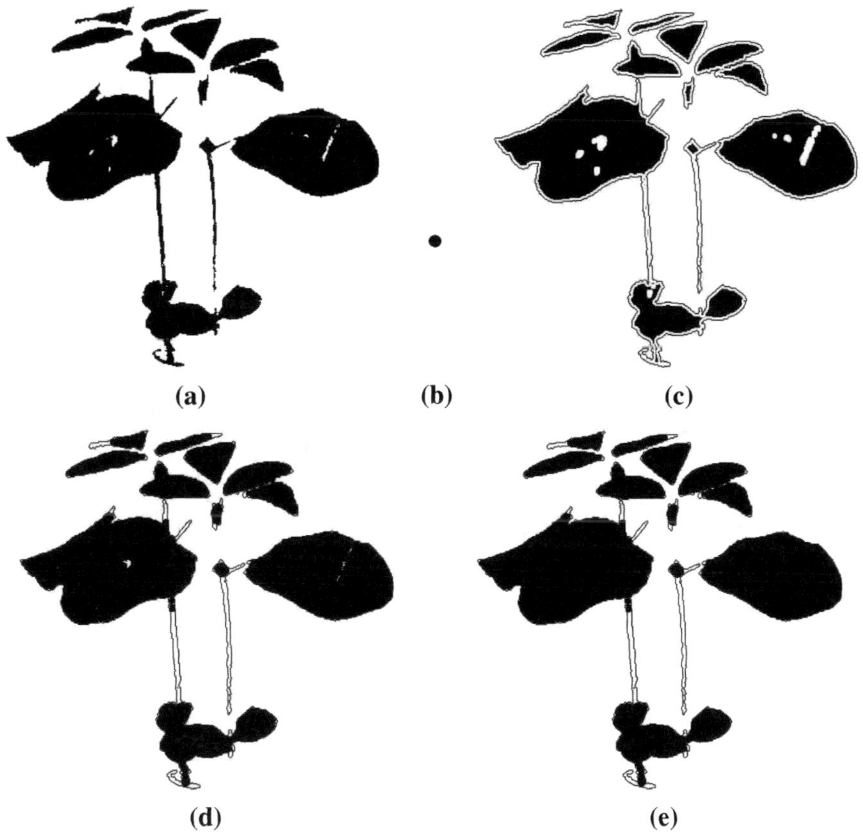

Figure 6. **(a)** Original image, **(b)** structuring element, **(c)** erosion of (a) by (b), **(d)** dilation of (c) by (b) and **(e)** elimination of holes in (d)

Figure 6 shows graphically the procedure used for the segmentation. To observe the effect of the operations, we assigned to the pixels that define the plant and the structuring element a gray value of 0 (black) and the background a gray value of 1 (white). We also superimposed to each result the original edges. In this way, we can see that the opening removed from the plant the stems and some leaves details. This does not significantly affect the final result because it still retains the bulk of the leaves. It can also be noticed that some leaves contain small holes that were eliminated after the opening.

2.2.5. Leaf area estimation
In this last stage, the identification of leaves was done using a similar procedure to the one used for the identification and subsequent elimination of small objects described in section 2.2.3. Once identified the pixels belonging to each of the leaves, using the connectivity 8 criterion, we proceeded to count them. The number of pixels of each leaf estimates the foliage area of it and can be converted to area units by multiplying by a conversion factor that depends, among other factors, of the distance between the camera and the plant.

Because in some cases there was an overlap of the leaves, especially in those images that contain a top view of the plant, the leaves overlapping were considered as a single object.

3. Results

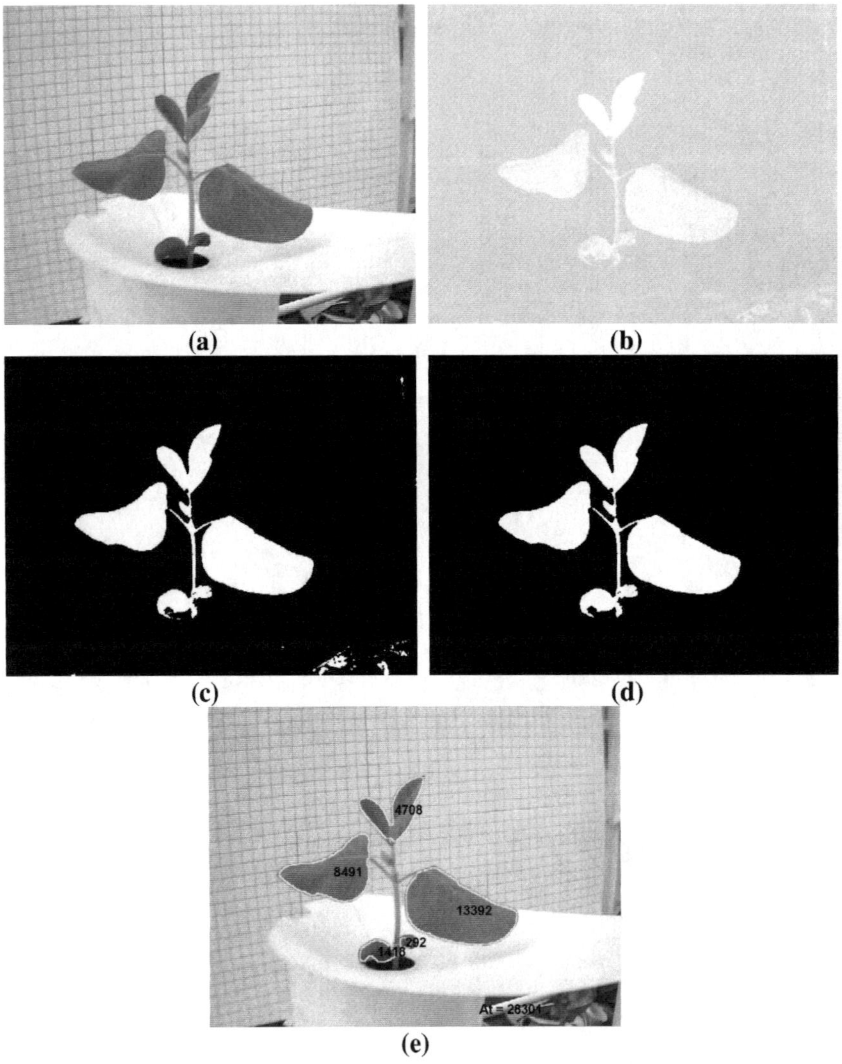

Figure 7. **(a)** Original image and resulting images from: **(b)** color filtering, **(c)** binarization of the G channel image resulting from the color filtering, **(d)** removal of objects smaller than 500 pixels and **(e)** final image with the area of each leaf and the total area in pixels.

In order to illustrate the steps used in the proposed algorithm, Figure 7 shows the results of the different stages of the algorithm applied to a side image of a soy plant, with high level of illumination. Figure (7a) presents the original image. Figure (7b) shows the result of applying the color filter in the HSI model. Figure (7c) shows the result of applying the binarization to the G channel of the resulting RGB image from the color filtering, with the selection of threshold by Otsu method. Figure (7d) shows the result of applying the noise filtering, through removing the objects smaller than 500 pixels and, finally, figure (7e) shows the result of the segmentation and subsequent estimation of area of each leaf. To improve the presentation of results, we superimposed the edges of the segmented leaves to the original image. Additionally, we added the value of the area in pixels in the center of each leaf and, the total value of area at the bottom right of image, which is the sum of all areas and corresponds to the foliage area estimation of the whole plant.

Figure 8. **(a)**, **(b)** and **(c)** are results for images with top view and different levels of illumination. **(d)**, **(e)**, **(f)**, **(g)** and **(h)** are results for images with side view and different levels of illumination.

Figure 8 shows some results obtained for several images of top and side views of soy plants, with different light levels and morphology. In images (a), (b) and (e) we can see that some leaves have not been detected due to excessive lighting in certain regions of the image, which deteriorates the color. This effect causes that during color filtering, some leaves are removed and considered as part of the background. In the case of images (c), (d), (f) (g) and (h) lighting is uniform, but varies from one image to another, however the algorithm works well in all these cases because the color filter is based on the hue and saturation and not at the level of illumination of the image. It should also be noticed that in some detected leaves, some details have been suppressed during the application of the morphological opening. This last fact is not significant because the estimated leaf area value is used only for comparative purposes.

Conclusions

This paper presents an algorithm to measure, automatically, the foliage area from an image of a soy plant. Because we used a color filter, based on the HSI model, together with the appropriate selection of the thresholds for hue, saturation and intensity, a correct detection of plants was achieved, regardless of the different levels or changes in brightness of the processed images.

Despite the wide variety of processed images tested, in many cases with significant differences, as a result of changes in lighting conditions during the acquisition stage, we were able to set general values for the parameters used in each stage of the algorithm, allowing us to process large groups of images, with a little processing time, and no human intervention, which, along with a proper presentation of the results, facilitates the analysis and comparison of growth for different soy genotypes.

References

[1] Biskup B., Scharr H., Schurr U., Rascher U.; 2007; A stereo imaging system for measuring structural parameters of plant canopies; Plant, Cell and Environment 30; pp. 1299-1308.
[2] Gonzalez R., Woods R.; 1992; Digital Image Processing, Addison - Wesley; pp. 221-248.
[3] Ordoñez C., Bouchet A., Pastore J., Blotta E.; 2011; Aplicación de filtros frecuenciales y cromáticos en tiempo real para la toma de biopsias dirigidas en endoscopias; V Congreso Latinoamericano de Ingeniería Biomédica (CLAIB2011).
[4] Chiho K.,Bum-Jae Y., Mun-Ho J., Hagbae K.; 2008; Color segmentation robust to brightness variations by using B-spline curve modelin; Pattern Recognition 41; pp. 22-37.
[5] Qazi I., Alata O., Burie J., Moussa A., Maloigne C.; 2011; Choice of a pertinent color space for color texture characterization using parametric spectral analysis; Pattern Recognition 44; pp. 16-31.
[6] Bouchet A., Pastore J., Abras G., Jury G., Jury H., Ballarin V.; 2007; Barrett esophagus: guided biopsies through digital image processing; Journal of Physics; vol. 90; ISSN 1742-6596 (On - Line); ISSN 1742-6588 (Print).
[7] Otsu N.; 1979; A Threshold Selection Method from Gray-Level Histograms; IEEE Transactions on Systems, Man, and Cybernetics; vol. 9; pp. 62-66.
[8] Serra J.; 1989; Image Analisis and Mathematical Mophology; Academic Pres; vol. 1; pp. 43-56.
[9] Gonzalez M., Ballarin V., Celín A., Sánche V., Rapacioli M., Flores V.; 2009; Análisis de axones en crecimiento mediante la función Granulométrica de Tamaños; XVII Congreso Argentino de Bioingeniería - VI Jornadas de Ingeniería Clínica (SABI); ISBN: 978-950-605-505-9.
[10] Soille P.; 2003; Morphological Image Analysis: Principles and Applications; Springer - Verlag; pp. 210.

Journal of Physics: Conference Series **332** (2011) 012035 doi:10.1088/1742-6596/332/1/012035

Algorithms evaluation for fundus images enhancement

V Braem[1], M Marcos[1], G Bizai[1], B Drozdowicz[1,2] y A Salvatelli[1]

[1]Grupo de Inteligencia Artificial, Facultad de Ingeniería, Universidad Nacional de Entre Ríos. [2]Facultad de Ciencia y Tecnología, Universidad Autónoma de Entre Ríos.
E-mail: gbizai@bioingenieria.edu.ar

Abstract. Color images of the retina inherently involve noise and illumination artifacts. In order to improve the diagnostic quality of the images, it is desirable to homogenize the non-uniform illumination and increase contrast while preserving color characteristics. The visual result of different pre-processing techniques can be very dissimilar and it is necessary to make an objective assessment of the techniques in order to select the most suitable. In this article the performance of eight algorithms to correct the non-uniform illumination, contrast modification and color preservation was evaluated. In order to choose the most suitable a general score was proposed. The results got good impression from experts, although some differences suggest that not necessarily the best statistical quality of image is the one of best diagnostic quality to the trained doctor eye. This means that the best pre-processing algorithm for an automatic classification may be different to the most suitable one for visual diagnosis. However, both should result in the same final diagnosis.

1. Introduction

In general, image pre-processing aims to improve the visual quality and basically prepare it for a subsequent automatic processing. Fundus images acquired with a fundus camera have artifacts such as noise and a characteristic effect of non-uniform illumination [1]. The illumination heterogeneity is due to the combined effects of the eyeball geometry, the dilation of the pupil, the alignment of the eye with the optical axis of the camera and the direction and shape of the lighting system.

A large number of algorithms that reduce the noise in the image, have been widely explored and developed. In contrast, the non-uniform illumination correction is a more difficult task that requires the combination of more complex techniques.

Both noise and illumination artifacts can hide information of diagnostic interest. In particular, diabetic retinopathy is a retinal complication that may lead to blindness. A fundus examination may provide an early diagnosis when the expert is able to detect the first signs, consisting of small aneurysms and exudates [2].

The visual results can be very dissimilar depending on the pre-processing technique applied. A pre-processed image can be seen as an improvement by an inexperienced observer, while an expert may consider it an image of a low diagnosis quality. The aim at this stage is, to standardize the images for later stages of unsupervised classification. It is also expected to provide the specialist with an image of better quality than the original for medical evaluation.

The performance of some pre-processing techniques was tested based on the results of two segmentation schemes: Fuzzy C-means and Fractal Dimension [1]. The diversity of the pre-process results (visual results) leads to the necessity of proposing a quantitative method to judge, according to the change of some parameters, which is the performance of the tested techniques.

2. Materials and methods

2.1. Materials

Images from the "DIARETB1-Standard Database Diabetic Retinopathy" database, collected as part of the "ImageRet" project, from Lappeenranta University of Technology Kuopio, Finland, were used. They are available online at no cost [3]. The database consists of 89 images which were acquired using a digital fundus camera in 50 degrees field, with different image settings (flash intensity, shutter speed, aperture, gain), controlled by the system itself. Images contain noise. Optical aberrations and photometric accuracy are the same for all.

Thirty images chosen randomly from the database were used in this work. In order to reduce the Gaussian noise present in the images, a Wiener adaptive filter was applied to all images. The window was selected as 3x3, so as not to affect the small structures of the retina.

2.2. Methods

From the literature, eight algorithms were selected and implemented. The algorithms deal separately with the illumination problems, contrast enhancement and color preservation. The main purpose of the chosen techniques is the correction of non-uniform illumination, but it was also considered the contrast improvement and color preservation.

It should be noted that the implemented algorithms use a mask that defines the region of interest (ROI), as proposed by Frank ter Haar [4]. The red channel is binarized and the mask with morphological operators is delimited. The mask eliminates undesirable border effects.

An overview of the proposed algorithms follows.

Algorithm 1

The first algorithm is based on the Equalization of Illumination technique [5]. Each pixel is adjusted according to the equation:

$$I_{Eql}(x,y)=I(x,y)+m-A(x,y)$$

where m is the desired average intensity (128 in a grayscale image of 8 bits) and A (x,y) is the average value of pixels in an NxN window W, centered on the pixel to be set .

Algorithm 2

It uses the Homomorphic Filtering technique, which considers the image as the product of an illumination component I (x,y) and the reflectance component R (x,y). I(x,y) has smooth variations and its spectrum is concentrated on the low frequencies, while R(x,y) shows sharp variations with its spectrum concentrated on the high frequencies. The idea is to isolate the reflectance R and remove the illumination I in order to obtain an image without illumination variations.

Algorithm 3

This algorithm applies an averaging filter to each color channel C in order to obtain an illumination estimate LC. The mean value mLC is subtracted from LC and the corrected channel CC is:

$$CC=C-(LC-mLC)$$

Later, the Local Contrast Enhancement (LCE) technique proposed by Xiaohui and Chutatape [6] is applied. Considering a sub-size image W NxN pixels size, centered on the pixel to be set, the mean $\langle f \rangle$ and standard deviation σ of the intensity is calculated. These local values are the parameters for the following transformation function:

$$\Psi_W(f)=\left[1+e^{\left(\frac{<f>_W-f(x,y)}{\sigma_W}\right)}\right]^{-1}$$

Algorithm 4
The mean value $\mu_{channel}$ of the logarithm of each channel is calculated and a high pass filter PA is applied. From the result of the high pass filter $PA_{channel}$ the mean value $\mu PA_{channel}$ is obtained, so that:

$$channel= \mu_{channel}+A.(PA_{channel}- \mu PA_{channel})$$

where A is an constant equal for the three channels. The aim is to obtain the reflectance alone, without illumination, while preserving the original image colors. Finally the LCE algorithm is applied.

Algorithm 5
Both a high pass and a low pass filter are applied to the logarithm of each color channel to obtain $PA_{channel}$ and $PB_{channel}$. Then the mean values $\mu PA_{channel}$ and $\mu PB_{channel}$ are calculated and the following formula is applied (A and B are weighting factors):

$$channel=channel+ A.(PA_{channel}- \mu PA_{channel})- B.(PB_{channel}- \mu PB_{channel})$$

By subtracting the low pass results, the luminosity variations are decreased. By adding the high pass results, the contrast of the retinal structures is increased., the original colors of the image are preserved by removing the value of the media. Finally the LCE algorithm is applied.

Algorithm 6
This one is similar to algorithm 5 except for the fact that instead of adding and subtracting the filtered images to the original channel, it is maken to its mean $\mu_{channel}$, and LCE is applied to the result.

Algorithm 7
This algorithm is based on the iterative Retinex technique developed by Frankle and McCann [7, 8] and applied to fundus images [9]. The original image in RGB color space is converted to YUV color model and the V channel only is used.

Algorithm 8
This technique was proposed by Jobson et al [10]. It is based on multiscale Retinex technique with some modifications [9] so as to keep color fidelity in the image.

3. Results
The gray level mean and standard deviation of the images before and after the Wiener filtering was evaluated. The results showed a variation average percentage of 0.0971% while the standard deviation decreased by an average of 10.522%. These results indicate that the image brightness was kept while the noise was decreased.

The 30 sample images were processed with each of the algorithms above, previously applying 3x3 window size Wiener filtering.

To assess the result of the algorithms on the illumination homogeneity and the contrast, each original image was processed and transformed into grayscale images. Then one hundred 20 x 20 pixels ROI randomly located in the grayscale image were selected. The mean value μ and standard deviation σ was calculated for every ROI

3.1. Algorithms performance to correct uneven illumination.
The average values of the original and processed ROI were used as illumination estimators in that area of the image. The standard deviation of the 100 samples was calculated to get an idea of their

dispersion, which gives an idea of the uniformity of illumination (greater deviation is related to less uniform illumination). Figure 1 shows the result for Algorithm 2 (the one of worst performance in homogenizing illumination) and Algorithm 3 (the one of best performance in this task.) Table 1 summarizes the obtained values for an image.

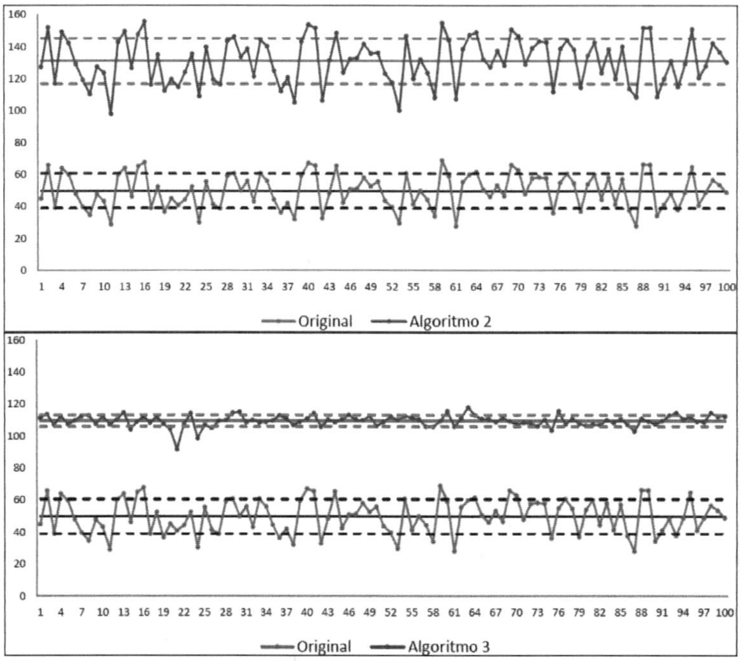

Figure 1. Brightness values for an image with and without the use of the algorithms. The Y axis represents the average value of each ROI, solid line indicates the average value and dotted line represents the deviation of the mean.

The 30 images were processed and the mean and standard deviation was obtained for each algorithm. Considering these results, the number of images for which the algorithms decreased or increased their illumination homogeneity with respect to the original was obtained. Algorithms 3, 4 and 7 increase the homogeneity a 100% every time (the homogeneity increased in all 30 analyzed images), algorithms 5, 6 and 8 show values above 80%; Algorithm 1 gets close to 50%. Finally, Algorithm 2 yields poor results because the 74% of the times worsens the homogeneity.

Table 1. Average illumination and standard deviation for each algorithm, for an image.

	Original	Alg. 1	Alg. 2	Alg. 3	Alg. 4	Alg. 5	Alg. 6	Alg. 7	Alg. 8
Average Ilumination	49.82	87.46	130.95	109.36	102.43	72.98	92.85	89.00	88.75
Standard D	10.82	12.84	14.25	3.61	4.20	7.22	10.42	8.74	9.84

Finally, the average value measured in which each algorithm improves or worsens the illumination homogeneity in regards to the original was quantified. The average value of the brightness deviation for each algorithm was calculated for all images and it was compared with the one obtained for the image without any processing. The algorithm showing better performance is Algorithm 3, which increased the homogeneity value in 78.74% regarding the original value, followed by Algorithm 4 with a 77.86%, Algorithm 8, 7, 5 and finally Algorithm 6. Algorithm 1 almost did not change and Algorithm 2 worsened the value 23%. These results are shown in Figure 2.

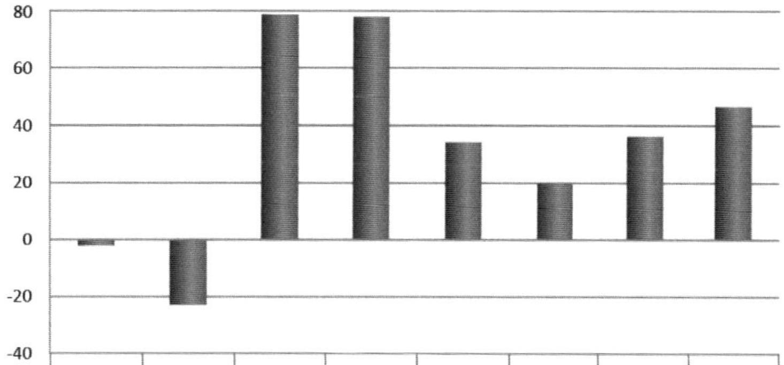

	Algoritmo1	Algoritmo2	Algoritmo3	Algoritmo4	Algoritmo5	Algortimo6	Algoritmo7	Algoritmo8
Porcentaje	-2,07	-22,86	78,74	77,86	34,08	19,89	36,13	46,57

Figure 2. Average value of improvement or worsening of illumination homogeneity regarding the original.

Figure 3 compares an image and the result of applying the algorithm 3, the one with best performance for this case.

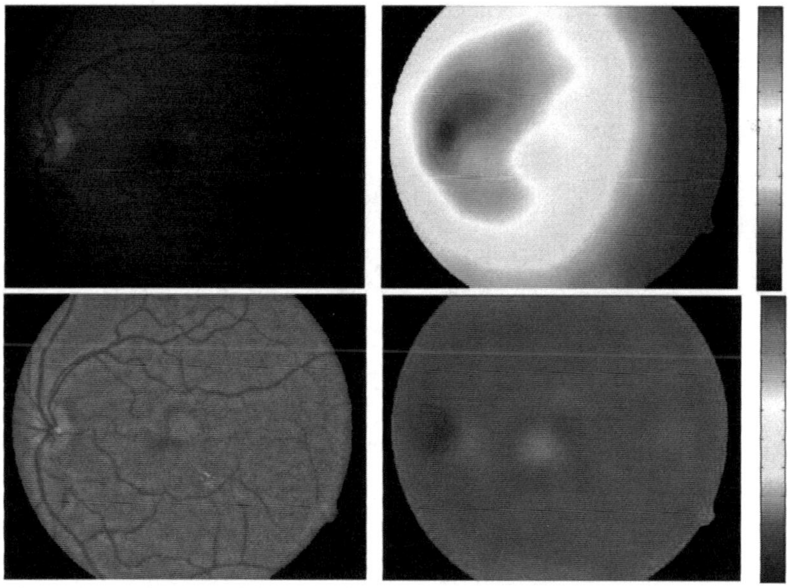

Figure 3. An image of the sample with light distribution: at the top shows the image without any processing and at the bottom the one processed with Algorithm 3

3.2. Contrast evaluation.

In order to analyze the change in contrast induced for each algorithm, a similar analysis was performed. In this case the 100 standard deviation data was used, which gave an idea of the image contrast.

Figure 4 shows an example of algorithms 5 and 8 applied to one image of the database. The figure shows the increasing or decreasing of the contrast of the processed image with respect to the original one.

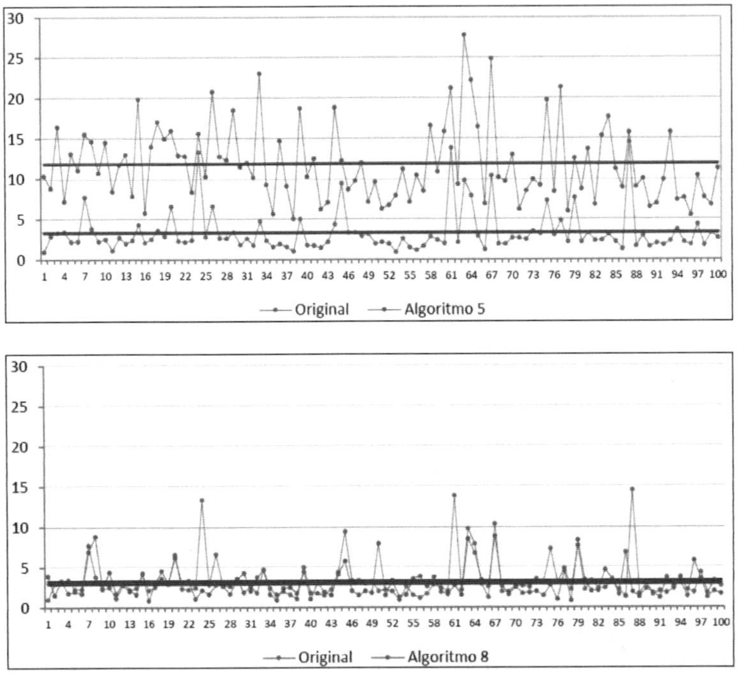

Figure 4. The original deviation in blue and the deviation of algorithms 5 and 8 in red, as example, for an image. The blue and red lines correspond to the average values of the deviations of the original image and the processed one respectively.

Then the average value of the ROI deviations for all original and processed images was calculated and the number of images in which the algorithms decreased and increased the contrast with respect to the original was calculated. Algorithm 3 and 5 always increased the contrast.

Finally the average value of deviations was compared with the average of the original images, in order to obtain the percentage of general contrast enhancement. This can be seen in Figure 5, which also shows that Algorithm 5 is the one that most increases the contrast, followed by Algorithm 4 and 3. Figure 6 shows an example of contrast increasing in an image of the sample.

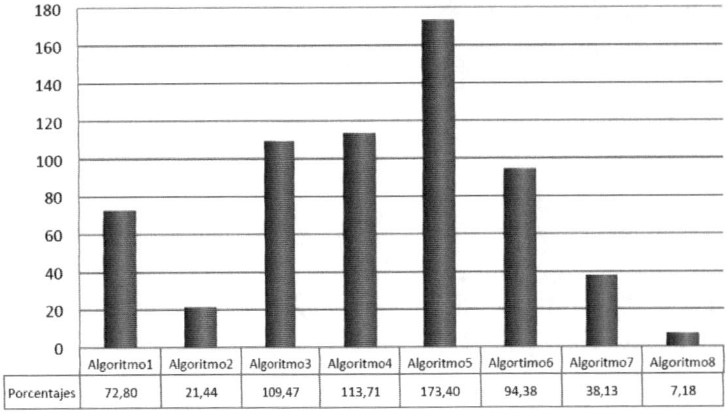

Figure 5. Increasing percentage of contrast obtained by each algorithm, regarding the average contrast of the original images.

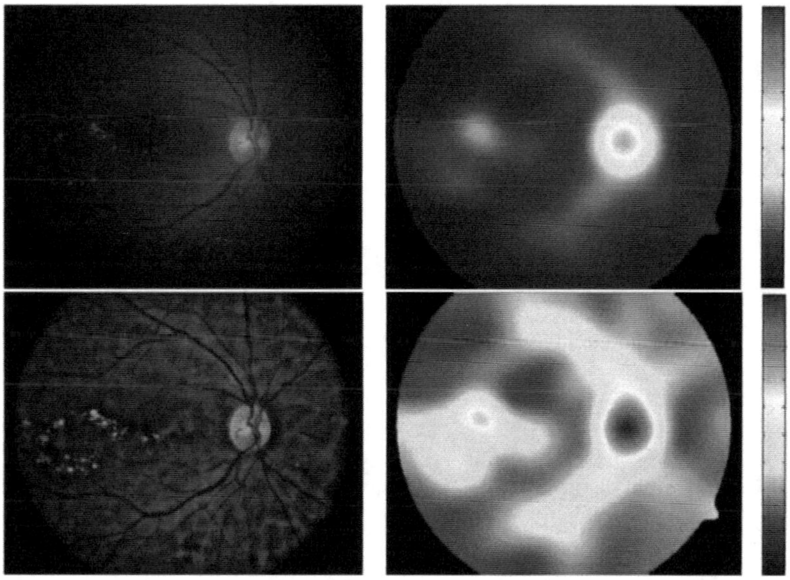

Figure 6. An image together with its contrast distribution: at the top, the image without any processing. At the bottom, the image processed with Algorithm 5.

3.3. Color preservation evaluation.
With the purpose of assessing if the chromatic features between original and processed images were kept, the R, G and B channels from both the original and the processed images were obtained.

A color normalization of each channel pixel by pixel is performed for all images as follows (for the red channel R (x, y)):

$$r(x,y) = \frac{R(x,y)}{D(x,y)} \qquad \text{with} \qquad D(x,y) = R(x,y) + G(y,y) + B(x,y)$$

Finally, whit these normalized values the Euclidean distance between the processed channel and the original one for each color is calculated, giving a general estimation about the preservation of the color ratio. In order to obtain a single general value for each algorithm, the results are averaged in the three channels:

$$P_{dist} = \frac{R_{dist} + G_{dist} + B_{dist}}{3}$$

The results of each algorithm can be seen in Figure 7. The smaller the distance, the more the color characteristics between the original and the processed image are kept. It can be noticed that the algorithm that best preserves color is 8 (multiscale Retinex), while the worst is 7 (iterative Retinex).

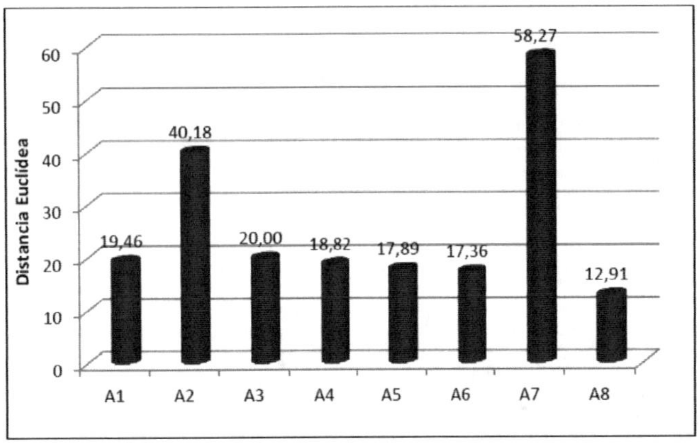

Figure 7. Average Euclidean distances of the color channels for each algorithm.

3.4. General Performance.

The results obtained in illumination homogenization, in contrast and in color preservation were normalized between 0 and 1. For illumination and contrast it was performed as follows:

$$V_N = \left(\frac{V - V_{min}}{V_{max} - V_{min}} \right)$$

A value of 1 represents the best performance and 0 represents the worst. In color preservation the best result is the smaller, so the scores are calculated as follows:

$$V_N = 1 - \left(\frac{V - V_{min}}{V_{max} - V_{min}} \right)$$

As shown in Table 2 and Figure 8, Algorithm 4 obtains the highest value, together with Algorithms 3 and 5. Then, with a slightly lower value are placed Algorithms 6, 8 and 1. Finally, the one of lowest score is Algorithm 2. It should be mentioned that the algorithms with better scores were proposed by the authors, that is to say, those containing the major modifications to standard processes known.

Table 2. Performance Scores for each algorithm.

	Alg1	Alg2	Alg3	Alg4	Alg5	Alg6	Alg7	Alg8
Illum. Homogeneity	0.20	0.00	1.00	0.99	0.56	0.42	0.58	0.68
Contrast	0.39	0.08	0.61	0.64	1.00	0.52	0.18	0.00
Color Preservation	0.86	0.40	0.84	0.87	0.89	0.90	0.00	1.00
Final Score	**1.45**	**0.48**	**2.46**	**2.50**	**2.45**	**1.85**	**0.76**	**1.68**

3.5. Expert opinion.

The proposed algorithms were pre-selected based on expert opinion. For this, a series of processed images was presented to an expert and he was asked to choose the one that he considered as the best for diagnosis. Based on this classification, algorithms 1, 3 and 6 were chose.

Another expert was provided with a graphic interface so as to evaluate which image was of his preference (original or processed) for diagnosis and which structure he considered as the best in the processed image: optic nerve, macula, vessels, none or another (to be specified by him). As a first result, in the 95.56% of the cases (86 of 90 images), the ophthalmologist chose the processed image instead of the original.

Regarding the performance of the algorithms, Algorithm 3 is the best in the opinion of the expert. Among the 30 images processed with this algorithm, 29 were chosen, generally standing out the three main features of the images (optic nerve, macula and arterial tree). Algorithms 1 and 6 also show good results in improving visualization of retinal vessels. In contrast, regarding optic disc visualization improvement, these algorithms were chosen a 50% of the cases.

The expert remarked the importance of maintaining the optic disc original color, as it is relevant while making a diagnosis and patient monitoring, as well as the characteristic orange color of the arteries compared to wine-red veins, because facilitates differentiation.

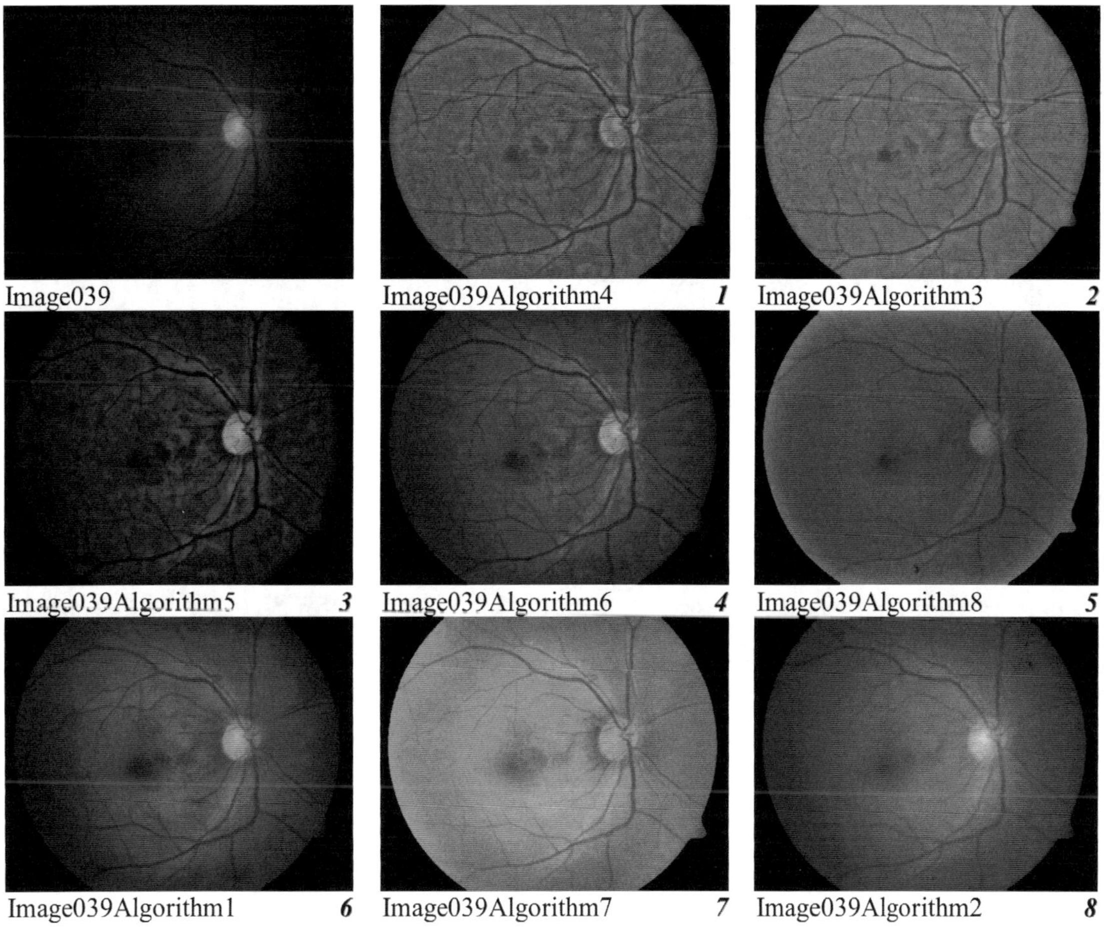

Figure 8. An original image and processed images by the algorithms in order of performance.

4. Conclusions and discussion

The proposed algorithms get different scores when evaluated for improvement of homogeneity of illumination, contrast enhancement or color preservation. In order to choose the one that produces the best result in general improvement of the image (which includes the three properties mentioned), Algorithm 4 is the one that achieves the best score. In turn, algorithms 3 and 5 get scores very similar to each other (2.46 and 2.45) and very close to the first (2.50). Figure 8 shows the results of the algorithms applied to the sample.

It should be noted that Algorithms 3 to 6 which are supported by the model where the image is the product of the illumination and reflectance, do not use standard methods, but combine results of high

Journal of Physics: Conference Series **332** (2011) 012035 doi:10.1088/1742-6596/332/1/012035

pass and low-pass filters with different weights given by the authors. The aim was to emphasize the reflectance component and attenuate the illumination.

When asked to choose the best algorithm for diagnoses, the expert did not choose the best qualified algorithms according to our score method. This supports the hypothesis that not necessarily the statistically best image is the best for diagnosis as evaluated by the experts. This means that the best pre-processing algorithm for an automatic classification may be different to the most suitable one for visual diagnosis. However, both should result in the same final diagnosis. To conclude, it is considered that the Algorithm 3 is the one that provides the best pre-processing performance obtaining a high overall score in the comparative scale. Hence, it is considered appropriated for the pre-processing of images that must be subjected to unsupervised processing. Besides, the experts considered the images pre-processed with Algorithm 3 as useful, indicating that it provides appropriate enhancement of the image and preserves the usual conditions expected by the expert for visual diagnosis.

References

[1] Salvatelli A, Bizai G, Barbosa G, Drozdowicz B and Delrieux C 2007 A comparative analysis of pre-processing techniques in color retinal images J. of Physics: Conference Series, Institute of Physics, http://www.iop.org/EJ/toc/1742-6596/90/1

[2] Drozdowicz B, Caropresi J, Salvatelli A, Bizai G 2007 Sistema multiagente para apoyo a la detección y análisis de patologías diabéticas en imágenes retinográficas X Simposio de informática en salud SIS2007

[3] Kauppi, Tomi et al. 2008 DIARETDB1 Diabetic Retinopathy Database and Evaluation Protocol. [On line] Medical Data Acquisition and Image Processing, doi=10.1.1.102.1922.

[4] Haar, Frank ter. 2005 Automatic localization of the optic disc in digital color images of the human retina. M.S. Thesis, Utrecht University

[5] Hoover A and Goldbaum M 2003 Locating the optic nerve in a retinal image using the fuzzy convergence of the blood vessels IEEE Transactions on Medical Imaging ISSN: 0278-0062 22(8) 951-958

[6] Xiaohui Zhang and Chutatape Opas 2004 Detection and classification of bright lesion in color fundus images. [On line] School of Electrical & Electronic Engineering, Nanyang Technological University

[7] Frankle J and McCann J 1983 Method and apparatus for lightness imaging. US Patent No. 4384336

[8] Funt B, Ciurea F and McCann J 2004 Retinex in MATLAB™ Journal of electronic imaging, 13 48-57

[9] Durango Londoño N, Bizai G and Drozdowicz B 2009 Implementación y aplicación de algoritmos Retinex al preprocesamiento de imágenes de retinografía color [On line] Revista Ingeniería Biomédica ISSN 1909-9762, 3(6) 36-46

[10] Jobson D, Rahman Z and Woodell G 1997 A multiscale Retinex for bridging the gap between color images and the human observation of scenes IEEE Transactions on image processing ISSN: 1057-7149 6(7) 965-976

SABI 2011 IOP Publishing
Journal of Physics: Conference Series **332** (2011) 012036 doi:10.1088/1742-6596/332/1/012036

Analysis and Implementation of Methodologies for the Monitoring of Changes in Eye Fundus Images

A. Gelroth, D. Rodríguez, A. Salvatelli, B. Drozdowicz, G. Bizai

Artificial Intelligence Group, Facultad de Ingeniería, Universidad Nacional de Entre Ríos (UNER).

E-mail: asalvatelli@bioingenieria.edu.ar,

Abstract. We present a support system for changes detection in fundus images of the same patient taken at different time intervals. This process is useful for monitoring pathologies lasting for long periods of time, as are usually the ophthalmologic. We propose a flow of preprocessing, processing and postprocessing applied to a set of images selected from a public database, presenting pathological advances. A test interface was developed designed to select the images to be compared in order to apply the different methods developed and to display the results. We measure the system performance in terms of sensitivity, specificity and computation times. We have obtained good results, higher than 84% for the first two parameters and processing times lower than 3 seconds for 512x512 pixel images. For the specific case of detection of changes associated with bleeding, the system responds with sensitivity and specificity over 98%.

1. Introduction

Today, the application of new medical technologies is a reality that does not cover all practices and does not extend to all professionals. Moreover, although many of these professionals have access to computer systems with large volumes of information, these systems can be underutilized if they are not assisted with tools, to deal in a practical, effective, efficient and synergic way for their business.

In the case of diagnosis and treatment of eye diseases, it is a routine practice to take photographs of the fundus, which are then analyzed in order to find characteristic signs and symptoms, thus evaluating the possible pathological development. In the analysis of the evolution of ocular pathology, by comparing images of the same patient taken at different time intervals, it is unusual for specialists to have computerized assistance to facilitate the observation of these images simultaneously and even less to assist in distinguishing changes.

We study methods that take advantage of digital images acquired at different time intervals, to change detection. Therefore we aim to assist professional ophthalmologists, through the computer system, by facilitating the monitoring of their patients and achieving a positive impact on treatment.

The selected methodologies are part of a test system that will allow us to analyze the results of its implementation on a series of images.

2. Materials and Methods

2.1. Materials

Based on online public databases (Neuro-Ophthalmology Collection Dr. William F. Hoyt, Atlas of eye disorders "RedAtlas.org - Recognizing Eye Disease: A Visual Review of Ophthalmic Disorders") pairs of images (each associated with a patient) were selected, taking into account clinically significant changes. These images were marked by specialists with medical practice, disregarding signs and diseases of low incidence. From this process, about 40 pairs of images were selected. Images generally come in two formats, JPEG and TIFF, RGB color with 8 bits for each of the 3 channels. The image sizes and resolutions range from 100 KBytes to 11 MBytes, with sizes of 128 x 128 to 2200 x 2200 pixels respectively.

We developed a graphical user interface that was necessary for the purposes of selection, combination and configuration of the methodologies to analyze.

2.1.1. Test images description

The set of images previously employed in the demarcation of the Gold Standart Masks (GSM) [1, 2] for parameter setting was grouped with a new set of pairs of images to form the "Test Group", comprising 9 pairs of images in total and more than 100 regions subject to change.

In the pairs of images belonging to the test group, changes associated with bleeding were predominant, following those associated with changes in the vessels, and other showed exudates. Most of these changes were found mainly in the area corresponding to the retina itself because of their greater size compared to the optic disc. Images with relative displacement greater than 20% in the photographic capture were discarded. At the same time changes in images containing edges of the optic disc were not considered, suggesting for these cases other methodologies with a different approach. Pairs of images associated with pathological signs that radically altered the nature of the retina (as can be the neoplasia and associated malignancies) were not considered.

2.2. Methods

The main considerations related to the selection of methodologies are directed toward a future development of an assistive system which can support ophthalmologists in the task of change detection in series of images.

2.2.1. Registration: Application of transformation functions with manual selection of point.

A set of pairs of points were generated manually, taking physical correspondence between the images that were part of each pair of fundus images [3]. One of the main advantages of this methodology lies in the simplicity of the method and the absence of problems, typical of automatic methods associated to identifying features when these are very distorted because of the pathological nature of the image.

From the pairs of points generated various transformation functions were applied [4-7], obtaining very good results with the affine and polynomial transformations. It was observed that the affine transformation required fewer pairs of points that the polynomial and was approximately 50% faster.

2.2.2. Masking relative to the size of the image

Masks were generated as circles centered on the center of the image, with diameters predefined by the user as a percentage of the dimensions of the image. This methodology allows us to separate the masking processes from the quality of the image and possible pre-masking.

2.2.3. Illumination correction - homomorphic filtering
The applications of this methodology [8-11] on the set of images gave good results. This methodology largely reduced the effects of nonuniform illumination distribution.

These good results were obtained for high-resolution images that were presented in Table I. However, this technique showed some problems associated with the presence of artifacts uniformly distributed in the image with high intensity values.

Table I: Duration in seconds of homomorphic filtering to pairs of images of size 256 x 256, 512 x 512, 1345 x 1134 and 2048 x 2048 pixels

Stage	256x256	512x512	1345x1134	2048x2048
Homomorphic Filtering	0,75	0,98	11,7	53,4

2.2.4. Illumination correction – Ratio of green channel and red channel (G / R)
The implementation of this methodology requires the preparation of images, to prevent zero division, and aims to override the lighting components that were present in green and red channels, generating a grayscale image in which pixels intensities were weighted that more differences exhibited between these channels [2.12 to 15].

Good results were obtained when the lighting components were minimized, with excellent processing times for this stage. These are presented in Table II.

Table II: Duration in seconds of the ratio G / R, for pairs of images of size 256 x 256, 512 x 512, 1345 x 1134 and 2048 x 2048 pixels

Stage	256x256	512x512	1345x1134	2048x2048
Ratio G / R	0,01	0,06	0,3	0,9

2.2.5. Global statistical normalization
This technique, based on the definition of identical mean values and standard deviation for all images used, produced good results for the following stages.

Here, it was observed that the use of this technique allowed the use of parameters such as threshold value for binarization stage, which were practically independent from the pair of images used. Thus, these parameters could be defined solely by the combination of methodologies used in different stages of processing [16-18].

2.3. Processing

2.3.1. Calculation of the Image Difference - Simple Difference
This technique consists of subtracting from a reference image a second image of the same scene, in which there may be changes. Algorithms were implemented to obtain the absolute value of subtraction and solve potential format problems.

2.3.2. Binarization Image Difference - Fixed threshold
The use of a fixed threshold is valid only when it is preceded by an appropriate standardization process. The process of defining the threshold was based, for each combination of methodologies, from masks generated by the expert, finding values that vary slightly for all test pairs.

2.4. Post Processing

2.4.1. Processing mask change - regions detection and filter properties.
We implemented the detection and segmentation of clusters of pixels in the changes mask [19]. Then features such as its centroid, number of pixels, bounding boxes, etc, were extracted.

Based on the number of pixels associated with each area it is possible to remove the masking areas that do not exceed a minimum. This threshold was defined based by experience and with the help of the MGS generated by the expert. In this way it was possible to remove isolated pixels associated with noise, without removing or modifying areas associated with significant changes, capacity that other methodologies, such as binary operators do not have [12.19-20]. Thresholds are defined in relation to the size of the images because the regions of change will grow proportionately to the size of the images to be processed and, therefore, the thresholds will adapts to those changes (for details see the final flow diagram).

2.5. Results presentation.

We have chosen to overlap the original images with the perimeter of areas with detected changes, so that we can compare these to the areas drawn from the MGS generated by the expert. Also, changes could be demarcated as areas filled through the use of bounding boxes [21-22]. The latter help to define areas of changes rather than changes, which would facilitate the specialist to evaluate in the same image the significant signs without obstruct her/his observation.

2.6. Workflow final

In Figure 1, we present the process of transforming the input images along the workflow, resulting in final change mask. This workflow embedded the image with the numbered regions from the mask generated by the system, which was provided to the expert.

2.7. Performance measurement methodology

Each region detected by the system was enumerated manually. Furthermore, for each pair of images, a template was generated. This, together with pairs of images and their corresponding masks demarcated by the system, was referred to the specialist to be completed, as indicated in Table III and IV.

Table III: Template Associated with each pair of images

Image N°	Region	**Sign** Hemorrhage (H) Exudate (E) Vessels (V) Optic Nerve (O) Retina (R) Otros	**Type of change** Appears (A) Disappears (D) Form (F) Texture	Subtype if there	Properly Change Detected	Improperly Change Detected	Change Detected
	1	H	A	Perif.	x		
1	2	E	F	.-.		x	
	3	.-.	.-.	.-.			x
	4	Otros	A	.-.	x		
Non Detect Change		H	E	V	O	R	Otros
		2	1	4	2	4	0

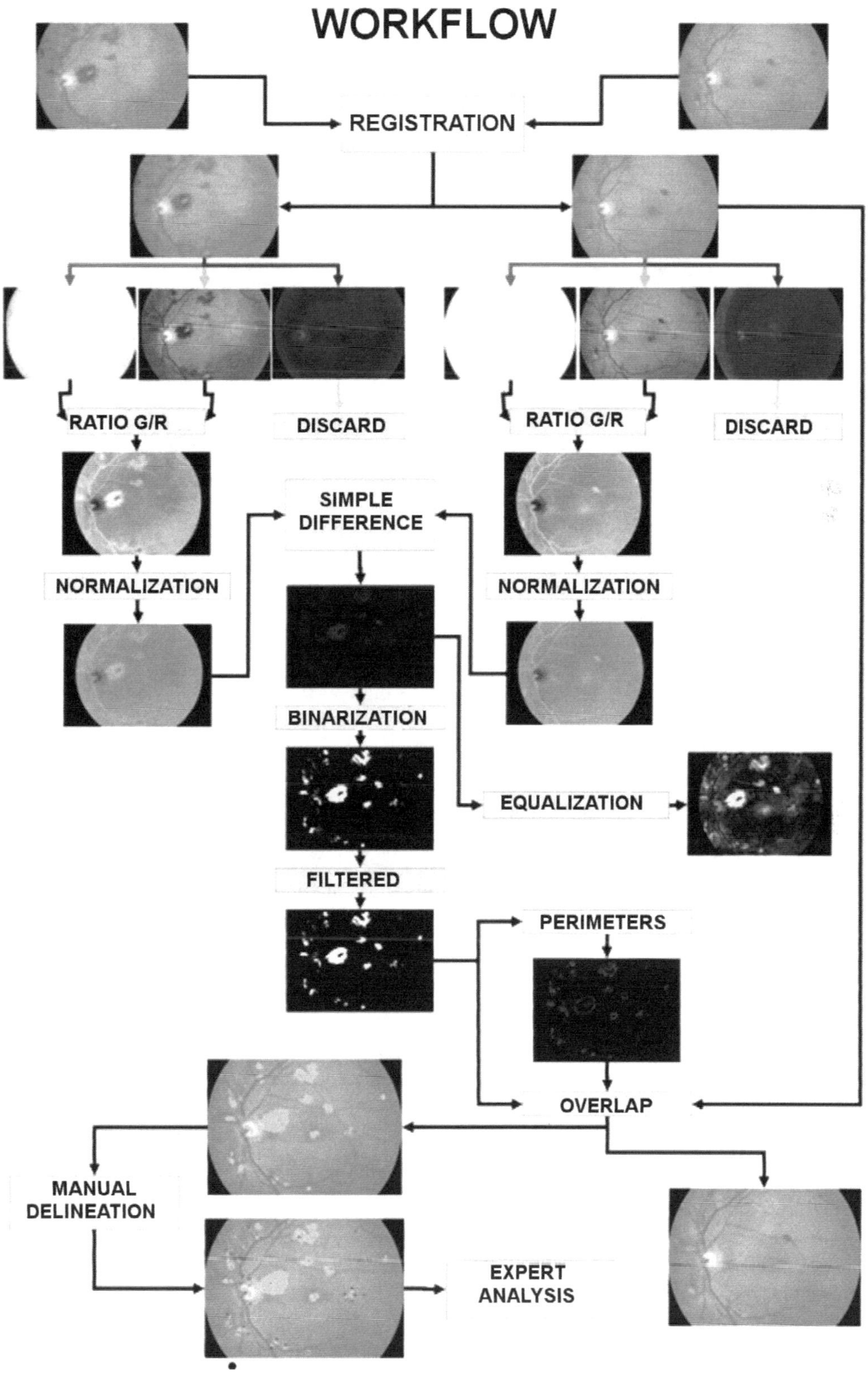

Figure 1. Workflow final imagery accompanied at every step.

Table IV: Summary data from all pairs

SIGN	RETINA	A	D	F	adequate	inadequate	not exsist	not detected	Total
H	54	33	20	0	52	1	1	0	57
E	9	7	2	0	8	1	0	7	14
V	13	2	0	9	9	2	2	3	0
O	0	0	0	0	0	0	0	0	4
R	1	1	0	0	0	1	0	0	1
Total	77	43	22	9	69	5	3	10	101

SIGN	Optic Disc	A	D	F	adequate	inadequate	not exsist	not detected
H	3	2	1	0	3	0	0	1
E	5	0	0	0	0	0	5	0
V	12	1	0	7	8	0	4	1
O	4	0	0	0	0	0	4	0
R	0	0	0	0	0	0	0	0
Total	24	3	1	7	11	0	13	2

First, the regions were separated according to their ownership to the retina or the optical disc. This division occurred as a result of an initial visual assessment showing a clear difference in the system's ability to detect changes in both regions. In each region, changes were divided by sign and classified first by whether they appeared (A), disappeared (D) or had associated changes in the form (F). In a separate set of columns the number of times that these changes were detected was counted as: properly changed, improperly changed, as changes that do not exist or changes not detected.

Thus, depending on the application, the amounts of TP (True Positive), FP (False Positive), TN (True Negative), and FN (False Negative) were specified. Their correspondence with the elements of the tables marked by specialist [12] and other tables generated by them were indicated.

2.7.1. Specification of amounts TP-FP-TN-FN

TP: it is considered true positive to any region, in Table III, marked by the expert as belonging to the column "properly detected". From the detection point of view, it also includes regions belonging to the column " inadequately detected" because, although the detection is inadequate, it exists as such. Figure 2 shows an example of TP which properly recognizes the occurrence of bleeding.

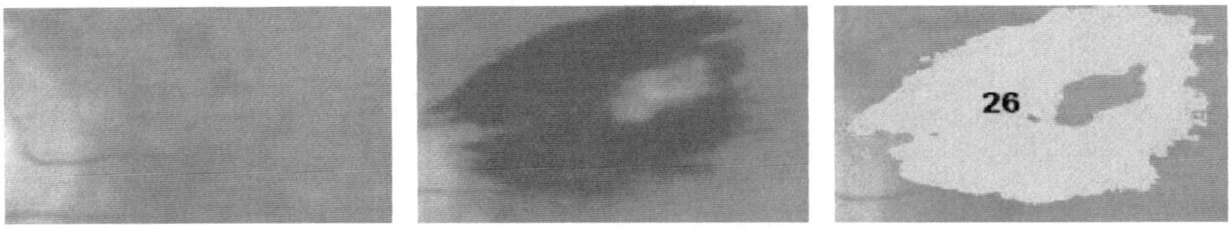

Figure 2. Example of True Positive (TP).

FP: this includes areas that have been marked by the specialist as belonging to the columns for "improperly detected" and "detected change where there is none?". From the standpoint of detection only, only the regions corresponding to the column mentioned in second place were taken. Figure 3 shows an FP example.

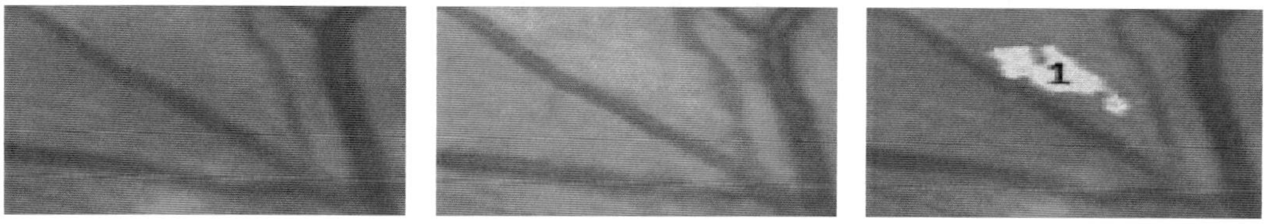

Figure 3. Example of False Positive (FP).

TN: For the determination of TN an estimation from the following equation is carried out:

$$TN=(AT - ATP - AFP - AFN)/AAR$$

Where: AT: area of region of interest of the analyzed image, ATP: total area corresponding to the regions TP; AFP: total area corresponding to the FP regions, AFN: total area corresponding to the regions FN; AAR: average area of TP and FN regions

FN: includes areas that have been marked by the specialist as belonging to the row "changes not detected". Figure 4 shows an example of an FN case where a reddish pigmentation possibly related to a punctate bleeding is not detected.

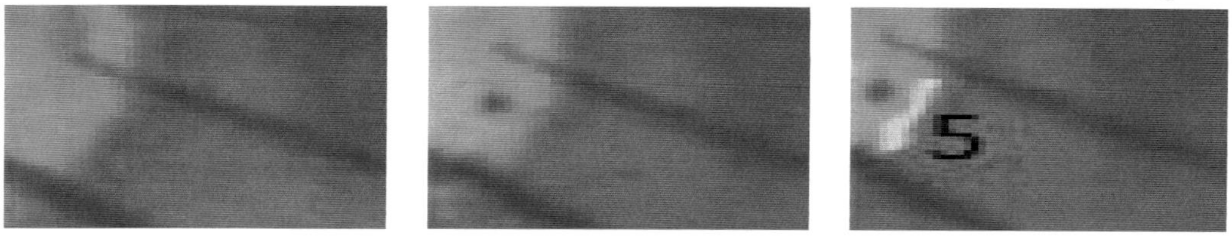

Figure 4. Example of False Negative (FN).

3. Results

The results are expressed by values of sensitivity and specificity of the system, according to different grouping data criteria as described below. The different ways of grouping the data intended to be used to determine strengths and weaknesses in system performance, depending on the region of the retina under study, the detected signs and changes detection.

Case A: From the general point of view of detection only: Taking into account all the signs, both in the region corresponding to the retina such as the optic nerve, consider all the changes, regardless of whether they are adequate or not.

Case B: From the general point of view of proper detection: Given all the signs, both in the region corresponding to retina as well as the optic nerve, consider only the proper detection and TP.

Case C: From the general point of view of only detection and considering only detected changes on the retina: Given all signs, related only to the region of retina, consider all the changes, regardless of whether they are adequate or not.

Case D: From the general point of view of appropriate detection considering only the changes on the retina: Given all signs, related only to the region of retina, consider only the proper detection and TP.

Case E: From the general point of view of detection only in the optic nerve: Considering all signs, related only to the optic nerve region, regardless of whether or not screening properly.

Case F: From the standpoint of just detecting changes associated with bleeding: Taking only bleeding into account, both in the region corresponding to retina and the optic nerve, consider all the changes, regardless of whether they are adequate or not.

Case G: From the standpoint of proper detection of changes associated with hemorrhage:

Taking only bleeding into account, both in the region corresponding to retina as the optic nerve, consider only the proper detection and TP.
In Table IV results are expressed in different cases in percentage values.

Table V: Results of system performance in terms of sensitivity and specificity. Data expressed in percentages

	A	B	C	D	E	F	G
Sensitivity	87.62	86.95	88.09	87.34	84.61	98.24	98.21
Specificity	99.70	99.61	99.93	99.83	99.72	99.98	99.96

3.1. Observations and analysis of results

The results showed that, for all grouping data criteria, the specificity of the system was above 99.6%.

The high proportion of true negatives with respect to total negative suggests a low probability that the system will detect changes where none exist. This is clearly evidenced in the results if we consider that, of all the regions analyzed, the system erroneously determined as corresponding to changes only 7 regions out of a total of 101 detected.

The overall sensitivity has values close to 87% for both the detection of changes as their appropriate detection. The small gap between the two values indicates that most changes detected by the system are carried out properly.

When data were grouped according to the region in which they are, there is an increase of approximately 5% of sensitivity to changes located in the region corresponding to the retina, with respect to those on the optic nerve. Furthermore, sensitivity is high when working independently with the optic nerve, suggesting the need to work with both regions in different ways, mainly with regard to correction lighting preprocessing stage. Figure 5 shows a case where the system detects changes in the optic nerve, while the expert recognizes such changes as false positives.

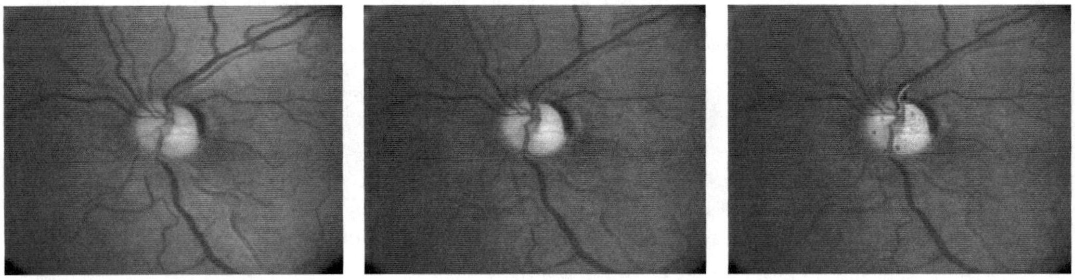

(a) (b) (c)

Figure 4. (a) First image of the retina. (b) Image of the same eye taken several months later. (c) Output of the system makes changes where there are no such changes (FP).

Independent analysis of the results of changes associated with bleeding with respect to other signs shows a sensitivity and specificity of 98.21% and 99.96% respectively, which corresponds to a sensitivity that is approximately 12% higher than for the general case.

3.2. Analysis of processing times

We conducted a study of the time taken for each stage of the workflow in order to recognize those with higher processing time. In turn, we used images with different resolutions in order to observe the variation that this factor produces on the results. Table V shows the times measured in seconds, which was obtained by applying the workflow end, through the generated interface.

We can see from the measurement of processing times that the stages of registration and masking generated the main delay, especially in large images. They have not been optimized so their processing time could be significantly reduced.

The stages which we worked on, tested and adapted showed minimal processing times, consistent with the purposes of the application. When working with images of standard resolution of 512 x 512, processing times which would make feasible its use in the dynamics of the professional consultation were required.

Tabla VI: Duration in seconds of each stage, for pairs of images of size 256 x 256, 512 x 512, 1345 x 1134 and 2048 x 2048 pixels

Stage	256x256	512x512	1345x1134	2048x2048
Registration	0,12	0,35	1,74	5,70
Initial masking	0,21	0,51	2,88	7,81
Ratio G / R for the pair	0,018	0,062	0,36	0,89
Statistical normalization for the pair	0,02	0,04	0,22	0,51
Masking Final	0,12	0,47	2,75	7,35
Simple Difference	0,0004	0,0014	0,0075	0,0044
Thresholds	0,002	0,003	0,008	0,02
Regions and properties	0,005	0,02	0,05	0,12
Image Overlay	0,10	0,12	0,20	0,85
Sum Total	1,19	1,58	8,21	23,30
Total Prototype	1.47	1.60	8.25	23.50

4. Conclusions and Discussion

In this work we have managed to acquire, standardize and characterize a set of images for further processing, by generation of MGS, selection and adjustment of methodologies and measurement performance of final workflow.

The selection, adaptation and implementation of methodologies successfully resulted in a robust workflow, which is simple, fast, clear and reliable. This is clearly seen on the results of the testing and evaluation of performance, where the sensitivity and specificity exceeded 80%, with processing times lower than 3 seconds for images of 512 x 512 pixels. This permits the inference that the workflow obtained has a high potential to be incorporated into a usable system in the dynamics of ophthalmologists.

The final results were suitable mainly for the determination of changes associated with signs of bleeding. There is a clear need for further testing and performance of the workflow for other signs present on the optical disc.

To adjust the parameters and the selection of methodologies, an MGS demarcated by a single specialist was used. Although this was considered ideal and plausible, we should consider the possibility of demarcation with a representative group of ophthalmologists.

The developed flow could be modified in order to extrapolate these results to other types of signs and fundus images.

Acknowledgements

Special thanks to the collaboration of Dr. Rodrigo Torres (ophtalmologist, Paraná, Entre Ríos), in his capacity as advisor and guide on the subject of monitoring changes in fundus.

5. References

[1] Muhammad-Amri Abdul-Karim et al. 2003 *Automated Tracing and Change Analysis of Angiogenic Vasculature from in vivo multiphoton confocal image time series.* Microvascular

Research 66. Págs. 113-125.

[2] Harihar Narasimha-Iyer et a, l2006. *Robust Detection and Classification of Longitudinal Changes in Color Retinal Fundus Images for Monitoring Diabetic Retinopathy*. IEEE Transactions On Biomedical Engineering, Vol. 53, Nº 6.

[3] A. Goshtasby. 2005 *2–D and 3–D Image Registration for Medical, Remote Sensing, and Industrial Applications*. Wiley – Interscience,

[4] G. K. Mastopoulos et at. 1999 *Automatic Retinal Image Registration Scheme Using Global Optimization Techniques*. IEEE-Trans. Inf. Technol. Biomed., Vol. 3, Nº 1, págs. 47-60.

[5] N. Ritter et al. 1999 *Registration of Stereo and Temporal Images of the Retina*. IEEE Trans. Med. Imag., Vol. 18, Nº 5, págs. 404-418.

[6] C.V. Stewart et al. 2003 *The Dual-bootstrap Iterative Closest Point Algorithm with Application to Retinal Image Registration*. IEEE Trans. Med Imag., Vol. 22, Nº 11, págs. 1379-1394.

[7] A. Can et al. 2002 *A Featured-based, Robust, Hierarchical Algorithm for Registratering Pairs of Images of the Curved Human Retina*. IEEE Trans. Pattern Anal. Mach. Intell., Vol. 24, Nº 3, págs 347-364.

[8] R.C. Gonzales y R.E. Woods. 1992 *Digital Image Processing*. Reading, MA: Addison-Wesley..

[9] B.H. Brinkmann et al. 1998 *Optimized Homomorphic Unsharp Masking por MR Grayscale Imhomogenity Correction*. IEEE Trans. Med. Imag., Vol. 17, Nº 2, págs. 161-171.

[10] R. Lillestrand. 1972 *Techniques for Change Detection*. IEEE Trans. Comput., Vol. 21, Nº 7, págs. 654–659.

[11] X. Dai y S. Khorram, 1998 *The Effects of Image Misregistration on the Accuracy of Remotely Sensed Change Detection*. IEEE Trans. Geosci. Remote Sens., Vol. 36, Nº 5, págs. 1566–1577.

[12] R. J. Radke et al. 2005 *Image Change Detection Algorithms: A Systematic Survey*. IEEE Transactions on Image Processing, Vol. 14, Nº 3.

[13] Li Chen. 2006 *Feature-Based Retinal Image Registration Using Bifurcation Structures.*

[14] A. Elgammal et al. 2002 *Background and Foreground Modeling Using Nonparametric Kernel Density Estimation for Visual Surveillance*. Proc. IEEE, Vol. 90, Nº 7, págs. 1151–1163.

[15] N. M. Salem y A.K. Nandi. 2007 *Novel and Adaptive Contribution of the Red Channel in Pre-Processing of Colour Fundus Image*". Journal of the Franklin Institute.

[16] L. P. Rosin et al. 2003 *Evaluation of Global Image Thresholding for Change Detection*. Department of Computer Science, Cardiff University, UK.

[17] L. P. Rosin. 2002 *Thresholding for Change Detection*. Comput. Vis. Image Understanding, Vol. 86, Nº 2, págs. 79–95.

[18] P. Smits y A. Annoni. 2000 *Toward Specification-Driven Change Detection*. IEEE Trans. Geosci. Remote Sens., Vol. 38, Nº 3, págs. 1484–1488.

[19] R. M. Haralick y L. G. Shapiro. 1992 *Computer and Robot Vision*. Reading, MA: Addison-Wesley, Vol. 1.

[20] E. Stringa. 2000 *Morphological Change Detection Algorithms for Surveillance Applications*. Proc. Brit. Machine Vision Conf.

[21] Tarkan Sevilmis et al. 2008 *Automatic Detection of Salient Objects and Spatial Relations in Videos for a Video Database System*. Image and Vision Computing, Vol. 26, págs. 1384–1396.

[22] 22.M. Tedder y C. Jin-Chung. . 2004 *Autonomous Robot Vision Software Design Using Matlab Toolboxes*. Lawrence Technological University.

SABI 2011 IOP Publishing
Journal of Physics: Conference Series **332** (2011) 012037 doi:10.1088/1742-6596/332/1/012037

An algorithm for automatic surface labeling of planar surgical resections

F E Milano[1,2]*, L E Ritacco[2], G Farfalli[3], L Aponte-Tinao[3], F González Bernaldo de Quirós[2] and M Risk[1,4]

[1] Instituto Tecnológico de Buenos Aires, Buenos Aires, Argentina
[2] Unidad de Planeamiento y Navegación Virtual, Hospital Italiano de Buenos Aires, Buenos Aires, Argentina
[3] Departamento de Oncología Ortopédica, Hospital Italiano de Buenos Aires, Buenos Aires, Argentina
[4] Consejo Nacional de Investigaciones Científicas y Técnicas - CONICET, Argentina

E-mail: fmilano@ieee.org

Abstract. Three dimensional (3D) preoperative planning and navigation in bone tumor resections have been used in the last five years with good results. The purpose of this study is to develop a method capable of detecting and labeling the nearly planar surface generated by the cutting saw in the surgical specimen taken off the patient during the resection procedure. This surface area labeling is fundamental to track the path that the cutting saw took during the surgery and compare it to the planned cutting plane. The algorithm presented here works by using a 3D reconstruction of the surgical specimen computed tomography (CT) scan, registered against the 3D reconstruction of the preoperative patient CT scan, and the cutting plane defined during surgical planning. The results show a high labeling accuracy (a matching mean of 98.5%) and a non significant accuracy variation for a range of distance and angle offsets.

1. Introduction

Three dimensional (3D) preoperative planning and navigation in bone tumor resections have been used in the last five years with good results [1]. These results were evaluated histologically, considering free margin from tumor. However, accuracy of preoperative planning [2] and navigation is not yet clear [3]. The purpose of this study is to develop a method capable of detecting and labeling the nearly planar surface generated by the cutting saw in the surgical specimen taken off the patient during the resection procedure. This surface area labeling is fundamental to track the path that the cutting saw took during the surgery and compare it to the planned cutting plane (PCP). The algorithm presented here works by using a 3D reconstruction [4] of the surgical specimen computed tomography (CT) scan, registered against the 3D reconstruction of the preoperative patient CT scan, and the cutting plane defined during surgical planning.

There have been reported algorithms for planar region detection on 3D laser scans of building facades [5] and for Lidar images roof detection [6]. Both algorithms work on point clouds, not taking advantage of information present in triangular meshes. In this work we present an

*This work was partially financed by a CONICET-UTN doctoral scholarship.

algorithm that mantains its accuracy even though the planar surface detected is surrounded by complex bone geometries.

2. Materials and Methods

2.1. Experimental Data and Preprocessing

The method presented here was developed, tested, and validated using a set of two patient CT images. Each patient comprises two image acquisitions: (1) The preoperative CT scan and (2) the CT scan of the surgical specimen. In this context the surgical specimen is the piece of bone containing the tumor, resected from the patient during the surgery. Both images are acquired using the same multi-slice CT scanner, applying the same acquisition protocol. Both images are 3D reconstructed using a custom medical image processing platform. The reconstruction is achieved by manually thresholding each tomographic slice to obtain a set of binary images that are then voxelised. This volume is then transformed into a closed surface represented by a triangular mesh. Both the preoperative mesh volume (PMV) and the surgical specimen mesh volume (SMV) are then set in a common frame of reference by a manual landmark registration [7], refined by an iterative closest points registration [8].

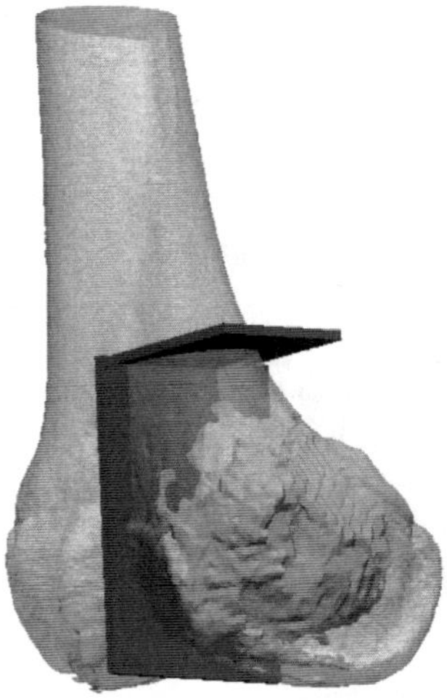

Figure 1: A planned surgery, showing two cutting planes in red and a bone tumor in green.

Our main objective is to evaluate the accuracy of the postoperative computed cutting surface (CCS) labeling.

2.2. Pipeline

The overall application of this method can be briefly described as follows: having a patient's PMV, the surgeons that will intervene in the procedure plan ahead where the bone affected by the tumor will be cut, taking into account medical factors such as the tumoral oncologic margin and the particular approach to the surgery. This plan is reflected in a set of PCPs located by the surgeons on the PMV. The plan together with the PMV is uploaded to a surgical navigator and the procedure is executed under intraoperative surgical guidance. After surgery the surgical specimen is CT scanned and reconstructed, obtaining a SMV. Now both PMV and SMV are registered in a common reference frame. The hereby presented tool is then used with the PCP and the SMV inputs to detect the computed cutting surface (CCS) created by the cutting saw. This way it is possible to compare what was planned in the PCP and what was actually executed in the CCS and measure the errors and discrepancies. This work focuses on the algorithm for the automated labeling of the CCS from the SMV.

2.3. Planar Resection Region Labeling Algorithm

The algorithm is divided in two phases:

(i) Each point of the SMV is projected onto the PCP.
Let the PCP be described by the plane

$$ax + by + cz + d = 0$$

with the normal vector $\mathbf{n} = (a\ b\ c)$ and $\mathbf{x} = (x\ y\ z)$.
Let SMV be a triangulated 2-manifold representing the surgical specimen. We look for the set of points K that belong to the SMV and minimize their normal distance to the PCP. Formally:

$$K = \left\{ \mathbf{k} \middle| \mathbf{k}, \mathbf{u} \in \text{SMV and } (\forall \mathbf{x} \in \text{PCP}) \left(\mathbf{k} = \operatorname*{argmin}_{\mathbf{u}} \frac{|\mathbf{n} \cdot (\mathbf{u} - \mathbf{x})|}{|\mathbf{n}|} \right) \right\} \qquad (1)$$

In practice, as the SMV is represented by a mesh built on an unstructured grid and it has potentially infinite points to project, the computation is performed in a different way. The PCP is used to generate a 2-dimensional uniform sampling grid, applying a homogeneus spacing in both plane's spanning directions. This grid is bounded by the projection of the bounding box of the SMV onto the PCP. Each point of the sampling space defined by this structured grid is projected onto the SMV in the line along the normal vector of the PCP. The closest point in SMV intersected by the projection of the sampling point is then added to the point set K that is going to be used in the second part of the algorithm. This is described in algorithm 1.

Algorithm 1 Projection

```
Input: SMV, PCP
Output: K, a set of points of the SMV that fullfill equation 1
UG ← Generate Uniform Grid from PCP
K ← empty
n ← n/norm(n)
threshold ← 20
for each u in UG do
  min_distance ← Largest Posible Number
  distance ← 0
  while distance < threshold do
    distance ← distance + 0.01
    k ← u + n · distance
    if Intersected(k, SMV) and distance < min_distance then
      min_distance ← distance
      min_k ← k
    end if
  end while
  Add(min_k, K)
end for
return K
```

(ii) The set of points generated in step (i) contains points belonging to the CCS, but it also contains other points projected from the SMV that are not part of the CCS. In this second step we propose a method to distinguish the points that belong to the CCS from the ones that do not. To this end we use a Random Sample Consensus (RANSAC) [9] algorithm to estimate the parameters of the ideal plane described by the cutting saw in the CCS. This algorithm generates, iteratively, plane models from three random points in the set, and then tries to fit the rest of the points to the model generated. The importance of this fitting stage is that the points are divided into two sets: the inliers and the outliers. If we adjust wisely the threshold that separate these two sets, the proposed method is robust enought to filter the outlier points, returning the largest inlier set of points found. In our case, we have been very stringent with this adjustment since we defined as inlier a point that is closer than 0.1mm from the model plane. The details are listed in algoritm 2.

Algorithm 2 Inlier Selection

```
Input: K
Output: largest_inlier_set
largest_inlier_set ← empty
max_iterations ← 10000
for i = 1 → max_iterations do
  p0 ← K[Random()]
  p1 ← K[Random()]
  p2 ← K[Random()] {Get three random points to model a plane}
  n ← (p0 − p1) × (p0 − p2) {Compute the normal}
  temp_set ← empty
  for each p in K do
    distance ← DistancePointPlane(p, n, p0)
    if distance < 0.1 then
      Add(p, temp_set)
    end if
  end for
  if length(temp_set) > length(largest_inlier_set) then
    largest_inlier_set ← temp_set
  end if
  i ← i + 1
end for
return largest_inlier_set
```

Each point included in the inlier set is then associated with the triangular face in the SMV that contains it. This triangular face is then labelled as belonging to the CCS.

At a post-processing stage, a connectivity filter is applied and all the small (one face) holes in the CCS are closed. The result of the algorithm is a set of labelled triangular faces that conform the CCS.

2.4. Validation protocol: Simulated clinical cases

A ground truth for this kind of algorithm would be a SMV with a labelling of the geometry touched by the cutting saw. Since we do not have such a ground truth to test the accuracy of the algorithm, we simulate the cutting process and generate SMVs directly from PMVs. In the simulation we take a PCP, push it along its normal and then vary this normal to a certain degree. This process simulates the offset and angle error that could occur in the operating room.

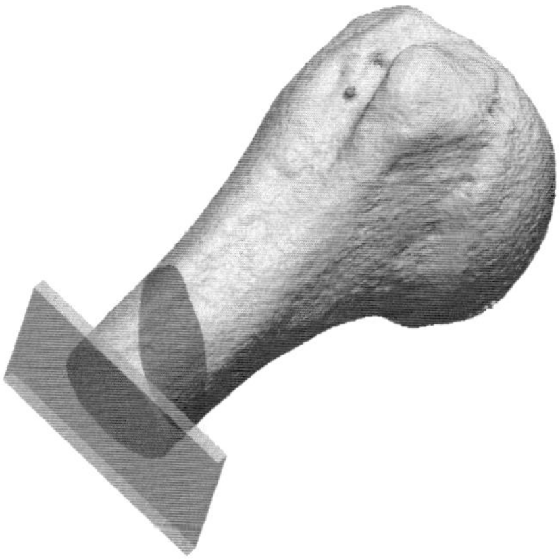

Figure 2: A simulated resection, showing in semitransparent gray the planned cutting plane and in green the simulated cutting plane (distance offset of 2mm and angle of 30°).

case	distance	angle	matched	not detected	not ground truth
		0	0.976	0.023	0.001
	0	15	0.979	0.020	0.001
		30	0.984	0.015	0.001
		0	0.981	0.018	0.001
1	2	15	0.978	0.021	0.001
		30	0.981	0.018	0.001
		0	0.982	0.017	0.000
	5	15	0.981	0.019	0.001
		30	0.981	0.016	0.003
		0	0.979	0.006	0.015
	0	15	0.992	0.006	0.002
		30	0.986	0.013	0.001
		0	0.994	0.005	0.001
2	2	15	0.988	0.011	0.001
		30	0.993	0.007	0.000
		0	0.994	0.006	0.001
	5	15	0.993	0.006	0.001
		30	0.990	0.008	0.001

Table 1: Algorithm performance the two cases. The **distance** and **angle** parameters are controlled in the simulation. The **matched** column shows the ratio between the ground truth area and the detected area that matches with it. The **not detected** column shows the proportion of the area that is in the ground truth but that it was no detected by the algorithm. The **not ground truth** column shows the proportion of the area that was detected as being part of the cutting surface but it was not part of the ground truth.

Then the PMV is cut generating a SMV and the triangular faces involved in the cut are labelled as belonging to the ground truth. Then our algorithm is run with the SMV and the PCP as inputs. The resulting CCS is matched to the ground truth following these rules:

(i) If a triangular face belongs to the ground truth and to the CCS then the area of that

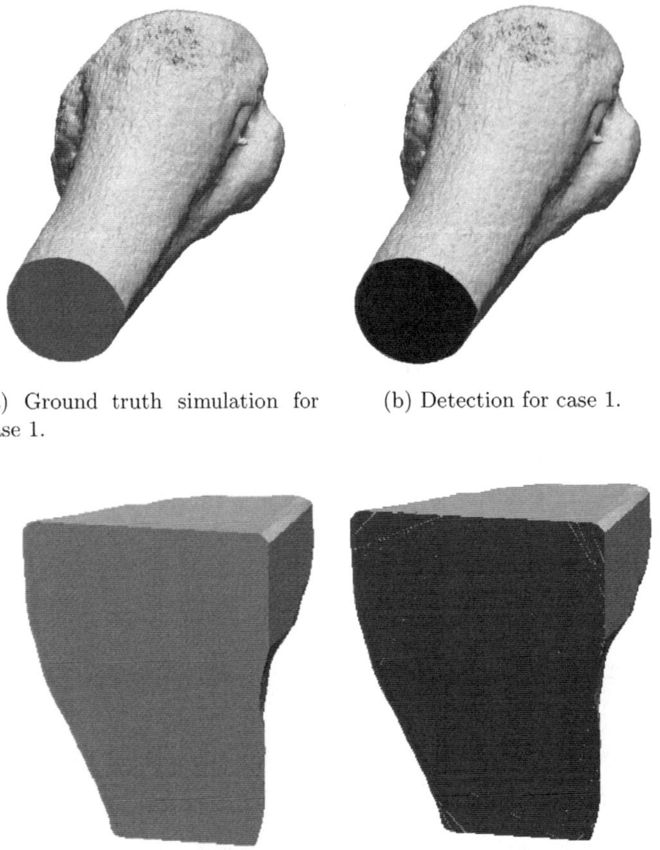

(a) Ground truth simulation for case 1.

(b) Detection for case 1.

(c) Ground truth simulation for case 2.

(d) Detection for case 2.

Figure 3: Results for different cases. In green, the simulated ground truth. In red, the detected cutting surface.

triangular face adds for the **matched** ratio (see table 1).

(ii) If a triangular face belongs to the ground truth but it was not detected, the the area of that face is added to the **not detected** ratio.

(iii) The last case occurs when a triangular face is detected as beloging to the cutting surface but it does not belong to the ground truth; in that case, it is added to the **not ground truth** ratio.

3. Results

The results of this work are shown in table 1. This table shows a comparison of the area detected by the algorithm and the ground truth generated by the simulation. We show that varying the distance offset and the angle of the simulated cutting plane in 0mm, 2mm, 5mm and 0°, 15°, 30° (for each distance offset all the angle variations are generated) does not affect the detection accuracy. The mean matched area detected is around a 98.5% of the ground truth area. The non significance (see figure 4) of the distance offset and angle variations was determined by multi way ANOVA testing, being the p-values for distance offsets of 0.29 and for angles of 0.69 (both $>> 0.05$). Some qualitative results can be visually assessed in figure 3.

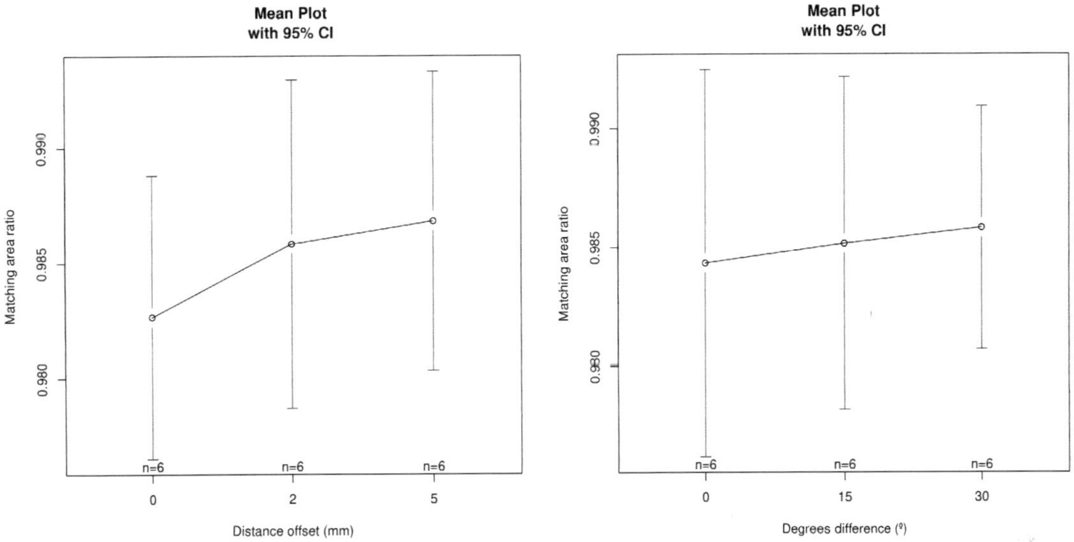

Figure 4: The matching area ratio variability analysis shows that variations in distance offset and angle of the cutting surface does not alter the algorithm detection capability.

4. Discussion

This work demonstrates the accuracy of an algorithm for cutting surface detection and labeling. The algorithm is useful for surgical intraoperative navigation evaluation, for surgery team performance evaluation and for surgical training. As future work we will simulate the cutting saw vibration noise and we will include real (non simulated) resection cutting path detections, visually validated by orthopaedic oncologic surgeons.

References

[1] Cho H S, Oh J H, Han I and Kim H S 2009 *Journal of Surgical Oncology* **100** 227–232
[2] Marmulla R and Niederdellmann H 1999 *Plastic and Reconstructive Surgery* **104** 938–944
[3] Shamir R, Joskowicz L, Spektor S and Shoshan Y 2009 *International journal of computer assisted radiology and surgery* **4** 45–52
[4] Wong K C, Kumta S M, Chiu K H, Antonio G E, Unwin P and Leung K S 2007 *The Journal of bone and joint surgery British volume* **89** 943–947
[5] Hammoudi K, Dornaika F, Soheilian B and Paparoditis N 2010 *2010 Canadian Conference on Computer and Robot Vision* **0** 122–129
[6] Tarsha-Kurdi F, Landes T and Grussenmeyer P 2007 *Science And Technology* **XXXVI** 407–412
[7] Wong K C, Kumta S M, Antonio G E and Tse L F 2008 *Clinical Orthopaedics and Related Research* **466** 2533–2541
[8] Zinß er T, Schmidt J and Niemann H 2003 *Image Processing, 2003. ICIP 2003. Proceedings. 2003 International Conference on* vol 2 (IEEE) pp II–695 ISBN 0780377508 ISSN 1522-4880
[9] Fischler M A and Bolles R C 1981 *Communications of the ACM* **24** 381–395

Identification of motility clusters per area in surfaces colonized by Pseudomonas aeruginosa

M Guzman[1,2,3], J Pastore[2,5], S Murialdo[4], V Ballarin[5] and M Trivi[3]
[1]Laboratorio Láser. Facultad de Ingeniería. UNMdP.
[2]CONICET.
[3]Centro de Investigaciones Ópticas (CONICET La Plata-CIC), UID Optimo, Departamento de Ciencias Básicas, Facultad de Ingeniería. Universidad Nacional de la Plata, Argentina.
[4]Grupo de Ingeniería Bioquímica. Departamento de Ingeniería Química. Facultad de Ingeniería. UNdMP.
[5]Laboratorio de Procesos y Medición de Señales. Dpto de Electrónica. UNMDP.

E-mail: jpastore@fi.mdp.edu.ar

Abstract. The biofilms are involved in pathogenesis and antibiotic persistence and resistance. The number of cells of a given species that will adhere to surfaces depends not only on the affinity of the cells but also on their number available for attachment. Therefore, the quick identification of the motile microorganism's area should be of great interest. The analysis of bacterial spatial patterns at the initial stage of biofilm formation is very important to know the success of the bacterial colonization. We propose a post processing method capable to distinguish motile microorganisms area of colonization in dynamic speckle images by applying Mathematical Morphologic techniques. The methodology would be effective for segmenting, detecting and describing patterns of colonization known not to be completely spatially random. The presented method is fast, reproducible, convenient, robust, and can be used to control the pattern, spacing, and orientation between colonies of different bacteria in order to prevent biofilm development.

1. Introduction

In natural environments, bacteria live mostly in biofilms [1], which are multicellular communities attached to surfaces. Motility gives access to surfaces suitable for biofilm formation, it is necessary for the recruitment of planktonic cells within the biofilm, and it is required for the spreading of biofilm on non-colonized surfaces [2]. Biofilm architecture was shown to be motility dependent in E. coli and in Pseudomonas aeruginosa [3,4]. These findings highlight the importance of discriminate clusters of motile bacteria in a sample. We recently developed a method which detects motile bacteria on a surface [5]. Analysis of bacterial spatial patterns at the initial stage of biofilm formation is very important to know the success of the bacterial colonization [6].
Chronic infections remain a major challenge for the medical profession and are of great economic relevance because traditional antibiotic therapy is usually not sufficient to eradicate these infections.

One major reason for persistence seems to be the capability of the bacteria to grow within biofilms that protects them from adverse environmental factors. In oral infections the number of cells of a given species that will adhere to oral surfaces depends not only on the affinity of the cells but also on their number available for attachment [7]. Therefore, the quick identification of motile microorganisms attached per area should be of great interest.

Pseudomonas aeruginosa is not only an important opportunistic pathogen and causative agent of emerging nosocomial infections but can also be considered a model organism for the study of diverse bacterial mechanisms that contribute to bacterial persistence [8]. In this context the elucidation of physiological characteristics, such as cell size (starved versus vegetative bacteria), and number of motile microorganisms per area, may influence their transport through porous media and should be considered. The identification of mechanisms responsible for the switch from planktonic growth to a biofilm phenotype and the pattern of motile bacteria aggregation per area in persistent disease should provide new insights in *P. aeruginosa* pathogenicity, and would contribute to a better clinical management of chronically infected patients.

The immediate goal of this work was to distinguish motile surface patterns per area of colonization by applying image processing techniques in particular Mathematical Morphology. The methodology would be effective for segmenting, detecting and describing patterns of colonization known not to be completely spatially random. The images of bacteria in biofilms are usually textured with fuzzy edges and nonhomogeneous gray level. Conventional image processing methods for analysis of shapes can not be applied in this case.

Unlike standard techniques, morphological techniques are based on concepts of geometry, algebra, topology and set theory, to characterize structural properties in images [9-12]. The central idea of these techniques is to examine the geometric structures in an image overlaying them with small patterns called structuring elements. Of all mathematical morphology analysis techniques, the most appropriate tool for segmented the bacteria in biofilms images.

This paper proposes a morphological segmentation in image of bacteria in biofilms in order to distinguish motile surface patterns per area. Disrupt these patterns of aggregation may help prevent development of biofilms. This method is rapid, reproducible, convenient, and can be used to control the pattern, spacing, and orientation between colonies of different bacteria in order to prevent biofilm development.

2. Materials and methods

2.1.1. Biospeckle patterns acquisition.

The proposed set-up for acquisition and storage of dynamic speckle patterns (biospeckle) is shown in fig. 1. An expanded HeNe laser (632.8nm and 30mW) illuminates the plate under study from the bottom through a ground glass diffuser. A CCD camera connected to a frame grabber registers a sequence of images (8 bits and 768 x 576 squared pixels) and stores it into the computer. The CCD height and objective were adjusted to focus the sample.

Using the experimental set up of Fig. 1, sequences images of speckle patterns were recorded. Time series corresponding to intensity level of each pixel were assembled (as many series as the image resolution = 768 x 576 pixels) to study the dynamics of the phenomenon.

To evaluate the dynamic within stationary periods, images sequences of 400 samples were registered using a 4 Hz sampling frequency, during 1min 40s. To distinguish motile surface patterns per area of colonization were applied techniques of Mathematical Morphology.

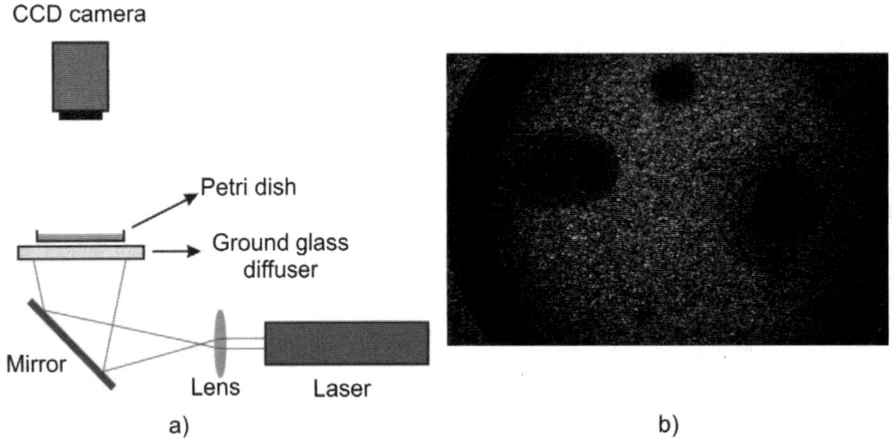

Figure 1. The dynamic laser speckle technique employed to easily discriminate filamentous fungi from motile bacteria in soft surface, such as agar plate a) Scheme of the setup for speckle imaging (the ground glass diffuser was removed and the mirror was replaced by a circular fluorescent tube over a dark surface for the white light imaging), b) example of speckle image.

2.1.2. Image processing

We used an RGB image obtained with the energy of filtered speckle images [5]. An RGB (red, green, blue) image is a three-dimensional byte array that explicitly stores a color value, between 0 (black) and 255 (white), for each pixel. A gray level image is obtained doing a weighted combination of each color channel in the RGB image ($I_{Gray} = 0.2989 \times R + 0.5870 \times G + 0.1140 \times B$).

Mathematical Morphology

The basic idea of Mathematical Morphology is comparing the objects contained in an image with a known object called structuring element [11,12].

The language of this important theory is set theory. Sets represent the shapes of the objects in an image. Every time the structuring element is superimposed on the image, a set operation is done for each pixel. The basic morphological operators of the MM are erosion and dilation [11]. A binary image can be modeled as a function X define on a subset of $G \subset \mathbb{Z}^2$ into the set $\{0,1\}$. For this work, X denotes a binary image.

Basic operators are defined, with respect to the structuring element B :

Erosion:

$$X \ominus B = \{ y \in X \,/\, B_y \subseteq X \} \tag{1}$$

Dilation:

$$X \oplus B = \{ y \in f \,/\, B_y \cap X \neq \varnothing \} \tag{2}$$

where $B_y = \{ b + y \,/\, b \in B \}$ and + represents the vector sum.

The remaining morphological operators are defined based on the combination of these two, such as opening and closing which in turn give rise to new filters [11,12]. Opening and closing are defined, respectively, as follows:
Morphological opening of the image X by the structuring element B is defined as:

$$\gamma_B(X) = (X \odot B) \oplus B \tag{3}$$

Morphological closing of the image X by the structuring element B is defined as:

$$\Phi_B(X) = (X \oplus B) \odot B \tag{4}$$

The shape and size of the structuring element play a key role in the detection and extraction of features resulting from the image shape and size.
Conditional dilation of Y by a set X ($Y \subset X$), is defined as:

$$\delta_X(Y) = (Y \oplus B) \cap X \tag{5}$$

where B is a structuring element.

Reconstruction, $\rho_X(Y)$ where X is named mask, and Y is called marker $Y \subset X$, is obtained by iterating conditional dilation as: $\rho_X(Y) = \bigcup_{n \geq 1} \delta_X^n(Y)$, where n is the number of conditional dilations necessary to obtain the idempotency. Reconstruction is a very useful operator provided by Mathematical Morphology. The Reconstruction transformation is relatively well-known in the binary case, where it simply extracts the connected components of an image which are "marked" by another image [10]. Figure 2 shows an example of morphological reconstruction.

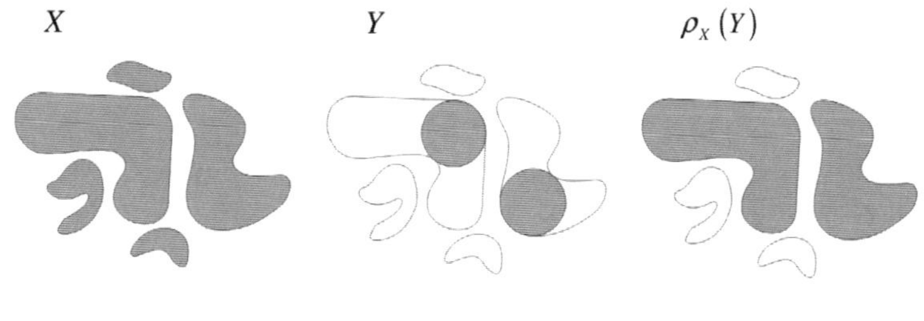

(a) (b) (c)
Figure.2: (a) Original image. (b) Marker image. (c) Reconstructed image.

The algorithm to distinguish motile surface patterns per area of colonization can be summarized as follows:

Step 1: Original RGB image is converted to gray level image (Fig. 3(b)). The weighted combination of each color channel (RGB) produce a gray - level image.

Step 2: Otsu binarization of the former resulting image. (Fig. 3(c))

Step 3: Area opening and area closing operators are applied to remove any pore (i.e., background connected component) with an area less than a threshold. Connectivity is given by the structuring element (square structuring element of size 3x3). (Fig. 3(d))

Step 4: Resulting image of the previous step is reconstructed from the border of the image. (Fig. 3(e))

Step 5: Image obtained in step four is subtracted from the one obtained is step three. (Fig. 3(f))

Step 6: Visualization. (Fig. 3(g))

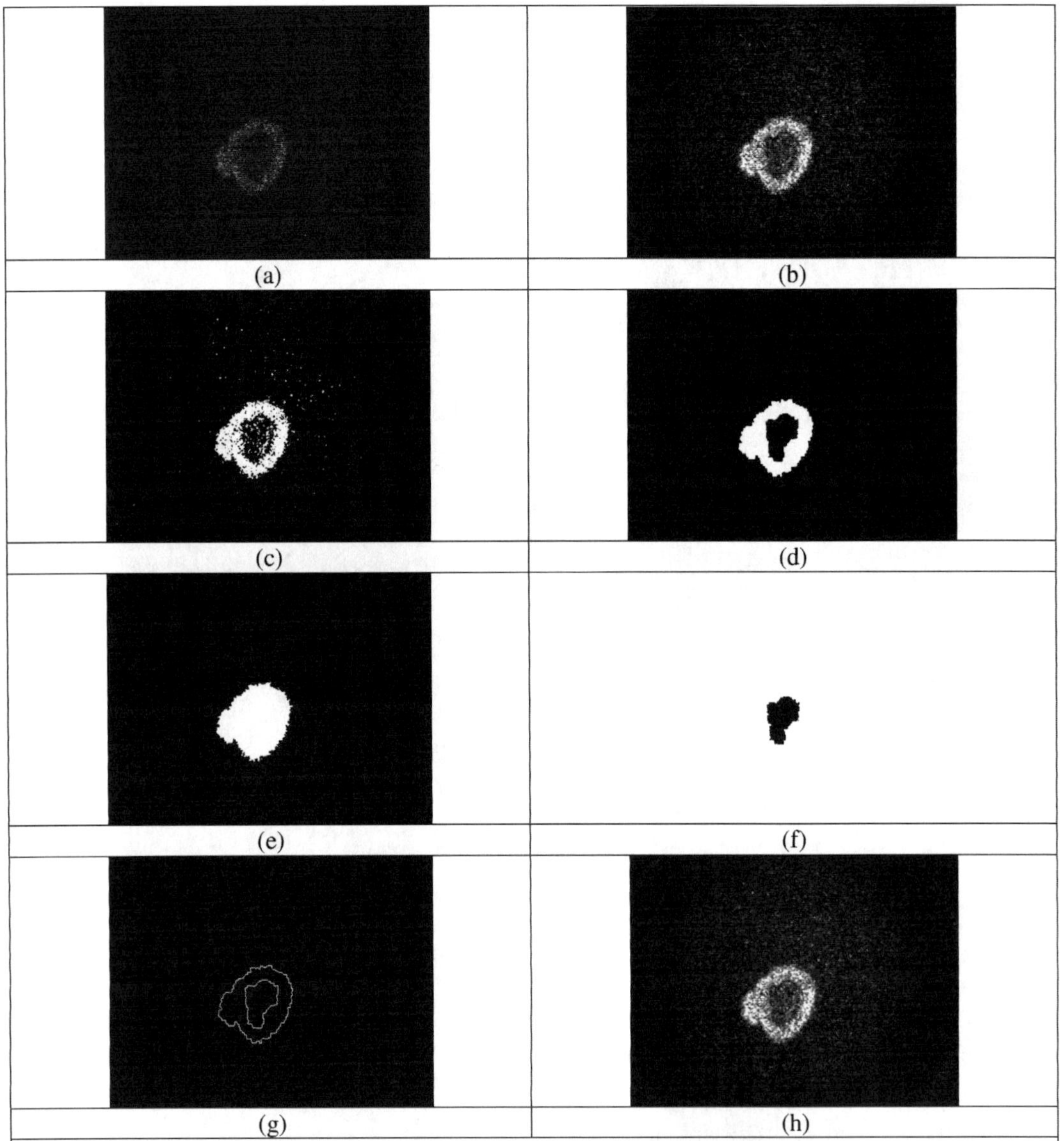

Figure. 3. a) Filtered RGB speckle images b) gray level images c) Binary image d) Filtered binary image e) reconstructed image f) g) h) visualization

3. Results and discussion

In figure 4 the first column are the resulting speckle filtered images [5] for chemotactic bacteria and for non-chemotactic and chemotactic bacteria in the same experiment. In the second column gray levels image are shown. These are obtained combining each color channel of the RGB image, and finally the last column shows in red the contours resulting after applying the calculation algorithm.
It can be observed that the algorithm locates the area of greatest bacterial activity. It is also shown that the geometry of growth does not affect the ability of the algorithm to discriminate the desired area.

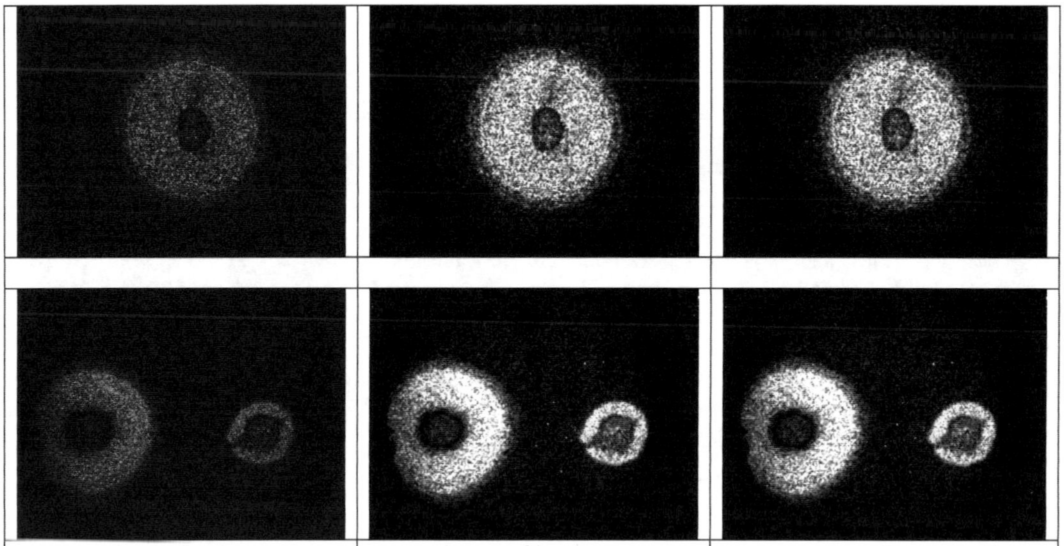

a)	b)	c)

Fig. 4: a) Filtered RGB speckle images b) gray level images obtained combining each color channel of the RGB image c) visualization, contours obtained after applying the calculation algorithm.

4. Conclusions

The analysis of spatial patterns of bacteria during biofilm formation is essential for understanding the success of bacterial colonization. Therefore the quick identification of motile microorganisms attached per area is essential.
We distinguished motile surface patterns per area of colonization by applying image processing techniques in particular Mathematical Morphology. The methodology was effective for segmenting and detecting patterns of colonization known not to be completely spatially random. The presented method is fast, precise, robust, and can be used to control patterns, spacing, and orientation between colonies of different bacteria in order to prevent biofilm development.

5. References

[1] Costerton, J. W., Lewandowski, Z., Caldwell, D. E., Korber, D. R. & Lappin-Scott, H. M. (1995). Microbial biofilms. *Annu Rev Microbiol* 49, 711–745.
[2] Houry A., R. Briandet,S. Aymerich and M. Gohar. Involvement of motility and flagella in Bacillus cereus biofilm formation. *Microbiology* (2010), 156, 1009–1018.
[3] Barken, K. B., Pamp, S. J., Yang, L., Gjermansen, M., Bertrand, J. J., Klausen, M., Givskov, M., Whitchurch, C. B., Engel, J. N. & Tolker-Nielsen, T. (2008). Roles of type IV pili, flagellum-mediated motility and extracellular DNA in the formation of mature multicellular structures in Pseudomonas aeruginosa biofilms. *Environ Microbiol* 10, 2331–2343.
[4] Wood, T. K., Gonzalez Barrios, A. F., Herzberg, M. & Lee, J. (2006). Motility influences biofilm architecture in Escherichia coli. *Appl Microbiol Biotechnol* 72, 361–367.
[5] Murialdo SE, Sendra GH, Passoni LI, Arizaga R, Gonzalez JF, Rabal H, Trivi M. 2009. *Analysis Of Bacterial Chemotactic Response Using Dynamic Laser Speckle. Journal of Biomedical Optics* 14(6): 064015-1-6. ISSN: 1083-3668, OCLC: 32795913.
[6] M. A. Hamilton' K. R. Johnson' A. K. Camper' P. Stoodley' G. J. Harkin' R. J. Gillis' P. A. Shope, 2007. Analysis of Bacterial Spatial Patterns at the Initial Stage of Biofilm Formation. *Biometrical Journal*. Volume 37, Issue 4, pages 393–408, 1995
[7] J. Van Houte and D. B. Green. Relationship Between the Concentration of Bacteria in Saliva and the Colonization of Teeth in Humans Infection and Immunity, Apr. 1974, p. 624-630.
[8] Cornelis P (2008). Pseudomonas: Genomics and Molecular Biology (1st ed.). *Caister Academic Press*. ISBN 978-1-904455-19-6
[9] González R, and Woods, R. 1996. *Tratamiento Digital de Imágenes*, Addison Wesley.
[10] Facon J. 1996. *Morfología Matemática. Teoría e ejemplos*. Editora Universitária Champagnat da Pontífica Universidade Católica do Paraná, Curitiba, Brasil.
[11] Serra J. 1982. *Image analysis and mathematical morphology*, Academic Press, London.
[12] Serra J. 1988. *Image analysis and mathematical morphology*, Academic Press, London.

SABI 2011 IOP Publishing
Journal of Physics: Conference Series **332** (2011) 012039 doi:10.1088/1742-6596/332/1/012039

Real time video processing software for the analysis of endoscopic guided-biopsies

C. Ordoñez[1], A. Bouchet[1,2], J. Pastore[1,2], E. Blotta[1]

[1] Laboratorio de Procesos y Medición de Señales/Facultad de Ingeniería/Departamento de Electrónica, UNMDP, Mar del Plata, Argentina.
[2] Consejo Nacional de Investigaciones Científicas y Técnicas (CONICET).

E-mail: eblotta@fi.mdp.edu.ar

Abstract. The severity in Barrett esophagus disease is, undoubtedly, the possibility of its malignization. To make an early diagnosis in order to avoid possible complications, it is absolutely necessary collect biopsies to make a histological analysis. This should be done under endoscopic control to avoid mucus areas that may co-exist within the columnar epithelial, which could lead to a false diagnosis. This paper presents a video processing software in real-time in order to delineate and enhance areas of interest to facilitate the work of the expert.

1. Introduction

Reflux oesophagitis takes place when the mechanisms existing in the gastro esophageal union fail. When enough time has passed, it can produce complications or turn into a chronic condition. One of the injuries that may appear is called Barrett esophagus, which is considered precancerous. Due to these reasons, the interest in this particular pathology has increased in the last couple of years.
Barrett esophagus is caused by the reflux of gastric juices or intestinal content, which is rich in bilis, towards the esophagus [1]-[4]. This produces an epithelial damage, generally erosions or ulcers, whose ulterior curative process using basal pluripotentials cells, gives place to the different types of columnar epithelials that are seen in such alteration [5].
Barrett esophagus is defined as the presence of columnar mucosa with intestinal metaplasia in the esophagus. In practice, the best identifying method is through endoscopic diagnosis. The importance of this point is that the metaplasia of Barrett is a histological criterion rather than an endoscopic one, because it is possible to identify the intestinal metaplasia only with a directed biopsy; that is, looking for areas of dysplasia presenting greater likelihood for neoplasia.
If the importance of Barrett esophagus lies on a high risk for malignant transformation, the treatments to induce its regression and the substitution of metaplastic mucosa by normal squamous epithelium may be the means to eliminate the risk of cancer. Because Barrett epithelium is a complication from chronic gastro-esophageal reflux, an efficient anti-reflux treatment could, theoretically at least, induce regression of the metaplasia. Such treatments include surgery and drugs to reduce gastric acidity.
Therefore, in order to not only verify the occurrence of Barret esophagus, but also to diagnose its possible complications, it is absolutely essential obtain adequate biopsies that need to be performed under endoscopic control, so as to avoid areas of squamous mucosa that may coexist inside from the columnar epithelium, which –in turn- could lead to a false diagnosis. It is of fundamental importance

that the specialist performing the endoscopic study be given the means for distinguishing on the image the correct area to draw the biopsy samples, because the borders that envelop the illness are typically not well defined.

Various guided techniques to increase accurate of the diagnoses and making it less intrusive and fast are studied by specialists. Among them are the chromoscopy, which involves staining of the mucosa with dyes; the acetic acid, due to its mucolytic effect, provides better eyesight; and the narrow band imaging (NBI), recently developed by Olympus, which is based on modifying the bandwidth of the emitted light. All of these techniques have their advantages and their weak points. The chromoscopy and the use of acetic acid require additional time to succeed, by the other way NBI equipment have a high cost.

The objective of this work is to implement a filtering software in real time to enhance endoscopic images, in order to delimit, with less error, the area affected by the transformation of the mucosa, typical of Barrett's disease. This would enable the expert differentiating the different surfaces, reducing the possibility of false analysis and, consequently, false diagnoses.

2. Experimental bank

The prototype was mounted on an endoscopic imaging equipment formed of an Olympus CV-160 processor and an Olympus CLE-145 halogen light source.

The proposed system consists of a desktop PC and a video capture card *Encore* ENMVG-USB model, which supports an input signal type RCA-composite or S-Video in NTSC or PAL format and uses a 10-bit analog-to-digital converter, with a maximum resolution of 720x480 pixels for NTSC or 720x576 for PAL. The proposed and designed software is capable of acquiring video frames from the video capture card and to apply various filters in real time.

3. Software Design

3.1. Initial considerations

We consider the following issues in the development of the software application,

according the professionals that perform the study:

- Real-time processing, ie, delay in the filtered images is indistinguishable.
- Good performance in any standard computer.
- Simple and intuitive graphical interface.
- Ability to store images for patient records and reports preparation, among others.
- Ability to store frequent use filter settings.
- The digital-analog conversion must not cause a significant loss in the quality of the video signal.
- Compatibility with different types of video signals.

according the developers:

- Easy to incorporate new processing algorithms.
- Supports multiple formats of image or video.

3.2. Software Development

We chose Visual C++ as programming language due to the following features:

- It is a very powerful object-oriented language, capable of generating simple and complex applications, such as operating systems or video games.
- It is widely used, it has a great support, so there are many online tutorials, books, open source codes, etc.
- Allows compiling and running C code.

- It has a very powerful development environment, that allows object-oriented programming (OOP) in conjunction with the development system SDK of Windows (also called API). Because the Windows SDK is a complex set of functions, Visual C++ includes the class library MFC (Microsoft Foundation Classes) allowing create and manage an intuitive way typical components of Windows.

For handling the video signal and images, it was considered the open source libraries from Intel OpenCV as the best option. OpenCV is a high-level library that implements algorithms for calibration techniques, feature detection and tracking, shape analysis, motion analysis, 3D reconstruction, segmentation and object recognition. This library is capable of manipulate almost any video format or image transparently to the programmer. The best feature of this library, in addition to its quality and functionality, is its performance [6].

3.3. Graphical environment

Figure 1 shows the proposed graphical user interface, which can command all capture and processing actions.

The graphical interface allows to:

- Set filter parameters to a preset value.
- Adjust the filter parameters to the desired value.
- Adjust the scales of the original and processed video.
- Capture and store areas of interest through the screen.
- Store filter parameters.

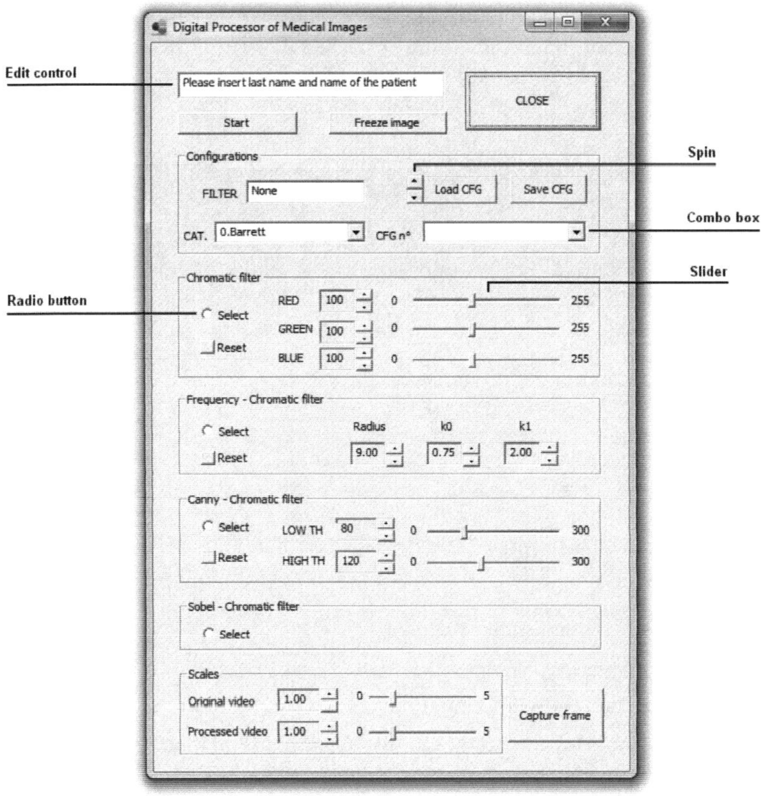

Figure 1. Graphical interface.

4. Implemented Filters

The filters implemented in the prototype were used successfully by the authors to delineate and enhance areas of interest to facilitate the work of the expert using the captured images from the video signal emitted by the endoscope [7][8].

4.1. Chromatic filter

A RGB color image is composed of three grayscale images of size $M \times N$, corresponding to the levels of red, green and blue.
The chromatic filter consists in multiply by a constant, selected by the user according to a criterion, each color in order to enhance different attributes in the area under study [7][8].

4.2. Frequency Filter

In order to enhance important changes of gray level in the image we proposed to use a spectrum analysis. As abrupt changes are related to high frequencies while moderate changes are related to low frequencies, we modify the spectrum of the image, enhancing high frequencies and attenuating low frequencies, ie performing a high pass filtering. By this way, a new image is obtained, enhancing the variations that exceed a threshold and making darker the ones that do not exceed the threshold.
The spectrum of the image is obtained performing the Fourier transform in two dimensions. Since the RGB images are three dimensional, they are decomposed into three components, resulting in three grayscale images in two dimensions. Thus, the spectrum of each of these images have been calculated and processed, and finally recombined to a filtered RGB image [7-13]. The calculation of the Fourier transform is performed using the algorithms from the free library FFTW 3.2.2 [14].

4.3. Canny edge detector

The edges of an image are the boundaries separating regions with different brightness or color. J. Canny proposed an efficient method for detecting edges [15]. The algorithm takes as input a grayscale image and returns a binary image wherein the pixels whose value is nonzero represent the detected edge. The algorithm can be described by the following four steps [15]:

Step 1- Image smoothing: The image is smoothed using a Gaussian filter whose band width is specified as parameter.
Step 2- Differentiation: After smoothing, the image is derived in x and y directions. Then, we obtain the magnitude and phase of the gradient values through hypotenuse and arctangent functions.
Step 3 - Non maximum suppression: Once calculated the gradient at each point of the image, the edges can be located at the points of maximum gradient magnitude. This is accomplished through the suppression of the "non maximum", ie eliminating the points of the magnitude gradient that are not maximum.
Step 4 - Edge thresholding: Finally we describe the edges by applying what is called "hysteresis threshold". We operate on two thresholds, a lower and an upper one. They consist in eliminate those gradient values that are below the lower threshold, and keep those values that exceed the upper threshold. Those gradient values which are found between the two thresholds are kept only if they have a neighbor that exceeds the upper threshold.

4.4. Sobel filter

Sobel filter is an edge detection algorithm based on the calculation of the gradient of the intensity function of an image. The gradient of an image is obtained by calculating the partial derivatives df/dx and df/dy in each of the pixels of the image. There are many ways to implement digital derivative operators, however, Sobel operators stand out by its smoothing effect. This feature is important because, typically derivatives operators tend to increase the noise level.
The implementation of the Sobel operator is performed through the use of masks [16]. These are applied to the image using a spatial convolution, moving through the image pixel by pixel until a

complete output image is generated. We use two dimensional 3×3 masks, one for the horizontal and one for the vertical direction.

5. Results

5.1. Barrett's images results

The images shown in this section reproduces the results obtained by using the developed software.
Figure 2 (left) shows how the chromatic filter applied on the original image (right) clarifies the area under study, giving another perspective of tissue change.
Figure 3 shows several examples for frequency filtering. The images of figures (a) and (b) were obtained using preset settings. Figure (c) was obtained by increasing the passband gain, achieving an image with sharper relief, and figure (d) was obtained by increasing the attenuation of the rejection band, achieving an image where the parts with little relief are darker.
Figure 4 shows several examples of Canny filter for different values of lower and upper threshold. Also, chromatic filter has been applied to improve performance. It is noted that by decreasing the lower threshold value, the lesions and the less pronounced edges stick out, however, by increasing it, only the injuries are detected and the edges results more defined.
Figure 5 shows how the Sobel filter provides a completely different perspective to the rest of the filters, enhancing and amplifying the edges in the area under study.

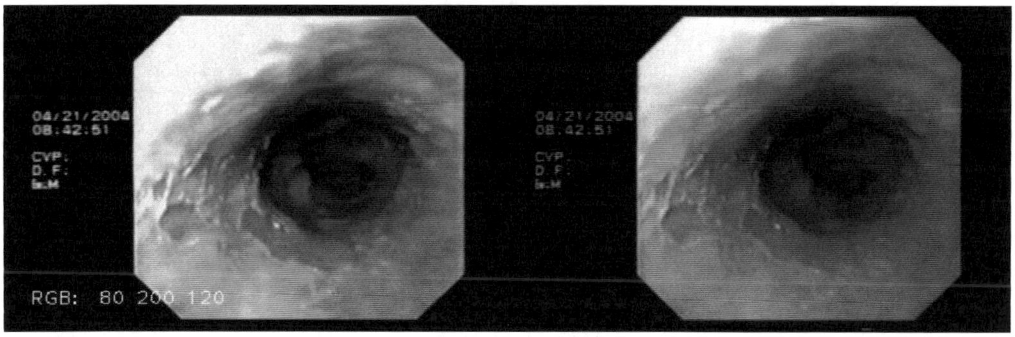

Figure 2. Original image combined with chromatic image

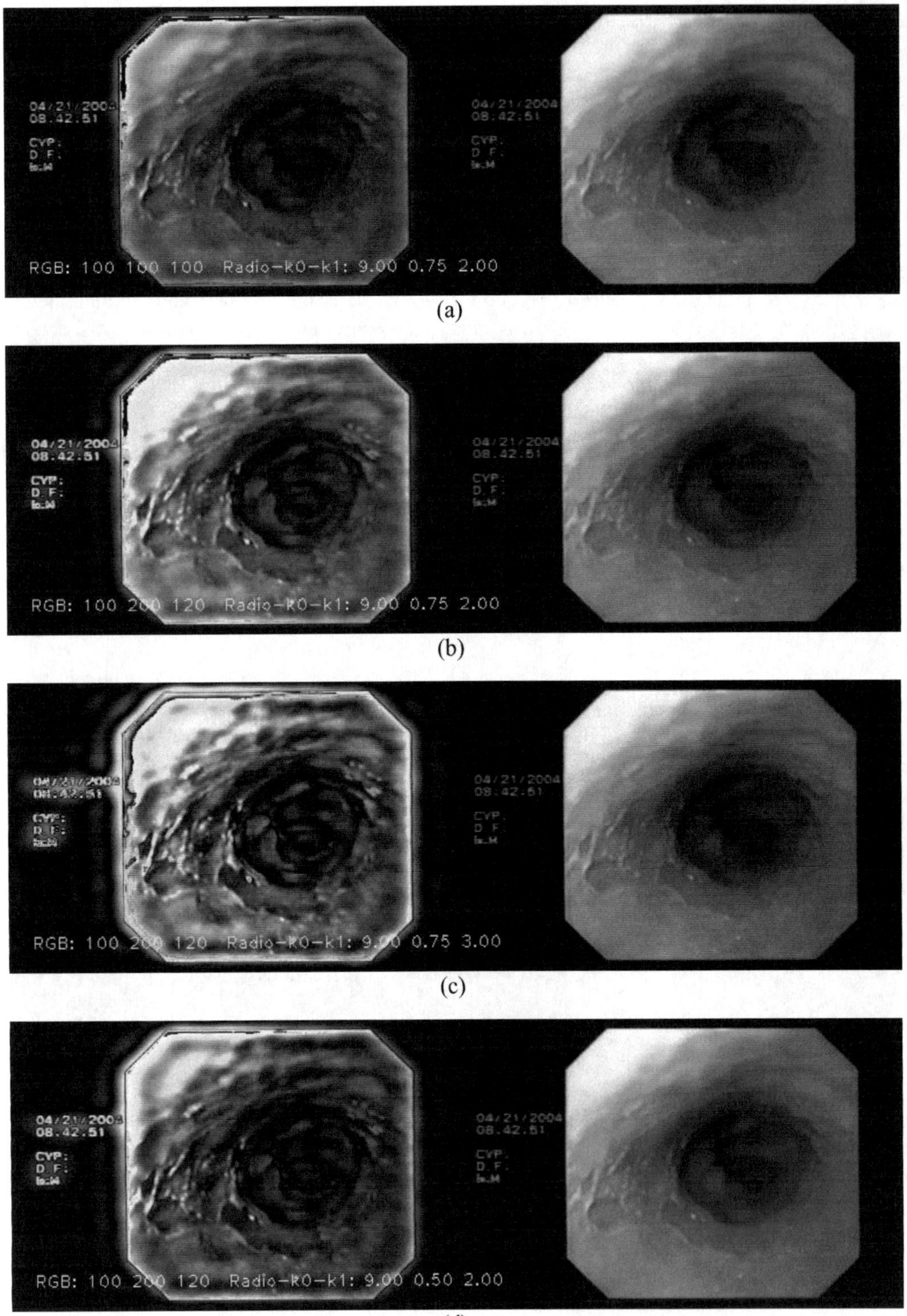

Figure 3. Original images combined with images resulting from the frequency filter for different configurations. (a) Preset settings. (b) Preset settings. (c) Increasing the passband gain. (d) Increasing the attenuation of the rejection band.

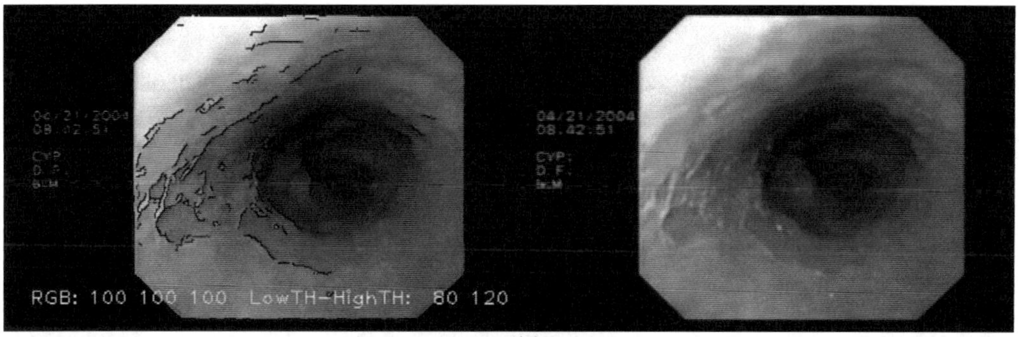

(a)

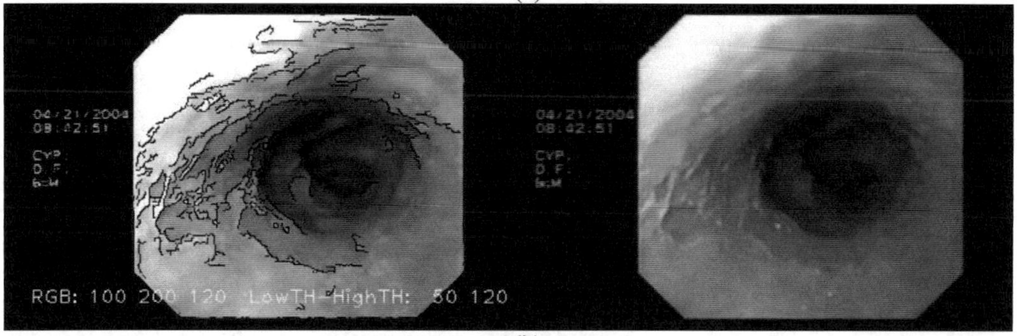

(b)

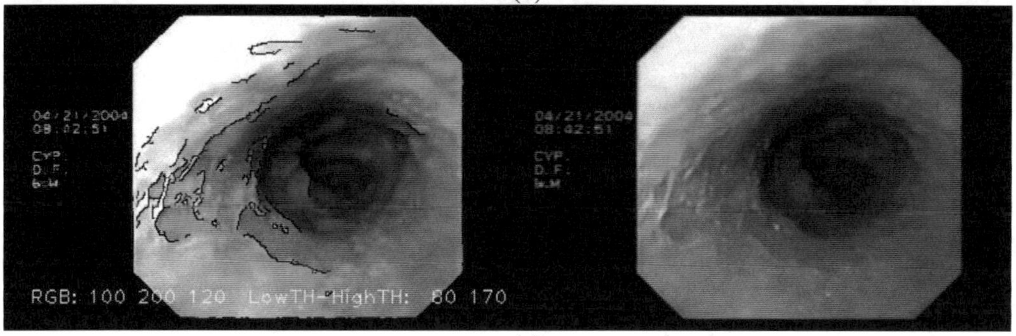

(c)

Figure 4. Several images by applying the Canny filter for different configurations. (a) Canny filter on the original image. (b) Canny filter on the chromatic image when threshold value decreases. (c) Canny filter on the chromatic image when increasing the lower threshold.

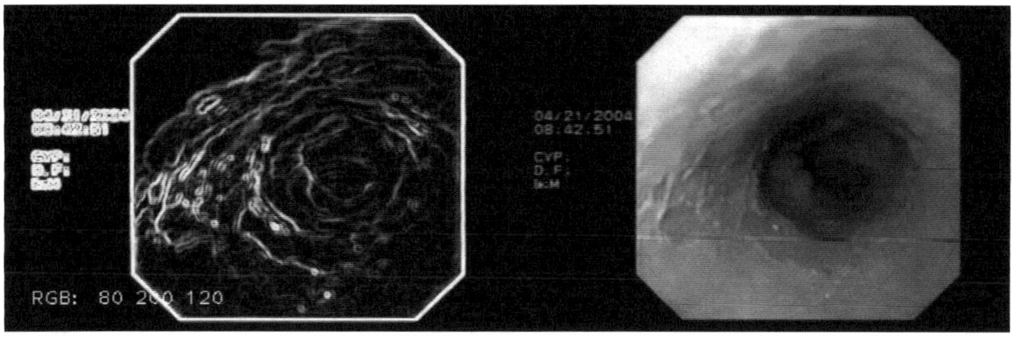

Figure 5. Original image combined with resulting image from the Sobel filter.

5.2. Other pathologies results

Currently the number of patients who come to the clinic with symptoms of Barrett's esophagus do not represent the majority of cases. Due to the characteristics of the implemented filters, the software can be used for any kind of symptoms under study. Following, there are two images of different patients, with a brief remark made by the doctor who performed the study.

Case 1: In figure 6, we can see that filtering enhanced the mucosal pattern (pit pattern) and the grooves of the lesion confirming the presence of a benign adenoma.

Case 2: In figure 7, the original image shows a diffuse gastropathy. By applying the proposed filter, a sharper image with more pronounced relief was obtained, allowing the specialist perform a biopsy in the most representative tissue with greater precision.

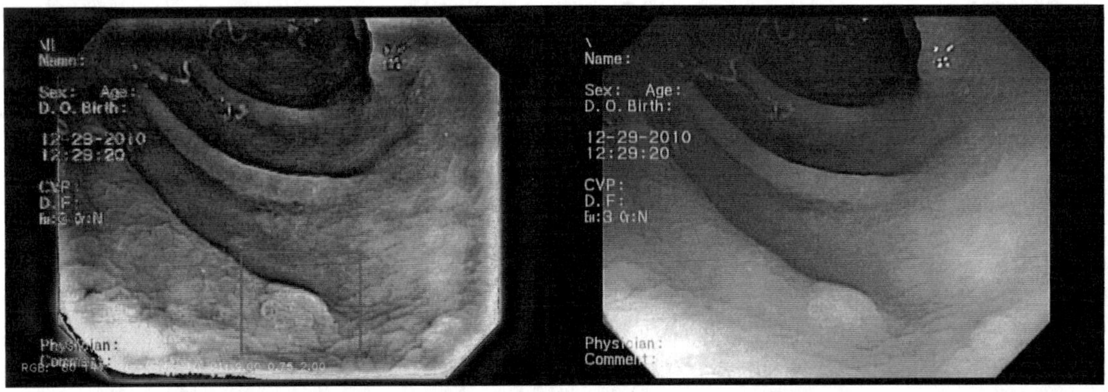

Figure 6. Processed image that allowed the diagnosis of a benign lesion.

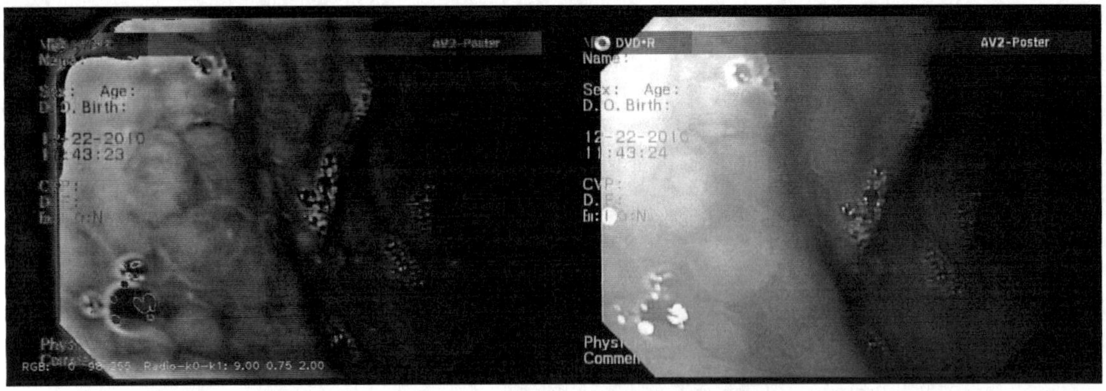

Figure 7. Enhanced image showing a diffuse gastropathy

6. Conclusions

Although digestive endoscopy has changed considerably in recent years, the images obtained by conventional endoscopic equipment face some technological difficulties. This is due in part to the fact that the white light that use these instruments does not allow identify areas of dysplasia. When identifying the affected area is not possible, the diagnosis lies in taking random biopsies. This adds complexity to the procedure and limits its performance, since sometimes the samples are not representative or are taken outside the lesioned area.

Select more precisely the pathological areas, it achieves a better diagnosis. The prototype developed has proved to be a simple and economic tool, allowing the delineation and enhancement of the areas of interest, the early detection of tissue changes and the errors reduction in the making biopsy. Its use does not require internal modification of the endoscopy equipment, simplifying the task of connecting and disconnecting the new image processing system.

7. Acknowledgements

The authors thank Dr. Gaston Jury and the CED (Centro de Estudios Digestivos de Mar del Plata) for their cooperation and the support provided for the development of the prototype.

8. References

[1] Bremer C, Lynch V and Ellis F 1970 Barrett's esophagus: congenital or acquired?. An experimental study of esophageal mucosal regeneration in the dog. *Surgery* **68** pp 209-216.

[2] Burgess JN, Payne WS, Andersen HA, Weiland LH and Carlson HC 1971 Barrett's esophagus. The columnar-lined lower esophagus. *Mayo Clin. Proc.* **46** pp 728–34.

[3] Crabb DW, Berk MA, Hall TR, Conneally PM, Biegel AA and Lehman GA 1985 Familial gastroesophageal reflux and development of Barrett's esophagus. *Ann. Intern Med.* **103** pp 52-54.

[4] Dahms R and Rothstein F 1984 Barrett's esophagus in children: a consequence of chronic gastroeophageal reflux. *Gastroenterology* **86** pp 318-323.

[5] Boch J, Shields H, Antonioli D et al. 1997 Distribution of cytoketatin markers in Barrett's specialized columnar epithelium. *Gastroenterology* **112** pp 760-765.

[6] Intel Corporation 1999-2001 *Open Source Computer Vision Library, Reference Manual.*

[7] Bouchet A, Pastore J, Abras G, Jury G, Jury H and Ballarin V 2007 Barrett esophagus: guided biopsies through digital image processing. *Journal of Physics* **90** pp 1-6. ISSN 1742-6596 (On-Line). ISSN 1742-6588 (Print).

[8] Bouchet A, Pastore J, Abras G, Jury G, Jury H and Ballarin V 2007 Esófago de Barrett: toma de biopsias dirigidas a través de Procesamiento Digital de Imágenes. In *XVI Congreso Argentino de Bioingeniería (SABI 2007)*, San Juan, Argentina.

[9] Serra J 1982 *Image Analysis and Mathematical Morphology* vol I (London: Academic Press)

[10] Jain A 1989 *Fundamentals of Digital Image Processing*, (London: Prentice-Hall International)

[11] Castleman K 1979 *Digital Image Processing*, (London: Prentice-Hall International)

[12] Pinilla C, Alcalá A and Ariza F 1997 Filtrado de imágenes en el dominio de la frecuencia, Revista de la Asociación Española de Teledetección **8**

[13] Gonzalez R and Woods R 1992 *Digital Image Processing* (Adison –Wesley)

[14] FFTW library for computing the discrete Fourier transform, www.fftw.org.

[15] Canny J 1986 A Computational Approach to Edge Detection *IEEE Transactions on Pattern Analysis and Machine Intelligence* **8**(6) pp 679-698.

[16] Pradabpet C, Ravinu N, Chivapreecha LS, Knobnob B and Dejhan K 2009 An Efficient Filter Structure for Multiplierless Sobel Edge Detection, *Innovative Technologies in Intelligent Systems and Industrial Applications* pp 40-44.

Colors Identification for Blind People using Cell Phone

A.L. Dominguez and J.P. Graffigna

Gabinete de Tecnología Médica. Facultad de Ingeniería.
Universidad Nacional de San Juan, San Juan, Argentina.

email: anadominguez_22@hotmail.com, jgraffig@gateme.unsj.edu.ar

Abstract. Assistive Technology (AT) is an interdisciplinary research area that allows finding solutions to the individual with disability [1] by easing or improving the functions or the skills for accomplishing daily activities. A technology can be considered "assistive" if it is fit for the needs, skills and capabilities of the person, taking into account mainly the intended activity and the limitations of the context and environs where the person performs such activity. The current work intends to solve the problems of vision impaired persons to recognize colors. To this aim, a Java application for cell phones has been made which lets complement the mobiles' technology with that of image processing. The means to obtain the colors from a view are based on analysing the different color models join to a mechanism to reduce the collected data. This paper describes preliminary experiences, methodology and results considering the user perception.

1. Introduction

AT has as a main objective to ease or allow for accomplishing a certain task within a given setting. The professionals that work with AT are in charge of priorly pondering the various factors involved so as to recommend the proper technical assistance that better fits to the needs and capabilities of the user in his/her specific environment [2]. Then, it is possible to state that AT is not aimed at solving the impairing condition; instead, it tries to attain acceptable functional results that the user seeks for his activities.

In the design of AT systems, devices are engineered to make the user functionally independent. In the case of sensorial impairment, ATs may offer the necessary assisting means to let the user acquire such sensorial information. To decide what kind of AT has to be developed for a given case calls for a prior analysis on the impairment state itself, its residual function and level. With this information, it is then possible to respond to whether the sensorial enhancement devices can be used or not. For example, if the residual sensorial function is insufficient, alternative sensorial ways have to be used [3]. Therefore, for persons whose vision cannot be used as the interface with the intended technology, other senses, like hearing and tactile functions may be studied as alternate ways.

For blinds, specifically, ATs offer help in various areas, such as obtaining visual information from the context, manipulating and controlling objects, bringing mobility and orientation of AT users, and other applications.

At present, AT for blinds looks for integrating the blind individual with his daily context by rendering the information means for visual recognition. In such a situation, a problem a blind has to overcome is to recognize colors. For example, it is important when deciding what to wear and how to combine clothing, or how to discern from different medicines, and similar situations. As is the case for anyone, it is very important for the blind to know they are dressed well, with a proper color combination when attending any social activity or event, so as to prevent scornful attitudes from others.

At present, blinds that need to know the color of some object, or to identify the object by just knowing its color, are able to do so only by asking it to another person, when this one is nearby. Therefore, the application proposed here allows the blind accomplish such a task independently.

This work has been motivated by observing the evolution and development experienced in recent years on the use of embedded programming of mobile telephone devices (MTD), along with the need to have concrete solutions for the visually impaired persons.

MTD technology offers a wide range of tools, along with ample advantages and salient features like a low cost of mobiles; a large amount of information offered by the manufacturers as regards the various functions and technical specifications that are relevant to develop the mobile's software for this application; the availability of numerous free public licence development environments and programming languages for the various phone models and operative systems available in the market [4]; the possibility of acquiring images; the ever-decreasing size of these electronic devices; its scarce need for maintenance, altogether with the possibility of getting communicated.

In Cook [5], a model is proposed to define an AT system by analysing the solution as a whole; that is, considering the activity to be done, the functional characteristics of the person and the context of use (see Fig. 1). For ATs devised for sensorial impairment and interface is required that is able to receive information from the context, plus a processor that obtains the useful information to be transmitted to the person through another interface. Specifically for this current work, the proposed design resorts to using the cellphone camera, colors processing software and the cellphone speaker used as the communicating interface with the person.

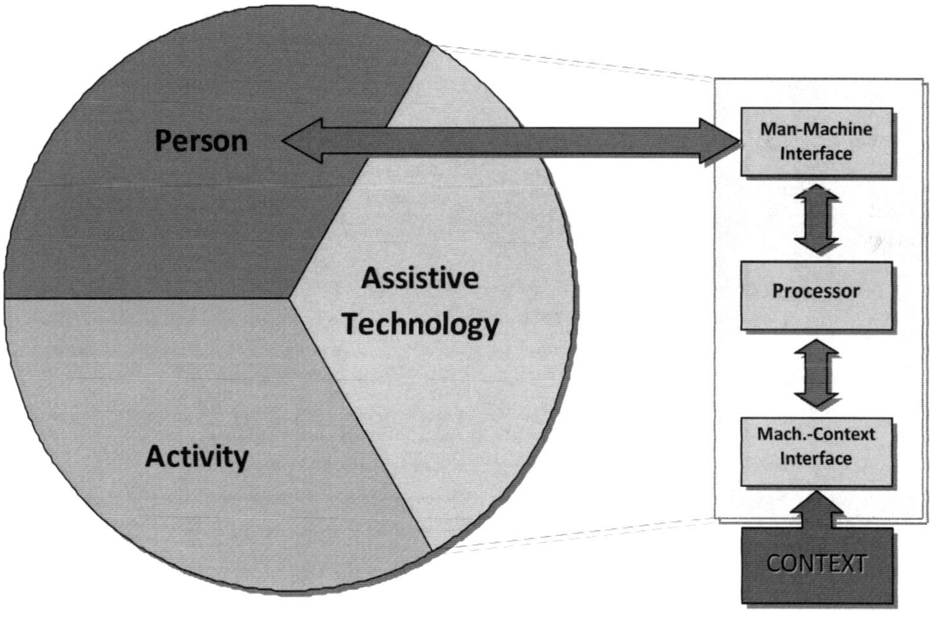

Figure 1. Interaction of the AT with the other components of the HAAT model.

Image processing requires that the various color models be priorly studied. The color information for each pixel can be specified in standard ways, and accepted through these color models. Each model is specified by a coordinate system, generally tridimensional, where each color is represented by a single point. The linear combination of coordinates generates the entire color space. These coordinates do not indicate which color it is from the viewpoint of the user's perception [6], but it rather objectively shows its location within a given color space. The more general spaces of color pursue to capture the largest number of colors visible to the human eye.

The chromatic models oriented to image management are: RGB, HSI, HSV, among other approaches [7]. The first one outstands from the rest by being the closest one to the hardware of artificial vision. The remaining two use perceptive attributes on luminance or intensity, saturation and color hue. They are defined as psychological, intuitive spaces, or user-oriented spaces, because they are the optimal ones for human interaction.

2. Methodology

The work proposed here consisted in obtaining the predominant color of a scene, and communicating it verbally to the user through a high-end cell phone (MTD). The methodology was divided into two parts involving, first, the color models study and, second, the implementation of application into the device.

2.1. Color Models.

The models allow identifying distinctly the colors of an image. There are more than five different proposals according to their application. In this work, the models RGB (red, green and blue) and HSI (hue, saturation and intensity) have been studied. The first model is based on the characteristics of the video hardware, while the second one is based on the human interpretation of color.

For each studied model, it was established a data reduction approach that allows reducing the number of colors in the view, to a quantity that can be verbalized by the device. This implies devising a univocal function that, for each pixel color of the domain space there is, correspondingly, a value in the image space. The domain is composed by 2^m colors, whereas the image space is composed by 2^n, with n<m. In the proposal for reduction, it is assumed that for each image space color c_i there is, correspondingly, $2^{(m-n)}$ colors in the domain (see Fig. 2) that conform a subdomain S_i.

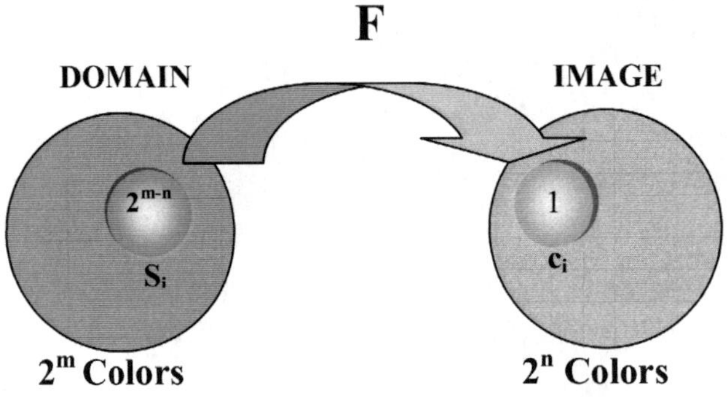

Figure 2. "F" transformation. Venn Diagram representation

The image obtained by the MTD is RGB 8:8:8, using 8 bits to represent each color. Therefore, the domain will have 224 possible colors (m=24). From this, four strategies of color number reduction are evaluated (see Table 1).

Table 1. Strategies of color reduction.

Representation	N	Number of colors: 2^n	Number of colors for each element of the image space: 2^{m-n}
RGB 2:2:2	6	64	262144
HSI 2:2:2	6	64	262144
RGB 3:3:3	9	512	32768
HSI 5:2:2	9	512	32768

Each strategy was evaluated according to the complexity to implement it and to the quality of reduction. The complexity is pondered according to the computational costs in working it by the MTD. In order to measure the image quality, for each value c_i, a color palette (CP) was made, that contained the entire subdomain S_i. Then, it was analyzed for each CP its capability to represent a given color. A palette of good quality is one that, for the observer, does not contain contrasting mixtures of colors.

After choosing the best strategy for data reduction, the color for each element c_i was determined. To this aim, the various S_i palettes were shown to different observers, and the color denominations were recorded for each palette. From this data, a definite color was assigned for each palette.

2.2. Application Development

An application development on a MTD necessarily implies to know the software programming language J2ME [8], which is widely used with this equipment type. Besides, the characteristics of the proposed development require a thorough study of the image acquisition hardware, data processing, the management of the audible information, and other aspects.

The requirements of the development are as follows:

- A simple algorithm
- A minimal time of execution
- A small memory for software execution due to MTD limitations
- A limited number of colors in order to have a simple verbalization
- Color characterization of the image as a function of the occurrence analysis of each color.

Basically, the system captures the image sent by the MTD camera and stores it in memory, to be later processed one pixel at a time. The analysis consists in taking the image's center, with a pre-established size of window, where the information corresponding to the RGB color is extracted.

For each RGB píxel, the representation levels were reduced in order to simplify the number of colors and to assign them to ranges experimentally determined in the previous stage. The software computes the number of pixels belonging to each range and, by means of an audible output, it tells the user the colors of the view. To attain this, it is necessary to complete a number of steps of image processing. Figure 3 shows the block diagram of the application.

For the audible output, the maximum values atained for each color are processed. This way, when there is a dominant maximum, the system automatically verbalizes the color name. When there are two dominant colors, both names are verbalized and, finally, when there are no relevant maximum counts, a color range analysis is performed.

3. Results

The results allow evaluating the preliminary tests and the final development of the system. From the preliminary tests, the comparison of the four proposals for color representation is made, along with the interpretation of colors by the observers. In the phase of final experiences, tests are performed on the implementation of the system.

3.1. Analysis of the representation models

The representation models listed in Table 1 were evaluated using the complexity and quality analysis of the CP obtained from Si. The study results for two 6-bit strategies and two 9-bit strategies are presented.

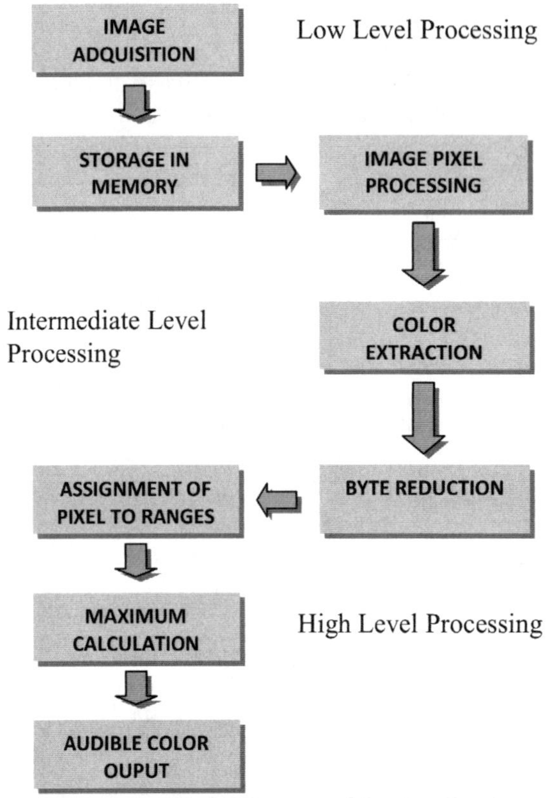

Figure 3. Block Diagram of the Application

3.1.1. RGB 2:2:2

In Figure 4 some CP are shown (represented as a color square) obtained through Matlab®, where each one represents 2^{18} possible colors for each element of the image space of the transformation. To the left, it can be noted those palettes that present some difficulty to differentiate some colors, and to the right there is another group of well-defined CPs that can be easily discriminated one from another.

In this model, most images were obtained with very definite colors and easy to discern, a situation that led to a low complexity in determining the color names. Besides, the reduction is simple, thus allowing a shorter software execution time due to the small number of colors.

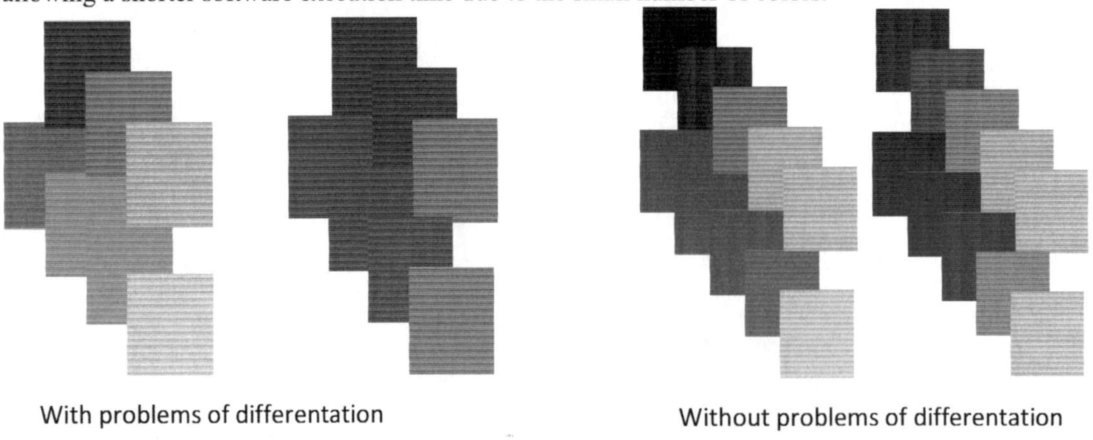

Figure 4. Color Pallet for RGB222 Image with and without problems of differentiation (Matlab)

3.1.2. HSI 2:2:2
In this model, by analysing the CPs obtained through tests, it was noted a better differentiation of colors, but also a significant lack of color hues. Its complexity and software execution time were greater due to the need to make an intermediate transformation between the native color model and this one. To the left of Figure 5, it can be noted the CPs that present difficulty to differentiate some colors; on the right, there is another group with well-defined CPs, where each one represents 218 possible colors. In this reduction strategy, colors like yellow do not have any representation in the image set.

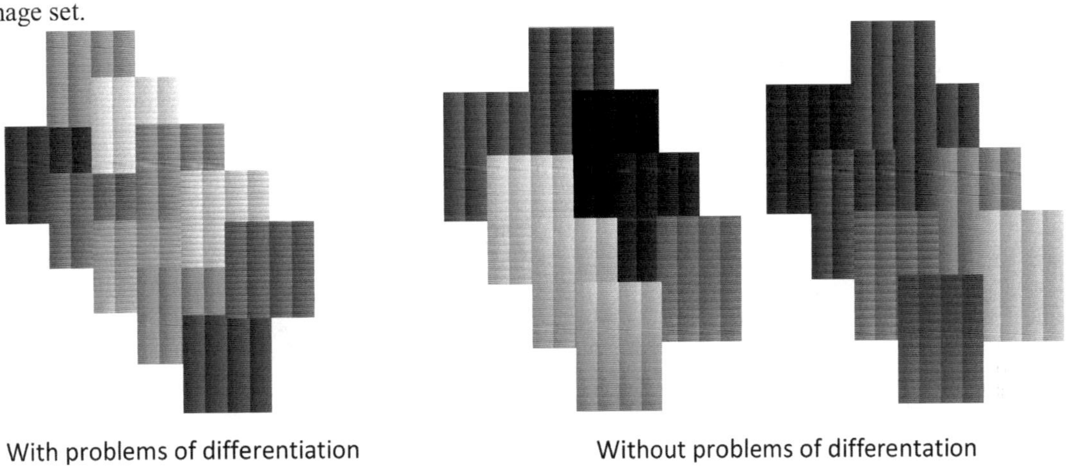

Figure 5. Color Pallet for HSI222 Image with and without problems of differentiation (Matlab)

3.1.3. RGB 3:3:3
In Figure 6, it can be observed that most CPs for this representation are well defined, but operative problems arise in the manipulation of 512 colors in the transformation, therefore causing more complexity in assigning names and ranges, and a longer software execution time.

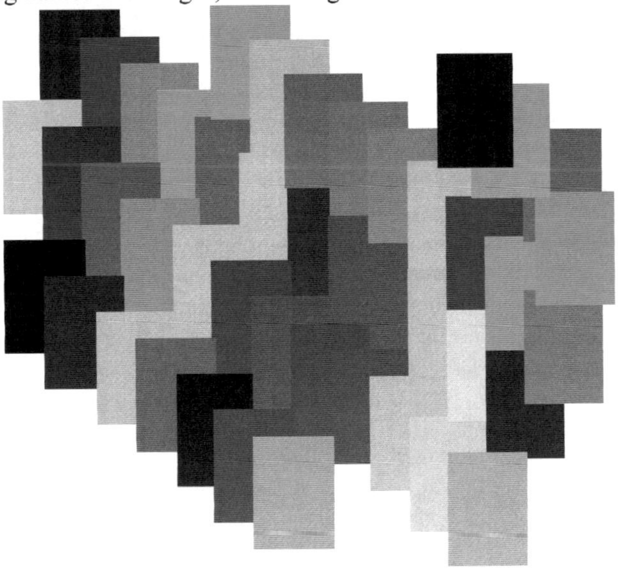

Figure 6. Color Pallet for RGB333 Image

3.1.4. HSI 5:2:2
This strategy was approached differently by interpreting the information that is stored for each model component. Since hue H contains all the information referred to the predominant color, S represents the saturation and I is referred to intensity, the function used for this model gives prevalence to the first parameter. This way, from the 9 bits used for the reduction, 5 bits are used for H, 2 bits for S and 2 bits for I. Even though the number of colors of the image space is equal to the number of RGB 3:3:3 strategy, the variation may be perceived in the representation of the color, because the modified parameters are different. To the left in Figure 7, there are CPs which show the difficulty for discerning colors, as opposed to the right-hand portion showing images with well-defined colors. In this model, a great percentage of non-defined images was obtained, which made cumbersome the task of defining names and ranges, in addition to a much longer software execution time on account of the large number of colors.

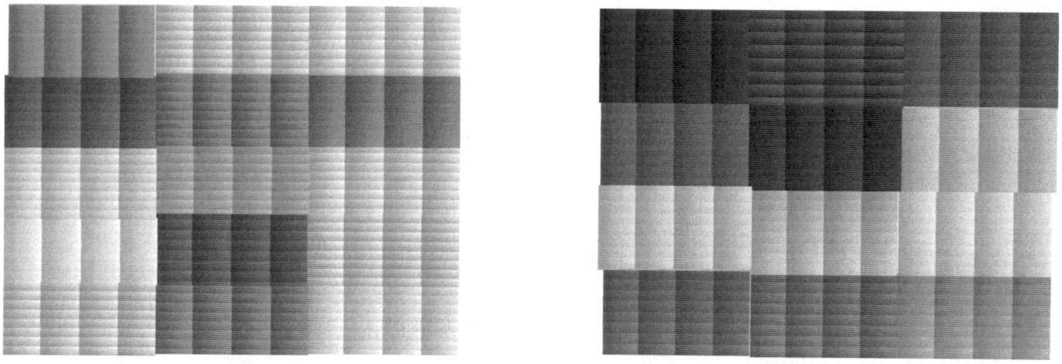

With problems of differentiation Without problems of differentation

Figure 7. Color Pallet for HSI522 Image with and without problems of differentiation.

3.1.5. Comparison among strategies
After performing the tests, the representation system chosen to carry out the application was the RGB 2:2:2 due to a number of better characteristics and advantages respecting the remaining strategies. A key feature that pointed to its selection is the fact that RGB is the native model of the hardware and of the J2ME language, in contrast with the HSI which needs a previous conversion step. This way, the RGB representation system allows to adequately discerning the colors as a function of in straight connection to the experience, while it also allows for an optimal processing within the system.

In addition, it was opted for reduction 2:2:2 due to its balanced characteristics which permit a proper identification of colors and a greater computational speed. This strategy makes it easier for the observers to group colors and defines the ranges.

3.2. Definition and interpretation of colors.
The determination of color names in order to allow for a proper verbalization was made by analysing each CP by 10 observers having normal vision. The perception of some of these observers was considered for defining the color naming, which resulted in 31 different names. The ten observers analyzed each CP and gave them a color name. Table 2 shows in part the result from such a task. For each CP, the chosen color is presented, along with the number of observers which opted for such a color name (name/ number of observers). When the CP has very different interpretations, it is referred as undefined. Only 1 CPs were classified as undefined.

Table 2. Analysis for color determination (color name/number of observers)

Color	1st Dominant Color	2nd Dominant Color	3rd Dominant Color	4th Dominant Color	5th Dominant Color	6th Dominant Color	Adopted Color
0	Black / 10	-	-	-	-	-	**Black**
1	Dark blue / 6	Bluel / 2	Navy blue / 2	-	-	-	**Dark blue**
2	Blue / 6	Navy blue/ 3	Dark blue / 1	-	-	-	**Blue**
3	French blue/ 5	Bright blue / 4	Light blue/ 1	-	-	-	**French Blue**
4	Army green / 5	Green / 2	Olive Green / 1	Moss Green / 1	Dark green / 1		**Military Green**
5	Oil blue / 3	Greenish blue / 2	Green / 2	Gray / 2	Blue / 1	Indefinite / 1	**Oil blue**
....	....	....	....	....	....	....	....
57	Orange /4	Light orange / 3	Light yellow / 2	Light brown / 1	-	-	**Orange**
58	Light pink / 2	Light orange / 1	Indefinite color / 2	Greenish purple /2	Light yellow / 1	Grayish pink / 2	**Indefinite color**
59	Light pink / 4	Llight purple/ 3	Purple / 3				**Light pink**
60	Yellow / 5	Dark yellow / 3	Lemon yellow / 2	-	-	-	**Yellow**
61	Yellow / 4	Light yellow / 4	Water green / 1	Yellowish green/1	-	-	**Yellow**
62	Light yellow/4	Yellow / 2	Water green / 2	Light brown/1	Cream / 1	-	**Light yellow**
63	White / 7	Gray / 2	Pruple / 1	-	-	-	**White**

For each of the 31 colors and its ranges, it were generated audio files with theirs names. This allows communicate color to a blind person using the audible output of the MTD, based on the processing of the dominant color.

3.3. System Test.
The experience tests were carried out into two stages. The first one was done by a normal-vison person, using as a tool two mobile devices (a Nokia 5800 and a N73), under different conditions of lighting (natural light, artificial light, etc.) to determine the optimal light conditions for work. The second stage was performed by three persons with visual impairment (blinds) who used a N73 mobile.

It can be noted in Table 3 for the first stage the coincidence percentage between the scene and the captured image, as well as that of the captured image and the audible output. Both values affect directly to the quality of the audible output respecting the scene. The main problem arises in the correspondence between scene and image. Particularly, it was noted that with recurrent image takes, when the MTD camera makes a self-adjustment (thus permitting a stronger or weaker input of external light), a significant change of color of the captured image was observed. It was also noted that when having objects or textures of glossy surface, a problem arises on lack of coincidence between the scene and the captured image. Reflex or gloss seen at various inclination angles caused gray tones in the captured image.

Table 3. Results obtained with Color Recognition Application

		% COINCIDENCES	
Cell Phone	**Light**	**Scene/Image**	**Image/Audible Output**
N73	Natural	36.66	73,33
Nokia 5800	Natural	50	86,67
N73	Natural Indoors	66,67	70
Nokia 5800	Natural Indoors	66,67	70
N73	Artificial	66,67	63,33
Nokia 5800	Artificial	50	50

The tasks for the blinds in the second stage were performed with the help of a person who guided them to capture the image and acquire the ability to use the MTD. The goal is to acquire capability to do the focusing and centring of the object by encouraging the autonomous usage of the application embedded in the MTD. The tests were carried out with lended MTDs, because there was no user owning high-end mobiles.

The results from these tests were highly satisfactory as regards the usefulness of the application to the MTDs. More practice is needed to increase the ability to focuse and use the MTD, as well as to handle the regulation of the room's light. The three test cases presented a high coincidence ratio between the captured image and the audible output, but there were some flaws between the scene and the captured image.

4. Conclusions

In this work, it was posible to design and implement an application on a MTD which allows determine and communicate audibly the predominant colors of an image taken by the cell-phone's camera. To this purpose, it was studied the problems associated to the activity and the technological aiding system that fits more properly.

Priorly to the development, the color models were studied, as well as some approaches for reducing the number of colors. Representation RGB 2:2:2 resulted to be the most convenient one when analysing the complexity and quality of results. From the perception of ten observers, it was possible to associate each color with a set of colors easy to verbalize.

The application thus developed allows capturing, determining and emitting an audible rendering of color names of an image taken by a MTD. The view that is wished to be captured gets modified by a number of factors external to the processing, which affect teh image to be analyised, and to the perception of the image by the human eye. Such factors can be noted in artifacts produced by the camera and artifacts stemming from external sources, such as shades, reflexes of natural light, the objects's own glow, the artificial lighting, and the like. These inconveniences are then translated into audible reproduction errors.

The implementation can be made on any high-end MTD, acondition set by the basic needs for dynamic memory and standardization. The application has been satisfactforily implemented on two mobile devices, Nokia models 5800 and N73.

In order to solve the inconvenience presented by the hardware, an analysis is under way to acquire the image in a controlled environment. For example, in the room of the user define distance to the object and fixed lighting conditions.

References

[1] OMS, ORGANIZACIÓN MUNDIAL DE LA SALUD: Clasificación Internacional del Funcionamiento, de la Discapacidad y de la Salud (CIF) 2001.

[2] GRAFFIGNA, Juan Pablo: "Sustitución Parcial de la Visión mediante Realimentación Táctil". 2009. Universidad Nacional de San Juan. ISBN: 987-05-7233-6.

[3] ZELEK, Jhon: The E. (Ben) & Mary Hochhausen Fund for Research in Adaptative Technology For Blind and Vissually Impaired Persons. 2002.

[4] GEORGE APPLING, Giulio Pappalardo: "The Rise of Mobile Applications Stores Gateways to the World of Apps". Booz and Company Consulting Firm. 2010.

[5] COOK A. HUSSEY .S.: Assistive Technologies, Principles and Practice. 2da Ed.Mosby. 2002

[6] MOJSILOVIC, Aleksandra. A Computational Model for Color Naming and Describing Color Composition of Images, *IEEE, 2005*

[7] GONZALEZ R, Y WOODS Richard: "Digital Image Processing (3rd Edition - 2007). Prentice Hall; ISBN-10: 9780131687288.

[8] GÁLVEZ ROJAS Sergio, ORTEGA DÍAZ Lucas "Java a Tope: J2ME (JAVA 2 MICRO EDITION)". Dpto. de Lenguajes y Ciencias de la Computación E.T.S. de Ingeniería Informática Universidad de Málaga. ISBN: 84-688-4704-6

Enhancement of medical images in HSI color space

E Blotta[1], A Bouchet[1,2], V Ballarin[1], J Pastore[1,2]
[1] Laboratorio de Procesos y Medición de Señales. Dpto de Electrónica. UNMDP.
[2] Consejo Nacional de Investigaciones Científicas y Técnicas (CONICET).

E-mail: eblotta@fi.mdp.edu.ar

Abstract. In this paper we use the HSI color space as an alternative to RGB space. The HSI space considers the image as a combination of the components: hue, saturation and intensity. In this paper we propose to design a chromatic filter in order to obtain improvements in the enhancement of medical images. In image processing systems it is usual specify colors in a form compatible with the hardware used. The RGB color model, where is computationally convenient, is not very useful in the specification and color recognition. The human being does not recognize a color by having an amount of red, green or blue components, but uses attributes perceptual of hue, saturation and intensity. The chromatics models HSI, HLS, HSV, and its variants, encodes the color with the above attributes and are defined as intuitive spaces.

1. Introduction

The color image processing is motivated by two important factors, by a similarity to human vision, fully chromatic, and second, by the increasing of the information that the chromaticity contributes to the analysis of images.

The digital image processing techniques proposed have been developed for binary images or gray levels [1-2]. For many years, this limitation has been imposed by the hardware. This is because the computational cost of the color processing image algorithms is very high, making it necessary to reduce the visual information to a unique two-dimensional map. While these images contains information of scenes captured, the reduced to two dimensions of a visual environment causes the loss of information to discriminate certain objects in it. This is because the chromaticity of the forms is not reflected and the luminance is homogeneous in this type of map. The growth of computing power, the storage capacity and the reduced costs of capture systems and printing images have made that the image acquisition is made in color, and that further analysis and processing is carried out in tridimensional maps.

Color spaces provide a method for specifying, ordering and manipulating colors. Usually it is determined by a base of n vectors whose linear combinations generate all elements of the space. There are numerous color spaces among which we can mention the gray scale, which is one-dimensional space, the planes RG, GB and BR, which are two-dimensional spaces and the spaces RGB (Red, Green and Blue), HSV (Hue, Saturation , Value), HSI (Hue, Saturation, Intensity) and YIQ, wich are three-dimensional spaces [3-4].

In the RGB model, the color is specified by red, green and blue amounts. In this model, a digital image consists of three planes of independent images, each of which stores the values of red, green and blue. This model has serious disadvantage when you want to perform different types of processing of the images such as enhancement, segmentation or classification [5]. To eliminate or reduce these

limitations be have created models that separate the color information of the lighting information such as the HSI space. As in the RGB space, a digital image in the HSI space consists of three matrices of $m \times n$ dimension, which saves the values of hue, saturation and intensity. Thus, each pixel of an image represented in this space has three data: hue and saturation which provide information of color, and intensity which describes the brightness.
The application of alternative color models to the medical image processing has as main objective improve the performance of algorithms for enhancement, segmentation and classification. Thus, it provides tools to the specialist to reduce the error in the performance of different types of diagnostic.

2. Color models or color spaces

Due to the nature of the human eye and due to the tri-chromatic theory, all colors that humans can recognize in an image are a combination of so-called primary colors red (R), green (G) and blue (B). The goal of a color model is to facilitate the specification of colors in a standardized way.
A color model is the specification of three-dimensional coordinate system and of one subspace of this system in which each color is represented by a unique point. The choice of a suitable space for color representation remains a challenge for scientists researching color image processing..

2.1. RGB Color Model

In RGB space, color is specified through red, green and blue amounts, forming in 3D space the cube depicted in figure 1. Each color is a point on the surface or inside this cube. The gray scale is located in the diagonal that joins the white color with black color, i.e. the diagonal joining the points (255,255,255) and (0,0,0). The range of each coordinate or chromatic RGB component is, usually, a value belonging to the interval [0,255]. Images in the RGB color model consist of three planes of independent images, each corresponding to a primary color. The individual processing of these three planes allows extract shapes and significant details of the images. However, there are times when this model is not suitable for image processing. Between the disadvantages of this model stands out the non-uniformity, i.e., is not possible to evaluate the observed difference between colors with a measure of distance from the primaries colors R, G and B.

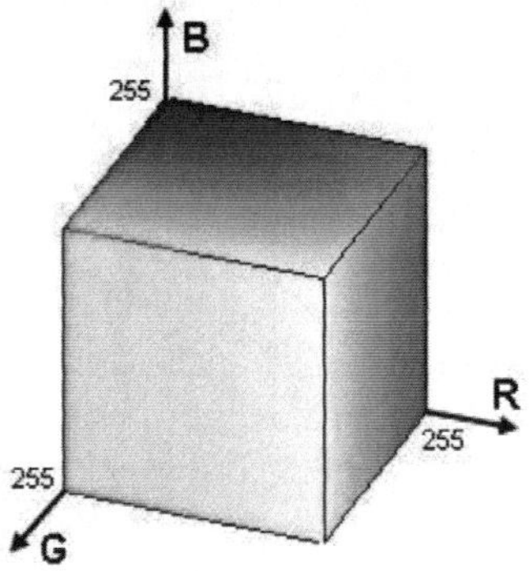

Figure 1. RGB model representation.

2.2. HSI color model

The HSI model defines a color model in terms of its components. This space has the ability to separate the intensity of the intrinsic information of color, which refers to the hue and saturation. This model is suitable for processing images that present lighting changes this is due to the fact that the colors of the environment are distinguishable from each other through the hue component.
The most common way of representing the HSI model is a double cone, as shown in figure 2. The center of the double cone is a circumference divided into equal size angles. Therefore, the value of the

hue component describes the color by its wavelength, and takes values between 0 and 2π, with 0 representing the red, $2\backslash 3\pi$ representing the yellow and $4\backslash 3\pi$ representing the green. The distance from the center to the outside of the circumference represents the color saturation and takes values between 0 and 1. Saturation refers to the mix of color with white light. Finally, the axis through the two cones corresponds to the intensity component. This will have a value between 0 (black) and 1 (white) and indicates the amount of light present in a color. Removing a small circumference of the figure formed by two cones, colors close to an intensity of 1 are lighter than those close to zero. When the saturation component is close to 0, colors only reflect a change between black and white. When this component is close to 1, the color will reflect the true value represented by the hue.

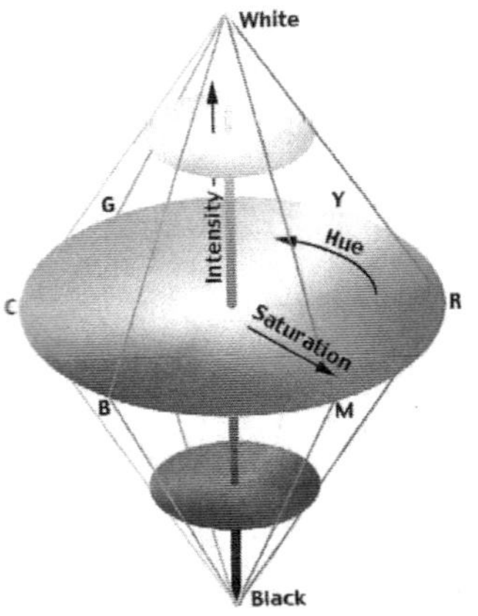

Figure 2. HSI model representation. The letters R, Y, G, C, B, M refer to = Red, Yellow, Green, Cyan, Blue and Magenta respectively.

2.3. RGB to HSI conversion

The spaces that represent the color in terms of hue, saturation and intensity, allow an intuitive description of the colors. The transformation of RGB space to HSI space is the conversion of a cartesian coordinate system to another in cylindrical coordinates, where the color is specified in terms of hue, saturation and intensity. A good representation of color must use norm or distance to make that the chromatic and achromatic components are independents. The equations that establish the change of coordinates between the components of both color spaces, using the semi-norm max-min, are given by:

$$\begin{cases} I = 0.23\times R + 0.715\times G + 0.072\times B \\ S = \max\ R,G,B\ - \min\ R,G,B \\ \theta = \arccos\left(\dfrac{2\times R - G - B}{2\left[R^2 + G^2 + B^2 -\ RG + RB + GB\ \right]^{1/2}}\right) \\ H = \begin{matrix} 360-\theta & \text{if } B > G \\ \theta & \text{if } B \le G \end{matrix} \end{cases} \quad (1)$$

The following section presents the proposed method for intensity enhancement.

3. Proposed Method: Selective Chromatic filter

The proposed algorithm for the selective chromatic filter has the following steps:

• First we extract the color of interest of the original image using manual selection. The algorithm select the correspondent color code on the matrix H (a value between zero and one).

• Based on the selected color and allowing a range of tones above and below it (color band pass filter), a mask with the image pixels that are within that range is obtained.

• At last, this mask is applied to get darker the pixels which colors are outside the filter passband in the intensity component.

To compare the quality of filtering in HSI space with respect to RGB space, we performed the same filtering, but now on each matrix of RGB space, red, green and blue. The final mask results the combination of a logic *and* function applied to the individual masks obtained. This chromatic filtering is more restrictive in RGB space because there is a high degree of correlation between the RGB components.

Figure 3 (a) shows a synthetic image that reproduces the color palette on a ring, while the inner triangle shows the variation of the intensity. Figure 3 (b) and (d) shows the results of applying the chromatic filter in the HSI space for two different tones, while figure (c) and (e) show the counterpart to the RGB space. It may be noted a further restriction of colors leaked in the latter space, being the passband of the filters used in both equivalents cases.

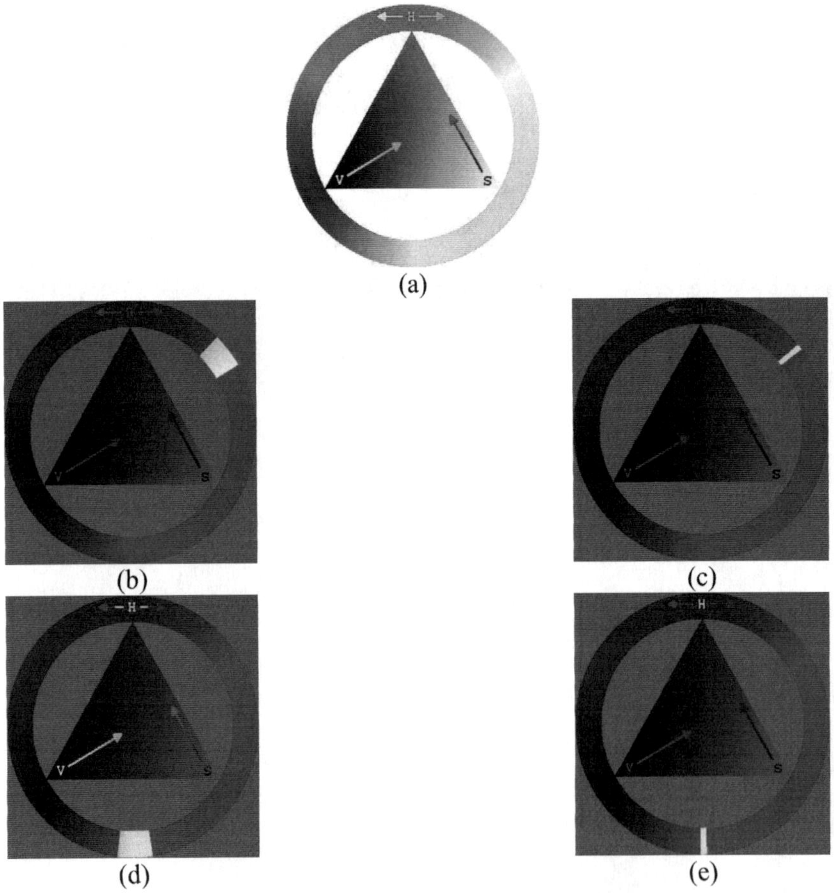

Figure 3. Processing a synthetic image. (a) Original image. (b) Enhancing the yellow component in the HSI space. (c) Enhancing the yellow component in the RGB space. (d) Enhancing the cyan component in the HSI space. (c) Enhancing the cyan component in the RGB space.

4. Some Bioengineering Applications

This section shows different chromatic filter applications implemented in HSI space in some medical images. To compare the results, we include the same segmented images in RGB space, using a similar chromatic filtering. In these images we can observe light and dark areas, the first corresponds to a passband filter, while the last correspond to the band reject filter.

Figure 4 represents an electron micrograph image showing a nerve demyelination [6]. The region coloured in red is a section of an axon- transmitter of nerve impulses to other neurons- which has lost its myelin sheath. This is a disorder seen in multiple sclerosis disease. Figure 4 (a) shows the original image, figure 4 (b), (d) and (f) show the enhancement of the blue component, the red component and the green component respectively, using the chromatic filter on the HSI space. In figure 4 (c), (d) and (g) we observe the same images filtered in RGB space.

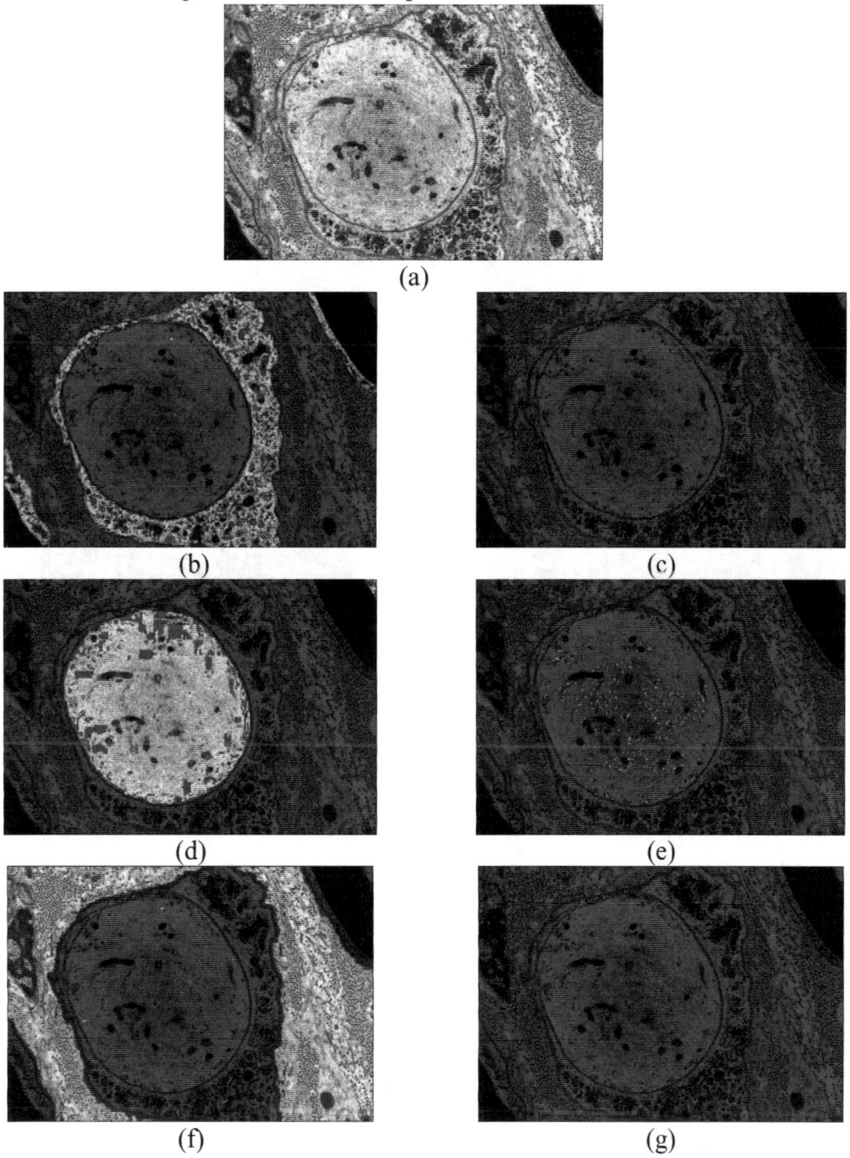

Figure 4. (a) Electron micrograph showing a nerve demyelination. (b) Enhancement of blue component in HSI space. (c) Enhancement of blue component in RGB space. (d) Enhancement of red component in HSI space. (e) Enhancement of red component in RGB space. (f) Enhancement of green component in HSI space. (g) Enhancing the green component in RGB space.

Figure 5 (a) corresponds to an endoscopy image with a Barrett's disease diagnose. This is a disease caused by reflux of gastric juices or intestinal content, which is rich in bilis, towards the esophagus. This occurs when the mechanisms existing in the gastroesophageal union for prevent it, fails [7-8]. Barrett's esophagus is considered precancerous.
To achieve a better diagnosis by observing this kind of lesion, in order to select more accurately the areas to biopsy, several techniques to enhance endoscopic images have been developped. These techniques are used in order to make a diagnosis more accurate, less intrusive. Figure 5 (b) shows the result of applying chromatic filter on the HSI space. The areas of interest where the specialist should take the biopsy were determined successfully. Figure 5 (c) shows the same segmented image in RGB space with very poor results.

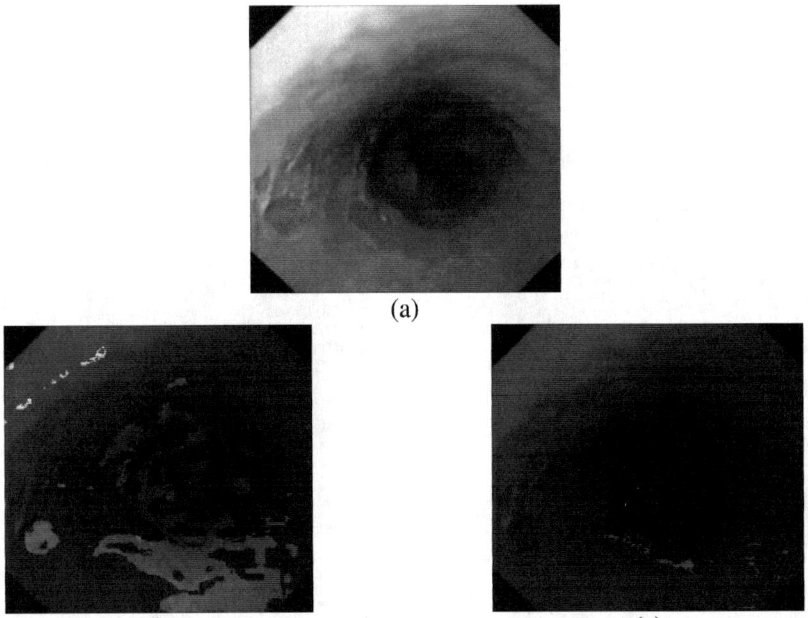

(a) (b) (c)

Figure 5. (a) Barrett's Esophagus image. (b) Enhancement of the affected area in HSI space. (c) Enhancement of the affected area in RGB space.

At present, studies of human identification using molecular biology techniques can be performed automatically. In third world countries horizontal electrophoresis in agarose gels is still used. In the last case the DNA fragments amplified by PCR (Polymerase Chain Reaction) or multiplex monoplex reactions are separated by horizontal electrophoresis [9-10]. The results are obtained analyzing the length of DNA strands present in the sample and its concentration on molecular weight markers. Making this old process automatic is a big step until an automatic instrument can be afford.
Figure 6 (a) shows an agarose gel produced by a process of electrophoresis. Figure 6 (b) shows how the application of the proposed filtering method to this kind of images. It can be observed the existence of two valid bands in the filtered image, optimizing the interpretation by the experts, whereas in figure 6 (c) that it is not possible because the bad results of the RGB filtering.

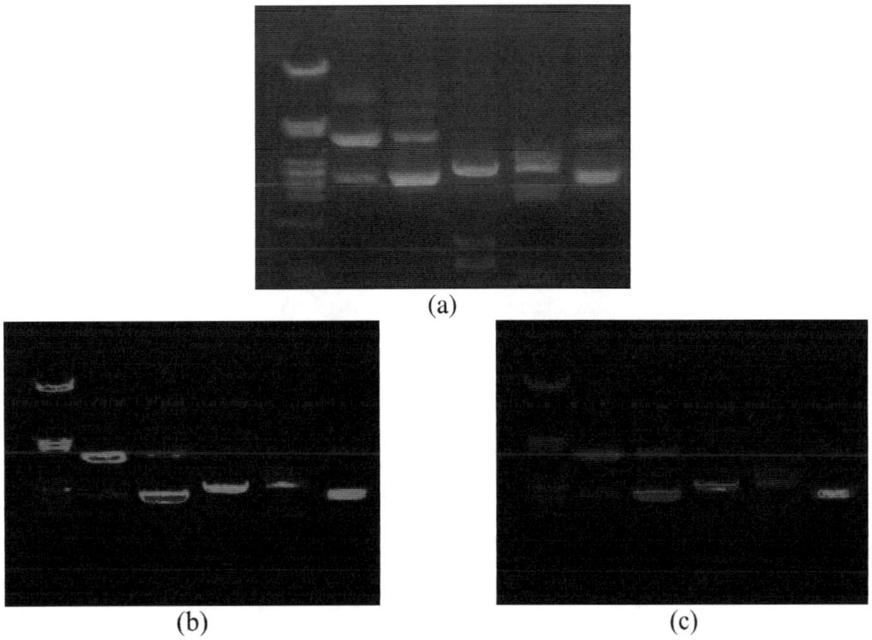

Figure 6. (a) Original image of agarose gel. (b) Enhancement of an agarose gel in HSI space. (c) Enhancement of an agarose gel in RGB space.

Figure 7 (a) shows an image of a brain scan with hypertensive encephalopathy in a woman of 23 years of age [11], which has suffered an acute arterial hypertension in association with cyclosporine. Figure 7 (b) shows the enhancement of areas of bright orange in HSI space and figure 7 (c) shows its counterpart in the RGB space.

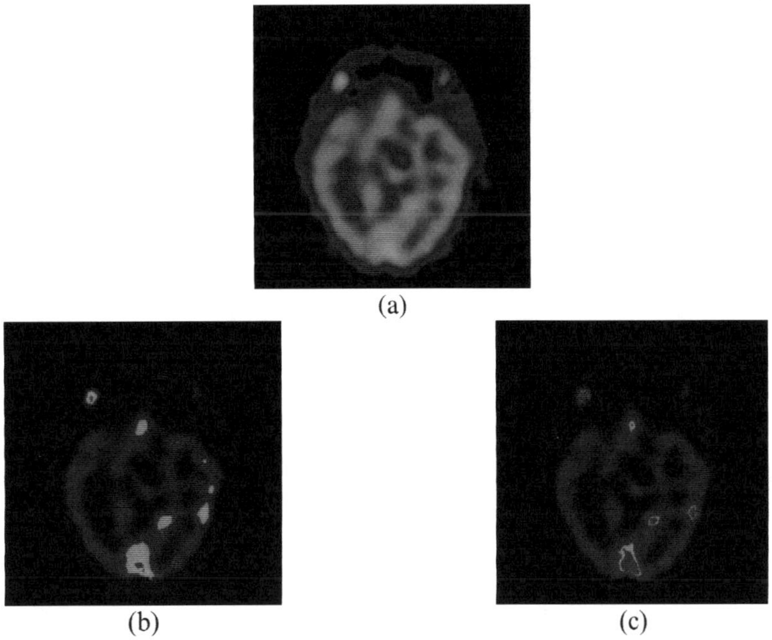

Figure 7. (a) Image of a brain scan: Hypertensive encephalopathy. (b) Enhancing the orange area in HSI space. (b) Enhancing the orange area in RGB space.

Journal of Physics: Conference Series **332** (2011) 012041 doi:10.1088/1742-6596/332/1/012041

5. Conclusions

Choosing a suitable color model is heavily dependent on the application and is the first step in the processing of medical images in color. Although the RGB model is best suited to display color images, the preliminary results obtained shows that this sapce is not suitable for analysis and processing imaging with a high degree of correlation between the components R, G and B. Furthermore, the distance between colors in this space do not represents color differences such as the human visual system perceives them. The HSI model have as key attribute the advantage of separating the intensity of the intrinsic color information, which is important when images are affected by changes in lighting. This paper proposed the HSI color space as an alternative to the RGB space and design a chromatic filter for improve medical image enhancement. It could be observed that in the HSI space is possible to integrate variations in lighting without compromising the classification significantly.

6. References

[1] González Woods 1992 *Digital Image Processing* AddisonWesley.
[2] Jain 1989 *Fundamentals of Digital Image Processing* Prentice Hall.
[3] H D Cheng X H Jiang Y Sun Jingli Wang 2001 Color image Segmentation: advances and prospects. Elsevier.
[4] Alain Tremeau, Shoji Tominaga, and Konstantinos N. Plataniotis. Color in image and video processing: most recent trends and future research directions. 7, January 2008, *Journal on Image Video Processing*, p. 26.
[5] Sierra, Angulo 2005 Segmentación de imágenes en color utilizando histogramas bivariables en espacios color polares HSV, *Computación y Sistemas*, **8** (4).
[6] Iglesias Rozas, J Rafael 2009 Desmielinización en fibras nerviosas causada por polineuropatía. Repositorio Documental de la Universidad de Salamanca.
[7] Bremer C, Lynch V and Ellis F 1970 Barrett´s esophagus: congenital or acquired? An experimental study of esophageal mucosal regeneration in the dog. *Surgery* 68 pp 209-216.
[8] Bouchet A, Pastore J, Abras G, Jury G, Jury H and Ballarin V 2007 Barrett esophagus: guided biopsies through digital image processing. *Journal of Physics* 90 pp 1-6. ISSN 1742-6596 (On-Line). ISSN 1742-6588 (Print).
[9] Lewin 1988 *Genes* Editorial Reverté.
[10] Martin 1996 *Gel electroforesis nucleid acids* Bios Scientific Publisher.
[11] Johnson, Keith A. M.D. and Becker, Alex J. Ph.D. The whole brain atlas. [Online] http://www.med.harvard.edu/AANLIB/home.html.

Analysis of chemical processes for the synthesis of magnetite for biomedical applications.

J Baena[1], J I Marulanda[1]
[1] School of Sciences and Humanities, Applied Electromagnetism Group, *EAFIT University*, Carrera 49 No. 7 Sur–50; Medellín, Colombia

E-mail: jbaenaro@eafit.edu.co

Abstract. This article demonstrates the evaluation of wet chemical routes to produce magnetic nanoparticles of iron oxide and surface chemistry characterization by infrared spectroscopy (IR). Its potential use in biomedicine as contrast agents or to deliver drugs in localized medical treatments, which reduce the toxicity associated with cytotoxic drugs, is also evaluated.

1. Introduction

In a broad spectrum of applications, nanotechnology increasingly innovates the biomedical field. Nanoparticles used to detect tumors, activation by alternate magnetic fields as a cancer treatment, and their accumulation, vectorization and functionalization for drug targeting, are some of the mote important applications in this field. But equally, their major limitations are studied to optimize their application; among them, the half–life in blood which, when injected into the bloodstream, is recognized and phagocytosed in the liver and then removed, a biological process, which–over long periods–prevents the action of the desired biomedical application [1].

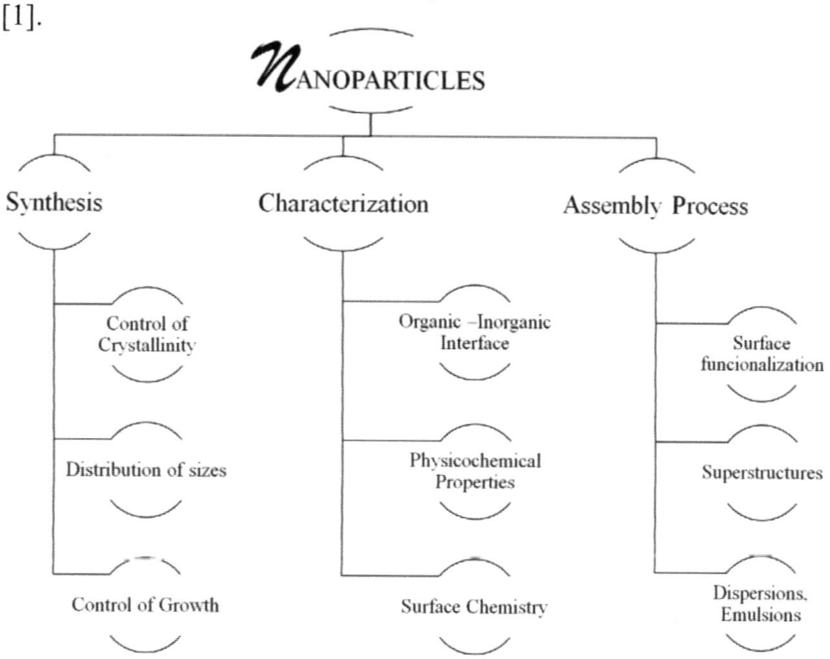

Figure 1. General scheme of synthesis

The half–life of the nanoparticles in contact with the blood depends on their size, the tendency to form aggregates and the nature of their coating. Hence, the importance of studying the feasibility and viability of the chemical pathways used for the synthesis of magnetite, since this will establish the shape, size, distribution, degree of crystallinity and surface chemistry, which determine the magnetic properties of the material and, thus, prolong its half–life to generate applications that impact current biomedical techniques [2].

These nanoparticles are widely used in the market, due to the possibility of their use in high–resolution biomedical images, implemented in the Nuclear Magnetic Resonance techniques, usually employed as contrast agents.

However, the synthesis process presents some problems, such as the lack of control of the average particle size and their broad size distribution.

2. Objectives

In this paper, we analyze and verify the chemical processes associated with the synthesis of magnetic nanoparticles for biomedical applications. It also raises the formulation and design of the experimental setup, the further characterization of the samples, their composition and superficial chemistry through the infrared–spectroscopy (IR) technique and the assessment of potential enhancements to the methods involved and the main variables, as shown in Figure 2.

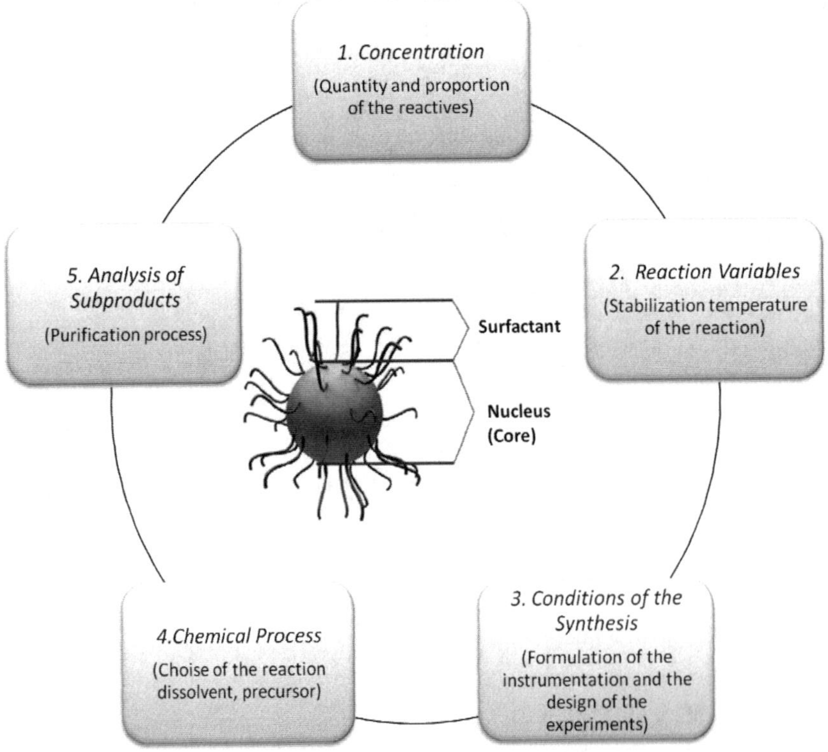

Figure 2. Diagram of the action mechanism in the synthesis of magnetite, with the necessary variables and parameters.

3. Methodology and Formulation of the Chemical Process

There are several methods used to obtain the nanoparticles, among which are physical, chemical, and gas–phase methods. The method used in this work is the synthesis of magnetite through thermal decomposition, described by Roca, A G; Morales, M P; Serna, C [3].

3.1 *Description of the Thermal–Decomposition Process in an Organic Iron Precursor*

The first stage of the synthesis of magnetite using iron oleate III (Fe $(ole)_3$) is formed. To do this, we add 2 g of $FeCl_3{\bullet}6H_2O$ and 4 g of sodium oleate to a mixture of dissolvents, formed by 8 ml of *EtOH*, 6 ml of distilled water and 14 ml of hexane. The mixture was heated to reflux during four (4) hours, as shown in Equation 1.

$$FeCl_3 + RCOO^-Na^+ \xrightarrow{hexano-EtOH-Agua-\Delta t} Fe\,(RCOO^-)_3 + NaCl \tag{1}$$

After the completion of the reaction, the aqueous phase was eliminated and the organic phase was washed several times with deionized water to remove all the ions that remained in the supernatant.

The experimental setup initially began with a heating mantle to 200°C. After characterizing it (thermal profile), it was decided that the assembly would be improved using a thermostatic bath to provide more precise temperature control. We used a glass agitator propeller tip at 200 rpm. A four–mouth, round–bottom reaction flask was designed and constructed to comply with the reaction conditions (hot injection of reagents and surfactants, as well as the conditions of inert atmosphere). A nitrogen–gas flow and a reflux condenser were available and the temperature measurements were obtained with a J–type thermocouple. All the experimental setup and the complete synthesis were performed in a fume hood, due to the volatility of the solvents, as shown in Figure 3.

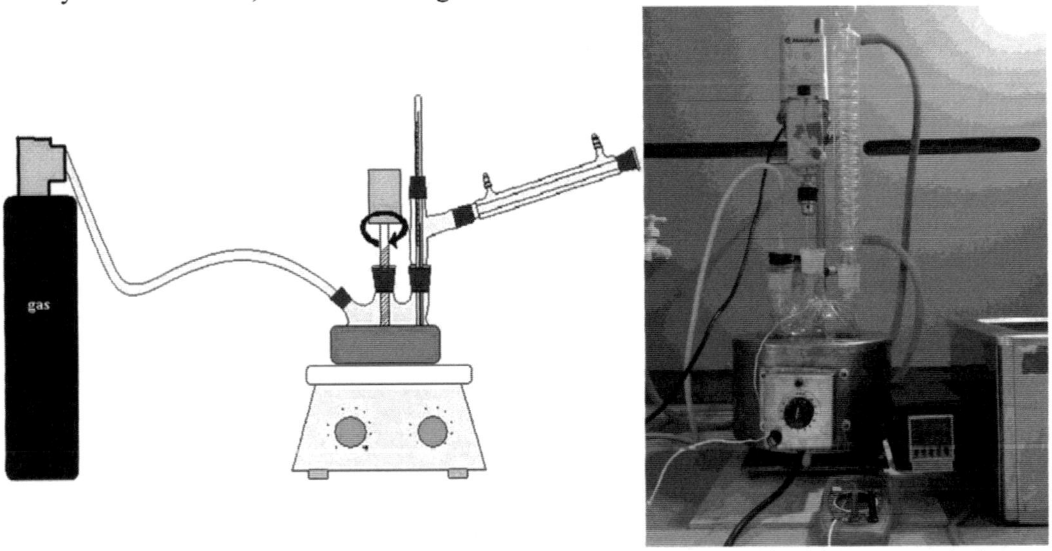

Figure 3. Experimental setup of the thermal–decomposition method in an organic iron precursor

3.2 *Description of the Salt Coprecipitation Method*

This method involves the addition of an Fe^{2+} and Fe^{3+} salt solution on a basic medium in excess. The advantages of this method are its simplicity and low cost. The method is described by García, L A; Rodríguez, O S; Betancourt, R; Saldívar, R [4].

For this reason, we performed an optimization of the method. The chemical–coprecipitation technique consists of mixing a 2.333 g ferric–chloride–hexahydrate solution ($FeCl_3.6H_2O$) and 0.8580 g of ferrous–chloride tetrahydrate ($FeCl_2.4H_2O$) with an 0.1 M concentration and a molar relation of 2:1, with continuous mechanical agitation. Later, this mixture was heated to a temperature of 70°C. A 10% volume solution of sodium hydroxide (*NaOH*) was added carefully from a burette. Instantly, we observed the formation of a dark precipitate as the base fell into the mixture, producing flocculation. The precipitate was washed several times with distilled water to remove the ions and the sodium hydroxide in the supernatant. The washed magnetite was kept in suspension to facilitate its peptized process.

The surfactant preparation was carried out through the peptization method. Oleic acid as the surfactant, hexane as the carrier liquid and suspended magnetite as the magnetic material were used.

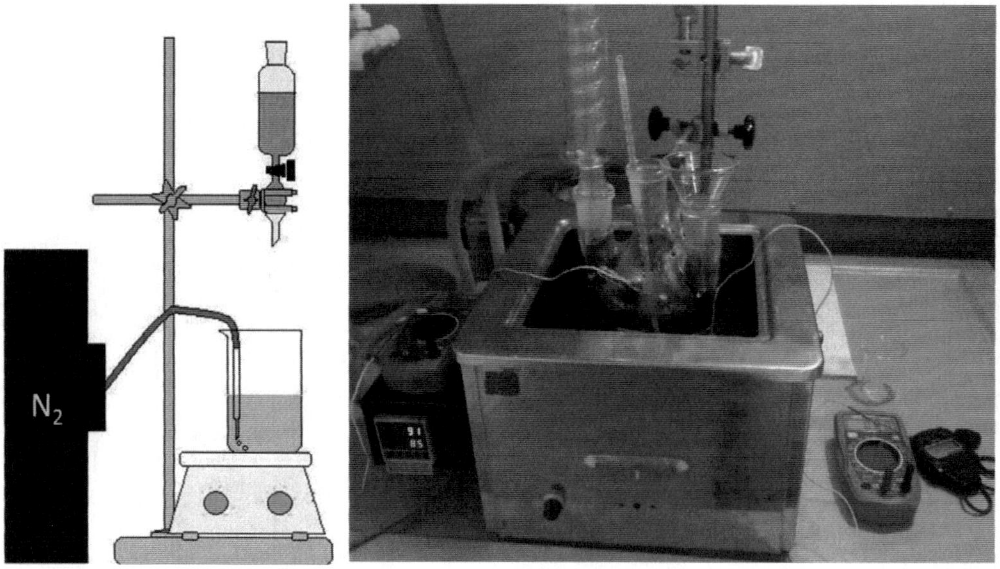

Figure 4. Experimental setup for the synthesis of magnetite, through the coprecipitation method

4. Characterization

We observed that the superficial chemistry of the magnetite presented an absorbance peak at 610 nm, which corresponds to a coordinated double bond (C=O) (Figure 5b), which means a coating of oleic acid to this substance, showing the steric isolation required. This also shows that the coprecipitation and thermal decomposition results belong to the bands of natural absorbance of magnetite (Figure 5a). Similarly, we compared two samples of magnetite powder, produced by the methods of coprecipitation and thermal decomposition, noting that the latter method is much closer to the values reported in the literature [5].

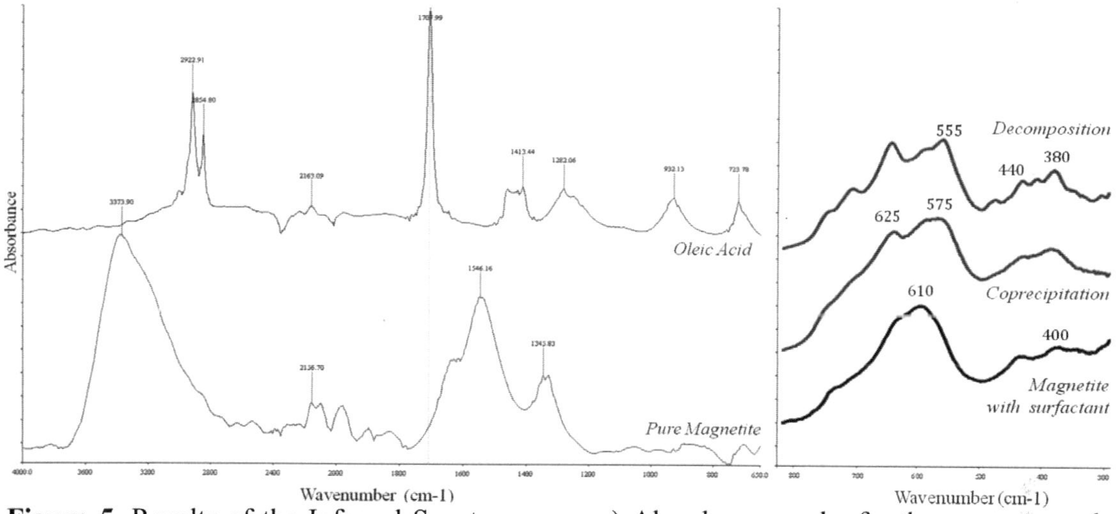

Figure 5. Results of the Infrared Spectroscopy. a) Absorbance peaks for the reactants and the products. b) Absorbance peaks of the products obtained through different chemical pathways.

5. Results and Discussion

Figure 6. Experimental results obtained: Magnetite in the carrier medium

Figure 6 shows the experimental results of the synthesis of magnetite through thermal decomposition in the middle of an organic precursor. It shows the hidrobo character of the colloid, due to an organic (hexane) carrier medium. This feature implies further exchange in an aqueous medium, since the suspension must be biocompatible. Additionally, it shows the peptization process of the magnetite, after six (6) hours of continuous agitation and constant temperature. As shown in Figure 5, the results of the Infrared Spectroscopy (IR) for these final products correspond to magnetite with surfactant on its surface [6].

6. Conclusions

In evaluating the viability and feasibility of the chemical pathways to synthesize magnetite, we present the following results:

In thermolysis with organic precursors, we observed that, although the method is viable, it requires production time of more than six (6) hours and the difficulty of maintaining a stable temperature of more than 300°C hinders its implementation. Similarly, we propose changing the solvent (hexane) in future tasks because of the difficulty of finding it in the local market, due to restrictions in handling.

7. Acknowledgements

Special thanks to Doctor *Alirio Jaramillo*, Mrs. *Norckzia Ortiz de Navarro* and Mr. *George F. Mobley*, for their unconditional support in every aspect and their generous financial aid, which was essential for the development of this work.

Thanks to *Álvaro Andrés Velásquez*, Ph.D. in Physics, for his assistance in the setup process, synthesis, lending of the laboratory and sharing his knowledge on the topic. Similarly, thanks to all the *EAFIT* University Process Engineering and Physics Engineering laboratory personnel (Medellín, Colombia), for providing the tools necessary to implement the project.

8. References

[1] Gómez, A R; Morales, M P; Serna, C J 2009 Preparation of uniform magnetic and high crystallinity nanoparticles for biomedicine. *Doctoral thesis* 4-9.

[2] De M; Arvizo, R R; Verma, A; Rotello, V M 2007 Tailoring nanoparticles for the recognition of biomacromolecule surfaces. *Chemical Analysis* **172** 91-117.

[3] Gómez, A R; Morales, M P; Serna, C J 2009 Preparation of uniform magnetic and high crystallinity nanoparticles for biomedicine. *Doctoral thesis* 23-44

[4] Garcia, L A; Rodríguez, O S; Betancourt, R; Saldívar, R 2003 Synthesis and properties of magnetite ferrofluids. *Superficies y Vacío* **16** (4) 28-29.

[5] Klostranec, J M; Chan, W C W, 2006 Quantum dots in biological and biomedical research: Recent progress and present challenges. *Advanced Materials* **18**(15) 1953-1964.

[6] Parak, W J; Gerion, D; Pellegrino, T; Zanchet, D; Micheel, C; Williams, S C; Boudreau, R; Le Gros, M A; Larabell, C A; Alivisatos, A P 2003 Biological applications of colloidal nanocrystals. *Nanotechnology* **14**.

Surface Electromyographic Onset Detection Based On Statistics and Information Content

Natalia M. López[1], Eugenio Orosco[2], Fernando di Sciascio[2].

1- Gabinete de Tecnología Médica, Facultad de Ingeniería, Universidad Nacional de San Juan.

2- Insituto de Automática, Facultad de Ingeniería, Universidad Nacional de San Juan.

nlopez@gateme.unsj.edu.ar

Abstract. The correct detection of the onset of muscular contraction is a diagnostic tool to neuromuscular diseases and an action trigger to control myoelectric devices. In this work, entropy and information content concepts were applied in algorithmic methods to automatic detection in surface electromyographic signals.

1. Introduction

The surface electromyographic signal (sEMG) is a common diagnostic tool to detect neuromuscular pathologies while it also finds applications in rehabilitation engineering, sports, ergonomy and research. Temporal and energy characteristics supply parameters to estimate force of contraction, so the accuracy in the detection of beginning and end allows the evaluation of posture and movement.

Differences of 20 ms in conduction velocity studies, or in alignment of movement-related potentials in the electroencephalogram (EEG) are significant to the diagnosis of several pathologies [1- 4]. In others applications, as prosthetic or assistive devices control, the correct detection of sEMG beginning is critical in applications where that event triggers an action, or in signal classification where the operator's intent is extracted in the first 300 ms of the muscular contraction [5].

Subjective visual determination by a trained operator is the most common way to evaluate contraction detection algorithms and used as reference [6-8]. However, precision and repeatability are low requiring a large number of well characterized EMG's if any meaningful statistical analysis is wanted. Also, this method is restricted to offline applications. For this reason, an algorithmic method for efficient onset detection is required to reduce the operator's induced errors and allow online processing. Several research groups propose algorithms based in lowpass filters [9], statistical considerations [10], whitening filters [11], energy operators [12] and others algorithms. A systematic comparison of the methods present in bibliography can be found in the work of Staude [13] or by Hodges and Bui [14].

In this paper we introduce an approach based in probabilistic considerations and information content. The concepts of information content, entropy and algorithmic complexity theory have found application in biological signal analysis [15-17], speech and audio recognition, and other fields.

Onset detection based in probabilistic models is a usual technique in the field of music analysis, where the onsets of significant events are perceived as surprising moments, based on the notion of observer's familiarity with the evolution of the signal. This observer generates a probability trace of the baseline and the occurrence of an event can be seen as an unexpected change in probability, building a temporal structure [18]. Thus, a dynamically evolving measure of surprise can be used as a detection function in the analysis of sEMG, since muscular contraction is a surprising event.

There is a close relation between information content and entropy. Complexity and information measures have been used in several applications in the field of biological signal processing, as speech recognition [19] and sEMG characterization using approximate entropy [20], so the Shannon Entropy of the signal was analyzed in this context. Shannon's entropy describes the uncertainty or the degree of disorder of the data [21, 22], that is, the information content of the signal.

In order to perform a statistical test of the method here proposed, a large amount of sEMG signals is required, with parameters such as start point, amplitude and signal to noise ratio (SNR) perfectly known. In this paper we use the same approach as the proposed by Staude et al [23], who use simulated EMG data with the possibility of changing the parameters that affect the sensivity and improve the reproducibility. The surprise and entropy algorithms were compared with other detection methods, with simulated and real EMG signals and the performance was evaluated under different levels of SNR.

In what follows, we will present the database used to validate the algorithms, the two methods proposed and the results obtained.

2. Materials

2.1. sEMG database

2.1.1. *Synthetic sEMG database.*

Automatic event detection from sEMG signals requires a model containing abrupt changes in signal properties and variable parameters in the baseline.
A zero mean steady stochastic process, of Gaussian distribution, band limited (10-500Hz) and contaminated with additive white noise appears as a good modeling choice [24, 25].
According to the model of Staude [23] an all-pole representation by an autoregressive (AR) filter provides an adequate bandwidth while the excitation x_k results from two statistically independent noise sources, m_k and w_k, dynamical mean signal and system noise, respectively (Fig. 1). The complete EMG digitized signal y_k can be model as:

(1) $$y_k = \tilde{y}_k + v_k$$

(2) $$\tilde{y}_k = -\sum_{i=1}^{p} a_i \tilde{y}_{k-i} + b_0 x_k$$

(3) $$x_k = m_k + w_k$$

where $\tilde{y}_k$ denotes the output of the AR filter of order p with coefficients a_i and static gain b_0 ,and v_k is the measurement and other biological signals acting as interference .
The onset of the contraction is flagged by an abrupt change in the profile of m_k at the arbitrary time t_1, modeled by a ramp between $t_2 - t_1$.

In the following, the onset time t_1 will be setting randomly among the interval of 3-5s. The final ramp time t_2, noise baseline and maximal contraction amplitude were setting in the same way, in a bounded span. To preserve the Nyquist considerations the sampling frequency (*fs*) is 1KHz. A typical simulated sEMG signal is shown in Fig. 2

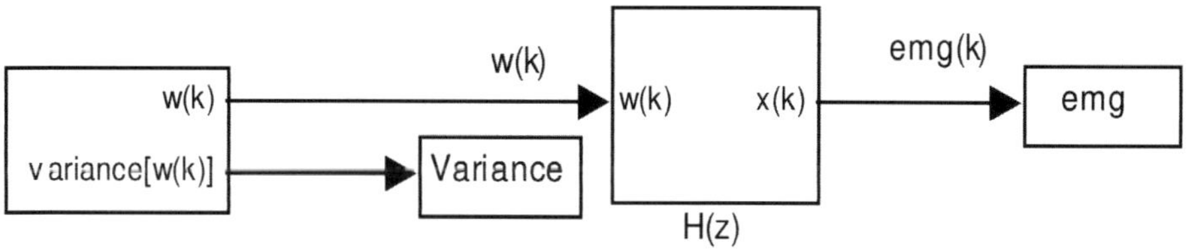

Figure 1: sEMG generation model. A gaussian white noise model and a linear filter *H(z)* model the emg_k signal. Internal and measurement noise are modeled by additional independent processes. Abrupt changes are driven by deterministic coefficients included in variance variable.

2.1.2. *Real sEMG database.*

To optimize onset detection algorithms, a sEMG real database was constructed with bipolar recordings made with Ag-AgCl electrodes (3M RedDot) fixed 2 cm apart in the biceps of 6 volunteers during right arm flexion. The signal was amplified, filtered (10 Hz/500 Hz for low/high frequencies cut-off, respectively), and digitized (*fs*=1KHz) on-line with BrainNet BNT36 technology (LYNX TECNOLOGIA ELETRÔNICA LTDA, SP, Brasil). Several acquisition conditions were tested using different noise sources and the analysis was performed off line.

3. Methods

Di Fabio's algorithm is most commonly used to detect onset time and it is also a common reference for a new technique [9], [14]. The method uses a fixed threshold (Th), calculated by adding a number of standard deviations above the mean value of the baseline of the EMG, in resting state. Besides, the signal must surpass the threshold during a whole period, to reduce false positive detections. Both conditions are selected empirically to obtain the best output.

(4) $$Th = \overline{emg}_{(rest)} + 3SD$$

We use this technique to compare with our results.

3.1. Surprise Function

In order to measure the degree of unexpectedness $s(E)$ of an event E, it is reasonable to assume the following conditions:

$s(E) = f(p(E))$, $s(E)$ only depend on the probability of the event E.

$s(E)$ is a monotone decreasing function of $p(E)$

$s(E) = 0$ for $p(E) = 1$, and $s(E) = \infty$ for $p(E) = 0$

The logarithm satisfies these conditions, thus the surprise can be defined as in [18]:

(5) $$s(E) = -\log p(E)$$

Using the concept of a negative log-probability as a measure of the surprise or unexpectedness of the contraction in the sEMG signal, it is possible to determine the onset in terms of an implicit probability model. This model can be unconditional or conditional, according to the behavior of the signal in relation with the past samples. In this work, the unconditional assumption was applied to reduce the computational cost.

Calculation of $s(E)$ function requires segmentation of the signal in disjointed windows ($v(k)$), independent and identically distributed. Now, if we assume that the elements of $v(k)$ are Gaussian distributed with zero mean and variance σ^2, the probability density function (*pdf*) is:

$$(6) \qquad P(v) = \prod_{i=1}^{n} \frac{1}{\sigma\sqrt{2\pi}} e^{-\frac{v_i^2}{2\sigma^2}}$$

Where v_i represents each of the n elements of the window elements. From (5) and (6) we obtain the measure of surprise, being c a constant. As can be seen, the surprise is closely related to the energy concept.

$$(7) \qquad s(E) = \frac{1}{2\sigma^2} \sum_{i=1}^{n} v_i^2 + c$$

In the work of Clancy and Hogan [25], the probability density function of the sEMG was revised theoretically and experimentally. They propose a model using a Laplacian density. Hence, a Laplacian distribution is assumed to compute the $s(E)$ function, that is,

$$(8) \qquad P(v) = \frac{1}{2} e^{-\lambda |v_i|}$$

$$(9) \qquad s(E) = \lambda \sum_{i=1}^{n} |v_i| + c$$

In this case, the algorithm gives the same results that rectification and smoothing, a procedure very usual in sEMG processing, related to the energy contained in the signal.

For both distributions, the success of the onset detection implies the correct fitting of the statistical model.

3.2. Entropy and information content

Entropy as a thermodynamic state variable was introduced into physics as a measure of system disorganization. Afterward, entropy was promoted to a generic measure of information content and Shannon [21, 22] formalized the concept applied to signal processing:

$$(10) \qquad S(m) = -\sum_{i=1}^{n} p_i \log p_i$$

Where p_i is an estimate of the *pdf* for sample i, n represents the window size and m is the window number. A proposed relation between entropy and sEMG processing is based on the hypothesis that the measurement and environmental noise is supposed to have the highest entropy value while the contraction signal has significantly lower entropy, as a large amount of information (i.e. muscular force) is present in the signal. To detect the contraction onset a modified version of Shannon equation was applied, computing the empirical histogram of a signal segment and counting the number of repetitions of each value in this section.

The EMG activity beginning was setting when at least 3 consecutive windows met the following criterion,

(11) $S(m) > \overline{S}(rest)$

That is, the window's entropy must be larger than the mean base line entropy.

For both algorithms, surprise function and Shannon's entropy, the window size was empirically set as a function of the sampling frequency recalling that the smaller the number of samples, the better the detection precision (meaning repeatability); however, if the window becomes too small, false detections are more likely to occur due to noise spikes. Keeping the aforementioned criteria led us to choose the window size n = 50 samples.

4. Results

Figure 2 shows onset detection results by the *surprise algorithms* and Shannon's entropy in a synthetic signal.

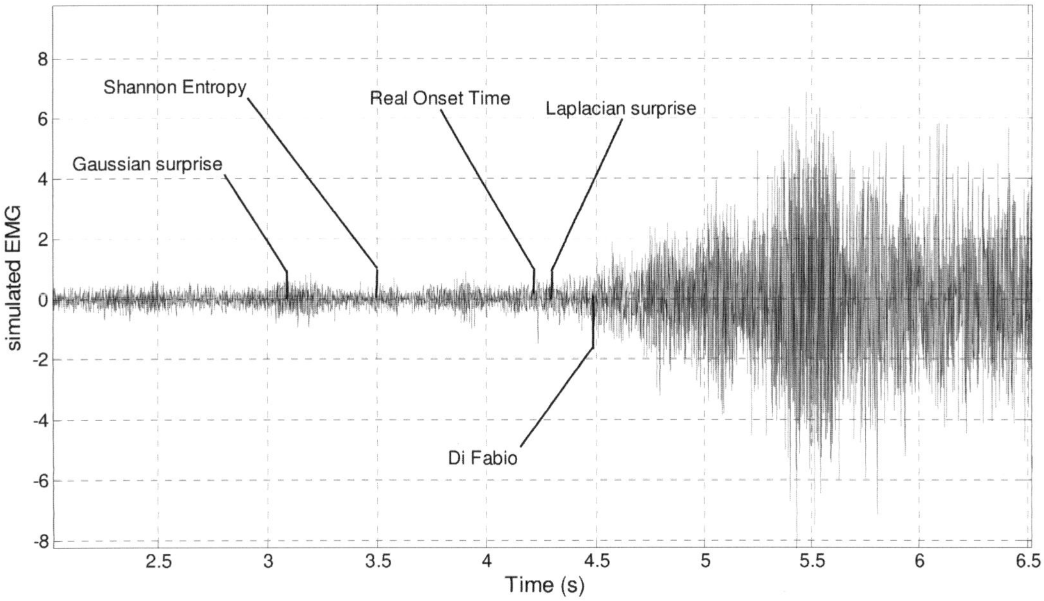

Figure 2: Results obtained with surprise algorithms, Shannon entropy and di Fabio method. Onset estimates are indicated by arrows.

A set of 1000 signals randomly created according to Section 2.1.1 was tested along a SNR range of 16 dB. To verify the algorithms, visual decision of an expert and Di Fabio's method were used as reference. Signal to noise ratio (SNR) in dB was computed as:

(12) $$SNR = 10 \cdot \log_{10} \frac{\sigma^2_{signal}}{\sigma^2_{baseline}}$$

The detection error was calculated as: $error = abs(t_1 - t_d)$, where t_d is the time of onset detected by the algorithms here proposed.

Table 1 depicts the statistical performance of the onset algorithms herein proposed, over the synthetic database. The *Gaussian surprise function* shows an unacceptable error and was discarded, while mean, standard deviation and mode in Shannon's entropy and *Laplacian surprise* are in the same order that di Fabio's method.

	Mean error	**Error STD**	**Mode error**	**Range error**
Di Fabio	1.219 ms	0.9995 ms	0.003756 ms	4.433 ms
Shannon Entropy	1.604 ms	1.478 ms	0.5507 ms	11.11ms
Gaussian Surprise	3.328 ms	0.5186 ms	3.34 ms	2.654 ms
Laplacian Surprise	1.392 ms	0.8479 ms	0.00766 ms	4.248 ms

Table 1: Statistics for onset algorithms, range of SNR 16dB.

Figure 3 illustrates the detection error at different levels of SNR. The *Laplacian surprise function* shows a slightly better performance at low SNR values and can be considered as an option to suitable onset detection. Shannon Entropy algorithm's presents a stepped behavior and needs to be improved, to be considered as a suitable choice.

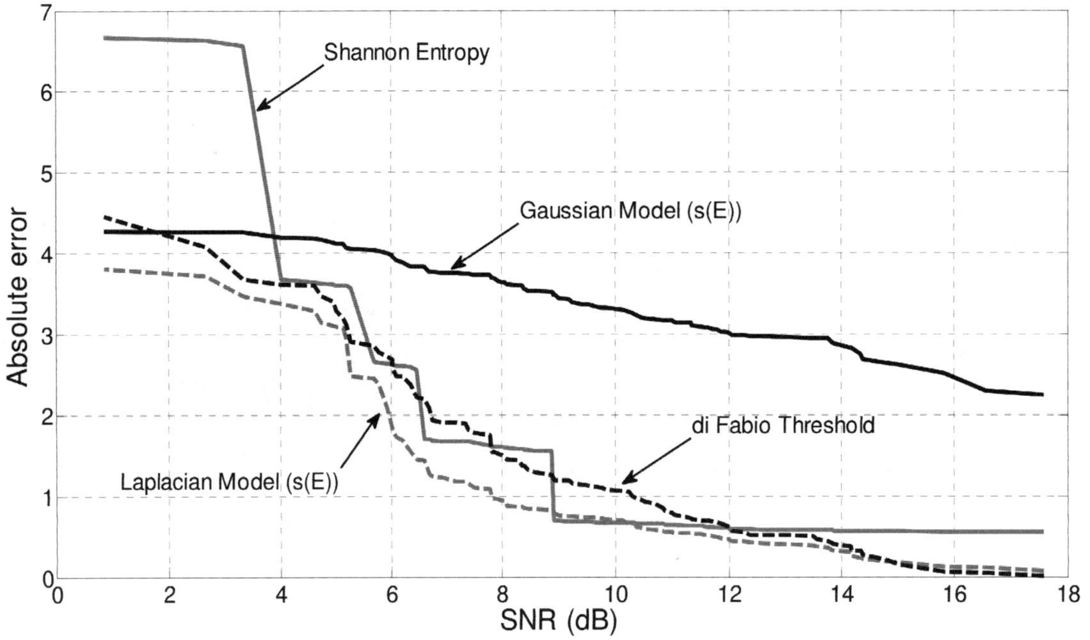

Figure 3: Dependence of error on SNR, with a 50 samples window's size.

In real EMG signals the performance of the algorithms is similar; *Laplacian surprise* is closer to the onset detected by the expert than di Fabio's algorithm. A statistical analysis is not suitable because the expert decision is not an exact parameter. A typical signal is shown in Fig. 4.

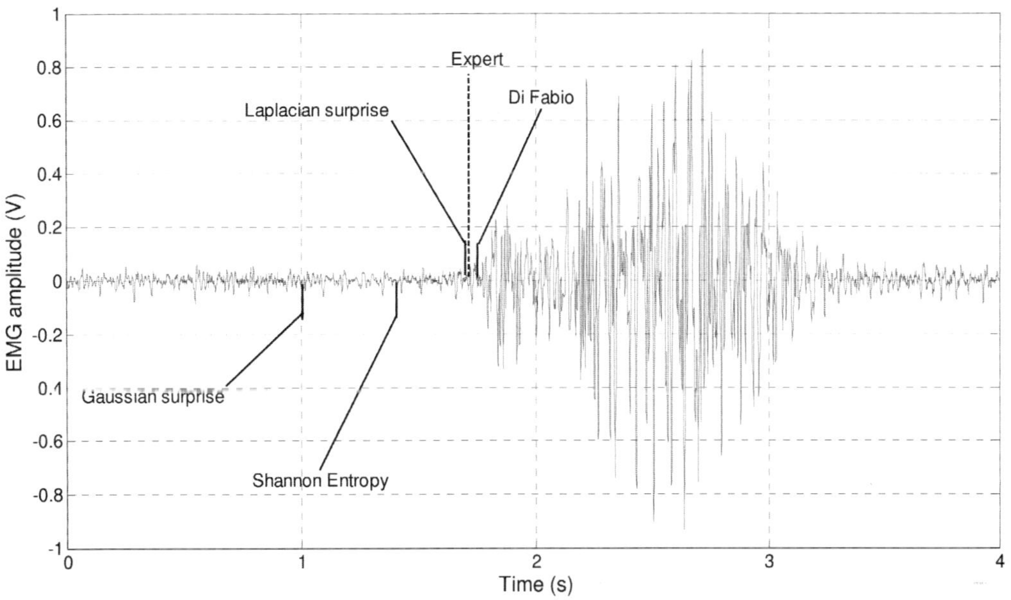

Figure 4: Real EMG and onset estimates. The event detection is marked with arrows and the expert reference is marked in dotted line.

5. Discussion and Conclusions

This paper proposes the use of information theory based techniques to automated detection of the onset sEMG contraction, based on statistical considerations and Laplacian distribution. Other techniques as Gaussian distribution and even Shannon Entropy were discarded as noisier, less reliable and more costly from a computational point of view.

The Gaussian surprise function is unable to detect the onset of contraction, because the statistical fitting supposes a change in distribution and both, baseline noise and sEMG, were simulated with a white Gaussian noise source. All detectors herein presented are related to signal energy estimation, which is consistent with an optimal detector under the assumption of Gaussian distribution for both signal to noise base.

On other hand, the Laplacian surprise detection error is in the order of the window's size, while a smaller window (i.e. 50 samples) yields to false detections. This algorithm shows better performance when high noise levels are expected, even better than Di Fabio threshold method.

References

[1] Adamovich SV, Levin MF, Feldman AG. "*Merging different motor patterns: coordination between rhythmical and discrete single-joint movements*". Exp Brain Res.99:325–37; 1994

[2] Takada K, Yashiro K. "*Automatic measurement of on/off periods of EMG activity*". Medinfo.8:751–4; 1995

[3] Rafal RD, Winhoff A, Friedman JH, Bernstein E. "*Programming and execution of sequential movements in Parkinson's disease.*" J. Neurol Neurosurg Psychiatry. 50:1267–73. 1987.

[4] Haagh S, Spijkers W, Van den Boogaart B., Van Boxtel A., "*Fractioned reaction time as a function of response force,*" Acta Psychol., vol. 66, pp. 21–35, 1987.

[5] Hudgins B., Parker P., Scott R. "*A new strategy for multifunction myoelectric control*". IEEE Trans. Biomed. Eng., vol. 40 pp. 82-94. 1993.

[6] Barrett G, Shibasaki H, Neshige R. "*A computer-assisted method for averaging movement-related cortical potentials with respect to EMG onset*". Electroenceph Clin Neurophysiol 1985;60:276–81.

[7] Badke MB, Duncan PW. "*Patterns of rapid motor responses during postural adjustments when standing in healthy subjects and hemiplegic patients*". Phys Ther 1983;63:13–20.

[8] Walter CB. "*Temporal quantification of electromyography with reference to motor control research*". Human Movement Science 1984;3:155–62.

[9] Di Fabio RP. "*Reliability of computerized surface electromyography for determining of onset of muscle activity*". Physical Therapy 1987;67(1):43–8.

[10] P. Bonato, T. D'Alessio, and M. Knaflitz, "*A statistical method for the measurement of muscle activation intervals from surface myoelectric signal during gait*," IEEE Trans.Biomed.Eng.,vol. 45, pp. 287–298, 1998.

[11] G. Staude,W.Wolf,, U. Appel, "*Automatic event detection in surface EMG of rhythmically activated muscles*," Int. Conf.IEEE BME Soc., vol. 17, pp. 1351–1352, 1995.

[12] Solnik S, DeVita P, Rider P, Long B, Hortobagyi T. "*Teager-Kaiser Operator improves the accuracy of EMG onset detection independent of signal to noise ratio*" Acta Bioeng Biomech 10(2), 65-68. 2008.

[13] Staude G, Flachenecker C, Daumer M, Wolf W. "*Onset detection in surface electromyographic signals: a systematic comparison of methods*". EURASIP Journal on Applied Signal Processing. 2001;2:67–81.

[14] Hodges PW, Bui BH. "*A comparison of computer-based methods for determination of onset of muscle contraction using electromyography*". Electroenceph Clinical Neurophysiology 1996;101:511–9.

[15] Abásolo D, Hornero R, Gómez C, García M, López M, "*Analysis of EEG background activity in Alzheimer's disease patients with Lempel–Ziv complexity and central tendency measure*". Medical Engineering & Physics 28, 315–322. 2006

[16] G. Barrett, H. Shibasaki, and R. Neshige, "*A computer-assisted method for averaging movement-related cortical potentials with respect to EMG onset*," Electroenceph.Clin.N europhysiol. vol. 60, pp. 276–281, 1985.

[17] P. Brodin, T. S. Miles, and K. S. Turker, "*Simple reaction time responses to mechanical and electrical stimuli in human masseter muscle*," Arch.Or al Biol., vol. 38, pp. 221–226, 1993.

[18] Abdallah S. , Plumbley M, "*Probability as metadata: event detection in music using ICA as a conditional density model*," Proc. 4th Int. Symp. Independent Component Analysis and Signal Separation (ICA2003), Nara, Japón. pp. 233–238, 2003.

[19] Torres M. E., Gamero L. G., "*Relative Complexity Changes in Time Series using Information Measures*", Physica A (ISSN: 0378-4371), Vol. 286, Iss. 3-4, pp.457-473, Oct. 2000.

[20] Chen W, Zhuang J, Yu W, Wang Z., "*Measuring complexity using FuzzyEn, ApEn, and SampEn*" , Med Eng Phys. 2009 Jan; 31(1):61-8.

[21] Shannon, C. E., "*A mathematical theory of communication*". Bell System Technical Journal, 27. p. 379-423.

[22] Shannon C., Weaver W. "*The Mathematical Theory of Communication*". Univ of Illinois Press, 1963.

[23] Staude G, Flachenecker C, Wolf W. "*A dynamic generator model for motor related biosignals*". Ann Int Conf IEEE BME Soc 1998;20:2374–7.

[24] De Luca CJ. "Physiology and mathematics of myoelectric signals". IEEE Trans Biomed Eng 1979;26(6):313–25.

[25] Clancy E, Hogan N. "Probability density of the surface electromyogram and its relation to amplitude detectors," IEEE Trans.Biomed.Eng., vol. 46, pp. 730–739, 1999.

Journal of Physics: Conference Series **332** (2011) 012044 doi:10.1088/1742-6596/332/1/012044

Electrical stimulation of mechanoreceptors

AM Echenique, JP Graffigna
Gabinete de Tecnología Médica. Universidad Nacional de San Juan
Av. Libertador 1109 (oeste). San Juan. Argentina

E-mail: amechenique@gateme.unsj.edu.ar

Abstract. Within the field of Rehabilitation Engineering, this work is aimed at identifying the optimal parameters of electric current stimulus which activate the nervous axons of mecanoreceptors found in the fingertip, allowing, this way, to resemble tactile senses. These sensorial feelings can be used by aiding technological means, namely, the sensorial substitution technology, in an attempt to render information to blind people through the tactile sense. The physical pressure on sensorial areas (fingertips) used for reading activities through the Braille System is the main effect that is imitated and studied in this research work. An experimental aiding prototype for Braille reading research has been developed and tested with blinds and reduced vision people, with highly satisfactory results.

1. Introduction

Assisting Technology (AT), when applied correctly, helps to overcome the limitations set by impairing conditions. Its core objective is to ease the execution of a task performed by an impaired person, within a given context for such an activity [1-2]. In the field of assistance to blind and visually impaired persons, whenever vision augmentation devices are not proper enough, the specialist resorts to using AT devices based on sensorial substitution [3]. This approach consists on capturing information that is generally perceived by normal vision and sending it to the user via another sense, typically the tactile or the hearing senses.

The devices used for sensorial substitution require three basic components: an interface that captures information of variables from the surrounding environment, a data processing system and, finally, an element that renders adequately the useful information to the user.

In order to capture and send the context information to the user, several tactile stimulation devices have been developed, the so called tactile displays, which operate by stimulating the perceptive nervous receptors of the skin. Several actuator types have been devised to stimulate the tactile sense. According to the research published in [4], there are three categories according to stimulus type: thermal, electrical and mechanical. Thermal actuators are heat pumps that stimulate the skin receptors that are sensitive to temperature changes. Electrical actuators are electrodes charged by an electrical variable (volts/amps) and which objective is to activate not the receptors but the associated afferent nerve terminals. Finally, the mechanical actuators translate the information into a force, a displacement and/or vibration that can be detected by the mechanoreceptors of the skin.

The tactile sense displays a series of advantages over auditive stimulation [3]. Moreover the electrical stimulation offers some advantages over the other approaches: it lacks thermal inertia, mechanical fatigue or wear of mobile elements, besides permitting manufacture compact designs with solid state component, all in a low-cost small-size package.

Main problems that visually impaired persons have to face are: to have access to reading material, to be restricted as regards mobility and orientation, and to manipulate and recognize objects.

As regards reading, the Braille reading/writing system is widely used by blinds and deaf/blind persons, for its universality and relatively easy usage. On this direction, this work presents a contribution to the selection of current stimulation parameters, in order to obtain a reliable tactile sensitivity in blinds so as to let them read in Braille via an adequately designed interface.

1.1. Human skin structure

Various mechanoreceptors of the fleshy fingertip were studied, mainly: (1) the Meissner corpuscle of rapidly adapting (RA) mechanoreceptors; (2) the Merkel cells, or slowly adapting type I (SAI) mechanoreceptors; (3) the Ruffini terminal, or slowly adapting type II (SAII) mechanoreceptors and, (4) the Pacinian corpuscles (PC). The number of SAII receptors is very small; hence, they were excluded from the study. Previous research works [5-8] have considered that it was possible to stimulate the mechanoreceptors separately and that the combination of these stimuli could produce different tactile sensitivity (responses). Figure 1 shows a cross section of human skin. It is assumed that RA and PC can perceive vibration, whereas SAI senses pressure [7-8]. Sometimes RA and PC (fastly adapting mechanoreceptors) are considered speed and acceleration sensors because they respond to strain with just one impulse, whereas SAI (slowly adapting mechanoreceptor) is called the position, or displacement, sensor because it responds to the intensity and duration of the stimulus.

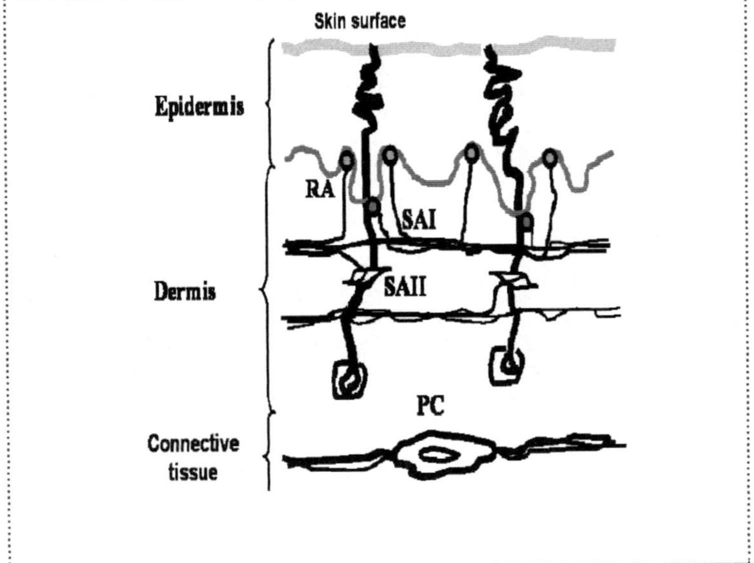

Figure 1. Cross section of human skin.

RA: Meissner corpuscle; SAI: Merkel cell; SAII Ruffini terminal; PC: Pacinian corpuscle.

Table 1[7] shows the main features of mechanoreceptors that are taken into account at the time of stimulation.

As a result of nerve size difference, the electrical stimulus of mechanoreceptor nerves is much easier to do than that of afferent nerves because, in general, the stimulation threshold increases with diameter size [7].

Another important characteristic of the fingertip is the thickness of the corneous layer of the skin, of approximately 600 μm, whereas in other body skin parts, it may be as thin as 15 μm [7-8].

Table 1. Depth, diameter and orientation of mechanoreceptor axons respecting the fleshy surface of the fingertip.

Mechanoreceptors	Depth	Diameter	Orientation
Meissner(RA)	0.7 mm	3 ~ 5 μm	Perpendicular
Merkel(SAI)	0.9 mm	7 ~ 12 μm	Parallel
Pacinian(PC)	2,0 cm	5 ~ 13 μm	Parallel
Pain, heat and cold	2.5 cm	1 ~ μm	-------------

1.2. Mechanism of selective electrical stimulation

Research papers on electrical stimulation mechanisms were reviewed, especially those where distribution control was performed of an electrical current source acting on the skin surface, and the indirect activation of nerves [7-8]. From these works, it is known that when the pulse of cathode (negative) current is applied to the skin surface, the axons of nerves oriented horizontally to the skin surface are activated. Conversely, when anode (positive) current is applied, the nerves oriented vertically to the skin surface are activated, such as the RA axon.

2. Development of the proposed prototype

In order to achieve a pressure feeling, a stimulus signal is defined as in Figure 2, where some restrictions are added in order to avoid possible lesions to the cutaneous tissue. First, the maximum output current of an electrode is limited to 2mA. Besides, even though the stimulation can be either anodic or cathodic, the typical applied current is two-phase (negative and positive) so as to be able to eliminate the accumulation of charges on the skin surface.

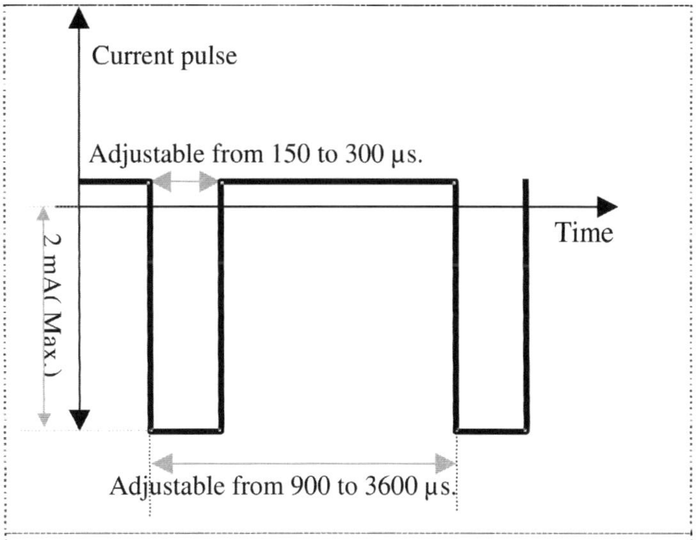

Figura 2. Waveform of the current stimulus.

Figure 3 shows a simplied block-diagram array of the prototype used in the experiences.

The electrocutaneous stimulus is controlled via a PC. A software developed for this work allows setting the parameters linked to the stimulation pulse (pulse frequency and pulse width). The selection of pulse width is made at 10 µs steps, in the 150-300 µs range.

When choosing the pulse width, it becomes readily available to the user a frequency selection set of three different values (F1, F2, and F3) ranging from 278 to 1111 Hz, as shown in Table 2. For each frequency value (F1,F2 and F3), a ratio Ap/T is kept constant, where Ap is the active pulse width and T is the period of the signal.

The interface (CP2102) allows adapting the PC's output information via USB to UART.

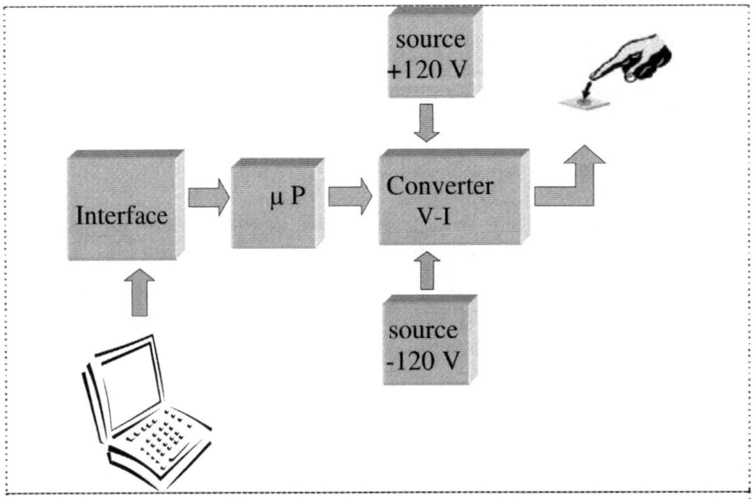

Figure 3. Simplied block-diagram array of the prototype used in the experiences.

Table 2. Frequency and Pulse-Width values permitted by prototype

Pulse width [μs]	Frequency [Hz]		
	F1 for a ratio Ap/T=0,16	F2 for a ratio Ap/T=0,12	F3 for a ratio Ap/T=0,08
150	1111	714	555
160	1042	685	521
170	980	658	490
180	926	633	463
190	877	610	439
200	833	588	417
210	794	568	397
220	758	549	379
230	725	532	362
240	695	515	347
250	666	500	333
260	641	485	320
270	617	472	309
280	595	459	298
290	575	446	287
300	555	435	278

The microcontroller (MC68HC908QY4) is programmed to receive the chosen parameters and to control the voltage-into-current converter (V-I converter). This last device was built with two current sources, Cascode type, with bipolar transistors adjusted for a maximum output current of 2mA in the negative phase and 0.5mA output in the positive phase of the signal.

The V-I converter is powered by ±120 V dc linear sources, with folded V-I characteristic and maximum output current limited to 2mA. These sources also present an isolation transformer.

The stimulation electrode was manufactured according to the knowledge gained in experiences made by this researchers team in previous projects [9-11], while altogether meeting the requirements of small size, simplicity and low cost for the device. This electrode is a 0.1cm dia. steel pin inserted in a polished acrylic pad.

The electrode for stimuli feedback was manufactured with a metal plate attached to a mesh fastening around the finger (Figure 4).

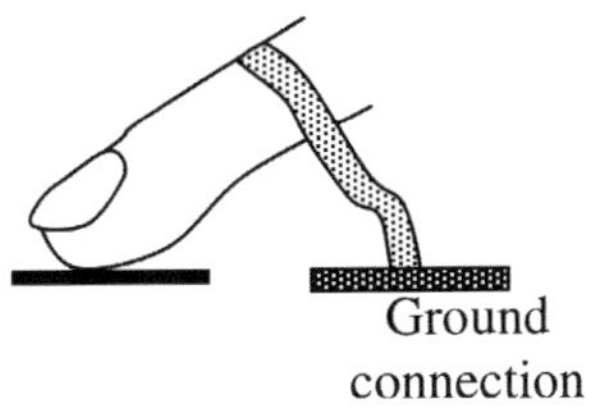

Figure 4. Stimulus return electrode

3. Experiences

3.1. Methodology
The experiences were oriented to use cathodic stimulation in order to mimic the tactile sensitivity (TS) by means of activating the various mechanoreceptors. To this aim, the following factors were evaluated:

- Sensitivity to perceive the first TS, with current intensity.
- Changes in TS perceived, with increases in current intensity.
- Range of the Ap/T ratio and active pulse width of the electric stimulus that is related to each TS.

A comparative experience was carried out with two groups of adult persons, so as to analyze the sensitivity while perceiving the TS with current intensity. The first group consisted of 10 normal-vision persons and the second group had 7 visually impaired persons.

The experience also tested the changes in sensitivity as regards current intensity value. This was specifically experienced for a 200Hz frequency and 200 μs pulse width in a visually impaired person and 3 blind ones.

Upon these experiences, tests were made to discern the variability range of the stimulus parameters (Ap/T ratio and pulse width), with two groups; the first one included four normal-vision adults while the second group had eight visually impaired and blind adults. The main objective of the experience with the first group is to study the stimulation mechanisms so as to reduce the number of parameters to include in the experiences of the second group.

The procedure of the tests consisted in stimulating at first with the least current intensity possible, and increasing it until reaching the first TS by the person. It was recorded only the data where TS was obtained, up to the current threshold acceptable for each individual, without coming into annoying sensation. The TS perceived were: Vibration, Pressure, Pricking, and Tickling.

- Vibration (Vi): sensation of intermittent tact.
- Pressure (P): sensation of permanent tact.
- Pricking (Pr): sensation of acute pressure, resembling that one caused by a pin.
- Tickling (T): sensation of smooth tact or high-frequency vibration.

3.2. Results.
Figure 5 shows the experience results on sensitivity of normal vision persons to the perception of the first TS with current intensity. It is clearly noted an activation threshold of about 1.2mA.

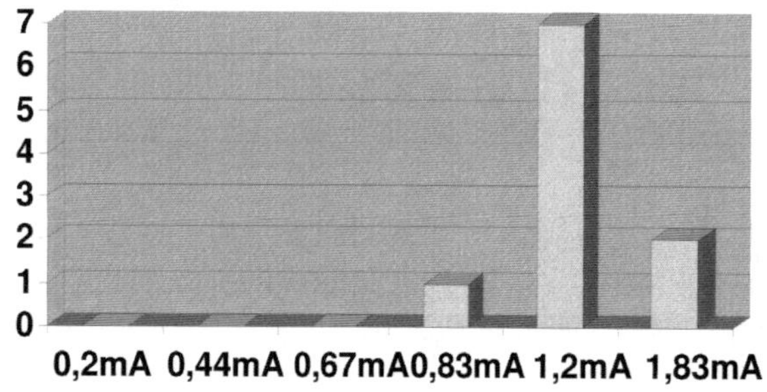

Figure 5. Number of normal-vision persons that perceive the initial tactile sensation vs. current intensity.

Figure 6 displays the results obtained for activation current thresholds for blind and visually impaired persons. Stimuli are perceived by most users from as early as 0.2mA. This value shows the high tactile sensitivity that blinds and visually impaired persons have, with very low current intensity thresholds. Due to this fact, moistening of fingertips was required by very few persons.

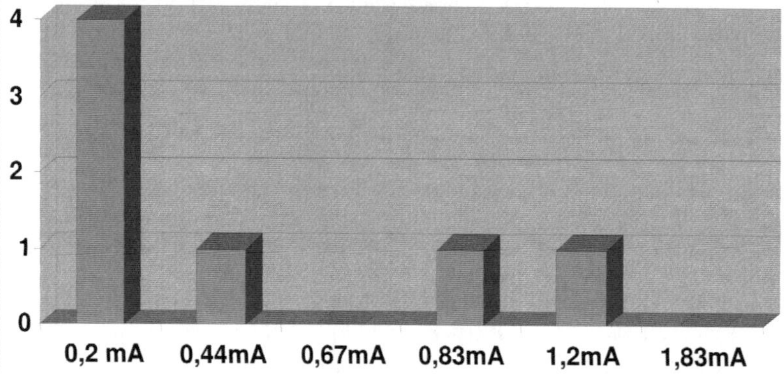

Figure 6. Number of blind and visually impaired persons that perceive the initial TS vs. Current intensity.

It is worth mentioning that in the group of blind and visually impaired persons, two persons presented high impedance in their fingertips due to the kind of daily manual work they do. This fact increases the activation threshold.

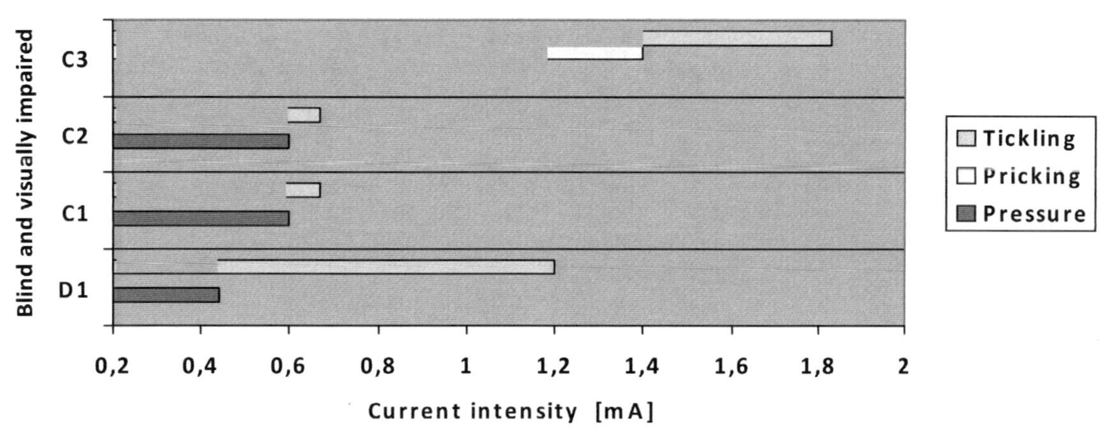

Figure 7. Change of sensation vs. Current Intensity.

In Figure7, it can be noted a change in TS perceived when the current intensity threshold is increased, that ranges from a pressure or pricking feeling (detected by the Merkel cells) to a tickling feeling. This can be explained by the fact that when the current intensity is increased, the mechanoreceptors lying deeper, as the Pacini corpuscles, can be activated as well, thus arising a tickling sensation.

Table 3. Number of normal-vision persons who perceive different TS, according to Ap/T ratio and Active Width Pulse.

ST	Ap/T	Active Pulse Width from 150 to 300µs, with 10 µs steps															
									220	230							
T	0,08	4	4	4	4	0	0	0	0	0	0	0	0	0	0	0	0
	0,12	4	4	0	0	0	0	0	0	0	0	0	0	0	0	0	0
	0,16	4	4	4	4	0	0	0	0	0	0	0	0	0	0	0	0
Vi	0,08	0	0	0	0	4	4	0	0	0	0	0	0	0	0	0	0
	0,12	0	0	4	4	4	2	2	0	0	0	0	0	0	0	0	0
	0,16	0	0	0	0	0	0	0	0	0	0	0	0	0	0	0	0
P	0,08	0	0	0	0	0	0	4	4	4	4	4	4	4	1	1	0
	0,12	0	0	0	0	0	2	2	4	4	4	4	4	4	3	3	3
	0,16	0	0	0	0	4	4	4	4	4	2	0	0	0	2	2	0
Pr	0,08	0	0	0	0	0	0	0	0	0	0	0	0	0	3	3	4
	0,12	0	0	0	0	0	0	0	0	0	0	0	0	0	1	1	1
	0,16	0	0	0	0	0	0	0	0	0	2	4	4	4	2	2	4

As regards the identification of the variability range of stimulus parameters, the experience results with the first group (Table 3) allowed to denote that the TSs for Vi and T are perceived with active pulse widths smaller than 200µs, with 0.08 and 0.12 Ap/T ratios. The TS of P can be perceived isolately from the remaining TSs, with pulse widths of 220 and 230μs, with the three Ap/T ratios. For this reason, to continue the experiences with the blind and visually impaired persons, the following pulse widths were chosen: 200, 250 and 300μs, and the three values of Ap/T ratios. The data of the experience performed with this group are detailed in Table 4.

Table 4. Number of blind and visually impaired persons that perceive the various TSs according to the Ap/T ratio and active pulse width.

ST	Ap/T	Pulse Width [μs]		
		200	250	300
T	0,08	0	0	0
	0,12	5	6	0
	0,16	0	0	0
Vi	0,08	0	0	0
	0,12	2	0	0
	0,16	0	0	0
P	0,08	8	8	2
	0,12	0	1	8
	0,16	8	8	2
Pr	0,08	0	0	6
	0,12	1	1	0
	0,16	0	0	6

4. Conclusions.

In the series of experiences performed with blind and visually impaired persons, it was possible to reach the perception of "virtual" TS starting from very low current thresholds (0,2mA), because of the high level of tactile sensitivity of these persons. This threshold, however, depends on the impedance of the fleshy fingertip of the individual under test, in strong relationship with the type of manual work performed by person.

It was noted a change in the resembled tactile sensitivity as there was a gradual increase in current intensity, which proved the theoretical base described in references.

The adequate stimulus to have a clear and well-defined pressure sensation is attained with Ap/T ratios of 0.08 and 0.16 and active pulse widths of 200 y 250μs.

5. References

[1] Gemperle F, Ota N and, Siewiorek D 2002 *Design of a Wearable Tactile Display*. Proceedings of the ISWC 02, IEEE Computer Computer Society.

[2] Cook A, Hussey S. (2003*) Assistive Technologies .Principles and Practice*. Second Edition. Publisher: C.V. Mosby.

[3] Lenay C, Gapenne O, Hanneton S. *Sensory Substitution: Limits and Perspectives*. 2003. Université de Technologie de Com-piègne.

[4] Vidal-Verdu F, Hafes M.(2007) *Graphical Tactile Displays for Visually-Impaired People*. IEEE Trans Neural S and Rehab Eng, V15.

[5] Asamura N, Tomori N and. Shinoda H. A *Tactile Feeling Dis-play Based on Selective Stimulation to Skin Receptors*, Proc. IEEE Virtual Reality Annual International Symposium '98, pp.36-42, 1998.

[6] N Asamura, N. Yokoyama, y H. Shinoda. *Selectively Stimulat-ing Skin Receptors for Tactile Display*, IEEE Computer Graphics and Applications. pp32-37 Nov- Dec 1998.

[7] Kajimoto H, Naoki K., Taro M. and Susumu T. *Tactile Feeling Display using Functional Electrical Stimulation*, In Proc. of The 9th Int. Conf. on Artificial reality and Telexistence, pp.107-114, 1999.

[8] Kajimoto H, Kawakami N, Tachi S 2002 Optimal Design Method for Selective Nerve Stimulation and Its Application to Electrocutaneous Display School of Information Science and Technology, The University of Tokyo, Bunkyo-ku, Hongo 7-3-1, Japan.

[9] Serra W, Lanzone M., Echenique AM., Graffigna JP: *Display Táctil Electrocutáneo*. Proceedings XV Congreso Argentino de Bioingeniería. Septiembre 2005. Pág.92. ISBN 950-698-156-6. 4 páginas.

[10] Rossini D, Pereyra PF, Echenique AM, Graffigna JP, Ensinck G: *Celda Braille por Electroestimulación*. IV Congreso Iberdiscap: Tecnologías de Apoyo a la Discapacidad – Vitoria – Brasil – Febrero de 2006. Vol 2 pág. CO103-CO107.

[11] Echenique AM, Graffigna JP, Mut V. (2010) *Electrocutaneous Stimulation System for Braille* Reading. 32nd Annual International Conference of the IEEE Engineering in Medicine and Biology Society, (EMBC 2010), 31/8 at 4/09/2010. 4pag 5827-5830 ISBN 978-1-4244-4124-2 : ISSN: 1557-170X

Design of a mechanical system in gait rehabilitation with progressive addition of weight

Ariel A. A. Braidot, Guillermo L. Aleman
Biomechanics Laboratory, FI-UNER
Ruta 11 Km 10. Oro Verde, Entre Ríos, Argentina
E-mail: abraidot@bioingenieria.edu.ar

Abstract. In this paper we designed and developed a mechanical device for gait rehabilitation based on the application of " partial body weight reduction therapy". An evaluation of the characteristics of devices based on this therapy currently available on the market was carried out obtaining information of the different mechanisms used in it. The device was designed to adapt to different height and weight of patients and to be used with additional equipment in gait rehabilitation, for example, treadmills, elliptical trainers and vertical scalers. It was envisaged to be used by patients with asymmetry in the lower extremities capabilities. We developed a stable structure in steel ASTM A36 which does not depend on the building conditions of the installation site. RamAdvanse software was used to calculate structural stability. A winch with automatic brake mechanism was used to raise/lower the patient, who was tied to a comfortable harness which provided safety to the patient and therapist. It was possible to quantify precisely, using counterweights, the weight borne by the patient during therapy. We obtained a small-sized and ergonomic low-cost prototype, with similar features to those currently considered cutting-edge devices.

1. Introduction

Walking is an essential daily activity carried out by any person to perform many activities independently. In normal conditions, gait is an automatic activity. Walking and weight bearing are the most important functions of the lower limbs, which must be considered in rehabilitation when gait is abnormal. Rehabilitation programs try to achieve the patient's recovery to a position of balance and ability to walk normally in a phased manner.

Patients with gait disorders due to different lesions often show great difficulty to support the body weight on the affected members, by abnormal patterns of motion, muscle weakness, or others.

In the past, the recovery of motor control was achieved by moving the segments involved with the help of a therapist who gradually introduced small resistances in order to enhance certain muscle groups which were already active. This traditional approach, which is still currently applied, was further enhanced after 1960 by neuromuscular and neurophysiological methods, which are based on the use of reflexes and sensory stimuli to stimulate or inhibit a particular motor behavior. Currently there are several methods that differ in the hierarchical CNS (Central Nervous System) approach and interactions with the periphery and muscles.

Today [1] the approach known as "task-oriented"[2], disconnected from the "practice of functional tasks" seeks to go beyond the concept of hierarchical organization of CNS and explain the control of all motor behaviors considering it like a systematic interaction in which there would be a multi-system control, exerting influence in parallel in order to perform a specific task. This means that it would not only be the CNS but also other systems such as muscle-skeletal and external environment which would influence motor behavior and therefore it would be taken into account in treatment. It also attempts to associate this idea of multisystem organization with the latest learning theory. Also it questions the assertion that "more training is better" and attests the fact that training is beneficial as long as it produces a beneficial change that addresses the functional purpose of diminishing some of the disabilities that the patient presents [1].

It has been suggested that "task oriented" rehabilitation should not be delayed, as in other methodologies, until the patient has got a good head, trunk in sitting and standing control, and also until abnormal reactions have disappeared and the spasticity has been controlled. Task-oriented techniques ensure that gait is an automated functional activity which can and should be trained at an early stage even in the absence of the conditions described above [1-2].

The most prominent gait rehabilitation in the current investigation into the technical "task-oriented" is "body partial weight reduction therapy" while walking on a treadmill. This is a relatively new technique that originated from the investigation of neural locomotion control of the vertebrates. Based on observations and the result of experimentation of this technique in cats [3-5], a number of research teams around the world began to test the therapy on humans [6-9].

This therapy is based on the combination of two elements:

1. A postural control device which reduces the amount of weight borne by the patient.
2. A treadmill which moves at a slow speed to retrain the appropriate patterns for better gait.

Several studies have shown that partial body weight support on a treadmill can improve walking in patients with spinal cord injury [10-14].

In recent years various devices that meet the requirements of this therapy have been developed. Some of those are the “biodex weight reduction system” [15], “pneumexpneu-walker” [16], “project walk equipment” [17] and “lokomat” [18], among others.

Some of these devices do not allow for a comfortable rehabilitation, nor do they have the ability to quantify the weight borne by the patient, and accessibility to patients is limited, which prevents the patient’s proper rehabilitation. Devices of this type that have all the appropriate features for gait rehabilitation have a high cost for the Latin American market.

In this work we designed a prototype for proper gait rehabilitation and for accurate quantification of the patient's weight which is supported on their lower limbs. Ergonomic and economic aspects are conditions that were considered in this device to ensure that these therapies are easily available to the society.

2. Design of the Device

The design of the device comprises gait rehabilitation of many disabilities, using the “body partial weight reduction therapy”. We developed a completely mechanical design, easy to manipulate and economically viable to start mass production.

The device was designed in a modular format that covers the structure, the counterbalance system of the patient, the lifting system and the patient’s harness, which act together to achieve the goal effectively. This feature of modularity allows for easy transfer to the installation site.

2.1. Structure Design

To keep the patient in an upright position to facilitate the rehabilitation activities, a horizontal beam that supports the weight of the patient was suggested. This beam has welded pulleys which are linked by a rope attached to a harness which holds the patient at one end and to the patient’s lifting device, on the other.

The first design proposed the use of a single cable tie. The drawback of this first design was that it did not limit the lateral movement of the patient, which can be sudden and involuntary and which may undermine the proper rehabilitation work.
The first design described required only a single horizontal beam associated with a vertical column. This unique longitudinal horizontal beam would be subjected to high stresses during lifting of the patient. The same situation of high stress would be generated on the lifting device and the only column in that design. In this sense we needed a larger section of the profiles to be used, significantly raising the cost of the device. For this reason, it was proposed to place two individual horizontal beams whereby there would be double support for the patient's weight. This enhancement allows the placement of different counterweights related to each half of the body [19]. This increases the functionality of the device with this new structure, considering that it may be useful for patients in whom the amount weight that can support by each side of the body is different. Mentioned below are some examples in which different discharges on each side of the body would be useful:

- Muscular atrophy in one member.
- Various operations on to a single member.
- Lateral or bilateral amputees.
- Hemiplegia.

The design of the device is adaptable to different rehabilitation clinic designs, where the device will be installed. The requirement is that the room has a flat supporting surface and is high enough for placement. Therefore, the beams described above are supported by columns associated with a solid base that is fixed to the ground.
Both columns should be associated with transverse beams to reinforce the structure against lateral forces generated by the patient on an involuntary basis during rehabilitation. Furthermore, this new design allowed the pulleys to be placed at different lateral positions welded to the new added beams, providing a comfortable and correct elevation (figure 1).
Longitudinal beams are welded to the columns at the end. The patient should be suspended from the front end of the longitudinal beams. This results in a bending moment determined by the product of patient weight and length of the beams. This moment can become large enough to compromise the rigidity of the structure. One solution used here is to place reinforcements in the structure (figure 2). Using a 45 degree beam between the columns and horizontal-longitudinal beams that keep the patient's weight, you get a more stable structure without affecting the ergonomics of the design.

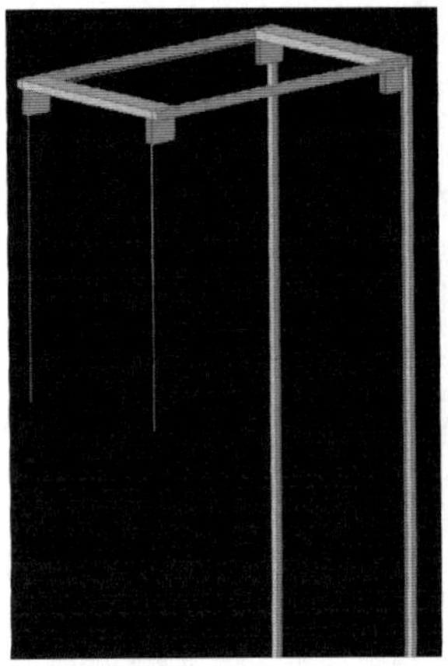

Figure1. Double column design linked by longitudinal and transverse beams.

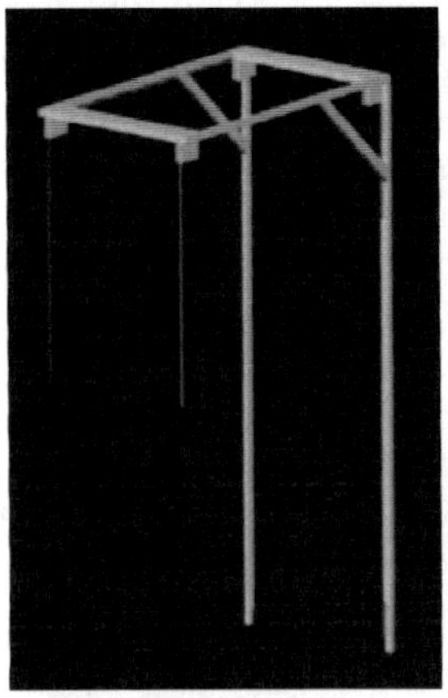

Figure 2. Reinforcements above 45degrees between side beams and columns.

As detailed above, the structure must be affixed to the ground and it is independent of walls, ceilings or walls of the room. It is necessary to build a stand that holds the various components of the structure and other mechanisms which belong to the device. This stand should be stable, stiff and ergonomic. Ultimately, the design takes into account the transfer of the patient and the mobility of the therapist around the structure.
For each column a horizontal beam was placed, which acts as a bracket attached to the floor, with reinforcements at 45 degrees on either side, which adds greater stability (figure 3).

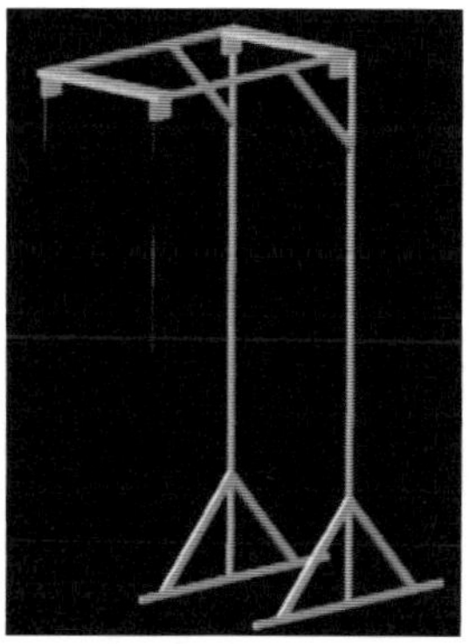

Figure 3. Base and lower reinforcements.

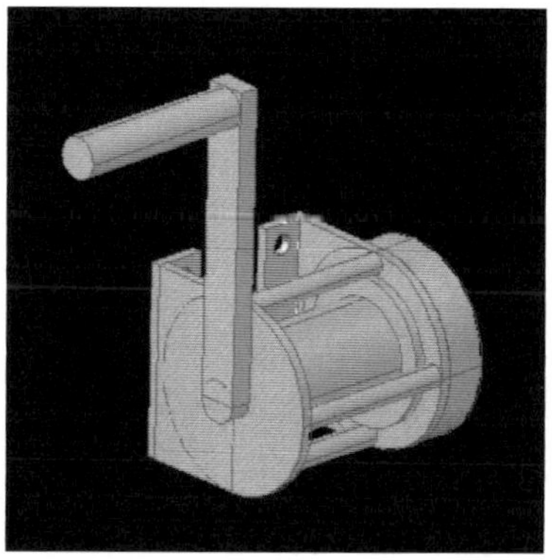

Figure 4. Winch scheme.

Figure 5. Photo of the winch mounted on the structure.

Each of the base beams have two anchors, one at each end, which fix the structure to the ground and act as constraints on all axes or directions. The reinforcements help maintain the stability of the structure and counter the torque exerted at the time of raising the patient.
The length of each beam, column or reinforcement is detailed in Section 3. This will take into account the average height of patient, size of the various aids to rehabilitation such as treadmills, elliptical walkers, among others, and cost and benefit relation between stability and ergonomic design.

2.2. Device for patient's elevation

At the beginning it was proposed to perform the elevation and descent of patient by means of a mechanical device. These devices are of easy manipulation and accessible price. We sought a device with the following characteristics:

- Elevation, descent and support of the patient's weight in suspension.
- Automatic brake system, both in ascent and descent.
- Safe and controlled Ascent / descent.
- Sturdiness
- Easy manipulation.
- Limited Size.
- Accessible Price.
- Easily available in the local or regional market.

Taking into consideration the aforementioned characteristics, the device of elevation that best adapts is the winch/capstan type. With this device the operator can raise a load of great magnitude exercising a small force. This device is available on the regional market in a great variety of sizes and forms.
In the winch considered ideal for this project the multiplication of force is obtained by means of a mechanism of reduction to planetary gears. It is a mechanism of long useful life by virtue of the sturdiness of all the elements. It possesses an automatic brake and that is why the operator's attention is not necessary. With this, the accidents that happen with a winch with a manual brake disappear.

The winches were fixed to profiles welded to the columns of the structure to a comfortable height for the operator (see figure 4-5).

2.3. Mechanism of Counterweight

The design considered the aptitude to quantify the load to know with certainty how much weight is supported for the patient, knowing the load of the counterweight. A system was designed for quantification of counterweight facilitation, of easy manipulation and accessible price.
The mechanism of counterweights is based on ingots of steel of 5 (Kg) by means of which it is obtained an exact quantification of the weight reduced to the patient, satisfying this one and other mentioned characteristics.
The ingots articulate vertically by means of two cylindrical axes which allow their correct ascent and decline (figure 6).
The addition of the ingots is done by applying a pin below them, which penetrates a cylinder guide which supports the weight. This cylinder guide is joined to a rope in the top part, which is tied to the rope from the patient's restraint system by means of a device of tie (sees section 2.4). This configuration allows the addition of the counterweight needed and controlled in precise form.
The height of the structure that contains the ingots must grant the quality of dynamic unload of weight. The body's mass center changes vertically in gait. The mechanism sought has to keep the same counterweight over the patient in the course of the therapy allowing the mass center to change vertically and the load on the low members to be constant (figure 7).
The counterweights rest on the floor when there are not patients doing therapy on the device. This minimizes the contribution of loads to the structure, preventing that the weight of the ingots and the structure that contains them relapse on the device.
The device was designed in such a way that it possesses two bodies of counterweight, offsetting the weight of the patient in right and left half.This disposition grants the possibility of training patients with asymmetric disability, who do not exercise the same distribution of load, force and neuromuscular activity in both members.

2.4. Tie of counterweights

It was sought an anchorage that possesses the aptitude to be tied to the rope from the element of subjection of the patient, from the beginning of the session and to cast off, at the end of the session a rapid and safe way. This mechanism must be small, light, safe and of easy operation. This tie remains associated to the counterweights in an uninterrupted way in the course of the therapy. The efforts that take place on the mechanism are of traction. This is due to the fact that the weight of the patient and the counterweights are opposed in direction.
After an exhaustive investigation, taking into consideration the previously mentioned characteristics, a device of tie with fist of ascension was chosen. The new element named JUMAR (see figures 8) possesses all the necessary characteristics.

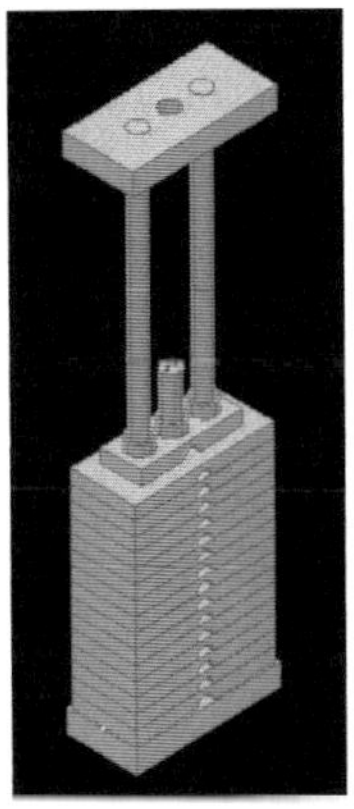

Figure 6. Diagram of the counterweights.

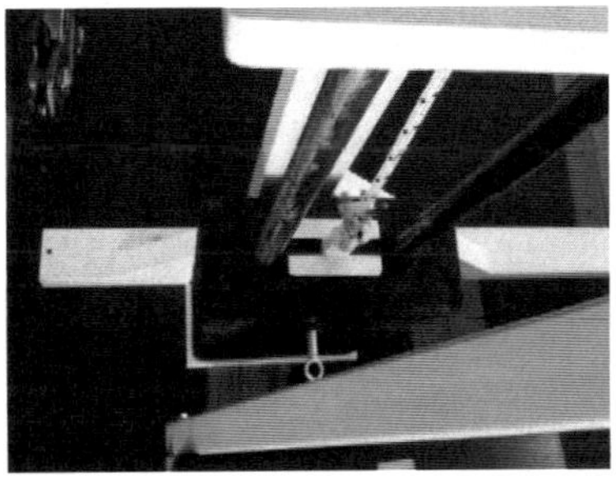

Figure 7. Photograph of the counterweights system.

Figure 8. JUMAR device.

2.5. Patient support

The patient's support must be at the expense of an element which redistributes the load of counterweights in a comfortable way in the body. Bearing this in mind, the chosen element was a harness which redistributes the load of the patient towards the pelvis and the trunk, diminishing the weight supported by the lower limb.
The harness must be comfortable, adjustable in size and safe, besides having the fastening elements over the shoulders. The harnesses which have the previously mentioned characteristics are: the harness of complete body and the parachute harness. In the latter, the elements for harnessing the patient are more comfortable. In addition, the parachute harness possesses a good fastening of the chest giving a correct raised position. This harness is of a high cost, which is a disadvantage.
The harness of complete body is composed of straps, nylon, polyester or of another type of tape, which are fastened around the body, distributing the load. Two elements of connection are necessary to allow the connection with the system of elevation by means of ropes. This type of harness is of a very low price.
The chosen harness is that of complete body with lumbar fastening, which ensures a correct position. Small modifications were performed on the fastening tapes, offering comfort similar to that of the parachute harness, adapted to carry out the therapy.

3. Dimensions

Below is a list of the desirable characteristics for the determination of the different parameters of the designed device.

- Possibility of patient's access to the device in a wheelchair.
- Dimension of supplementary devices. For example, a treadmill.
- Patient's maximum Height: 2 (m).
- Patient's maximum Weight: 150 (Kg) (hundred fifty kilograms).
- Construction of structure by means of steel ASTM A36 profiles.
- Dimensions of the ingots.
- Ergonomic Design.

3.1. Height of the structure

The height of the training supplementary devices, such as a treadmill was taken into consideration. Therefore, the dimension of the structure columns depended of the device whose work base has the maximum height. To define that dimension, the training, vertical climber and elliptical climber were considered like standard devices in rehabilitation. The major height corresponds to the vertical climber, which is 300 (mm) high. Bearing in mind a patient's maximum height of 2 (m), columns of 2.5 (m) long were chosen, giving a safety margin of 200 (mm).

3.2. Width of the structure

The possibility of patient's access in wheelchairs, both through the front and the back of the structure was taken into account. This offers an advantage as for the orientation of the device, adapting it to the infrastructure of the available building.

A great percentage of patients with severe gait disorders cannot walk on their own, at least in the first rehabilitation phases. Having this in mind and allowing for the different possibilities of access to the device, the width of a standard wheelchair was taken as the minimum access length. This dimension is 630 (mm).

The ingots of counterweights possess a width of 130 (mm). The structures that contain them were located in the interior side of the columns. Bearing the detailed dimensions in mind, the width of the structure, or distance between columns, which gives comfortable access of the patient on wheelchair is of 1.1 (m).

3.3. Length of the structure

For the correct choice of the lengths of the structure girders, the following points were considered:

- The comfort of the patients and the professionals working in gait rehabilitation.
- The length of the supplementary devices (treadmills, climbers, elliptical good walkers) in rehabilitation.
- Material, section and the profiles length used in the structure.

The longest supplementary device is the treadmill. Its center is an ideal position for patient's work, obtaining great freedom of movement, proximity to the armrests and easy access of the therapist to the patient. The treadmill length, in average, is 1.6 (m). Bearing this in mind, the selected length dimension was 1 (m).

The base of the structure, in its front was decided with the idea of minimizing the falling moment determined by the product of the patient weight and the length from the nearest end of the base to the patient's position. This moment has void value if the length of the previous section of the base is equal to the dimension of the longitudinal girders. Therefore, it was decided that they should be of the same length.

The base in his back part was 300 (mm).

3.4. Winch position

The correct position of winches was conceived with the idea of providing comfortable access and easy manipulation. It was chosen to place the device at elbow height, a comfortable position to realize the ascent/descent of the patient of a constant and safe way.

1.7 (m) was taken as the average height of a person in charge of manipulating the structure. Considering anthropometric tables [20] the winch was placed at 1.1 (m) high.

3.5. Dimensions of the structure which contains the counterweights

The ingots of counterweight used are 25 (mm) high. Bearing in mind the patient's maximum weight of 150 (Kg), every system of counterweights possesses 75 (Kg) in ingots. Therefore, their total height is 375 (mm).

Vertical margins of movement were given in order for the counterweights to be able to articulate while the patient does the therapy. The reason of this margin was explained in paragraph 2.3. This length is 460 (mm). This is a great margin of vertical movement, which allows the patient to move the joints of the limbs and hip. The structure that contains the counterweights has a top ceiling, which limits the movement of the ingots preventing a possible fall of the patient in case of stumbles, decompensation or imbalance of gait. The brake of the winch also avoids possible falls.

3.6. Dimensions of the reinforcements

Bearing in mind that the base must be fixed to the floor in its ends and the dimension of his back part is 400 (mm), the ends of the reinforcements were placed to 300 (mm) from the intersection node between columns and base. This reinforces the structure granting better stability.
The top reinforcements were placed bearing the same criterion in mind and they were located at a similar distance.

4. Results and Discussion

The structural calculations were made using the RamAdvanse software. The calculations took into account:

- Maximum weight of patient: 150 (kg).
- Design using steel profiles (ASTM A36). Elements on the local market which has the required performance.
- Square or rectangular profiles. Through these a good final of the structure may be obtained
- Profile weight.
- Winch weight.
- The weight of strings and washers was not taken into account.
- Distance between washers.
- Structure bulwarks to the ground by 4 points.

The patient's weight is transmitted through a system of pulleys and counterweights. Each node location of the pulleys has horizontal and vertical force components which are transferred to the structure as those are welded to the it.
The separation distance of the pulleys must be the same as the distance of medial separation of the counterweights, 960 (mm), considering that these can move only vertically. Consequently, the pulleys are welded to the upper transversal beams with a distance between the pulleys of 960 (mm).
RamAdvanse software has a database of both materials and profiles. There were no features of different profiles available on the local market, so we proceeded to create and load it in the program.The obtained information satisfies the conditions of stability with a profile of 80x40 (mm). Figure 9 shows in range of color the efforts on the designed structure.

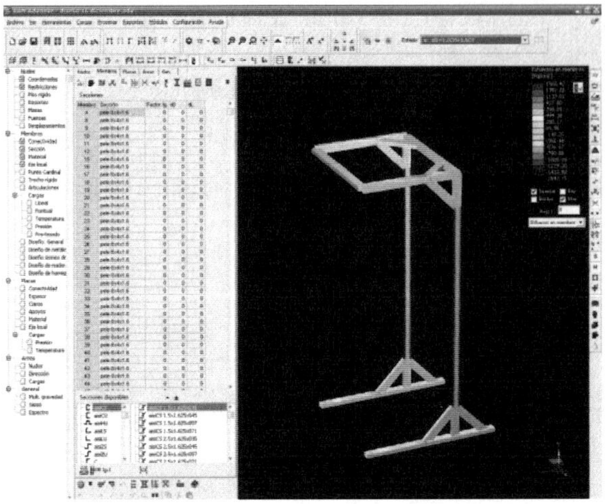

Figure 9. Tensions in the structure.

5. Conclusions

The design criteria used for the developed device provide adaptability to almost any standard rehabilitation clinic, and even the patient's home. The patient can access the device from the front and the back, a quality that allows that its installation is not be limited by any specific direction.

The shape and the dimensions used in the design generated wide lateral working margins, allowing the therapist to have free access to patients. This point is very relevant considering that some patients have a high degree of gait dysfunction, so therapists must simulate it through the mobilization of the lower limb segments. Moreover, the device is comfortable for both the therapist to manipulate the equipment and the patient in therapy.

The implemented device was installed at the Centro de Atención Integral y Cuidados Especiales (CAICE) from Paraná City - Entre Ríos – Argentina and it is now in operation (see details in figures 10 and 11).

Figure 10. Device installed in CAICE.

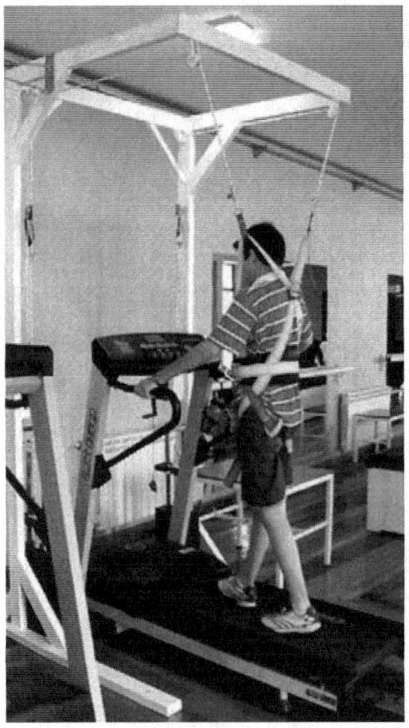

Figure 11. Patients through their therapy with the device.

The device is suitable for the rehabilitation of patients with different weights and heights, children or adults, considering the possibilities offered by the adjustable harness. The device has the ability to quantify the weight of the patient offset, allowing to decide when to increase the weight bearing in the rehabilitation process.
The prototype is ergonomic; the length and width dimensions are a few inches larger than the dimensions of treadmills, increasing the space required in a small proportion. The device was also adequate to put the treadmill at both the front and the back.
The correct sizing of the structure facilitates the use of the device with different additional equipment to help the gait rehabilitation, such as a treadmill, a stair climber and an elliptical vertical and others.
The large area of patient handling, together with the visual granted allows to study the different characteristics of gait using videography, as well as the possibility of using FES (Functional Electrical Stimulation) [21-22].

6. Bibliografia

[1] Beudoin N, Fleury J, Boulanger YL. 1994 *Reeducation des accidents vasculairescerebraux.Bilanetprise en chargue.Encycl Med Chir* (Elsevier, Paris-France)
[2] Malouin F, Potvin M, Prévost J, Richards CL, Wood-Dauphinee S. 1992 Use of an intensive task-oriented gait training program in a series of patients with acute cerebrovascular accidents. *PhysTher.***72**(11) 781-9
[3] Shurrager PS, Dykman RA. 1951 Walking spinal carnivores.*J Comp Physiol Psych.* **44** (3) 252–262
[4] Lovely RG, Gregor RJ, Roy RR, Edgerton VR. 1986 Effects of training on the recovery of full-weight-bearing stepping in the adult spinal cat. *Exp Neurol.* **92**(2) 421–435
[5] Barbeau H, Rossignol S. 1987 Recovery of locomotion after chronic spinalization in the adult cat. *Brain Res.* **412**(1) 84–95
[6] Barbeau H, Wainberg M, Finch L. 1987 Description and application of a system for locomotor rehabilitation. *Med BiolEngComput.***25**(3) 341–344
[7] Dobkin BH, Harkema S, Requejo P, Edgerton VR. 1995 Modulation of locomotor-like EMG activity in subjects with complete and incomplete spinal cord injury.*J Neurol Rehab.***9** 183–190
[8] Dietz V, Colombo G, Jensen L. 1994 Locomotor activity in spinal man. *The Lancet***344** 1260–1263
[9] Wernig A, Muller S. 1991 *Improvement of walking in spinal cord injured persons after treadmill training. Plasticity of Motoneuronal Connection.*Ed Wernig A, (Elsevier; Amsterdam). pp. 475–485.
[10] Renato Grasso1, Yuri P. Ivanenko1, MyrkaZago, Marco Molinari, Giorgio Scivoletto, Vincenzo Castellano, VelioMacellari and Francesco Lacquaniti. 2004 Distributed plasticity of locomotor pattern generators in spinal cord injured patients. *Brain***127**(5) 1019–1034.
[11] Behrman AL, Harkema SJ. 2000 Locomotor training after human spinal cord injury: a series of case studies. *PhysTher.***80**(7) 688–700
[12] Wirz M, Colombo G, Dietz V. 2001 Long term effects of locomotor training in spinal humans. *J NeurolNeurosurg Psych.* **71**(1) 93–96
[13] Barbeau H, Norman K, Fung J, Visintin M, Ladouceur M. 1998 Does neurorehabilitation play a role in the recovery of walking in neurological populations *Ann New York Acad. Sci.* ;**860** 377–392
[14] Wernig A, Nanassy A, Muller S. 1999 Laufband (treadmill) therapy in incomplete paraplegia and tetraplegia. *J Neurotrauma***16**(8) 719–726
[15] The Biodex Unweighing System http://www.biodex.com/rehab/unweighing/unweighing_pres.htm

[16] PneumexPneu-Walker™ Adult. http://pneumex.com/id53.html[PubMed]
[17] Project Walk Equipment. http://www.projectwalk.org/center/equipment.htm
[18] Lokomat® - Enhanced Functional Locomotion Therapy http://www.hocoma.com/en/products/lokomat/
[19] Waagfjörd J, LevangieP K, Certo C, 1990 Effects of Treadmill Training on Gait in a Hemiparetic. *Physical Therapy*; **70** (9) 549-559
[20] Tilley A. R. ,Dreyfuss H. 2001. *The measure of man and woman*. (ed. Wiley, John & Sons)
[21] Lindquist A RR, Prado Ch L, Barros R ML, MattioliR, Lobo P H, Salvini T F 2007 Gait Training Combining Partial Body-Weight Support, a Treadmill, and Functional Electrical Stimulation: Effects on Poststroke Gait. *Physical Therapy***87** 1144-1154
[22] Edelle C 2001 Combined Use of Body Weight Support, Functional Electric Stimulation, and Treadmill Training to Improve Walking Ability in Individuals With Chronic Incomplete Spinal Cord. *Archives of Physical Medicine and Rehabilitation***82**, (6) 818-824

Journal of Physics: Conference Series **332** (2011) 012046 doi:10.1088/1742-6596/332/1/012046

Construction of an automated table standing

R E Neira, M D Barbero, E A Di Giulio, J F Folco and G F Jiménez

Universidad Tecnológica Nacional, Facultad Regional San Francisco, 501 University Avenue, (2400) San Francisco, Córdoba, Argentina

E-mail: rneira@arnet.com.ar

Abstract. In this paper, the construction of an upright stretcher designed for patients in rehabilitation. The standing back to patients' expectations of therapeutic improvement, allowing correct all the troubles of a passive long. It has been shown that this table favors not only physically but also has a psychological reach beyond the scope of physical therapy and strongly affects the recovery. At the same time, the use of an upright stretcher greatly decreases the biomechanical disorders of hospital staff in the process of recovery. Thus the problem of rehabilitation of trafficking in a comprehensive way which not only focuses on the patient's undivided attention but also includes medical and auxiliary personnel.

1.Introduction

The axis travel is finding a solution to a major health problem is the lack of proper equipment necessary in the treatment of patients suffering from immobility associated with neuromuscular disorders and / or cardiovascular diseases, which involve evolution disabling sequelae. This is why there is a need to regain standing in mechanically assisted as part of the successful outcome of such treatment [1].

Defined from the rehabilitation aspect, the standing engine is a major milestone, which returns the expectations patients therapeutic improvements, allowing correct any disorders passive supine posture long as they are not anti-gravity postures appropriate.

A neurophysiological level the human nervous system is programmed motor sequences of movements and postural coordination program, which are the settings in different muscle groups to facilitate implementation. These adjustments help in the planning and the intent of the movement [2, 3].

Reaching the standing position by any means or method, allows treatment to begin rehabilitation in the patient a great stage, whose scope is beyond the scope of psychological and physical therapy impact on its position vigorously address the problems imposed by their position [4 -8].

Empirically analyzing variables of an adult male patient population of an average weight of 82 kg, with involvement of cardiovascular disease and nursing staff (nursing, kinesiology, speech) of female majority, result: an immobilized patient beyond expected and nursing staff working conditions based on biomechanical disorders, such as diseases of backbone, arthritis of multiple joints, among others. These are health problems associated with highly relevant to the quality of institutions working with these patients.

Since the 60's to date there has been a major shift in approach to the problem of rehabilitation: the interest of nation states, provincial and municipal is no longer exclusively focused on patient care but is addressed a more integrated way, which includes medical and auxiliary personnel. They are making great economic efforts to address this problem more comprehensively.

This brings greater responsibility in the area of health, to provide rehabilitation services appropriate technological and human capacity for.

In return, to deal with this new technical necessity, shall be studied and relate new materials, conduct a thorough systematization of production, resulting in better products, more affordable and more competitive

2. Materials and methods

By working with patients who need to recover the standing, we chose to work with an element commonly used in health institutions, such as the horizontal fixed table. These are used for the transfer and management of patients with many limitations. To assist affected staff, we designed a special table with a base in tubular welded metal and painted with epoxy material to meet the rigorous standards applied to these elements. This database is based on wheel slip and equipped with emergency brake. In addition, the base material has rolling environments that support health, even with spilled liquids. For stress analysis supported by the structure, we used a design and simulation software as SolidWork, to visualize the areas of greatest demand in the structure so we can strengthen them, taking into account the most unfavorable position of the table (vertically).

The bed is built on a tubular metal base secured by screws flat wooden structure with high density polyester material 100 mm thick and covered with a special leatherette fabric visible seamless for easy cleaning and replacement.

For lifting the stretcher from the horizontal position at the bottom of the base metal, placed an electronic actuator taking into account the minimum maintenance time. Other options were analyzed lift (air, water), which were ruled by the requiring special maintenance and high initial cost. The actuator selected was WB10-24-650-9-304, 8 provided by the Company for Mechanical Industries Welter, it has an operating voltage of 24 volts, a maximum load of 650 kg, a maximum load speed of 9 mm / sec and a piston stroke of 304.8 mm.

The union of the stretcher to the metal base was made using a non-skid device mounted on bushings that allow the rotation of the table in a horizontal position to vertical.

The stretcher has footrest, supports tape to prevent knee flexion and trunk inclination, and wheels that allow mobility within the hospital environment.

For the electronic control of standing, was added to the base keyboard commands an electronic board that contains the elements to store the instructions where medical personnel program the number of positions up the table (required angle) and the time spent at each station stop. This task can be performed by medical personnel without difficulty due to the simplicity of the commands sent to the electronic actuator and allows you to change your routine, your stop and back flat. This system provides protection against overload stress and emergency system to return to the horizontal position.

3. Results

Through a study of existing elements applied to improve the patient's condition in the process of standing, it appears that they are poor, inadequate and high cost, usually through outside sources. With the design and development of this automated table we can improve patient mobility and reduce substantially the problems of caregivers, reducing their working conditions and improving the quality of life of patients and medical personnel and support staff.

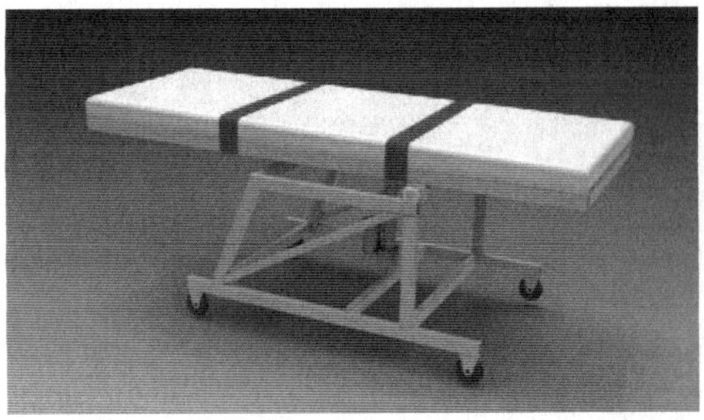

Figure 1. Table standing automated

The design and development of this table, see Figure 1, is made entirely in the local, and industrial production will bring many benefits in this area, by leveraging the technical and intellectual capacity of their community, gaining recognition regionally and nationally.

The cost of building a prototype is estimated to amount close to $ 8,000 pesos, an amount that is substantially reduced if the production is done on a large scale.

4.Conclusions

No longer discusses the need to monitor the patient in all stages of rehabilitation, nor the work of the staff concerned, but we must approach this task so that the effort is directed to patient care and spend time at a routine process, mechanical and slavish times.

The physical consequences they would suffer serious medical and support staff negatively affects the provision of the service.

Regarding the automated table is noticeably better than others on the market because it allows the management of a person regardless of size or physical condition of the therapist. In addition to regular digital control allows programming of time standing, angles required and can be performed by medical personnel without difficulty due to the simplicity of the commands sent to the electronic actuator.

By treating the patient's rehabilitation process in a comprehensive way, the improvements obtained with the use of automated litter standing are designed not only the patient but also to medical and auxiliary personnel.

Acknowledgements

The authors acknowledge the collaboration of the Regional Rehabilitation Hospital José Bernardo Iturraspe; to the company Frontier Living, the company, Daniel J. Aimaretti and Machining Workshop Di Giulio, all of the city of San Francisco, Cordoba, Argentina.

Referencias

[1] Farfalli L 2010 *Hip fracture in older adults*. Geriatric Clinical Guidelines CMPC Cap. 12
[2] Cardinali D 1995 *Manual of Neurophysiology*. Control of Posture, Balance, Movement Generation (UBA) Cap. 10
[3] Bobath B 1973 *Reflects abnormal postural actvidad caused by brain damage*. Normal postural reactions (Ed. Panamericana) Cap. 5
[4] Maltinskis R 1993 *Rehabilitation-functional recovery of the Disabled*. 4º edition (Ed. Medical Technical Science)
[5] Miralles R I 1998 *Clinical Biomechanics of the locomotor*. Stance (Editorial Masson) Cap 18.
[6] Montoto Marques, A 2006 *Manual SERMEF de Rehabilitación y Medicina Física*. Lesión Medular (Ed.Panamericana) Cap. 42
[7] Chae J, Triolo R, Kilgore K, "Y" Creasey G H 1998 *Rehabilitation Medicine.Principles and Practice*. Functional Neuromuscular Stimulation (Ed. Lippincott Williams & Wilkins) Cap. 24.
[8] Cibeira Z 1991 *Cerebral palsy*. Prescription Media (Ed. El Ateneo) Cap. 9

SABI 2011 IOP Publishing
Journal of Physics: Conference Series **332** (2011) 012047 doi:10.1088/1742-6596/332/1/012047

Development of a voice database to aid children with hearing impairments

M G Kuzman[1], P D Agüero[1], J C Tulli[1], E L Gonzalez[1], A J Uriz[2] and M P Cervellini[1]

[1] Laboratorio de Comunicaciones, Facultad de Ingeniería, Universidad Nacional de Mar del Plata, Juan B. Justo 4302, Mar del Plata, Argentina
[2] CONICET - Laboratorio de Comunicaciones, Facultad de Ingeniería, Universidad Nacional de Mar del Plata, Juan B. Justo 4302, Mar del Plata, Argentina

E-mail: {melisakuzman,pdaguero}@fi.mdp.edu.ar

Abstract.
In the development of software for voice analysis or training, for people with hearing impairments, a database having sounds of properly pronounced words is of paramount importance. This paper shows the advantage that will be obtained from getting an own voice database, rather than using those coming from other countries, even having the same language, in the development of speech training software aimed to people with hearing impairments. This database will be used by software developers at the School of Engineering of Mar del Plata National University.

1. Introduction

Hearing impairments approximately correspond to 18 % of all disabilities in Argentina. This statistic has been published by the CAS (Argentinan Federation of Deafs) [1], which keeps records of thousands of people with hearing impairments.

Early childhood is a particularly critical period for language acquisition and children development. In general, hearing losses cause difficulties for social adaptation [2]. Therefore, it is important to increase the opportunities of communication for preschool children with hearing losses, who have poor pronunciation skills due to the lack of auditory feedback.

Hence, it is essential to assist them as soon as possible in order to palliate this situation. For this purpose, creating properly designed applications for computers is very important.

Currently there are several programs that are used for oral speech training and rehabilitation of people with hearing impairments. Among them, "Dr. Speech"[3], which is a software for voice training. Another development is carried out at the University of Zaragoza. It is named "Alborada 13-A"[4],it is a software application for recognition and evaluation of the voice, which is able to detect pronunciation errors. This application is designed to help people with severe hearing losses in the speech training.

These applications have been created with the aim of improving communication capabilities, contributing in this way to the social interaction of people with hearing impairments

Software for voice recognition and evaluation is generally formed by three main components: a **voice database** to estimate statistical models or templates, a **voice recognition engine**, and finally, a user interface that allows for an **evaluation** of the utterance or gives users a visual feedback of their pronunciations.

A critical aspect of speech databases used at the laboratory of this research group at Mar del Plata National University, is the Spanish origin. This fact brings problems among acoustic analysis, since the linguistic forms of some words and their corresponding pronunciations often significantly differ with respect to the Argentinian dialect.

Considering that, the first goal that arises is the building up of a local database, it means the argentinian dialect. Its construction is divided into two groups: the first contains information about children with hearing impairments, while the second group contains voices of children with normal hearing abilities.

Once the voice database is done, it is intended to be used to train the SPHINX recognition engine [5] (a voice recognition tool written for Java). This application makes possible the evaluation of deletions, insertions, or substitutions of phonemes, a clinically phenomenon known as dysarthria [2]. Finally, the data and the recognition engine will be used to develop programs for people with severe hearing losses, such as SPO [6].

The paper is organized as follows. Section 2 describes several possibilities for data voice acquisition and analysis. Section 3 shows the achieved diphoneme coverage for the Spanish voice database, and how it could be improved. Finally, Section 4 details the conclusions of the study, analyzing future issues on the Spanish database.

2. Methods

In order to develop software projects applications that help people with severe hearing losses in their oralization practices, it is necessary to use a voice recognition system [7], also known as **ASR**[1]. It is a computer tool used to process voice signals and produce the transcription of the content. Because the ASR system uses statistical models, it is necessary to obtain appropriate amounts of data in order to get precision in the acoustic models.

2.1. Selection of the data to record

Depending on the application, there are several possible groups of phonemes that may be included in any voice database, which are listed below:

- Phonemes: it is the minimum unit used in voice recognition systems[2]. Spanish language has 18 consonants and 5 vocals [8]. In this paper work an inventory of 24 phonemes and 7 allophones is used, which are defined by Alarcos (1950) [9]. Because of the contextual variations, these units must be recorded in the context of words or sentences.
- Diphonemes: these units consider the coarticulation between speech sounds, by taking into account the left or the right context. There are 961 different combinations, but not all of them exist in Spanish.
- Triphonemes: these units take into account the central phoneme and the left and right context. In this case there are 29791 combinations, a higher amount than the previously case. However, many combinations must be deleted because they are not used in Spanish.
- Words and sentences: these are the only units with semantic meaning. Nevertheless, they are not useful to be assembled for recognizing any word.

For the development of the voice database, words and sentences are going to be chosen according to their diphoneme coverage. These units give flexibility to the system, and require a smaller number of words to be recorded. In order to have a good coverage of Spanish words,

[1] Stands for Automatic Speech Recognition.

[2] The phoneme is the minimum unit without semantic meaning in the spoken language.

the recorded data will be carefully selected to minimize resource needs (mainly time, people and storage). The resulting database will be useful to develop different aid applications.

2.2. Recording conditions

Recordings will be performed with open-source software tools capable of handling and manipulating sounds. The capabilities that are of interest in this work are: sound recording, speech analysis and phonetic annotation.

The data recording is going to be made with a dynamic microphone (SPK Pro-92 XRL, unidirectional, with a frequency response between 50- 16000Hz, output impedance of 600 Ohms and a sensibility of -52dBV $\pm$ 3dB).

All recordings must be done with the supervision of a voice therapist in order to help children with their utterances. The recording will be performed with the support of a professional in order to instruct the children. It is thought to generate an appropriate environment for the children's performance, and it will give naturalness to the speech [2].

Voice database recordings must take into account several specifications. Some of them are:

- Signal-to-Noise Ratio (SNR).
- Sampling frequency.
- Number of quantification bits.
- Reverberation time.
- Dynamic range.
- Presence of saturations (mostly in plosive sounds).

In the next paragraphs each specification is going to be explained, and the strategies to improve them will be described.

Signal-to-Noise Ratio. The recording room must be of appropriate size and with soundproof walls (irregularities on the walls, curtains, etc.) in order to avoid inconvenient acoustics reflections. To improve the **SNR**, recordings will be done with a laptop working on battery mode (without connection to the electric network), to avoid low frequency electric interferences. The speaker will be far from external noise like cellphones, air conditioning or computer's fans (including the laptop computer). The SNR is a very important specification of a system. It provides the relationship between the power of the signal under study and the noise power from other sources.

Sampling frequency and number of sampling bits. This parameter must be appropriate according to the required data quality. Sampling frequencies are limited by the capabilities of the computer's sound card . The sampling frequency of 16 Khz is chosen following the specifications of the project "Vocaliza". [4].

Reverberation time. RT is defined as the time it takes for the signal to fall 60 dB, since the speaker finished the phonation. Thus, it is considered that reflections finish when the intensity of the perceived signal correspond to 0,1% of the original signal. Minimization of this parameter results in a more accurate information obtained from the data.

Use of dynamic range. When the voice is recorded, some sounds may be louder than others, for example plosive sounds. These are produced when the vocal tract is closed, stopping all airflow and pressure is built up behind the occlusion. If this air is suddenly released, for example when pronouncing the occlusive /p/, the abrupt change of air pressure may cause a saturation in the recording. A good compromise between dynamic range and signal level is necessary in order to produce reliable recordings.

Voice recognition engine
The goal of the automatic speech recognition system is to produce a word sequence (or labels) which are the textual representation of the acoustic signal. This task is not trivial, because the acoustic signal not only contains words, pauses, but also disfluences , environmental noise, articulatory noises (lips, breathing), etc.

In order to perform this task, the automatic recognition system uses tools of statistical modeling and sequence decoding. It applies a set of simplifications in the approach to get an implementable solution.

The limitation of the vocabulary is the first simplification made by ASR systems. Its size could be a few words (e.g. recognition system of verbal orders) or tens of thousands of words (large vocabulary recognition system). It is not possible for the ASR to recognize words that are not in its dictionary, because it will not find the boundaries of them. For example, the sentence *"la casa de la pradera está habitada por un ermitaño"* could be pronounced *"lacasadelapradera estáhabitada porunermitaño"*. As seen, there are pauses before the words *pradera* and *habitada*, but the others are pronounced without any silence between words. Therefore, it is necessary to know the words to find their boundaries.

Nowadays, automatic speech recognition systems made their task based on statistical models of the acoustic signal (acoustic modeling of the phonemes) and the language (language modeling using words) [10]. The parameters of the statistical models are estimated using collected data, such as acoustic signals, and their corresponding phonetic and orthographic transcriptions. The reliability of the statistical models will be bounded by the amount of available corpus.

In the development of software for hearing disabled people, the functioning of the automatic speech recognition technology is partially modified to detect mispronunciations. In this case, the ASR does not have to recognize the word, but it should analyze the word that the speaker was asked to pronounce. In these conditions, it is possible to improve the analysis capabilities of the system incorporating possible alternative pronunciations due to dysarthria.

To be consistent with the use of JAVA in the development of systems to aid people with hearing impairments in their oralization practices, the ASR SPHINX was chosen to segment the recorded voice into phonemes to perform the analysis.

SPHINX-4 is an updated recognition system which was programmed in JAVA language. This system was created by the joint collaboration of the Carnegie Mellon University, Sun Microsystems Laboratories, Mitsubishi Electric Research Labs (MERL) and Hewlett Packard (HP), with the contributions of the University of California of Santa Cruz (UCSC) and the Massachusetts Institute of Technology (MIT).

3. Experiments

In order to know the word coverage of phonemes of the Alborada's database, the automatic phonetic transcription of the 57 recorded words was performed. It was done by a software that converts the word into the canonical phonetic transcription. Then, the phonetic transcription was used to get the phonemes of the words. The results show that the Alborada's database contains 147 phonemes. Figure 1 shows the phonetic transcription of a group of words made by the previously mentioned software.

The analysis of the coverage of the Spanish words with some specific phonemes was made using the written CREA corpus. The reference corpus of Spanish is a set of texts of different sources, which are stored in a computer support. The last one has the information of the words, their meanings and their context [11]. The Real Academy of the Spanish Language explains

Table 1. Words split into phonemes

Words	Phonemes
boca	_+b b+o o+k k+a a+_
bruja	_+b b+r r+u u+x x+a a+_
cabra	_+k k+a a+B B+r r+a a+_
campana	_+k k+a a+m m+p p+a a+n n+a a+_
casa	_+k k+a a+s s+a a+_
caramelo	_+k k+a a+r r+a a+m m+e e+l l+o o+_
casa	_+k k+a a+s s+a a+_
clavo	_+k k+l l+a a+B B+o o+_
cuchara	_+k k+u u+tS tS+a a+r r+a a+_
dedo	_+d d+e e+D D+o o+_
ducha	_+d d+u u+tS tS+a a+_
escoba	_+e e+s s+k k+o o+B B+a a+_
flan	_+f f+l l+a a+n n+_
fresa	_+f f+r r+e e+s s+a a+_
fuma	_+f f+u u+m m+a a+_
gafas	_+g g+a a+f f+a a+s s+_
globo	_+g g+l l+o o+B B+o o+_
gorro	_+g g+o o+rr rr+o o+_
grifo	_+g g+r r+i i+f f+o o+_
indio	_+i i+n n+d d+j j+o o+_
jarra	_+x x+a a+rr rr+a a+_
jaula	_+x x+a a+w w+l l+a a+_
lápiz	_+l l+a a+p p+i i+T T+_

that the corpus is a representative state of nowadays Spanish, because the text it contains has been selected according to the current parameters.

The 1000 words most used in Spanish according to CREA have been transcribed into diphonemes. The results show that there are 440 phonemes in the CREA Corpus. Therefore, the 147 diphonemes of the Alborada's database are not enough to cover the group of 1000 most frequent words.

In order to get a list of words that cover the 440 phonemes, a greedy algorithm was used. It selects a set of words to minimize the number of words needed to cover the 440 diphonemes. The final total was 320 words.

Applying the same procedure, it was possible to find the number of triphonemes necessary to cover the CREA Corpus. The results indicate that it is necessary 1778 words to comprehend the 2919 triphonemes in the Corpus.

Therefore, the use of diphonemes implies a positive trade-off between the reliability of statistical models and the amount of data that is necessary to record to get a good coverage.

4. Conclusion

In this paper has been mentioned that people with hearing impairments may experiment some problems when social interaction is required. Also, there are mentioned some applications that are useful for these people, in order to improve their speech abilities using specific softwares. Besides, it was emphasized the importance of having a suitable voice database for the development of these applications.

The origin of the voice database used at the laboratories of the School of Engineering is Iberic Spanish. The main problems are the different linguistic forms compared with the Argentinian Spanish. An own voice database will give the chance of improving the applications developed by this research group, and made it freely available to the community, like the project "Sistema de Práctica de la Oralización" (SPO) [6, 12, 13, 14].

The goal of this project is, on one hand, to get an Argentinian voice database and improve the results of voice reeducation with respect to those obtained with the Iberic Spanish database. On the other hand, it is proposed a different method to select the phonetic content that will give flexibility to the voice recognition system for many existing or new applications.

The results of the experiments show that the Spanish database has not enough coverage of the 1000 most frequent Spanish words. In order to get that coverage, it is necessary to record at least 320 carefully chosen words. This gives the chance to create and develop new applications, using a reliable and versatile voice database.

References

[1] "http://www.cas.org.ar," Tech. Rep.
[2] R. L. Belloto, *Voz y pronunciación en el niño discapacitado auditivo.* Ediciones ARES, 1974.
[3] "http://www.drspeech.com," Tech. Rep.
[4] O. Saz, E. Lleida, C. Vaquero, and W.-R. Rodríguez, "The alborada-i3a corpus of disordered speech," Tech. Rep., 2010.
[5] W. Walker, P. Lamere, P. Kwok, B. Raj, R. Singh, E. Gouvea, P. Wolf, and J. Woelfel, "Sphinx-4: A flexible open source framework for speech recognition," CMU, Tech. Rep., 2004.
[6] J. M. Garin, P. D. Agüero, J. C. Tulli, E. L. Gonzalez, and A. J. Uriz, "Spo: una ayuda en java para la oralización de hipoacúsicos," Tech. Rep., 2010.
[7] H. van den Heuvel, K. Choukri, C. Gollan, A. Moreno, and D. Mostefa, "Tc-star: New language resources for asr and slt purposes," in *Proceedings LREC 2006*, 2006, pp. 2570–2573.
[8] J. Llisterri and J. B. Mariño, "Spanish adaptation of sampa and automatic phonetic transcription," Tech. Rep., 1993.
[9] E. Alarcos, *Fonología española.* Gredos, 1950.
[10] S.-C. Yin, R. Rose, O. Saz, and E. Lleida, "Verifying pronunciation accuracy from speakers with neuromuscular disorder," in *In INTERSPEECH-2008*, 2008, pp. 2218–2221.
[11] "http://www.rae.es," Tech. Rep.
[12] J. M. Garín, P. D. Agüero, J. C. Tulli, and E. L. Gonzalez, "Sistema de práctica de la oralización en plataforma java," in *SICA*, Noviembre 2009.
[13] J. M. Garin, P. D. Agüero, and J. C. Tulli, "Nuevos aportes al entrenamiento de personas hipoacúsicas sistema de entrenamiento para personas hipoacúsicas en plataforma java," in *TISE*, Diciembre 2008.
[14] J. M. Garin, P. D. Agüero, and J. C. Tulli, "Nuevos horizontes - aportes al entrenamiento de personas hipoacúsicas," in *SICA*, Septiembre 2008.

A reconstruction platform for coronary arteries, finite element mesh generation and patient specific simulations

Sabina Salles [a], Fernando P. Salvucci [a] and Damian Craiem [a,b]

[a] Universidad Favaloro. FICEN. Av Belgrano 1723, C.P. C1093AAS, Ciudad de Buenos Aires, Argentina

[b] CONICET, Argentina

e-mail: sabinasalles@gmail.com

Abstract. The left coronary artery (LCA) is one of the most important sites of atherosclerotic plaque formation and its progression may lead to stroke. There are many systemic risk factors that are related to the appearance and development of atherosclerosis. It has been observed, however, that the lesions occur in specific regions of the arterial tree. Nowadays it is accepted that these regions are the ones with low abnormal values of Wall Shear Stress (WSS). The aim of this work is to generate a platform to obtain a three-dimensional (3D) reconstruction from an LCA, generate a finite element mesh and run a patient specific blood flow simulation in it.
The 3D model of the LCA reconstructed was based on a computed tomography (CT) study from a healthy patient. The method used for this procedure consisted basically on three stages: (1) segmenting the LCA principal branches from TC images, (2) extracting several contours from the arteries so as to get the essence of the geometry, applying a self-developed software, and (3) using this outlines to generate a *Non-uniform rotational basis spline* (NURBS surface).
A finite element unstructured volume mesh was generated and refined and a simulation was run on it using a Computational Fluid Dynamics (CFD) application. The values of pressure and velocity were analyzed and WSS was computed. As expected, regions of low WSS were found in the outer walls of the bifurcation and in the inner wall of curvatures.

1. Introduction

Atherosclerosis is currently one of the leading causes of mortality worldwide. It is a chronic, inflammatory, fibroproliferative disease primarily of large- and medium-sized conduit arteries (Chatzizisis et al., 2007). It is well known that there exist many risk factors that may contribute to the appearance of the atherosclerotic plaques, such as age, gender, hyperlipidemia, cigarette smoking, hypertension, diabetes mellitus, chronic infections, stress and genetic predisposition (Naghavi et al., 2003). Despite the systemic characteristics of these factors it was found that atherosclerotic lesions form in specific regions of the arterial tree, like in the vicinity of branch points, the outer wall of bifurcations and the inner wall of curvatures (Malek et. al., 2007).
This seems to be related with the hemodynamic forces that depend on the individual geometry of each patient's vessels. The most influencing of these forces is, by far, the tangential one that generates the blood flow in the inner arterial wall, and can be synthesized as the wall shear stress (WSS). Currently,

it is well accepted that the regions where plaques are more likely to emerge are those in which the flow bifurcates and there is low WSS (Davies et. al., 2009; Caro et. al., 2009; Steinman et. al., 2002). Moreover, for the particular case of the coronary arteries, recent studies showed that bifurcation angles in the in these arteries might influence atherosclerotic processes in the coronary Junctions (Chaichana et. al., 2010; Craiem et. al., 2009).

The fact that low WSS has an important role in the development and progression of atherosclerosis lies in the endothelial response: it attenuates nitric oxide (NO)-dependent atheroprotection; it promotes low-density lipoprotein cholesterol (LDL) uptake, synthesis, and permeability; it promotes inflammation and oxidative stress; it promotes vascular smooth muscle cell (VSMC) migration, differentiation, and proliferation; it promotes ECM degradation and attenuates its synthesis in vascular wall and plaque fibrous cap; it has a potential role in plaque neovascularization and calcification; it increases plaque thrombogenecity (Chatzizisis et al., 2007).

Computational Fluid Dynamic (CFD) modelling using the Finite Element Method (FEM) has been proved to be a practical and reliable tool for studying time-varying, three-dimensional (3D) blood flow patterns in complex arterial geometries. It has been shown to provide prognostically relevant hemodynamic data based on 3D reconstructions of vessels from medical images (Steinman et. al.,2002). It is imperative to use patient-specific reconstructions of arteries and make them as reliable as possible. The FEM is a numerical technique for solving partial differential equations (PDE's). Its first essential characteristic is that the continuum field is subdivided into cells, called elements, which form a grid. (Wendt et. al., 2009)

In this paper we present a method to reconstruct human coronary arteries from Computed Tomography (CT) images using a self-developed software. We use then the results of this reconstruction platform to generate a finite element volume mesh from a Left Coronary Artery (LCA) to run a FEM simulation of the blood flow in it.

2. Methods

2.1. 3D surface reconstruction

A critical issue to obtain reliable simulations is the reconstruction procedure used to model the vessels. We propose a method that consists basically on three steps: (1) segmentation of the principal LCA branches from CT images; (2) extraction of several contours from the arteries in order to characterize the geometry; (3) use of these outlines to generate a *Non-uniform rotational basis spline* (NURBS) surface.

2.1.1. *CT segmentation.* In the present work we used a multislice CT scan performed to an asymptomatic and healthy patient in the Hospital Universitario, Fundación Favaloro, Buenos Aires, Argentina. The scan was done using a 64-row MSCT scanner (Aquilion 64, Toshiba, Japan). Images were acquired during diastole by means of ECG-gating using a 400–960 ms time window in order to reduce motion facts. Transaxial images were reconstructed with 0.415 mm slice thickness and 0.3 mm increments.

All digital imaging processes were done using a custom software designed in the Favaloro University. A detailed explanation of this processing is explained elsewhere (Craiem et. al., 2011). We imported with the mentioned software 40 consecutive images (512 by 512 pixels, 256 gray levels) so as to reconstruct the whole LCA, beginning in the ostium, including the left main artery (LM), the left circumflex (LCX) and the left anterior descending (LAD) arteries until an important secondary branch in any of them appeared (Figure 2) (Salvucci et. al., 2010).

The LCA extraction proceeding is based on two main stages: Heart isolation and coronary artery segmentation.

The first phase, where the heart is isolated from the images, starts with the user selecting a volume of interest (VOI) where the whole heart must be included. The heart isolation consists in an iterative process of 5 steps and concludes in the clustering of every pixel between Heart-Region and Not-Heart-Region, being the values of the last group set to the minimum value of the whole volume in Hounsfield Units (HU).

The second stage (the segmentation of the coronary artery), involves 6 steps which include a 3D median filter, a 3D morphological opening filter, digital substraction, contrast increase, a windowing process and a volumetric region-growing algorithm. If at this point some elements as coronary veins, ventricle blood pools regions near papillary muscles or spurious noise connected to the LCA remain visible, the software provides a manual correction 3D erasing tool to remove them.

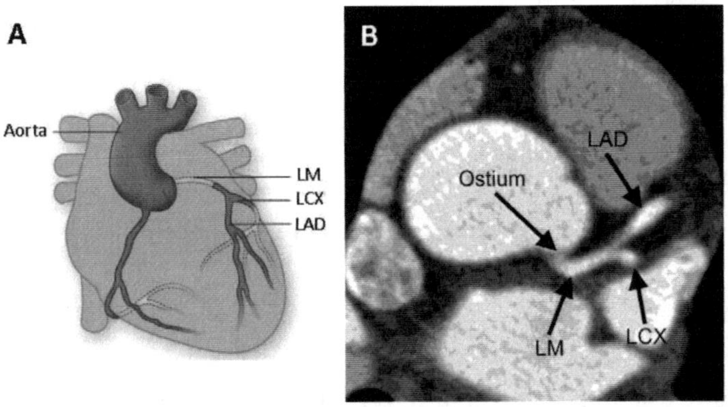

Figure 2: (A) Aorta and LCA anatomy. (B) Example of a CT slice image showing the coronary ostium, the left main artery (LM), the left anterior descending (LAD) and the left circumflex (LCX) branches.

2.1.2. *Contour extraction.* The idea is to obtain the contour of the LCA, with planes orthogonal to the midline (skeleton) of the vessels, in order to get the artery 3D geometry. For this purpose we used a self-developed software implemented in Matlab®. This routine has 5 algorithms: skeletonization, branch division, surface triangulation, contours computation and IGES exportation.

The **skeletonization** method is a sequential 3D thinning algorithm adapted from the one developed by Palágyi K. (Palágyi et. al., 2001). It is basically an iterative process that iteratively eliminates certain points form the solid object until a single pixel skeleton is found. It analyzes the 26-neighbourhood of every voxel and determines thereby if it belongs or not to de surface of the remaining solid. Each iteration in the algorithm itself is based on 6 subiterations that allow classifying the border points. Figure 3 (A) shows a solid reconstruction of the LCA used in this investigation and (B) its skeleton.

The **branch division** algorithm implemented in this platform begins by automatically classifying points of the skeleton in bifurcation-points, end-points and middle-points (Figure 3 (C)). This is done by counting the points in the immediate vicinity of each skeleton point. The user is then asked to select the beginning-point and the end-point of each branch. Then, the software compares every point of the solid LCA geometry and associates each point to the nearest branch as shown in Figure 3 (D).

For accomplishing the **surface triangulation,** the software implements the *Marching Cubes* algorithm. This procedure is used to render isosurfaces in volumetric data (Bourke et. al. 1994). A result of applying this algorithm is shown in Figure 3 (E).

The **contours computation** is done with the intersection between planes orthogonal to the vessel centerline and the marching cubes volume. The centerline is estimated using a spline curve. The user manually sets the spline order and the number of points to be analyzed in each segment. A representative artery is shown in Figure 3 (F)

Finally, the **IGES exportation** step is simply an algorithm that uses the information from the contours to generate an *.iges* file that can be recognized by any CAD software as spline curves in 3D.

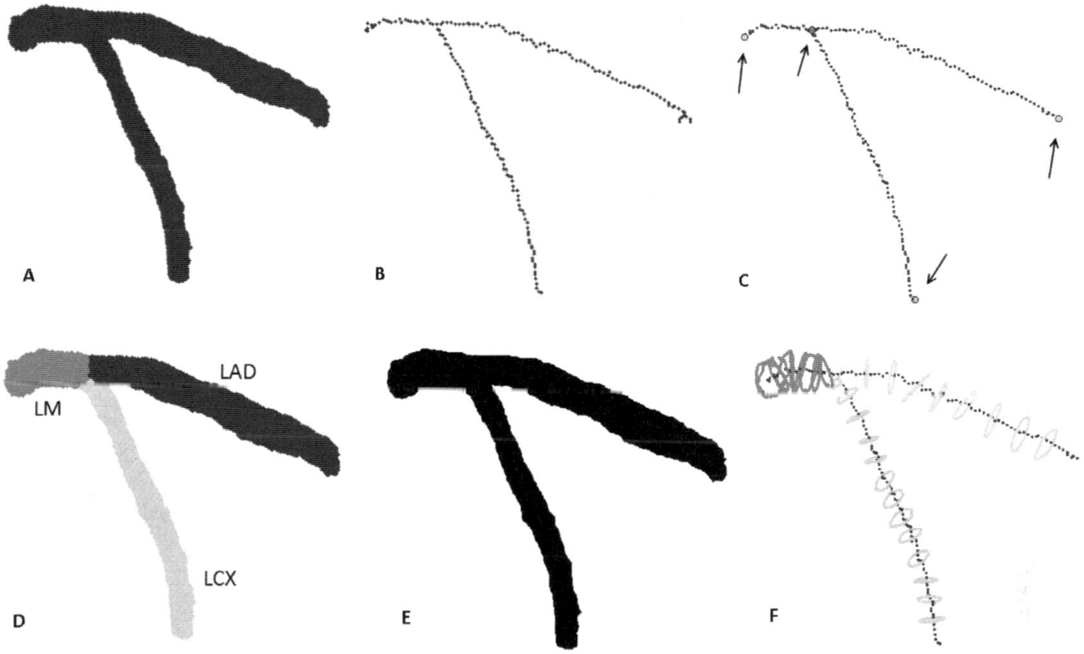

Figure 3: (A) Group of points in 3D representing a solid LCA. (B) Skeleton of the LCA obtained by implementing the Palágyi algorithm. (C) Detection of the bifurcation- middle- and end-points of the skeleton. (D) Branch division of the solid LCA. (E) Surface Triangulation obtained with the Marching Cubes algorithm. (F) Contours estimation.

2.1.3. NURBS surface generation. This phase was done using a custom developed CAD software. First of all the *.iges* file with the splines of the arterial contours are imported (Figure 4 (A)). The two branches of the bifurcation have to be lofted separately, sharing the curves from de main branch (LM) as shown in Figure 4 (B) and (C). Finally, these two surfaces are joined into the final surface, including the inlet and outlet patches (Figure 4 (D) and Figure 4 (E)).

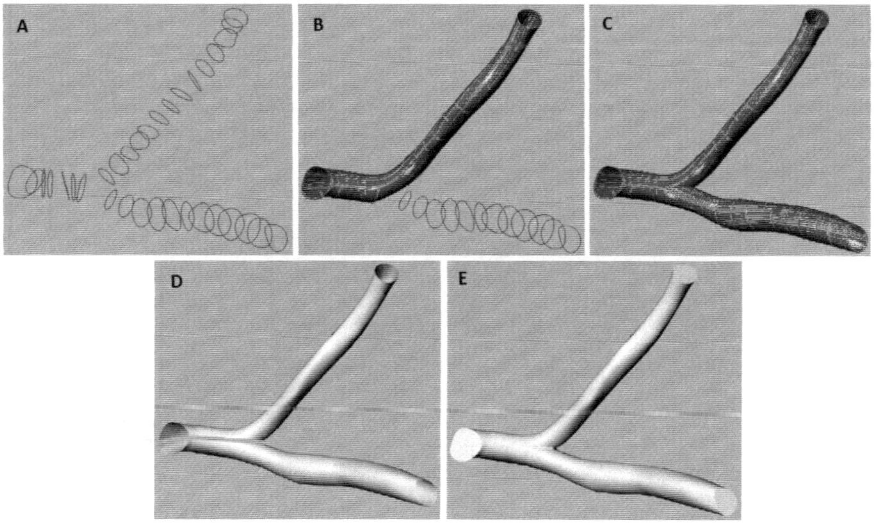

Figure 4: (A) Spline Curves representing the contours of the LCA. (B) Lofting of one side of the arterial bifurcation. (C) Lofting of the other half. (D) Trimming the surfaces leaving an empty space between both halves. (E) Blending the bifurcation and generating a closed polysurface.

2.2. Mesh generation and refinement

For generating the mesh we used Netgen® software designed by Schöberl J.. The mesh size and granularity were set so as to obtained a useful mesh for our investigation. In particular, the mesh size was set smaller in the proximity of pronounced curves. The mesh was then exported to enGrid® software. Here we created a prismatic element boundary layer in the vicinity of the arterial wall and we divided it in order to get a final outcome with a very thin boundary layer. This is the key point for an accurate calculation of the WSS. The obtained mesh was exported to the OpenFoam format.

2.3. Computational Fluid Dynamics

Blood flow through the LCA was simulated using the finite element code OpenFoam (OpenCFD Ltd.). A SIMPLE solver was used to solve the governing three-dimensional Navier-Stokes equations to obtain the velocity vector $\vec{v}$ and pressure p in every point of the mesh. Flow was assumed to be steady and laminar. Blood was considered as a homogeneous Newtonian fluid with a dynamic viscosity of μ=0.0035 Pa×s and incompressible with a density of ρ=1050 kg/m3. The vessel wall was modeled as rigid (Salvucci et. al., 2010).

The simulation was done setting p in the LCX outlet patch to zero and imposing $\vec{v}$ in the inlet and LAD outlet patches with a surface normal planar velocity profile. Flow splitting was calculated according to Murray's cubic law (Soulis et. al., 2006). Free traction condition was set to the LCX outlet patch and $\vec{v}$ was set to zero at the vessel walls according to the non-slip condition.

Once the simulation was performed we computed WSS from the gradients of the $\vec{v}$ fields near the wall:

$$WSS = \left| \mu \frac{d\vec{v}}{d\vec{n}} \right| \tag{1}$$

where $\vec{n}$ is the vector normal to the wall in each point.

3. Results and Discussion

In Figure 5 we show the mesh obtained following the procedure described in the previous section. Particularly, in (A) we present the mesh as it was created and refined with Netgen® and in (B), the final result, after generating and dividing the prismatic boundary layer in the vicinity of the arterial wall with enGrid®. This was the mesh used for running the simulation.

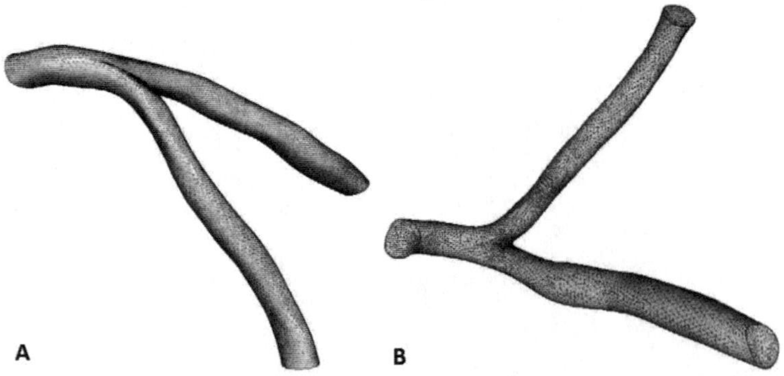

Figure 5: (A) Mesh obtained with Netgen®. (B) Final mesh used for the simulation with a thin prismatic boundary layer generated with enGrid®.

The results of the simulation run with OpenFoam are exhibited in Figure 6. The first panel (A) shows the pressure values computed in this LCA reconstruction. OpenFoam uses a "cinematic pressure", this is the pressure divided by the fluid density. These relative values are all calculated taking into account that pressure was set to zero in the LCX output as the boundary conditions were defined. In panel (B) a velocity vector graph is shown. Finally, in panel (C), the WSS distribution is presented. As expected, regions of very low WSS were found in the outer walls of the bifurcation (Caro et. al., 2009; Soulis et al., 2006). Furthermore, other zones of low WSS were detected in the three branches in the inner wall of curvatures as reported by others (Malek et. al., 2007).

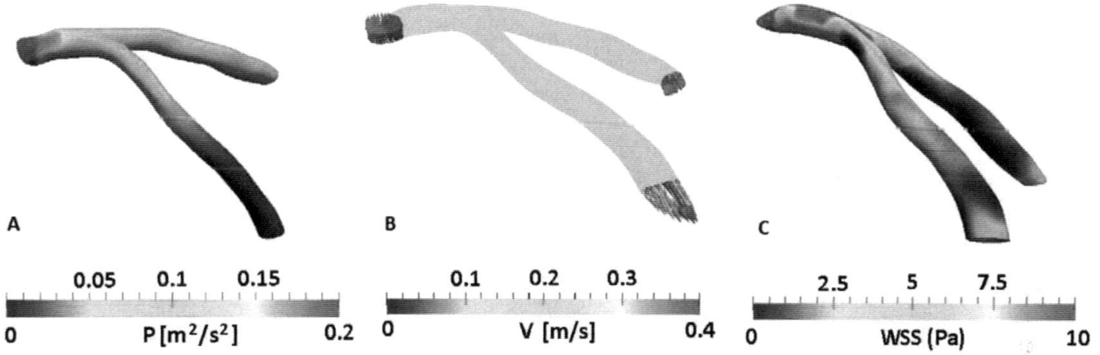

Figure 6: Blood Flow simulation results: (A) Pressure, (B) Velocity and (C) WSS.

One of the main limitations of this work is that it considers the flow as stationary and, in consequence, it doesn't include the effects caused by the oscillations. An accurate determination of the WSS is obtained when combining the pulsatile (unsteady) nature of the arterial blood flow with the complex geometric configuration of the coronaries (Chatzizisis et al., 2007). Furthermore, this investigation doesn't take into account the interaction of the blood flow with the vessel's wall, which plays an important role in the appearance and development of the plaques (Malek et. al., 2007). Besides, here were consider only a small segment of the coronary arterial tree, not looking at the effects that may bring the surroundings.

In spite of these limitations, we show that coronary arteries can be reconstructed in vivo from MSCT images and flow profiles can be simulated with realistic results. User intervention was minimized and computational cost was reduced. The time spent for the segmentation and 3D surface reconstruction was about 30 minutes, and for configuring the simulation, 5 minutes. The time that takes the simulation to run is highly variable and depends on the particular features of the computer, the characteristics of the mesh and the CFD configuration. In this case, using an average laptop, it took about 3 hours. All these results indicate that we walk towards a true 3D individual characterization of the coronary tree geometry and WSS profile that will be applied to better understand atherosclerosis progression.

4. Conclusion

In the present work we present a platform and a methodology to perform a 3D surface reconstruction from CT images of LCA and generate a finite element volume mesh of it. The whole process itself has many automatic steps but it requires also a lot of user interaction. This is what guaranties a reliable geometry.

We studied a specific case, performed the corresponding reconstruction and used the resulting mesh to perform a FEM simulation. The pressure, velocity and WSS obtained correspond, qualitatively, to the expected according to the bibliography. Moreover, regions of low WSS where found in the predicted zones.

Concerning the limitations of this work, further investigations should be done considering a pulsatile flow, it's interaction with the vessel's wall and taking into account a bigger portion of the coronary arterial tree, not just the first bifurcation.

5. Acknowledgment

We would like to thank Dr. Juan Gabriel Barra for his support through the *PICTO project 31355 (Préstamo BID 1728 OC/AR).*

References

[1] Bourke P., "Polygonising a scalar field", 1994, http://paulbourke.net/geometry/polygonise/.

[2] Calvo Plaza F.J., Simulación del flujo sanguíneo y su interacción con la pared arterial mediante modelos de elementos finitos - Tesis Doctoral, *Universidad Politécnica de Madrid*, 2006.

[3] Caro C.G., Discovery of the role of wall shear in atherosclerosis, *Arterioscler Thromb Vasc Biol*, 29, 158-161, 2009.

[4] Chaichana, T., Zhonghua Sun, Jewkes, J., Haemodynamic Effect of Coronary Angulations on Subsequent Development of Coronary Artery Disease: A Preliminary Study, *2010 Sixth IEEE International Conference on e-Science Workshops*, 39-43, 7-10, doi: 10.1109/eScienceW.2010.16, 2010.

[5] Chatzizisis Y.S., Coskun A.U., Jonas M., Edelman E.R., Feldman C.L. and Stone P.H., Roleof endothelial shear stress in the natural history of coronary atherosclerosis and vascular remodeling: molecular, cellular, and vascular behavior. *J Am Coll Cardiol*, 49: 2379–2393, 2007.

[6] Craiem D., Casciaro M.E., Graf S., Glaser C.E., Gurfinkel E.P., Armentano R.L., Coronary Arteries Simplified with 3D Cylinders to Assess True Bifurcation Angles in Atherosclerotic Patients, *Cardiovasc Eng,* DOI 10.1007/s10558-009-9084-1, 2009.

[7] Craiem D., Casciaro M.E., Graf S., Gurfinkel E.P., Armentano R.L., Non-invasive assessment of allometric scaling laws in the human coronary tree, *Artery Research*, 5, 15-23, 2011.

[8] Davies P.F., Hemodynamic shear stress and the endothelium in cardiovascular pathophysiology, *Nat Clin Pract Cardiovasc Med*, 6, 16–26, 2009.

[9] Malek A. M., Alper S. L. and Izumo S., Hemodynamic shear stress and its role in atherosclerosis, *Journal of the American Medical Association*, 282, 2035-2042, 1999.

[10] Naghavi M. et al., From vulnerable plaque to vulnerable patient: a call for new definitions and risk assessment strategies: Part I, *Circulation*, 108(14), 1664-72, 2003.

[11] Palágyi K., Sorantin E., Balogh E., Kuba A., Halmai C., Erdohelyi B., and Hausegger K., A sequential 3D thinning algorithm and its medical applications, *Lect Notes Comput Sci*, 2082: 409–415, 2001.

[12] Salvucci F.P., Perazzo C.A., Gurfinkel E., Armentano R.L. and Barra J.G., A patient-specific method for the evaluation of wall shear stress in human coronary arteries. *Conf Proc IEEE Eng Med Biol Soc*, 3788–3791, 2010.

[13] Salvucci F.P., Perazzo C.A., Salles S., Barra J.G., Armentano R.L., Influence of inlet conditions in wall shear stress distributions of left coronary arteries in patient specific simulations. *Asociación Argentina de Mecánica Computacional*, 29, 5935-5960, 2010.

[14] Soulis J.V., Farmakis T.M., Giannoglou G.D. and Louridas G.E., Wall shear stress in normal left coronary artery tree, *Journal of Biomechanics*, 39, 742–749, 2006.

[15] Steinman D.A., Image-Based Computational Fluid Dynamics Modeling in Realistic Arterial Geometries, *Annals of Biomedical Engineering*, 30, 483–497, 2002.
[16] Wendt J.F., Anderson Jr. J.D., Degroote J., Degrez G., Dick E., Grundmann R. and Vierendeels J., Computational Fluid Dynamics, An Introduction, *Springer-Verlag*, 2009.

SABI 2011 IOP Publishing
Journal of Physics: Conference Series **332** (2011) 012049 doi:10.1088/1742-6596/332/1/012049

Model of avascular tumor growth and response to low dose exposure

J.M. Rodriguez Aguirre[1,2] **and E.R. Custidiano**[1]
[1]Departamento de Física. Facultad de Ciencias Exactas y Naturales, y Agrimensura. Universidad Nacional del Nordeste, Avda. Libertad 5600, 3400 Corrientes, Argentina
E-mail: juakcho@gmail.com , cernesto@exa.unne.edu.ar

Abstract. A single level cellular automata model is described and used to simulate early tumor growth, and the response of the tumor cells under low dose radiation affects. In this model the cell cycle of the population of normal and cancer cells is followed. The invasion mechanism of the tumor is simulated by a local factor that takes into account the microenvironment hardness to cell development, in a picture similar to the AMTIH model. The response of normal and cancer cells to direct effects of radiation is tested for various models and a model of bystander response is implemented.

1. Introduction

The main goal of radiotherapy is the targeting of malignant cells with ionizing radiation to kill them without affecting the neighboring normal tissue. In this sense computational simulations can help to increase the effectiveness of radiotherapy. To achieve reliable results the response of cells to direct and indirect effects of ionizing radiation should be taken into account, and the growth and invasion of tumors should be accurately reproduced, simultaneously.

Mathematical modeling of tumor growth and tumor response to radiation has increased our knowledge in this field. In this path several models have been proposed to simulate early avascular tumor growth [1, 2]. The discrete approach of cellular automata have been developed using one[3], two [4] or more levels [5, 6]. The main reason behind the multi level cellular automata simulation rests in the disparity in characteristic diffusing time of the chemical species involved in the cells metabolic processes. Nevertheless a one level cellular automata model is able to simulate observed growth behavior of avascular tumors with a low number of parameters [3]. This simplified model requires more attention in the interpretation and meaning of the parameters involved, but his simplicity gives some insights in the underlying invasion mechanisms. The Acid mediated tumor invasion hypothesis (AMTIH) is based on the acidification of the microenvironment that the proliferating cancer cells generate [5]. The *growing potential* factor used by [3] can be interpreted as a crude approach of the AMTIH mechanism.

In recent years the radiobiological attention has been focused in the observed radiation induced bystander effect (RIBE). The development of microbeam irradiation techniques [7] have expanded our knowledge in this phenomenon. In this work we proposed a new model of CA that can reproduce the expected growing behavior of avascular solid tumors and, simultaneously reproduce the low dose region response of cells to ionizing radiation.

[2] Scholarship supported by CONICET

2. Model

The model is set as a bidimensional grid with constant grid size, and one to one correspondence between physical cells and sites. Each grid element is represented by a vector state: $(ns, np, g)_{i,j}$. Where i and j are the position coordinates of the cell in the grid . ns is the cell state, that can be in one of four possible states: empty, normal, cancer or necrotic. np is the cell phase in his normal cycle, the normal and cancer cells can be in one of the four possible states (i.e. G_1 , S , G_2 M). And g is the growth factor. This later value is intended to represent the influence of the medium to cellular viability as a global weight that takes into account all the chemical fields involved in cellular development. In this sense the g factor is interpreted as single factor enclosing the acidification of the microenvironment by lactose production of active cells and the depletion of useful vital resources (i.e. glucose and oxygen) in the neighborhood of the cell. Because of the correlation between waste products and consumption of resources, these two fields are treated in one single factor. Normal cells are autoregulated and are less active than the cancer ones, so his g factor remains in the basal value $g_0 = 10^{-H_0}$ (with H_0=7.4), but the cancer cells with increased activity consume more resources and originates a rise in the waste products, so the sites occupied by cancer cells continually increases the local g.

2.1. Cell cycle

The CA cycle is set to last one hour. And the cell cycles comprised: gap 1 (G_1), Synthesis (S), gap 2 (G_2) or mitosis (M). The last three phases have constant duration and the G_1 phase duration is probabilistic. Because of the deterministic nature of the cell evolution beyond G_1 phase, the probability of transition from G_1 to S phase is identified with the probability of mitosis.

The probability of mitosis is calculated as a global measure of the number of cancer and normal cells, according to a logistic like equation:

$$p_m = p_0 \left(1 - \frac{N_t}{K}\right) \tag{1}$$
$$N_t = 2N_c + N_n$$

Where: p_m is the mitotic probability, p_0 is the basal probability and K is the carrying capacity that is set as the total number of cells in the grid. To take into account the greater activity of the cancer cells, for which the glucose consumption rate is observed to be of the order of 5 to 10 times the normal cells consumption rate [8], the total number of cancer cells N_c is double weighted, with respect to the number of normal cells N_n. This mechanism depends on the dynamics and on the number of cancer and normal cells, but in order to enhance the effect of this global feedback inhibition mechanism, a local influence of the cancer cells is considered through the g factor. In every site the mitotic probability is obtained by the expression:

$$p_{i,j} = p_m.e^{(H_o - H_{i,j})} \tag{2}$$
$$H_{i,j} = -log(g_{i,j})$$

Where i, j are the index numbers of the site in the grid, and $g_{i,j}$ is the growth factor of the i,j site.

2.2. Cellular automata rules

All the elements of the grid are updated in each cycle of the simulation, in random order to avoid artifacts that can arise from the scanning method. Each cell can change its vector state according to the next local rules:

(i) Normal cells: the mean growth factor $\bar{g}$ is calculated in the Moore neighborhood [9] of the cell $\bar{g} = \sum_{i,j} g_{i,j}/N$, with i,j index changing in all the neighbor sites. If the value $H_{i,j} = -log(\bar{g})$ is greater than the surviving threshold value for normal cells (H_n) then the cell remains in normal state, with $g_0 = 10^{-H_0}$. Otherwise the cell is incapable of sustain his vital activity and becomes unviable, so the site changes to empty site with $g = g_0$. If the local $H_{i,j} > H_n$ and there is at least one empty site in the Moore neighborhood of the cell, then his phase state is increased.
The phases in the cycle have constant duration with the only exception of the G_1 phase that has a minimum duration, but the transition to S phase is probabilistic, with the probability given by equations 1 and 2. If there is any free site then the normal cell is autoregulated and remains in G_1 state and keeps his local value of g_0. When the cell reaches the end of his cycle, and there is at least one empty site, then mitosis occur and one randomly chosen empty site is used for replication. In mitosis the mother cell is eliminated and one of the daughters is placed in the mother site, and the other in the empty adjacent site, both of them at the beginning of G_1 state.

(ii) Cancer cells: if there is, at least, one neighbor site in *empty* state, then the cancer cell cycles in the same manner as the normal cells. Mitosis proceeds in the same way as in the normal cells, previously described. The local g value is increased in every loop of the simulation. In order to take into account the nutrients depletion, and the related waste production rate, the local $H_{i,j}$ is increased. The increment is given by: $\Delta H_c = \Delta H_{c0}.p_m$. Where p_m is given by equation 1, and ΔH_{c0} is the maximum increment allowed for cancer cells.
If all the neighbors are occupied and the local $\bar{g}$ is lower than a given threshold H_c then the cell becomes necrotic. In the same situation of all the neighborhood filled when the local $\bar{g}$ reaches a value below a threshold H_q but is above H_c, then the cells enter in a quiescent state and his increment of g is lowered to $\Delta H_q = \Delta H_q/100$ [5].

(iii) Necrotic cell: In this state the local $g_{i,j}$ remains constant (his value depends on the g-value of the cell that originated the necrotic site). If a necrotic cell has one or more neighbors that are empty site, then it can change to empty site with probability $1/2T$, where T is the total period of cell cycling.

(iv) Empty site: an empty site can be filled with mitotic cancer or normal cells in the mechanisms described above. In the case of competing normal and cancer cells, the cancer cells have priority to fill the empty site.

2.3. Radiation response

In order to take into account the hyper radio-sensitivity (HRS) effects and the increased radio resistance above 0.5 Gy [10] two set of parameters in the lineal quadratic model are used. The values are the same as used by [11] for glioma cells and are reproduced here only for completeness. Each phase of the cells have his surviving curve given by the parameters of the table 1.

Table 1. Parameters used to calculate the surviving fraction of irradiated glioma cells from reference [11]. Mean value and standard deviations refer to threshold that triggers repair process.

Phase	α_s	α_r	β	Mean	Standard deviation
G_1	0.615	0.03	0.038	0.508	0.33
S	6.803	0.266	-0.012	0.113	0.264
G_2/M	0.329	0.048	0.043	0.429	0.03

The surviving fraction is computed with:

$$Sf(d) = e^{-(\alpha_\mu + \beta d)d} \tag{3}$$

Where the μ index refer to the low or high dose response. The region of dose response depends on the particular threshold value that every cell in the population has. This threshold is distributed among the population with mean and standard deviation given in table 1. As a consequence of radiation damage the cells affected with the probability given by equation 3 can be killed with two different mechanisms: a) if the DNA damage is severe enough the cell suffer apoptosis in a process known as Radiation induced interphase dead (RIID), in this case the cell changes to empty site immediately, with probability 0.1 . b) The cell DNA can be affected in a way that his mitotic daughters are not viable any more, this happens in Radiation induced mitotic dead (RIMD) with 0.9 probability. To account for the later mechanism of cell killing an extra set of parameters should be added to the vector state of each cell. The cells marked to be killed in RIMD are treated as usual in the simulations, but at the end his M phase booth daughter cells suffers apoptosis and are replaced for empty sites in the grid.

2.4. Bystander

Non-targeted effects of ionizing radiation are caused by intercellular chemical signaling that diffuse in the media; they include production of DNA damage and alterations in cell apoptosis, differentiation, senescence or proliferation. The bystander phenomenon results from signaling molecules that rapidly propagate from irradiated cells and decrease in concentration as distance increases [12]. The bystander effect is modeled as a global effect that can kill any cell in the grid that have survived, or have never been exposed to radiation. The signal variation of intensity with distance is not modeled in this work, because of the distances involved in our simulations. Exposed cells becomes sources of the bystander signal, that can last in the medium 4 days in our simulations [13, 14]. If a cell survives to the first bystander effect it becomes immune to the signal. This mechanism was simulated according to the next rules:

(i) Cells that have been directly exposed to radiation and survives can emit an adimensional bystander signal with probability:

$$p_b = 1 - e^{-d/d_l} \tag{4}$$

with d_l chosen to be 0.1 Gy in accordance to the observations from [7] for 278 eV C_k x-ray beam over V79 cells. The bystander signal is approximately linearly dependent from the dose and saturates at 0.3 Gy of exposure.

(ii) The proportion of cells exposed to the signal that were killed is calculated by the expression:

$$df(s) = df_{max}(1 - e^{10.s(t)}) \tag{5}$$

where s(t) is the concentration of the signal in the medium and df_{max} is the maximum proportion of killed cells, for the current simulations this parameter where chosen to be 0.1 for all cell phases.

(iii) The cells exposed to the bystander signal that survives becomes immune to it. But can become secondary sources of bystander signal with probability given by equation 4.

3. Results

The parameters for the model were chosen in accordance with the typical tumor growth observations of spheroids in vivo and in vitro. The cell cycle duration widely vary between different kind of cells, for the purposes of this work we consider a typical duration of cell cycle in the order of experimental observations [15]. G_1 duration is of 12 hours, S phase duration is of 7 hours, G_2 phase duration is of 4 hours and M phase last 1 hour, to give a 24 hs cell cycle period. The same cycle phases duration is used for normal and cancer cells [15] as a valid simplification for the purposes of this work .

3.1. validation of the model and parameters: tumor growth

The general model for the ontogenetic growth of living organisms of West et al [16, 17] have been successfully translated to the growth of solid malignant tumors in-vitro and in-vivo [18]. This general or *universal* model rests in the simple expression:

$$r = 1 - e^{-\tau} \tag{6}$$

Where r and τ are the adimensional variables:

$$r = \left(\frac{m}{M}\right)^{0.25} \qquad \tau = \alpha.r_0.t - ln(1 - r_0) \qquad r_0 = \left(\frac{m_0}{M}\right)^{0.25}$$

Where m_0 is the initial tumor mass and M is the final mass of the tumor. In this work we are assuming the following equivalence between mass rate and cell number rate as a valid approximation:

$$r = \left(\frac{n}{N}\right)^{0.25} = \left(\frac{m}{M}\right)^{0.25}$$

Where n is the number of cancer cells and N is the asymptotic final number of cancer cells. The parameter α is approximately constant in a given taxon and proportional to the organism metabolic rate across taxa [16]. In the case of solid tumors α is supposed to be related with tumor particular characteristics as its invasion ability [18]. This parameter is obtained from the linear regression:

$$ln(1 - r) = t.\alpha.r_0 - ln(1 - r_0) \tag{7}$$

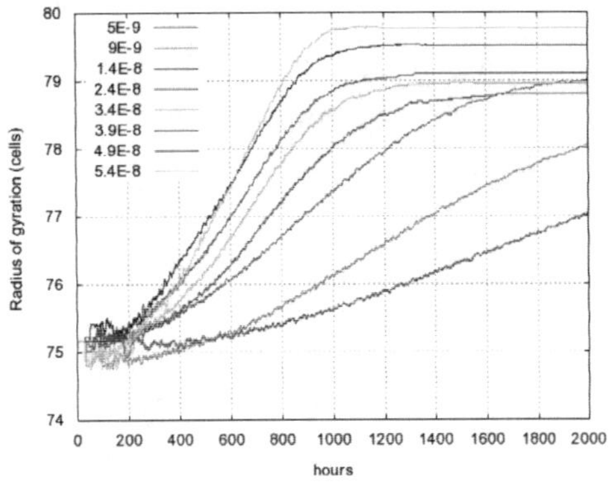

Figure 1. Radius of gyration of tumor for different values of ΔH_{c0}. The reference of colours is shown in the picture.

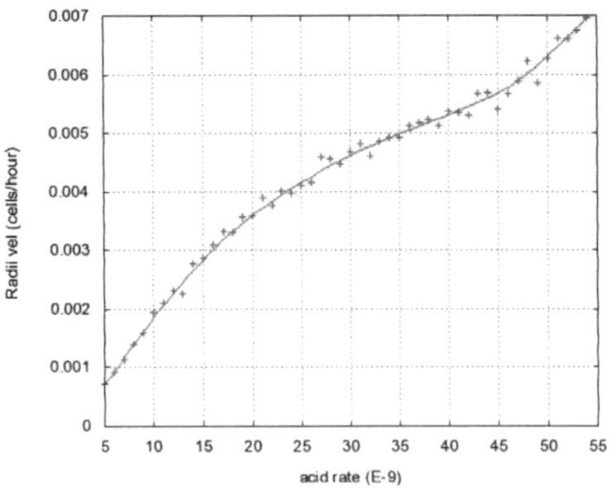

Figure 2. Mean velocity of the radius of gyration vs ΔH_{c0}. Simulations are the red cross and a smoothing line is set as visual reference.

This growth behavior is used as selective criterion for adjusting the parameter ΔH_{c0}. This parameter is directly related with the growing rate of the radius of gyration of the tumor in

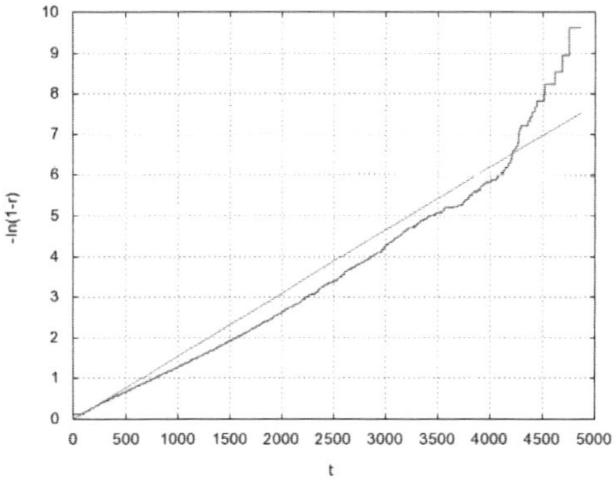

Figure 3. Linear regression (green line) of the reduced variables r and τ (red line) to obtain the α parameter of equation 7.

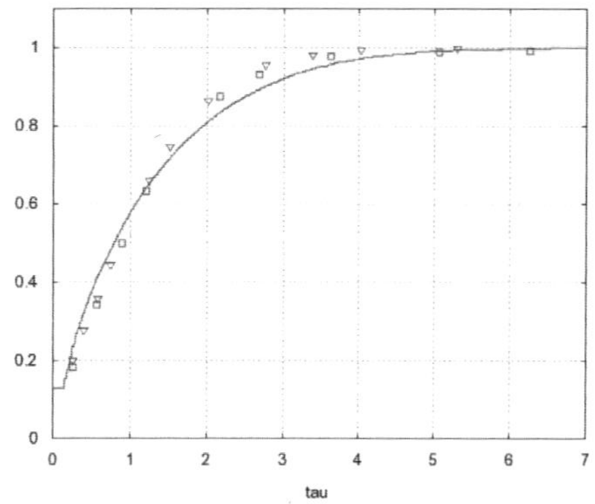

Figure 4. Relation of reduced variables r and τ of the present simulations. Experimental *in vivo* data from reference [19] (green ∇) and [20] (purple $\square$) (extrapolated values from [18]).

our simulations (see figure 1 and 2). The radius of gyration is defined as the square root of the second moment of the spatial distribution.

The best value for the parameters where chosen to be $\Delta H_{c0} = 8.10^{-9}$ and the values for quiescent threshold $H_q = 6.04$. The values of the remaining parameters where set according to the pH of serum acid concentration and threshold surviving values of pH of normal and cancer cells [5] $H_0 = 7.4$, $H_c = 6.0$ and $H_n = 7.1$. These values where used in all the present simulations.

3.2. validation of the model and parameters: surviving fraction of V79 cells
In order to test our simulations against available experimental data the radiation response of V79 cells to 250 KeV X rays were simulated with the multitarget model:

$$df(d) = \left(1 - e^{-d/d_0}\right)^n \tag{8}$$

Where the parameters are n=2 , and $d_0 = 1/0.44$ [21]. Simulations where carried out with randomly distributed cells in an empty grid. The entire grid where irradiated once and the cells where allowed to cycle during 72 hours. Then the surviving fraction were calculated as the rate $s_f = n_s/n_0$ of the surviving population n_s against the control unirradiated population n_u. Simulations were run with and without bystander response. The increased surviving fractions of the series without Bystander effect separates the simulated results from the experimental data (see figures 5 and 6).

4. Discussion and concluding remarks
The present model is the first one level cellular automata model that can reproduce cellular cycling and tumor invasion mechanisms of normal tissue. A simplified model of tumor invasion through resource depletion and acidification of microenviroment is proposed, inspired in the AMTIH hypothesis. Is expected that original vascularization of invaded tissue and neovascularization of growing tumor were crucial for tumor growth in more advanced stages, so the present model should be limited to avascular tumor growth stages. The response of cells to low radiation exposure is well reproduced and a simpler bystander mechanism is proposed. Further investigations and experiments are needed to fully incorporate RIBE for

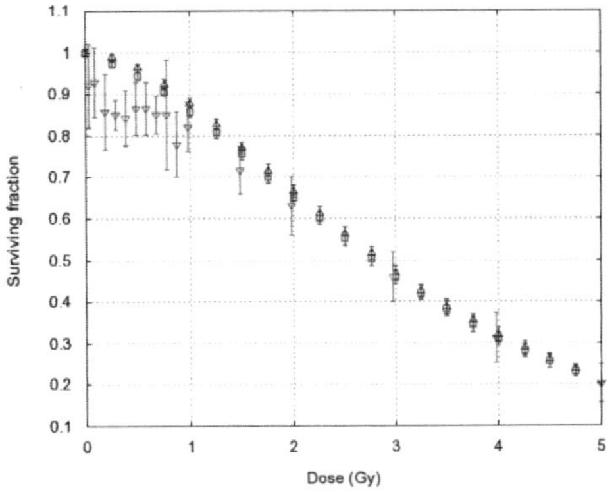

Figure 5. Surviving fraction of simulated V79 cells cultured in vitro, irradiated with 250 KeV x-rays using the multitarget model (eq. 8). With RIBE : purple □, without RIBE: blue △. Experimental values from [22] (green ▽).

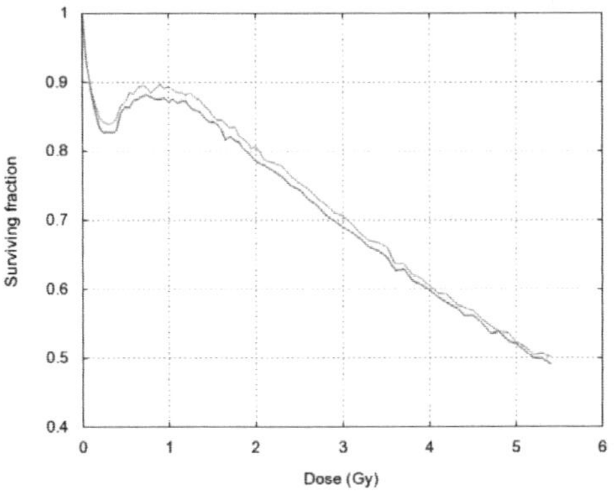

Figure 6. Surviving fraction of simulated Glioma cells using equation 3 with parameters given in table 1. With RIBE (red line) and without RIBE results (green line).

different radiation types and modulation of Bystander effect with cell cycle. The model allows to taking into account the cell cycle response to non direct and direct radiation effects.

Acknowledgments

The authors are gratefull with *Secretaria General de ciencia y Tecnica (U.N.N.E.)* for his financial support. One of the Authors (JMRA) acknowledges support from Consejo Nacional de Investigaciones Científicas y Técnicas (CONICET), Argentina.

References

[1] Jiang Y, Pjesivac-Grbovic J, y Charles Cantrell and Freyer J P 2005 *Biophysical Journal* **89** 3884
[2] Junior S F, Martins M and Vilela M 1998 *Physica A* **261** 569
[3] Reis E, Santos L and Pinho S 2009 *Phisica A* **388** 1303
[4] Richard M, KJKirkby, Webb R and Kirkby N 2007 *Nucl. Instr. and Meth. in Phys. Res. B* **255** 18
[5] Patel A A, Gawlinski E T, Lemieux S K and Gatenby R A 2001 *J. Theor. Biol.* **213** 315
[6] Alarcon T, Byrne H M and Maini P 2003 *J. Theor. Biol.* **225** 257
[7] Schettino G, Folkard M, Michael B D and Prise K M 2005 *Radiation Research* **163** 332
[8] Kallinowski F, Vaupel P, Runkel S, Berg G, Fortmeyer H, Baessler K, Wagner K, Mueller-Klieser W and Walenta S 1988 *Cancer Res.* **48** 7264
[9] Wolfram S 2002 *A New Kind of Science* (Canada: Wolfram media)
[10] Joiner M, Marples B, Lambin P, Short S and Turesson I 2001 *Int. Radiat. Oncol. Biol. Phys.* **49** 379
[11] Richard M, KJKirby, Webb R and Kirby N 2009 *Applied Radiation and Isotopes* **67** 443
[12] Prise K M, Schettino G, Folkard M and Held K D 2005 *Lancet Oncol* **6** 520
[13] Belyakov O V, Mitchell S A, Parikh D, Randers-Pehrson G, Marino S A, Amundson S A, Geard C R and Brenner D J 2005 *Proc. Natl. Acad. Sci. USA* **102** 14203
[14] Xia J, Liu L, Xue J, Wang Y and Wu L 2009 *Nucl. Instr. and Meth. in Phys. Res. B* **267** 1015
[15] Gordon R E and Lane B P 1980 *Cancer Research* **40** 4467
[16] West G B, Brown J H and Enquist B J 2001 *Nature* **413** 628
[17] Brown J H, West G B and Enquist B J 2005 *Functional Ecology* **19** 735
[18] Guiot C, Degiorgis P G, Delsanto P P, Gabriele P and Deisboeck T S 2003 *J. Theor. Biol.* **225** 147
[19] Yorke E, Fuks Z, Norton L, Whitmore W and Ling C 1993 *Cancer Res.* **53** 2987
[20] Norton L 1988 *Cancer Res.* **48** 7067
[21] Richard M, Webb R, KJKirby and Kirby N 2009 *Applied Radiation and Isotopes* **67** 440
[22] Schettino G, Folkard M, Prise K M, Vojnovic B, Held K and Michael B D 2003 *Radiat. Res.* **160** 505

Estimating RASATI scores using acoustical parameters

P D Agüero, J C Tulli, G Moscardi, E L Gonzalez and A J Uriz
Facultad de Ingenieria - Universidad Nacional de Mar del Plata
E-mail: pdaguero@fi.mdp.edu.ar

Abstract. Acoustical analysis of speech using computers has reached an important development in the latest years. The subjective evaluation of a clinician is complemented with an objective measure of relevant parameters of voice. Praat, MDVP (Multi Dimensional Voice Program) and SAV (Software for Voice Analysis) are some examples of software for speech analysis. This paper describes an approach to estimate the subjective characteristics of RASATI scale given objective acoustical parameters. Two approaches were used: linear regression with non-negativity constraints, and neural networks. The experiments show that such approach gives correct evaluations with ± 1 error in 80% of the cases.

1. Introduction

In the latest years the acoustical analysis of speech has reached an important development thanks to the progress of computers. The main advantage of computer analysis of speech is the non-invasive and objective assessment of the voice.

The human auditory system is one of the main obstacles in the perceptual diagnostic of voice by the clinician ear. Humans are fundamentally prepared to perceive the voice as a whole, which is particularly advantageous from the point of view of linguistic communication. However, this ability is limited when it is necessary to individualize relevant aspects from a clinical perspective.

It is often difficult to determine the origin of certain anomalies of the voice using a perceptual procedure. For example, Baken et al. [1] show that some aspects of the pitch are more related to resonant frequencies of the vocal tract rather than to the frequency of vibration of vocal chords. The hypernasality of voice can be a consequence of the desynchronization in the timing of velar occlusion instead of an incomplete occlusion. Hence, the same attribute or alteration of the vocal quality may have its origin in different subsystems which can not be easily isolated with the audition of an expert.

In other cases, an adequate perception can not be quantized with the degree of precision of a numerical measure. For example, it is possible to measure the degree of breathiness of a breathy voice through the corresponding speech parameter, the index of turbulence of voice (or VTI). In this way, the subjective evaluation of a clinician is complemented with an objective measure of relevant parameters of voice. As a consequence, the objectivity of the report is enhanced, and it is possible to measure the degree of progress more accurately.

Validity and reliability of acoustic analysis performed with different tools is affected by many factors. These include microphone type, noise levels, data acquisition system, sampling rate and software used for analysis [4, 5]. Ostensibly, the values of the commonly used frequency and amplitude perturbation measures should not be dependent on the software used to obtain them. Jitter and shimmer, for example, are defined by relatively simple and standardized formulas [2]. The differences observed between numerical values obtained for these measures using different softwares apparently stem from the raw fundamental frequency (f_0) data on which these calculations are based. Despite the basic nature of this parameter, there is no standardized algorithm to calculate f_0, which has been adopted and implemented by all programs.

While different methods for calculating f_0 may yield relatively small differences in the mean value of f_0, they may influence the perturbation measures to a far greater extent. This introduces a difficulty for the clinical voice specialist, because different programs which are available for conducting voice analysis could report different values when analyzing identical voice samples. Moreover, it is not clear whether normative data which are presented by specific software (e.g., the data used for the radial graph in Multi-Dimensional Voice Program, or MDVP) are comparable with values obtained in other programs. This possible discrepancy between the results obtained by different programs was previously noticed and addressed by various researchers [8, 4, 12, 6].

In the voice analysis of speech disorders exists a gap between the subjective and objective measures. On the one hand, objective measures are based on time and frequency calculations. On the other hand, subjective scores are based on scales with more complex concepts that have a correlate with objective parameters. One example of a subjective criterion is the RASATI scale, which considers aspects such as hoarseness(R), rough (A), breath (S), asthenic (A), strain (T) and instability (I). In that scale the severity of a pathology is measured using a zero to three discrete scale.

This paper explores the relationships between the perturbation measures and the subjective scale RASATI. The main goal is to estimate the values of the subjective evaluation given objective measures, such as jitter, shimmer, HNR, etc. In this way, it would be possible to trace the progress of a patient using more human perceivable factors, such as hoarseness or strain.

This paper is organized as follows. Sections 2.1 and 2.2 describe the objective and subjective approach to assess voice quality in speech disorders, and Section 2.3 depicts the proposed methodology to find the relationships between objective and subjective scores. Section 3 shows the experimental results with the acoustic parameter calculated by SAV and PRAAT. Finally, conclusions and future work are drawn in Section 4.

2. Evaluation of speech disorders

The evaluation of speech disorders can be performed using both subjective and objective scores. The former use some ratings to measure different aspects of the voice quality. On the other hand, objective measures use acoustic parameters obtained with different computer algorithms to grade the voice pathology.

2.1. Subjective evaluation of speech disorders

Proposed by Hirano [7] and accepted as standard by the Japanese Society of Logopedics and Phoniatrics and the European Group on the Larynx, the GRBAS scale comprises five qualitative characteristics: Grade of dysphony (G), Roughness (R), Breathiness (B), Asthenicity (A), and Strainess (S). For each one, a value in the range 0-3 is considered, where 0 corresponds to healthy voice, 1 to light disease, 2 to moderate and 3 to severe. Despite some limitations, GRBAS is simple and fast, and has a good correlation with some acoustic parameters [11].

The severity of hoarseness is quantified under the parameter G (Grade) integrating all deviant components. Two main components of hoarseness can be identified: Breathiness (B), which is the audible impression of turbulent air leakage through an insufficient glottal closure, and it may include short aphonic moments (unvoiced segments); and Roughness (R), which is an audible impression of irregular glottic pulses, abnormal fluctuations in F0, separately perceived acoustic impulses (as in vocal fry), and includes diplophonia and register breaks [11].

These two parameters have shown sufficient reliability (inter and intra observer reproducibility) when used in a current clinical setting [3]. The behavioral parameters A (Asthenicity) and S (Strain) are commonly less reliable and sometimes are omitted from the basic protocol. R and B features are associated to organic lesions in which there is a lowering of vibration (R) and default of closure (B), whereas features A and S are associated to functional disorders, related with vocal tiredness (A) and hyperphonic emission (S) [11].

The GRBAS evaluation is usually carried out based on continuous or conversational speech. However, sometimes it is approached by means of sustained vowels, although there are studies demonstrating that the results might differ depending on the material used [10]. They conclude that the evaluation from sustained vowels is less severe (i.e. dysphony is underestimated) than that carried out from continuous speech, especially in those patients with severe dysphony. The same study calls the attention over the variability of each of the five GRBAS parameters. The most consistent parameter is G, whereas scales A and S demonstrated a strong variability, due to the fact that these concepts are more complex to evaluate, even by a human expert [11].

The RASATI scale is the acronym proposed by Pinho et al. [9] to replace the English acronym GRBAS, and incorporates another factor named instability. Some authors consider that Instability corresponds to the tremor of the structure of the vocal tract, and must not be included in the analysis of alterations in the glottal source. The RASAT scale (not including Instability) is the standard adopted by the "Sociedad Argentina de la Voz" to measure voice quality.

2.2. Objective evaluation of speech disorders

The literature on voice analysis reveals that one or two voicing parameters alone, such as jitter and shimmer, are not sufficient to accurately describe an aberration in a patient's voice. Jitter values may be within normal limits in a patient who demonstrates a breathy voice quality, and periodic modulation over many glottal periods (tremor) should be differentiated from cycle-to-cycle modulation. Similarly, turbulence caused by incomplete glottal closure can contribute a different type of "noise" compared to noise from aperiodic vibration; and, longterm periodic modulation of amplitude (amplitude tremor) may have physiological causes that differ from those of long-term periodic modulation of frequency. The analysis of voice requires a multi-dimensional approach which is followed by many software applications. Examples of applications that agree with this direction are the Multi-Dimensional Voice Program (MDVP) of KayPentax, and the Software for Voice Analysis (SAV).

Some of the parameters estimated by MDVP are: Jita (Absolute Jitter), Jitt (Jitter percent), RAP (Relative Average Perturbation), PPQ (Pitch Perturbation Quotient), sPPQ (Smoothed Pitch Perturbation Quotient), vFo (Fundamental frequency variation), ShdB (Shimmer in decibels), Shim (Shimmer percent), APQ (Amplitude Perturbation Quotient), sAPQ (Smoothed Amplitude Perturbation Quotient), vAm (Peak-to-Peak Amplitude Variation), NHR (Noise Harmonic Ratio), VTI (Voice Turbulence Index), SPI (Soft Phonation Index), FTRI (Fo-Tremor Intensity Index), ATRI (Amplitude Tremor Intensity Index), DVB (Degree of Voice Breaks), DSH (Degree of Sub-harmonics), and DUV (Degree of Voiceless). Many of these parameters have become standards in the analysis of voice, and several papers about the study of speech disorders are based on the results of this software.

The Software for Voice Analysis developed estimates several parameters (some of them available in MDVP), such as jitter (jittr, jitta, jittrap,jittppq5), shimmer (shimr, shima, shimrap, shimppq5), HNR and SPI. The algorithm to estimate the values of pitch period is similar to the one used by PRAAT, which is based on the autocorrelation method. MDVP pitch estimation model is based on peak picking, and such approach is under some controversy in the paper of one of the authors of PRAAT: "Should jitter be measured by peak picking or by waveform matching?".

2.3. Relationships between objective and subjective evaluations

The study of the relationships between objective and subjective evaluations can be carried out with several statistical techniques, such as logistic regression and correlation studies. This paper proposes to estimate the values of the subjective evaluation given the objective parameters using two approaches: linear regression with non-negativity constraints and feed-forward neural networks.

Linear regression is a modelling approach to estimate a linear relationship between a scalar variable y and one or more variables denoted X. In linear regression, data are modeled using linear functions ($y = \beta_0 + \sum_i^N \beta_i x_i$), and unknown model parameters are estimated from the data ($\beta_0, \beta_1, ..., \beta_N$). y is called the dependent variable, and x_i are called predictor or independent variables. In this case the dependent variable y is one of the RASATI values that ranges from zero to three, and the independent variables x_i are the parameters of the objective evaluation, such as relative jitter or HNR.

The main drawback of linear models is the lack of a non-negativity constraint in the weights (β_i). Some β values may be negative, and such linear function will have a distorted behaviour if it is considered each weight as a measure of the importance of each parameter. For example, a negative weight would mean that the higher the jitter the lower the severity of the pathology, which is completely unlogical. Because of that the linear regression uses a non-negativity constraint in the least squares optimization of the weights.

$$y = \beta_0 + \sum_i^N \beta_i x_i, \beta_k \geq 0 \tag{1}$$

The neural network approach is also explored in this paper. An artificial neural network (ANN), usually called neural network (NN), is a mathematical model that is inspired by the structure and functional aspects of biological neural networks. A neural network consists of an interconnected group of artificial neurons, and it processes information using a connectionist approach to computation.

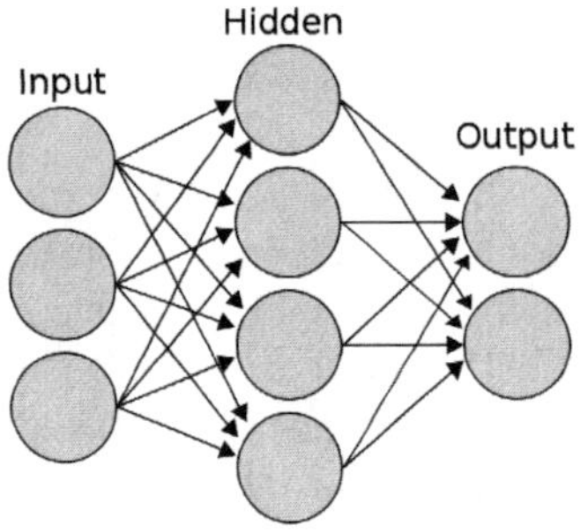

Figure 1. Example of a neural network with three layers: input, hidden and output

Mathematically, a neuron's network function $f(x)$ is defined as a composition of other functions $g_i(x)$, which can further be defined as a composition of other functions. This can be conveniently represented as a network structure, with arrows depicting the dependencies between variables, as shown in Figure 1. A widely used type of composition is the nonlinear weighted sum, where $f(x) = K\left(\sum_i w_i g_i(x)\right)$, where K (commonly referred to as the activation function) is some predefined function, such as the hyperbolic tangent. This work in this paper uses feed-forward networks, because their graph is a directed acyclic graph, as shown in Figure 1.

The inputs of the neural network are the parameters calculated in the objective evaluation, and the outputs is the value of the objective evaluation in the RASATI scale. The first neuron corresponds to the value zero, and the last one to the value three.

3. Experiments

This section shows the experimental results of the estimation of the subjective values of the RASATI scale with the parameters calculated from the audio signal with two different acoustical analysis applications: SAV and PRAAT.

3.1. Experimental setup

The experiments of the two approaches to estimate the subjective values of the RASATI scale given the objective parameters were performed using a database of 105 samples. Each recording has a sampling frequency of 44100 samples per second, and 16 bits of amplitude resolution. Each vocalization was evaluated by the speech therapist to score each qualitative characteristic of the RASATI scale from zero to three, with a rounding to plus infinity when some score fall between two possible values. For example, a value of one to two was scored as two.

Each patient uttered a sustained vowel (/a/ or /e/) with an approximate duration of eight seconds. The patients range from the age of 16 to 88, and there are patients from both sexes with different pathologies, such as disfunctional dysphonia, nodules, laryngitis, polypus or papilloma. Some recordings correspond to several sessions of the same patient.

The experiments were performed using the two approaches: linear regression with non-negativity constraints in the weights, and feed-forward neural networks. In both cases the leave-one-out cross-validation technique was used because of the limited available data in the experiments. Leave-one-out cross-validation involves using a single observation from the original sample as the validation data, and the remaining observations as the training data. This is repeated such that each observation in the sample is used once as the validation data. Leave-one-out cross-validation is usually very expensive from a computational point of view because of the large number of times the training process is repeated. The small amount of training data in the experiments is suited to be used with this cross-validation approach.

The linear regression with non-negativity constraints was calculated using the *LSQNONNEG* command of MATLAB to estimate the weights using the training data. LSQNONNEG returns the vector X that minimizes NORM(C*X-d) subject to $X \geq 0$. The non-negativity constraints are necessary to estimate the weights in order to avoid negative values that would indicate an inverse behaviour. For example, if the weight of the jitta parameter is negative, it would mean that the greater this value, the smaller the pathology, which is clearly wrong.

Feed-forward neural networks were trained using the backpropagation optimization algorithm. The neural networks have multiple inputs and four outputs, each one corresponding to a possible value of the RASATI evaluation. The hidden layer consisted of 10 neurons with a tansig activation function.

Two applications to estimate the objective parameters were used: SAV and PRAAT. PRAAT was chosen because is one state-of-the-art free software for the analysis of speech in phonetics written by Paul Boersma and David Weenink of Phonetic Sciences of University of Amsterdam (The Netherlands).

The acoustical parameters estimated with the Software for Voice Analysis in this experiment are jittr, jitta, jittrap,jittppq5, shimr, shima, shimrap and shimapq5. The parameters estimated with PRAAT were jittr, jitta, rap, ppq5, ddp, shimr, shimdB, apq3, apq5, apq11 and dda.

3.2. Experimental results

The experimental results with the linear regression with non-negativity constraints approach is shown in Figure 2. The columns that end with the letter S correspond to the results obtained with SAV, and the columns with an ending P correspond to the results obtained using PRAAT. Each column shows in different colours the proportions of cases were the estimation of the RASATI values using the objective parameters was lower than the value decided by the speech therapist (-1, -2 or -3), higher (+1, +2 or +3), or equal. The similar results obtained with the parameters estimated with PRAAT and SAV shows that the linear regression with non-negativity constraints achieves an error around ±1 in the 80% of the cases. Such errors are not so severe because RASATI scale is a subjective assessment procedure with only four possible values, and any slight difference of perception produces a ±1. In these experiments, fluctuations in the order of ±1 are possible due to the sensitivity of objective measures. Nevertheless, future work should be devoted to better match the subjective opinion or uncover their origin in each case.

Figure 3 shows the experimental results using the feed-forward neural networks. The distributions of the errors are similar to the results obtained with the linear regression with non-negativity constraints approach. The first approach shows a smaller error than the neural networks for all conditions, but such difference is not significant due to the small amount of data used in the experiments.

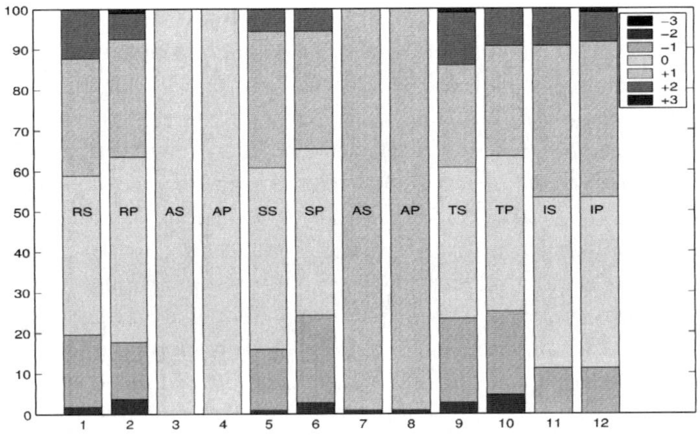

Figure 2. Distribution of the absolute errors for the linear regression with non-negativity constraints approach

The analysis of the weights shows that the important parameters to estimate RASATI scores by means of SAV parameters are (ordered by relevance):

- R : jittr, jitta, shimr and shimapq5.
- A : too few cases available.
- S : jitta, jittr, shimr and shimapq5.
- A : too few cases available.
- T : jittr, shimapq5 and shimr.
- I : jittr, shimapq5 and shimrap.

In the case of PRAAT acoustical parameters the order of relevance are:

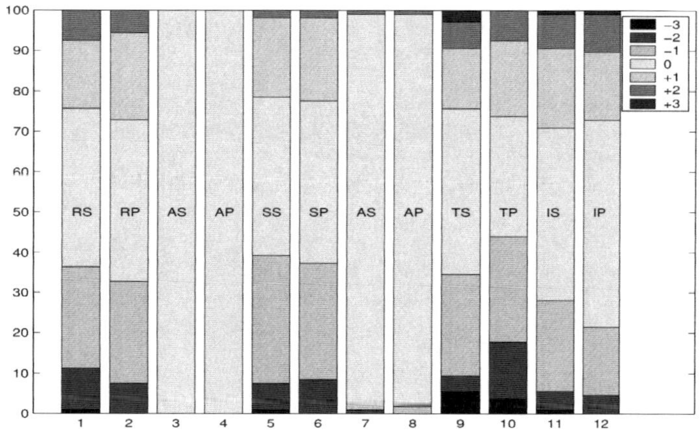

Figure 3. Distribution of the absolute errors for the feed-forward neural network approach

- R : jittr, apq11, shimdB and jitta.
- A : too few cases available.
- S : apq11, ppq5 and shimdB
- A : too few cases available.
- T : apq11, jitta and shimdB.
- I : jittr, apq11 and ppq5.

In both cases is observed changes in the acoustical features and also in the order of relevance. This fact is important to model each RASATI characteristic individually.

4. Conclusions

In this paper it was made a set of comparative experiments to study the relationships between objective and subjective evaluations of speech disorders. The subjective measure is the RASATI scale, the standard chosen by the "Sociedad Argentina de la Voz" to measure voice quality.

Experimental results shown that linear regression with non-negativity constraints and neural networks achieve similar estimation performances. The error in the estimation of the RASATI value is around ± 1 in the 80% of the cases. Such errors are not so severe because RASATI scale has only four possible values, and any slight difference of perception produces a ± 1. These fluctuations are possible due to the sensitivity of objective measures.

Future work will focus in the evaluation of additional objective acoustical features to improve the estimation of the six qualitative characteristics of RASATI scale, and to uncover the origin of the differences between the estimation and the opinion of the speech therapist in each case.

References

[1] Baken, R., Orlikoff, R.: Clinical measurement of speech and voice. In: Second Edition. San Diego, CA: Singular Publishing Group (2000)
[2] Baken, R.: Clinical measurement of speech and voice. In: Allyn and Bacon, Needham Heights, MA (1987)
[3] Dejonckere, P.: Effect of slightly louder voicing on acoustical measurements in different etiological categories of disphonia. In: Proceedings de Voicedata. pp. 86–91 (1998)
[4] Deliyski, D., Shaw, H., Evans, M.: Influence of sampling rate on accuracy and reliability of acoustic voice analysis. In: Logopedics, Phoniatrics, Vocology. vol. 30, pp. 55–62 (2005)
[5] Deliyski, D., Shaw, H., Evans, M.: Regression tree approach to studying factors influencing acoustic voice analysis. In: Folia Phoniatrica et Logopaedica. vol. 58, pp. 274–288 (2006)

[6] Godino-Llorente, J., Osma-Ruiz, V., Saenz-Lechon, N.: Acoustic analysis of voice using wpcvox: a comparative study with multi dimensional voice program. In: European archives of otorhinolaryngology. vol. 265, pp. 465–476 (2008)
[7] Hirano, M.: Psycho-acoustic evaluation of voice. In: New York: Springer-Verlag (1981)
[8] Karnell, M., Hall, K., Landahl, K.: Comparison of fundamental frequency and perturbation measures among three analysis systems. In: Journal of Voice. vol. 9, pp. 383–393 (1995)
[9] Pinho, S., Pontes, P.: Musculos intrinsecos da laringe e dinamica vocal serie desvendando os segredos da voz. In: Revinter (2008)
[10] Revis, J., Giovanni, A., Wuyts, F.: Comparison of different types of vowel fragments for the evaluation of voice quality. In: Proceedings de Voicedata. pp. 80–85 (1998)
[11] Saenz-Lechon, N., Godino-Llorente, J., Osma-Ruiz, V., Blanco-Velasco, M., Cruz-Roldan, F.: Automatic assessment of voice quality according to the grbas scale. In: International Conference of the IEEE. pp. 2478–2481 (2006)
[12] Smith, I., Ceuppens, P., Bodt, M.D.: A comparative study of acoustic voice measurements by means of dr. speech and computarized speech lab. In: Journal of Voice. vol. 19, pp. 187–196 (2005)

Medical Data Transmission Using Cell Phone Networks

J Voos, C Centeno, G Riva, C Zerbini, E Gonzalez

Grupo Ingeniería Clínica – Universidad Tecnológica Nacional, Facultad Regional Córdoba, Maestro López esquina Cruz Roja, Ciudad Universitaria, X5016ZAA Córdoba, Argentina.

jvoos@sdct.frc.utn.edu.ar

Abstract. A big challenge in telemedicine systems is related to have the technical requirements needed for a successful implementation in remote locations where the available hardware and communication infrastructure is not adequate for a good medical data transmission. Despite of the wide standards availability, methodologies, applications and systems integration facilities in telemedicine, in many cases the implementation requirements are not achievable to allow the system execution in remote areas of our country. Therefore, this paper presents an alternative for the messages transmission related to medical studies using the cellular network and the standard HL7 V3 [1] for data modeling. The messages are transmitted to a web server and stored in a centralized database which allows data sharing with other specialists.

1. Introduction

Due the availability of new data transmission technologies and new standards for medical studies development, e-health systems have had a sustained adoption in recent years. However, one of the issues for a standard adoption during the medical data transmission is given by the increase in the message size needed to represent the acquired medical data following the data models defined by the selected standard.

In the case of HL7 V3, the requirement to support the reference information model (RIM) for the entities participating in one medical study and the XML [2] tags used for data representation implies the need of high bandwidth for the available communication channel to allow HL7 V3 well formed messages transmission. For this reason, many e-health systems use proprietary representations during transmission and offer HL7 interfaces to represent medical studies already stored at the server side; it means when the transmission from the remote site is completed.

Because of this, the integration between different e-health systems using standards occurs mainly via broadband networks provided by ISPs, where there are no limitations related to the message size during the transmission.

2. Methodology

In order to transmit medical studies represented by the HL7 V3 standard using the cellular network, we developed a customizable telemedicine system to generate, transmit, validate and store partial or complete medical studies according to a particular diagnosis.

The cellular network as a transmission channel was selected based on its availability in remote locations in our country. While cell phone providers are incorporating new technologies (GPRS,

EDGE, HSDPA, etc.) allowing different data transfer rates, the system could be configured to improve the available bandwidth for the area where the medical study is performed.

2.1. Architecture

The system is implemented using a service-oriented architecture (SOA). This architecture evolution provides from previous telemedicine projects executed by our team [3] which implies adding new functionalities during each project enhancement. It means adding new services provided by software components written in Java, to manage topics like message transmission, security, HL7 messages processing, front end and database management for the medical studies stored in the web site (Figure 1).

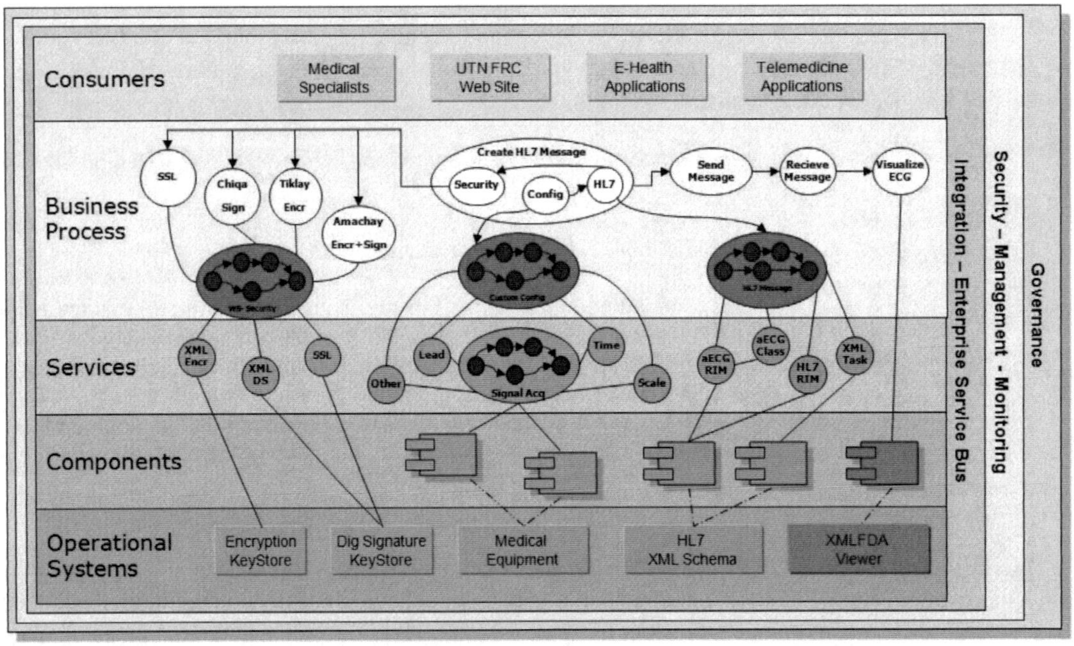

Figure 1. Service Oriented Architecture (SOA) diagram.

The main objective behind this system design is to offer different execution alternatives, providing the necessary adaptability to deal with the problem of having diverse execution environments for telemedicine applications, considering that each region has different needs and available technologies.

This objective is simple achieved through a specific system configuration according to the technical resources available where the medical study is performed and transmitted. During transmission request creation, the system specifies the parameters needed to validate and process messages with the web site module (Figure 2).

```
- <soapenv:Envelope xmlns:soapenv="http://schemas.xmlsoap.org/soap/envelope/"
    xmlns:teleMed="http://teleMed.utnfrc.gic.com">
    <soapenv:Header />
  - <soapenv:Body>
    - <teleMed:customConfig>
        <teleMed:leads>I,V1,V2,AVR,AVL,AVF</teleMed:leads>
        <teleMed:scale>25</teleMed:scale>
        <teleMed:seconds>10</teleMed:seconds>
        <teleMed:security>Amachay</teleMed:security>
      </teleMed:customConfig>
    </soapenv:Body>
  </soapenv:Envelope>
```


Figure 2. Transmission request using SOAP protocol.

Regardless of the selected configuration, all messages are transmitted following the HL7 V3 standard (Figure 3). This architectural decision let us reuse the existing software components associated with XML files processing and HL7 V3 schema validation in order to add new medical studies representations.


```
- <sequenceSet>
  - <component>
    - <sequence>
        <code code="TIME_ABSOLUTE" codeSystem="2.16.840.1.113883.5.4" codeSystemName="ActCode" displayName="Absolute
          Time" />
      - <value xsi:type="GLIST_TS">
          <head value="20030730091733.000" />
          <increment value="0.0010000" unit="s" />
        </value>
      </sequence>
    </component>
  - <component>
    - <sequence>
        <code code="MDC_ECG_LEAD_I" codeSystem="2.16.840.1.113883.6.24" codeSystemName="MDC" />
      - <value xsi:type="SLIST_PQ">
          <origin value="0.000" unit="uV" />
          <scale value="2.500" unit="uV" />
          <digits>36 35 35 35 36 35 33 33 32 31 30 30 30 30 31 31 31 31 31 32 32 31 31 31 31 31 31 31 31 32 32 32 32 31 31 31
            31 33 35 35 35 35 36 36 36 35 35 33 31 32 32 33 35 35 35 35 35 35 35 34 33 33 33 34 35 33 32 32 32 32 32 31 31 31
            31 31 31 31 31 31 31 31 31 31 31 31 31 32 32 32 32 31 31 30 30 31 30 30 31 30 29 29 29 30 31 34 34 34 33 34 33 32
            31 31 31 29 31 30 31 30 29 28 30 30 30 30 28 29 32 32 32 29 30 28 28 27 27 27 27 28 28 28 29 31 30 31 29 29 29 29
            29 30 30 30 31 31 30 30 30 31 33 32 33 31 32 30 29 30 30 30 30 30 30 30 30 31 30 30 30 30 28 27 26 26 28 29 30
            30 31 32 33 32 33 32 33 34 34 33 35 35 35 35 33 34 32 30 30 30 28 29 30 29 29 29 29 29 29 30 29 30 29 29 30 29 29
            29 30 30 29 30 29 29 29 29 28 28 29 29 30 32 31 31 31 30 31 30 31 31 32 33 33 32 31 30 30 29 30 30 30 30 30 30 30
            30 30 30 31 32 31 30 30 30 30 29 29 29 28 28 27 27 29 32 31 33 34 34 35 34 34 34 34 34 35 35 35 34 34 37 35 35 35
            35 35 35 33 34 34 33 34 33 33 35 33 36 35 34 32 30 29 29 27 26 27 26 27 28 30 31 30 30 28 28 29 29 30 30 31 33 33
            33 34 33 33 32 31 29 31 29 30 30 30 30 30 30 32 30 33 34 34 34 35 35 36 34 37 37 37 37 37 35 36 38 36 35 37 37 34
            34 32 33 35 35 36 37 38 38 39 39 39 40 43 43 44 46 46 45 46 48 48 47 49 51 51 51 52 52 53 54 54 54 55 55 55 55 57
            58 57 58 58 58 58 57 56 55 54 54 54 53 51 47 47 45 43 40 40 39 39 37 37 36 36 36 35 36 33 33 35 35 37 35 37 35 34
```


Figure 3. ECG signal representation using the FDA XML standard.

The system is based on a producer and a consumer module, while the communication between these modules is implemented by web services using SOAP protocol for message transmission. The producer module includes libraries that require a Java Runtime Environment version 1.4, 1.5 or 1.6, using Windows or Linux operating system. The consumer module runs on a web site hosted at the university using a Java application server version 1.5.

2.2. Security

Most telemedicine applications provide security at transport level, basically to keep the message size reduced and avoid encryption algorithms and digital signature complexity at both communication channel ends. The main issue associated with this security level is that the message could be modified at one intermediate point without detection, ensuring security only in the communication channel implementing a HTTPS session between the requester and the provider.

In our system the communication is protected by message-level security implementation using WS-Security and W3C XML Encryption [4] and XML Digital Signature [5] recommendations to encrypt and sign messages respectively. This approach ensures message integrity end to end because the messages are encrypted and/or digitally signed before transmission. Only the recipient has the valid credentials to decrypt the messages using a key store that contains the certificates exported by the message provider for message decryption. It facilitates detection of message changes during the transmission.

The message level security ensures confidentiality, integrity and authentication for patient data and medical parameters with a more effective method than transport level security using SSL.

3. Results

For the field tests, two medical studies represented by the HL7 V3 standard were used: laboratory report and electrocardiogram. For laboratory reports alternatives were checked with 4, 12 and 20 components; while electrocardiogram signals used during the tests have the following leads combinations: 3, 6 and 12 with 20 seconds as duration and 25 mm/sec as scale.

In all the variants mentioned above, the transmitted messages contain all the mandatory fields required by HL7 V3 standard to represent the reference models (RIM) associated with entities like patient, specialist, place/date for the medical study and data acquisition device.

These medical studies were saved in one laptop to be transmitted through GPRS network using two different GSM modems. One of them uses 2G technology (Motorola G24 with RS-232 connectivity), while the other one supports 3G technology (Huawei E166 with USB connectivity). Each of them uses cellular phone networks from different providers. Relating security, four options were implemented: no security, digital signature, encrypted and signed and encrypted.

Fifteen tests were executed at different times of day for each of the alternatives associated with the medical studies to validate. These tests were conducted in 3 remote areas located in the province of Cordoba with the objective of evaluating the system behavior in response to different cellular network bandwidths. Despite the increment of the transmission time associated with the GPRS network conditions, all of the messages were transmitted successfully. Each value for the figures 4 and 5 represents the average times in seconds using 2G and 3G technologies for each of the transmitted alternatives associated with medical studies.

	3 leads		6 leads		12 leads	
	2G	3G	2G	3G	2G	3G
Plain Text	8,43	2,27	8,62	5,72	12,78	8,15
Signed	14,67	6,78	17,21	7,89	26,31	10,53
Encrypted	28,33	11,89	38,29	12,47	97,01	18,01
Enc + Sign	29,39	13,47	36,25	13,51	90,22	25,03

Figure 4. Average transmission times in seconds for electrocardiogram alternatives.

	4 components		12 components		20 components	
	2G	3G	2G	3G	2G	3G
Plain Text	6,17	2,02	8,12	4,16	9,34	5,01
Signed	13,46	4,12	17,34	8,92	19,12	8,13
Encrypted	15,12	8,43	22,18	10,21	23,46	13,12
Enc + Sign	20,12	11,16	27,19	9,16	29,18	15,81

Figure 5. Average transmission times in seconds for laboratory report alternatives.

Figure 6 shows a graphical comparison between the average times for each security combination using 2G technology.

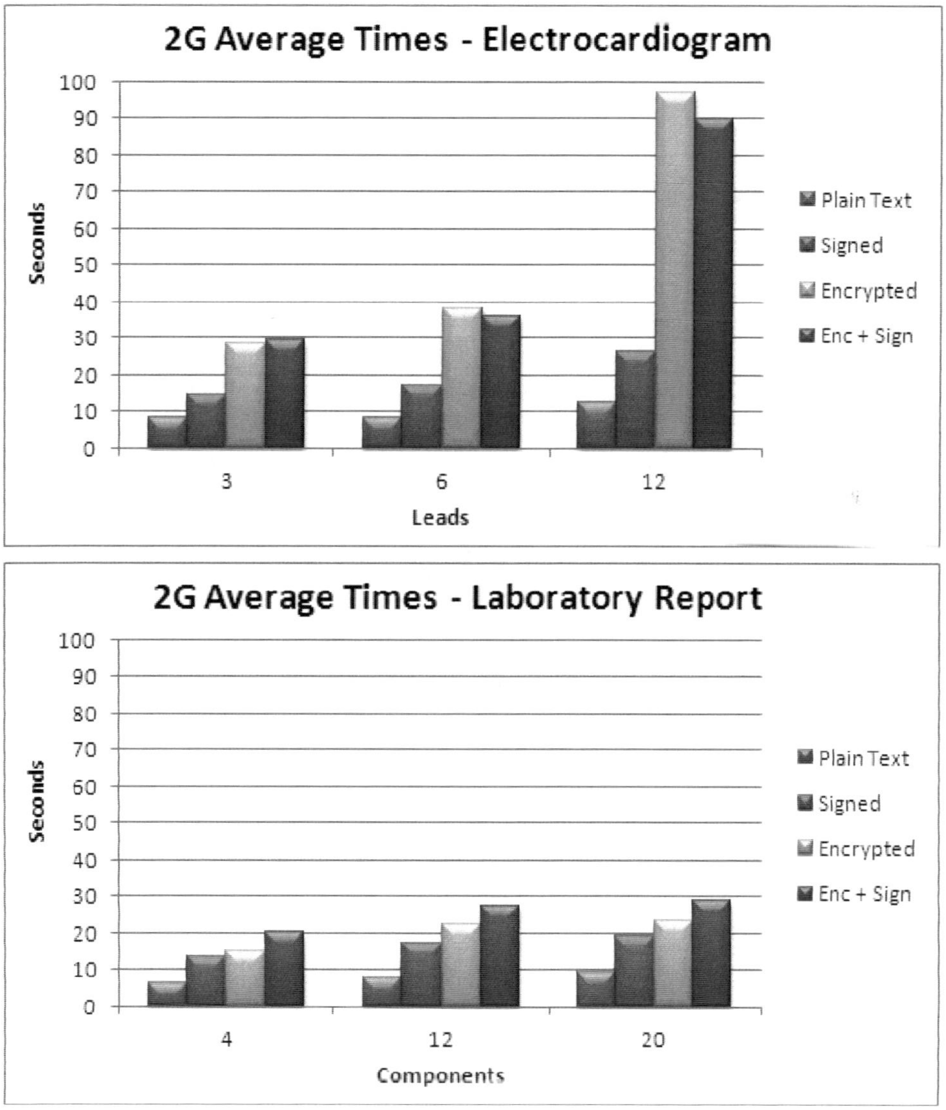

Figure 6. Average times using 2G technology.

Figure 7 shows comparative data related to 3G technology average times.

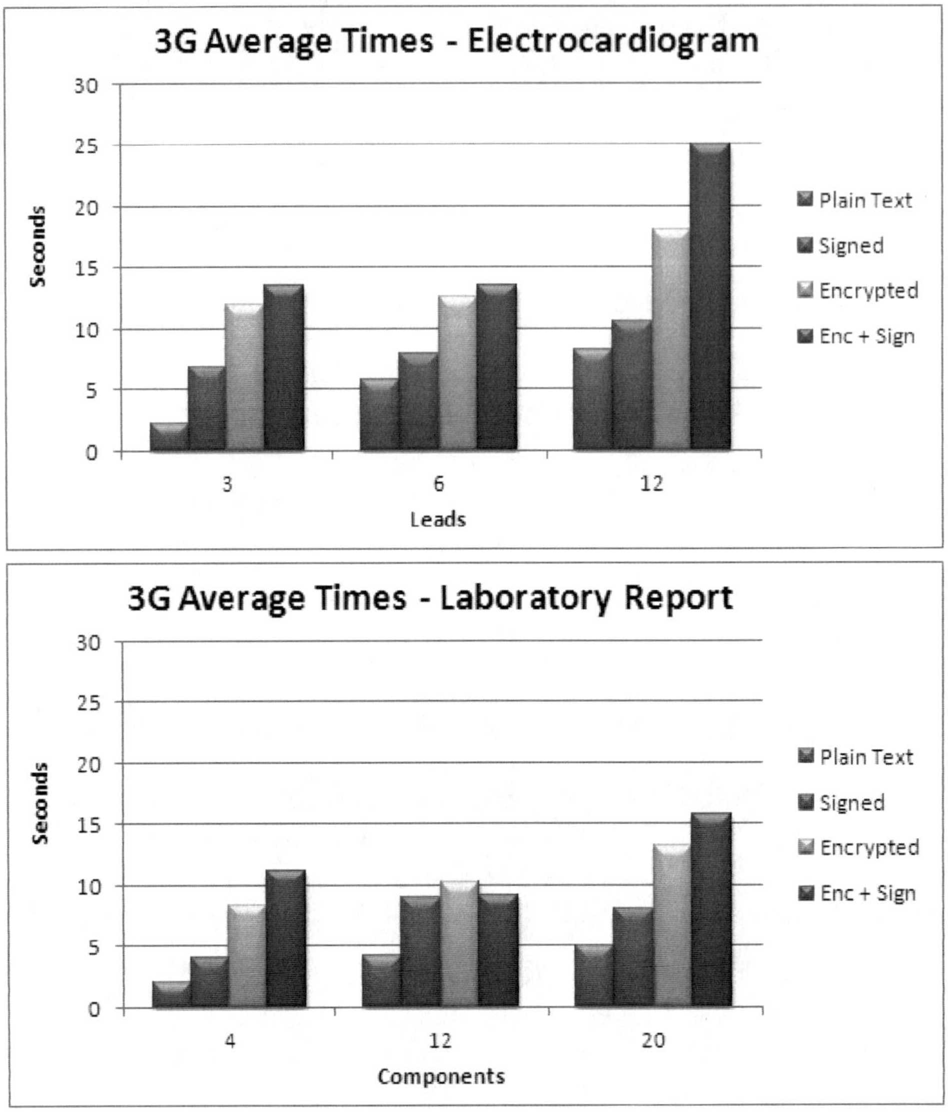

Figure 7. Average times using 3G technology

4. Conclusion

This initiative validates the technical feasibility of enabling the HL7 V3 standard for laboratory reports and electrocardiograms transmission using the cell phone network, through field tests in remote areas where the available medical infrastructure demands using of telemedicine. In this way any telemedicine system with a HL7 V3 interface can receive these studies without transformation and/or integration efforts for their processing.

With the purpose of managing available bandwidth issues which affect the medical study transmission; a software services group was developed in order to build HL7 V3 messages containing only critical data needed to formulate a study first diagnosis, together with the possibility to choose different levels of security during the transmission. Through the implementation of new software services; the system architecture allows new interfaces in response to different diagnoses and adoption of additional HL7 standards for other medical studies.

References

[1] V3 Messaging Standard, Health Level Seven (HL7) Standard, 2009. http://www.hl7.org/implement/standards/v3messages.cfm
[2] Extensible Markup Language (XML), W3C Note, 2001. http://www.w3.org/XML/
[3] Voos J, Vigliecca N and Gonzalez E 2007 Web based aphasia test using service oriented architecture (SOA) *J. Phys.: Conf. Ser.(2007) 90 012003*
[4] XML Encryption Syntax and Processing, W3C Recommendation, 2008. http://www.w3.org/TR/xmlenc-core/
[5] XML Digital Signature Syntax and Processing, W3C Recommendation, 2008.

Low level communication management for e-health systems

Guillermo Riva, Carlos Zerbini, Javier Voos, Carlos Centeno, Eduardo González

Clinical Engineering Group, National Technological University, Córdoba Regional Faculty
Maestro M López St and Cruz Roja Argentina St, Córdoba, Argentine

E-mail: griva@scdt.frc.utn.edu.ar, czerbini@electronica.frc.utn.edu.ar

Abstract. The heterogeneity of e-health systems encourages the use of standards such as Health Level 7 (HL7v3) to ensure interoperability. Many actual implementations address this problem by unoptimized high level programming of top-range portable computing platforms. However, this approach could pose excessive demands on battery-powered mid-range terminals. In this work, we propose low-level support for portable HL7v3-compatible embedded systems in order to better exploit their limited processing and communications capabilities. In particular, we present our experience in mobile communication management through two different approaches, which proves the feasibility of this proposal.

1. Introduction

E-health systems have been broadly adopted in last years to take advantage of technology for medicine applications. Particularly, *telemedicine* systems are used for supplying of sanitary attention, prevention, research and education wherever geographic location is a critical factor. Telemedicine is of particular interest in our country due to long geographical distances between medical centres. A generic e-health scenario is shown in Figure 1.

Two main technology advances are driving actual developments on telemedicine. On the one hand, *computing resources* enable engineers to design standard, smart, and secure data processing and management applications for embedded devices. On the other hand, *communication technology* enables both increasing bandwidth and mobility, which paves the way for innovative e-health systems. While much work focuses on high-level abstraction implementation of telemedicine systems, we rather discuss low-level design for efficient use of locally available computing and communication resources.

The rest of the paper is organized as follows. In Section 2, we compare communication and digital processing platforms suitable for our purposes. Based on this analysis, in Section 3 we propose approaches to cope with our particular needs. Section 4 presents experiences based on these proposals. Finally, Section 5 reviews obtained results and its application in future work.

2. System Analysis

The presented ideas emerge as enhancements of medical data acquisition and transmission system previously implemented in our laboratory [1]. In order to enable interoperability, on the one hand, adoption of the HL7v3 Standard was proposed [2]. On the other hand, connectivity

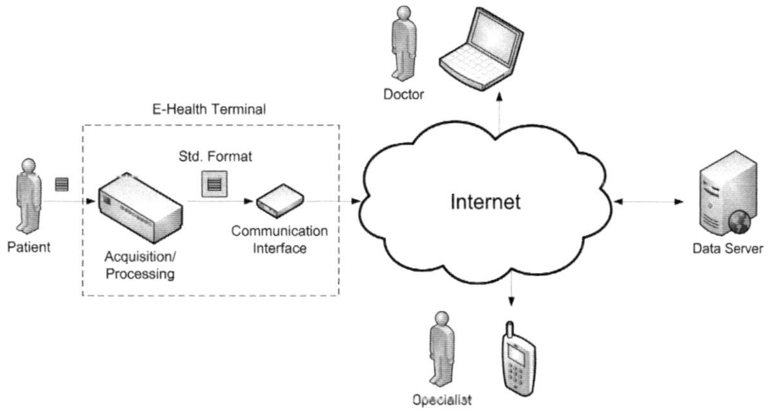

Figure 1. Generic e-health architecture

improvements through additional communications technologies are addressed. In the following, we survey available choices for this implementation.

2.1. Communication Technologies
At the moment, our existing platform supports Electrocardiogram (ECG) data transmission to a remote server over Ethernet links. Available communication technologies are discussed in this subsection.

2.1.1. Dial-up Access. Because of its wide availability and robustness to outdated physical transmission media, dial-up access to Internet through the analog Plain Old Telephone System (POTS) is still an option to consider. In this case, a permanent link is first established with a DSL provider through the point-to-point protocol (PPP), which provides an IP address to the user for subsequent data transaction with time-based billing approach. Its very narrow bandwidth restricts its application to low volumes of data, while fixed access to the transmission line restricts mobility.

Our actual application, namely transmission of 12-lead ECG studies, deals with files in the 100 Kilobytes range, therefore dial-up looks like a suitable communications technology. However, adoption of standards such as HL7v3 adds considerably to the volume of transmitted data. As a consecuence, Dial-up could turn into an inadequate technology in the near future. From this reasoning, we nowadays consider dial-up only as a last resource.

2.1.2. Ethernet Networks. Ethernet (IEEE 802.3) family of standards, has been for years the preferred technology for local-area networks (LANs), providing relatively high bandwidth, robustness and easy of use. Embedded solutions increasingly include Ethernet as standard communications interface, which enables portable telemedicine devices to attach to any Ethernet socket without the need of additional hardware. In the last years, enhanced Ethernet versions with up to 40-100 Gbps bandwidths are in development for local, metropolitan and wide area networks, which makes this technology a very promising option for wired, nomad Internet access.

2.1.3. Wi-Fi Networks. Wi-Fi (IEEE 802.11) wireless networks enables high-bandwidth interconnection between embedded devices and the Internet using unlicensed bands.

Actually, no open access everywhere is available, even though low-cost network devices are making it increasingly feasible. For example, an ambicious project is in development to provide open access to Wi-Fi in 66 cities of Córdoba province [3], which pretends to be one of the most important of our country.

Wi-Fi is suitable for medium-capacity battery powered devices. In the case of ultra-low-power mobile e-health devices, however, the IEEE 802.15.4 standard [4] arises as the best option due to its smart modes and enhanced energy management.

2.1.4. Satellite Systems. Satellite-based systems are suitable for areas which lack of any terrestrial communication infraestructure, as well as aircrafts and ships. This communication system is also the only available platform in case of natural disasters such as earthquakes, tornadoes, tsunamis, volcanic eruptions, etc. Main advantages of satellite systems are worldwide coverage and relatively high data transfer speeds [5].

Drawbacks of satellite communication are expensive equipment, dedicated RF links, and need of skilled operators. In addition, high RF power is required in satellite systems with the consequent size and autonomy impact, which restricts its application to very special cases.

2.1.5. Mobile Phone Networks. The General System for Mobile (GSM) technology enables digital communication in not only urban but also rural areas. While 1800/1900 Megaherz (MHz) band is used for short-distance high-bandwidth communication in cities, long-reach 850/900 MHz band is used in rural areas. Available GSM data services are:

- *Circuit-switched Data (CSD).* The CSD service is very similar to a voice call, requiring a dedicated digital channel with a remote system. Its implementation is quite straightforward and requires no digital interfaces which makes it possible to use any mobile phone. However, bandwidth is low (i.e., 9600 bps) and billing is time-based, which restricts it to special applications.
- *Short Message Service (SMS).* SMS service has been widely deployed in recent years for systems such as billing systems and event alert. It can be deployed between mobile stations (MSs) inside the provider's network or between a MS and external networks by means of a SMS server. It is quite easy to implement and cheap for low data volumes. Drawbacks are no delivery time guarantees and need for multiple messages for mid-sized data volumes [6].
- *General Packet Radio Service (GPRS).* GPRS is the specific GSM service for broadband mobile data transactions. MSs first establish a connection to a Gateway, which serves as interface between GPRS network and internet. From this point, application data can be transparently carried on using the TCP/IP protocol stack, as illustrated in Figure 2.
  This service was introduced after second generation (2G) GSM, but it wasn't until recently that it was broadly adopted and implemented as mobile Internet platform in Argentina. Its main advantages are data volume-based billing and enhanced bandwidth, i.e., 56-114 Kbps. Drawbacks are, service quality is strongly time-and place-dependant, latency is high, and cost is not low enough yet in our country.

As demand and competition increases, it is expected that GPRS system is going to evolve to better services and coverage with decreasing costs. In addition, several development platforms include GPRS interface, that's why it was adopted as communications platform for our work.

2.2. Digital Processing Technologies
Two main requirements drove a review of our processing platform. On the one hand, adoption of the HL7v3 standard demands parsing of Extensive Markup Language (XML) structures requiring dedicated processing and memory. On the other hand, additional processes for communication management are required. As a consequence, the natural question arises about whether to add new tasks to the existing platform or executing them on an external one.

Previous software was based on a PIC18F4550 microcontroller from Microchip, which takes care of user interface, visualization and formatting of raw ECG samples. It also manages Serial Peripheral Interface (SPI) with programmable amplifiers and serial A/D converter, as well as raw data transfer through the isolated serial V.24 port to a Personal Computer (PC). The PIC microcontroller runs 12 Mega Instructions per Second (MIPS) at 40 MHz, and featuring 2 Kbytes of RAM and 32 KBytes of FLASH program memory, which may be enough for HL7 and GPRS modem management. However, in the near future a real-time operating system (RTOS), additional memory and enhanced speed could be needed for additional processing tasks such as ECG storage, compression, encryption, and new protocols support.

Advanced microcontrollers such as ARM cores with appropriate peripherals, meanwhile, can fulfill current needs while providing extra processing power for future applications. In particular, the TS-7200 Single Board Computer (SBC) from Technologic Systems [7] features a powerful Cirrus EP9302 ARM9 processor running 200 MIPS at 200 MHz; Ethernet, SPI, Keypad, USB, V.24, GPIO, and Compact FLASH interfaces; as well as on-board 32Mb DRAM and 8 Mb Strata Flash memories. This processor includes a Memory Management Unit (MMU) which supports high level operating systems such as Linux and Windows CE. Despite all of these features, drawbacks such as cost and power comsuption call for a deeper analysis.

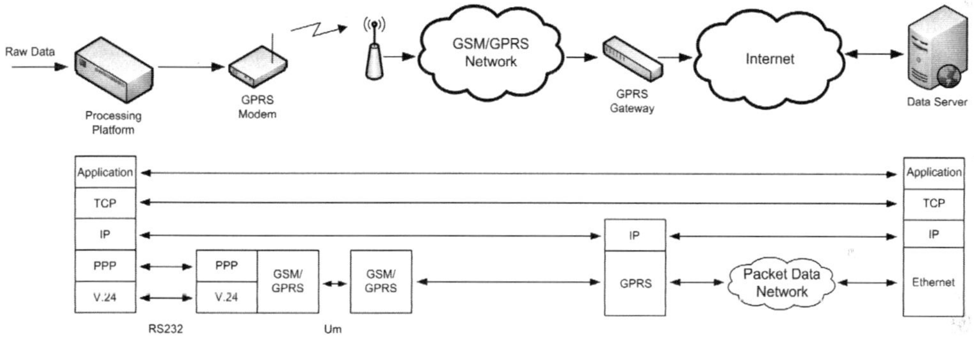

Figure 2. GPRS-based e-health architecture

3. Implementation

3.1. Discussion

Our ECG acquisition platform consists of a 12-lead analog processing stage, followed by a microcontroller which provides extra processing, user interface and transmission to a PC through an isolated standard RS232 serial port. Software running on this PC is in charge of transmitting raw ECG data to a server through Ethernet interfaces. Custom storage and visualization applications, developed in C++, are required for management and later access to ECG studies.

Even if this platform proved to be useful for our previous goals, it suffers of some drawbacks. First, it does not support any standard for medical information, which makes it very difficult to interoperate with larger systems. In second place, transmission of raw (i.e., not-compressed) data is bandwidth-inefficient and does not provide enough confidentiallity, which is essential while handling patient data. Third, the transmission equipment is based on a PC, which is far much complex and costly hardware than is demanded for our case. Finally, Ethernet technology could not fit mobility requirements for applications such as field data collection, remote consulting and emergencies. We focus on solving these problems while keeping in mind the cost-efficiency tradeoff.

As mentioned, the most appropriate communication solution seems to be GSM/GPRS. Communications infrastructure is already installed and managed by external players, while increasing demand pushes for better service and reduced cost. Multimedia applications are already present in most devices, and some development platforms are suitable for our work. Despite these advantages, developers must cope with some problems before achieving a functional module. On the one hand, development platforms sometimes lack of enough documentation or are launched to market without being completely debugged. On the other hand, interacting with such platforms involves many firmware-specific details.

3.2. GPRS Connectivity Platforms

Main discussion arises about whether to use mobile phones or modems for transmission of medical studies. In the following, we compare the use of mobile phones with respect to two different modems available in Argentina.

3.2.1. Mobile Phones. The first considered platform due to its availability, low cost and proper user interface was a mid-range mobile phone. At the time of our survey, interfaces and protocols used to be proprietary. In addition, coverage could no be enhanced by external directional antennas, which limits its application for e-health services in remote areas.

Altough many models offer standard Wi-Fi, Bluetooth and USB interfaces nowadays, they are not intended as interfaces for custom hardware development, and they still do not offer detailed hardware and programming documentation. Therefore additional effort is needed in order to get familiar with every available model.

3.2.2. Siemens TC65T Modem. TC65T combines the Siemens TC65 GSM/GPRS module with a powerful ARM9 microcontroller, which enables applications running on Java 2 Micro Edition (J2ME) to establish so-called Machine-to-Machine (M2M) communications. Quad-band operation is included, which is appropriate for roaming in different countries, and multiple general-purpose I/Os (GPIOs) serve as analog and digital interfaces to external platforms.

Despite all its features, this modem is strongly oriented to stand-alone use and therefore offers poor documentation and difficult programming guidelines for interaction with external devices. Finally, it is rarely adopted in latin-american e-health projects, and its under-utilized features do not justify its high cost for our project.

3.2.3. Motorola G24 Modem. This modem offers Quad-band operation for international use, CSD, SMS, and GPRS services. The G24 modem also includes an internal TCP/IP stack for FTP, SMTP, SSL, or HTTP data transfer, and Java support is provided in most recent versions. The internal G24 module offers two UARTs, one USB and some general-purpose interfaces, of which only one UART is available in the modem pinout.

Both, our previous PIC-based platform, and a new TS-7200 SBC were used with GPRS modems. Altough a TC65 modem was first tested for our implementation, the G24 was finally preferred due to good support from local suppliers, ease of use and low cost. Mentioned features are briefly compared in Table 1.

For transactions between any terminal device (e.g. a PC, our embedded platform, etc) and communication device, namely the GPRS modem, the *AT command set* is commonly used. These commands are formed by a set of ASCII characters starting with the *AT* prefix, which stands for *attention*. By using these commands, both configuration and information requirements are possible. It is to note that most of these commands are standard, while a small subgroup is proprietary of each modem provider [8].

Table 1. GSM/GPRS Modems features

Feature	Modem G24	Modem TC65T
Cost	Moderate	High
Documentation	Good	Poor
Assembly	Argentine	Germany
Interfaces	V.24	V.24, Audio, I2C, SPI, GPIO
Power Cons.	2.5 mA - 3.3-4.2V (Sleep)	3 mA - 8-30V (Sleep)
GPRS Class	Multi-slot Class 10	Multi-slot Class 12
Embedded TCP/IP	Yes	Yes
Processor	MC9S08GT32 (G24MCU)	ARM7

3.3. Methodology

We tested transmissions of ECG studies to a HTTP server hosting a web service application through the well-known port 80. For the purpose of this work, previously generated HL7v3 messages were used. We note that the inclusion of additional studies to the existing XML scheme is possible without substantial software modification.

Two different implementation approaches were evaluated during our tests, as shown in Figure 3. The first one uses the TCP/IP stack of the Linux kernel, which we call *external* stack; while the second one takes advantage of the *embedded* TCP/IP stack of the G24 modem. The first option is easier to implement, because many open development tools are available for Linux, which raises the abstraction level. On the other hand, the second option allows stand-alone applications, suitable for constrained-resources portable devices. It is worth to note that the TCP/IP stack could be also embedded in the microcontroller included in our acquisition platform [9], however, it makes no sense to do so considering that the modem includes this functionality.

Both approaches (external and embedded stacks) were evaluated by using three testbenches:

- *Modem connected to PC.* This setting enhances fast testing but it is the more expensive solution. The external (i.e. Linux) TCP/IP stack is used, while any PPP dialer such as *wvdial* can be applied. This software takes care of modem management, hiding AT commands from the user. For communication through the created socket, any Telnet client can be used, such as *Telnet, Putty*, etc, see Figure 3(a). We mention that embedded TCP/IP stack could also be tested with this testbench by using any available serial terminal (e.g. *Hyperterminal, Putty*, etc.), to send application data to the modem, emulating the processing platform behaviour.
- *Modem connected to SBC.* In this case, the Linux kernel must be compiled and ported to the TS-7200 board. The general approach is similar to the previous case (Figure 3(a)), but the limited resources of the SBC imposes the use of command-line tools, which demands additional development.
- *Modem connected to custom acquisition platform.* This is the most optimized testbench regarding use of hardware resources, but it requires most development efforts and work at hardware level (Figure 3(b)). Prior to hardware implementation, simulation tools (e.g. *Labcenter Proteus and Microchip MPLab*) and C compilers (e.g. *Hi-Tech C Compiler*) are valuable for this testbench.

While the first testbench was used for fast initial application test, the last two solutions are candidates for our final hardware implementation due to their strong optimization.

Two application-layer protocols were used during tests, *Simple Mail Transfer Protocol* (SMTP) and *HyperText Transfer Protocol* (HTTP). SMTP is used for electronic mail

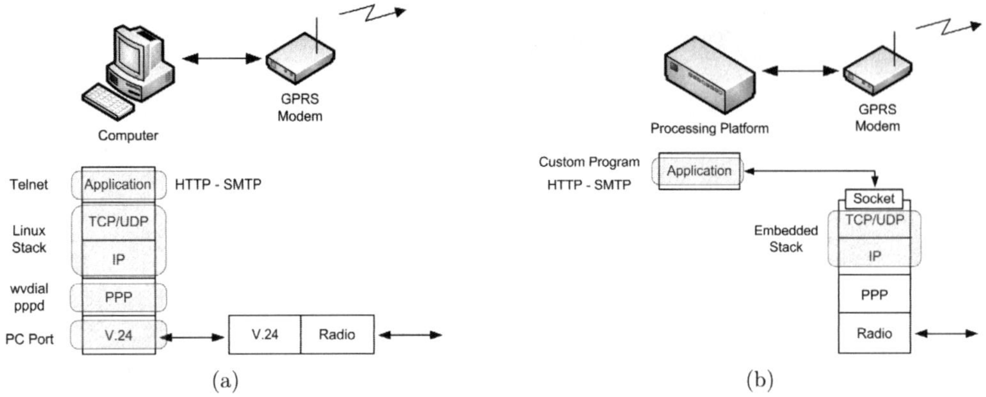

Figure 3. Evaluated testbenches: (a) Linux TCP-IP stack, (b) Modem TCP-IP stack

transmission across Internet Protocol (IP) networks using TCP port 25. Most of the documentation about the G24 modem is based on SMTP. HTTP, meanwhile, is a networking protocol for distributed, collaborative, hypermedia information systems comunications in a client-server model. Although there are nine methods in HTTP, the two most used are GET and POST. GET requests a representation of the specified resource, and POST submits data to be processed. This protocol will be used in our final implementation for storage of ECG records in a central database, therefore our results are based on it.

4. Results

4.1. Test of the External TCP/IP Stack

In this implementation the *wvdial* software is executed in a host PC on the Linux operating system. First, wvdial reads the file *wvdial.conf*, which stores information such as modem inicialization commands, Packet Data Protocol Context (i.e., packet data protocol and Provider IP Address), modem interface to be used, etc. After that, it dials the Provider's Gateway and starts the *pppd* daemon, which thereafter manages transactions between both ends. Finally, a Telnet client is executed in the host PC in order to send HTTP commands to the remote process (i.e., a web service) running on the remote server. The following sequence illustrates this procedure:

```
griva@localadmin-desktop:~$ sudo wvdial
--> WvDial: Internet dialer version 1.60
--> Initializing modem.                                  // Modem Inicialization
--> Sending: ATZ
ATZ
OK
--> Sending: ATQ0 V1 E1 S0=0 &C1 &D2 +FCLASS=0
ATQ0 V1 E1 S0=0 &C1 &D2 +FCLASS=0
OK
--> Sending: AT+CREG=1                                   // GPRS Network attachment
AT+CREG=1
OK
--> Sending: AT+CGDCONT=1,"IP","internet.gprs.unifon.com.ar"
AT+CGDCONT=1,"IP","internet.gprs.unifon.com.ar"
OK
--> Modem initialized.
--> Sending: ATDT*99#
--> Waiting for carrier.
ATDT*99#
CONNECT
```

```
~[7f]}#@!}!}!} }8}"}&} } } } }#}$@#}%}&} })};k}'}"}()"+s~
--> Carrier detected.  Waiting for prompt.
~[7f]}#@!}!}"} }8}"}&} } } } }#}$@#}%}&} }),[7f]}'}"}()"e)~
--> PPP negotiation detected.                          // pppd daemon starting
--> Starting pppd at Fri May 27 17:22:37 2011
--> Pid of pppd: 22483
--> Using interface ppp0
--> pppd: [1f]q[10][06][08]
--> pppd: [1f]q[10][06][08]
--> pppd: [1f]q[10][06][08]
--> pppd: [1f]q[10][06][08]
--> pppd: [1f]q[10][06][08]
--> local  IP address 10.82.131.137
--> pppd: [1f]q[10][06][08]
--> remote IP address 192.168.100.101
--> pppd: [1f]q[10][06][08]
--> primary   DNS address 200.49.193.140
--> pppd: [1f]q[10][06][08]
--> secondary DNS address 200.63.128.203
--> pppd: [1f]q[10][06][08]                            // Socket ready
```

In this approach, AT commands are used only for modem configuration and PPP negotiation in order to establish the connection with the server's application. From this point on, application data is transferred transparently using the Linux TCP/IP stack, so no further AT commands are necessary.

4.2. Test of the Modem-embedded TCP/IP Stack
This approach, unlike the previous one, makes intensive use of AT commands in order to both establish the PPP negotiation and exchange data with the server application. Two main modes are available for data exchange using the modem-embedded TCP/IP stack:

- *Buffered Data Mode*, using AT+MIPOPEN, AT+MIPSEND, AT+MIPPUSH commands
- *Online Data Mode*, which is enabled by using the AT+MIPODM command

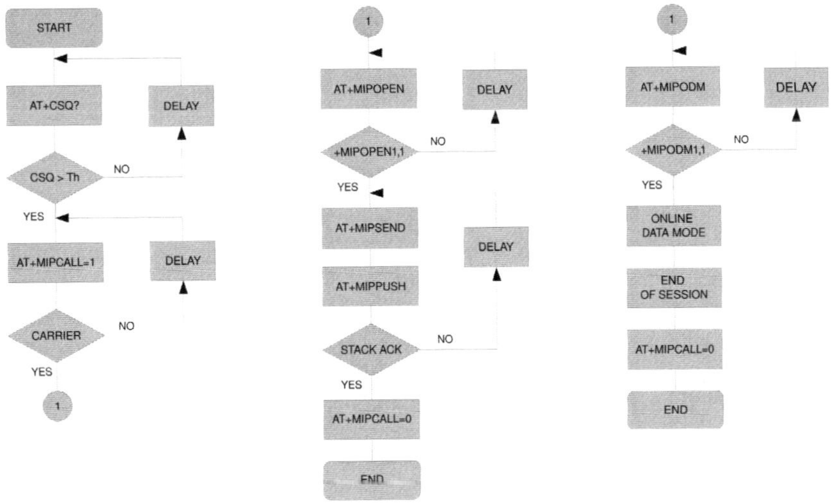

(a) (b) (c)

Figure 4. Command sequences: (a) GPRS attachment, (b) buffered mode, (c) online mode

As we appreciate in Figure 4(a), an initial negotiation stage with the GPRS Gateway is neccesary in both modes. AT+CSQ checks signal quality, before connecting with the Gateway through the AT+MIPCALL command. After successfull connection, the Gateway assigns a dynamic IP address to the modem.

In the first mode, AT+MIPOPEN is used to initialize a new socket and open a TCP or UDP connection to a remote server through a specific port. After that, AT+MIPSEND command is required to store application data into the modem buffer. Through AT+MIPPUSH command, data is sent from this buffer to the protocol stack (Figure 4(b)). Drawbacks of this approach are that data must be encoded in hexadecimal format, and segmented in small packet sizes before transmission using AT+MIPSEND. As advantage, this mode allows opening up to 4 simultaneous TCP/IP or UDP/IP connections. The following sequence of commands, which posts a HL7v3 message to our server, illustrates this fact:

```
at+mipcall=1,gprs."internet.gprs.unifon.com.ar","wap","wap" // Create a wireless link
at+mipopen=2,3322,"neurotests.frc.utn.edu.ar",80,0          // Open a TCP socket and connect to server

at+mipsend=2,"504f535420687474703a2f2f6e6575726f746573"     // POST http://neurotes
at+mipsend=2,"74732e6672632e75746e2e6564752e61722f5465"     // ts.frc.utn.edu.ar/Te
...                                                         // Keep storing data on 1372-bytes buffer
at+mipsend=2,"0a3c2f736f6170656e763a456e76656c6f70653e"
at+mipsend=2,"0d0a0d0a"                                     // End of HTTP POST method
at+mippush=2                                                // Push stored data into TCP/IP protocol stack
```

In the second mode, AT+MIPODM is used to initialize a socket and open a connection with a remote server. This command simplifies data transmission because neither hexadecimal enconding nor segmentation is required (Figure 4(c)). Disadvantage of this case is that only one remote connection can be implemented at a time. The following sequence demonstrates the ease of use of online mode:

```
at+mipcall=1,"internet.gprs.unifon.com.ar","wap","wap"      // Create a wireless link
at+mipodm=1,2343,"neurotests.frc.utn.edu.ar",80,0           // Open a TCP socket and connect to server

POST http://neurotests.frc.utn.edu.ar/TeleMed/services/WSTest HTTP/1.1
Content-Type: text/xml;charset=UTF-8
SOAPAction: "sendMessage"
User-Agent: Jakarta Commons-HttpClient/3.1
Host: neurotests.frc.utn.edu.ar
Content-Length: 271

<soapenv:Envelope xmlns:soapenv="http://schemas.xmlsoap.org/soap/envelope/"
xmlns:test="http://test.ws.gic.com">
<soapenv:Header/>
<soapenv:Body>
<test:sendMessage>
<test:message>HL7_message</test:message>
</test:sendMessage>
</soapenv:Body>
</soapenv:Envelope>
```

GPRS communication management with our PIC microcontroller-based platform was simulated before firmware download to target hardware. The environment used for this work is the *Labcenter Proteus* software, which enables direct access to the physical RS232 port of a PC for transmission to the G24 modem (Figure 5). In this way, the exact behavior of the microcontroller was evaluated.

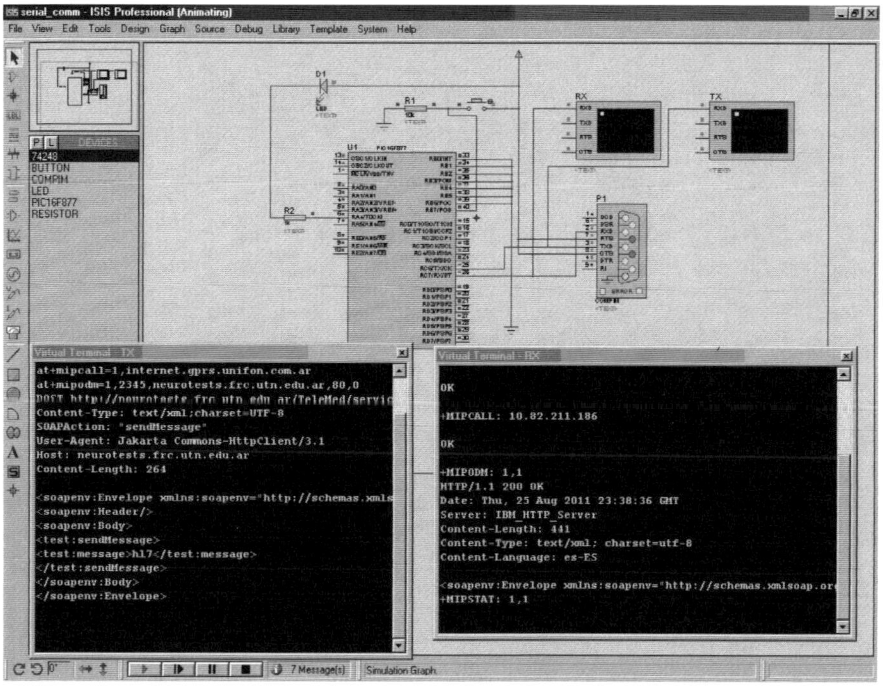

Figure 5. Simulation of the target PIC-based platform

5. Conclusions

In this work we propose taking advantage of actual data processing and communications technologies available in our country for mobile e-health applications. Special emphasis is put on efficient use of constrained resources through low level communication management. We demonstrate the viability of this approach through simulations of our target hardware platform. Future work includes integration to our medical data acquisition system, and debugging of application-level transactions with the process executing at the web server.

Acknowledgment. This work is partially funded by the Universidad Tecnológica Nacional - FONCyT IP-PRH 2007 Posgraduate Grant Program and 25/E134 Project of Ministerio de Educación de la Nación Argentina.

References

[1] Gonzalez E, Cagnolo F, Olmos C, Centeno A, Riva G and Zerbini A 2007 *Medical Data Transmission System for Remote Healthcare Centres* J. Phys.: Conf. Ser., Vol. 90, No. 1. (2007)

[2] Voos J, Riva G, Zerbini C, Centeno C and Gonzalez E 2010 *Custom HL7 V3 Message Provider using Web Services Security Features* Conf Proc IEEE Eng Med Biol Soc.,pp. 3899-902 (2010)

[3] Epec, Conectados Magazine, No. 30, available at: www.epec.com.ar (2010)

[4] IEEE Computer Society: IEEE Standard 802.15.4, IEEE Standard for Information Technology, Part 15.4: Wireless Medium Access Control (MAC) and Physical Layer (PHY) Specifications for Low-Rate Wireless Personal Area Networks (2006)

[5] Dario C, et al, Opportunities and Challenges of ehealth and Telemedicine via Satellite, Eur J Med Res Supplement, Proceedings of ESRIN-Symposium, Italy (2004)

[6] Triventi M, Mattei E, Censi F, Calcagnini G, Strano S, Bartolini P, A SMS-based Platform for Cardiovascular Tele-monitoring, World Congress on Medical Physics and Biomedical Engineering (2009)

[7] Technologic Systems, TS-7200 Hardware Manual (2010)

[8] Motorola G24 Developer's Guide, AT Commands Reference Manual (2007)

[9] Microchip, AN833: The Microchip TCP/IP Stack (2002)

Teledermatology: an experience in Tucumán

V. I. Rotger[1], P. F. Solarz[2], L. M. Ruiz[2,3], A. Salas[2], M. C. García Mena, J. M. Olivera[1]

[1]Departamento de Bioingeniería, Fac. de Cs Exactas y Tecnología, Universidad Nacional de Tucumán.
[2] Instituto de Bioelectrónica, Fac. de Medicina, Universidad Nacional de Tucumán.
[3]Sistema Provincial de Salud, Ministerio de Salud Pública

E-mail: vrotger@herrera.unt.edu.ar

Abstract. Since 2006 projects have been carried out in the Universidad Nacional de Tucuman to introduce Telemedicine in the Provincial System of Health (SIPROSA) in Tucuman, Argentina. These projects aim at improving accessibility to remote populations, providing equity and avoiding costly and unnecessary transfers as well as helping early diagnosis and treatments, timely and appropriate referrals. On the other hand, this consultation has already been set up within the SIPROSA with the application of the system of reference and counter-reference for several years. By applying Telemedicine it is possible to accelerate the times and make this more efficient. This paper presents a prototype Web application for performing consultation in dermatology. These were conducted by professionals in Ranchillos, a small town within the Province, together with specialists from the capital city. After the experience, it can be said that the teledermatological application of the platform has been well received by physicians and dermatologists of rural areas, especially because the interfaces could be adjusted to their requirements. The physicians consider it a useful tool to complete the process of referral and counter-referral.

Introduction

Public health is one of the main challenges for developing countries. Having both material and human resources to solve major health problems of the population is a difficult goal to achieve. In many cases the application of cost-effective technology, easy to use and accessible both to patient and doctor can help. This is the case especially in rural and marginal urban areas. Telemedicine has been used for over 50 years for this purpose. [1]
Already in 1998, the WHO defined it as "The provision of health care services where distance is a critical factor for professionals who appeal to information and communications technology to exchange data for diagnosis, to improve and indicate treatment, for the prevention of diseases and injuries, as well as for the training of health care professionals and for research and evaluation in order to improve the health of people and the communities they live in..."
In Tucumán, Argentina, the Department of Bioengineering of the Faculty of Science and Technology and the Institute of Bioelectronics of the School of Medicine, both from the Universidad Nacional de Tucumán, have undertaken projects to introduce Telemedicine in the Provincial System of Health

(SIPROSA) since 2006. Such projects aim at improving accessibility to remote populations, providing equity and avoiding costly and unnecessary transfers as well as helping early diagnosis and treatment, timely and appropriate referrals.
Among them, a system has been developed that includes information storage, acquisition, distribution, modification and display of images in JPG, PNG, TIFF and BMP formats, performance management, remote communication through a Web page for the exchange of images and security through access control. You can use these elements in conferences between health workers allowing, at the same time, the exchange of experiences and information between them.
Moreover, the consultation has already been set up within the SIPROSA, with the application of the reference and counter-reference system for several years. By applying telemedicine it is possible to accelerate the times and make this more efficient.
In a previous study we described a first prototype used in cardiology. [2]
One of the specialties in which they began using the Web developed prototype is Dermatology. The use of telemedicine in dermatology is based on the fact that its queries are based primarily on thorough medical records and a proper physical inspection, so if the records and the images sent are right, the professionals will be able to issue a diagnostic opinion and a treatment appropriate to the patient's condition. [3]
In the rural areas of Tucuman and other regions of Argentina, it is very common to develop dermatological diseases, but the visit to a specialist's office is difficult because of inaccessibility and lack of Human Resources. To cover this deficit, a network was organized, where doctors or Health Care Workers in rural centres, when consulted for skin disorders, could use digital photography, chat and standardized default medical records, thus connecting peripheral health facilities with specialized services and the Department of Dermatology using our Web application. [4]
In Tucumán, specialists far from the capital do consultation with those belonging to the Department of Dermatology at the UNT, located at Hospital Avellaneda. Until the implementation of the new system, the consultation was performed by the specialist himself who had, to the extent of his possibilities, to take the photos of the patient and the corresponding medical records and tests to the capital. Though the proceedings are set, they are not available and the course of action is left to the initiative of each professional.
The experience was performed with professionals from Ranchillos. This town has 19,000 inhabitants and is located 25 km. from the capital. This population suffered from common and frequent dermatological disorders. The professionals agreed to use the system and they even suggested some changes. A medical record was included more in agreement with that used in the service.
The aim of this paper is to show aspects of specialization of a Web system already developed as a general telemedical platform to be used in dermatology and to establish a synchronous and asynchronous consultation protocol.

System Features

We used the web platform prototype developed and described in a previous paper to record multimedia clinical information, that is, the storage of both text data from entry forms from typical medical records, including multimedia information from electrocardiography, electroencephalography, evoked potentials, ultrasound, photographs, etc. The information can be used through the conference module, where the different actors can communicate via chat. All information generated during the conference is kept in the database, which allows speakers to interact at different times. [5] [6]

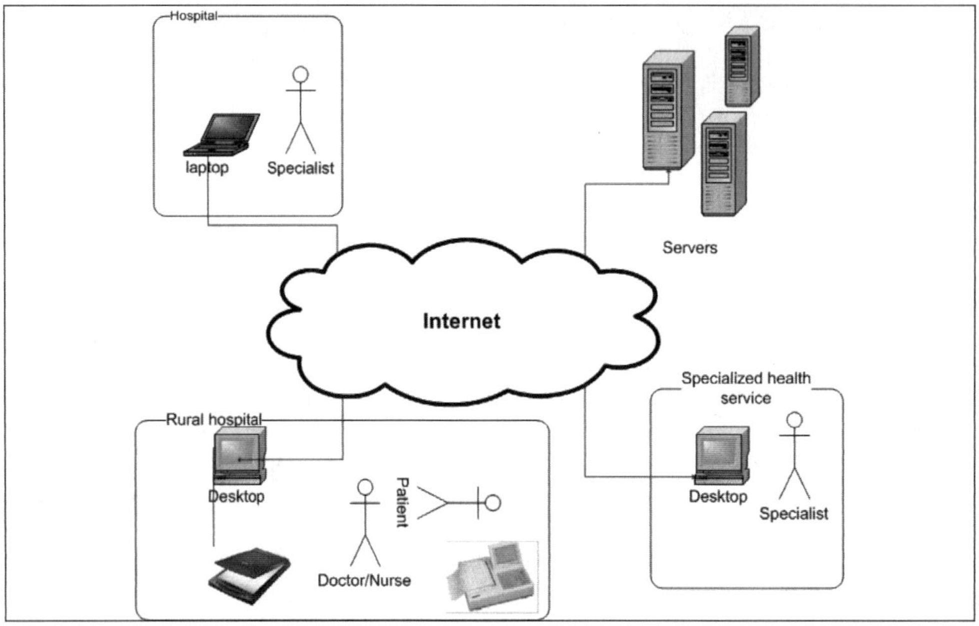

Figure 1. System Functional View

From a functional point of view, Figure 1 shows a diagram of the computer system components. Internet is the central element that allows intercommunication between the different actors from different types of terminals, both fixed and mobile including exchange of text, voice, images, videos, etc. that support complementary diagnostic methods

Internet, applications and databases servers are necessary for the operation and are the means to carry text, images, etc. into the web. But at least at an experimental stage, where the volume of data and transactions is low, the sizing of these resources does not require too many considerations, contrary to what is planned for a period of extensive use, when the safety aspects, efficiency, etc. should be considered as distributing applications and data on properly sized servers.

For users of the system, PCs, notebooks, PDAs, etc., are elements of access to all functionalities of the system. Typical system actors are, on one side, rural doctors and health workers at the rural hospital, who require a system support service that provides information on the condition of the patient; and on the other, a specialist who, from central hospitals or from any device that can access the web, brings expertise to provide solutions to the original request.

The information sent comes mainly from the output of diagnostic equipment, so it has different formats depending on which testing equipment it comes from. The reading of the source files is made using a format analyser that checks the input format with the format the application has previously registered (jpeg, avi, mpeg, etc., additional modules to interpret other formats can be added later). This is integrated with patient's information (personal and clinical data, etc.) that will constitute the basis of the exchange between actors, that is, the conference.

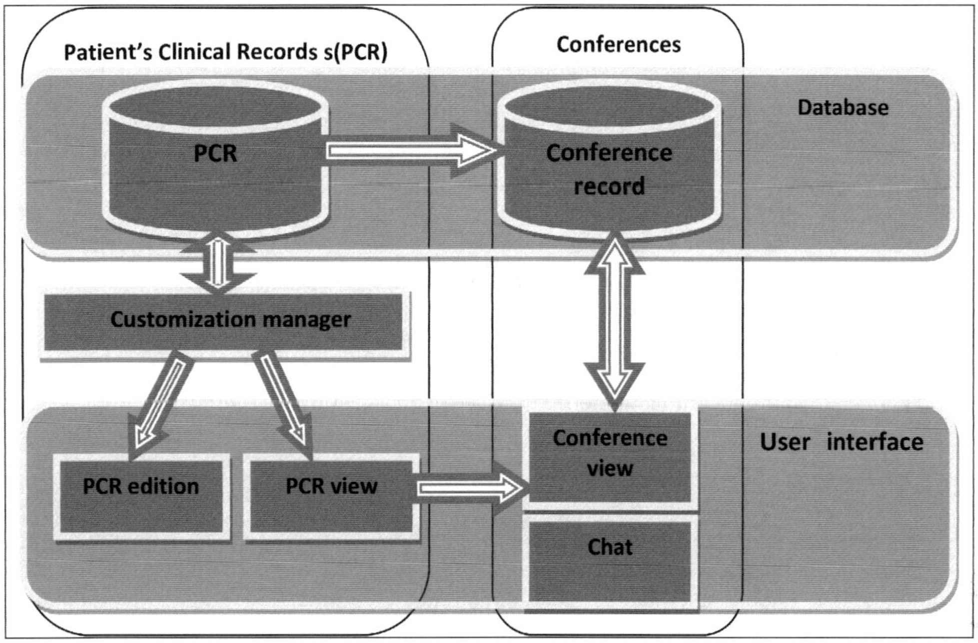

Figure 2. Design scheme

In Figure 2, we see the two most important aspects of the system: the management of multimedia patients' clinical data on the left and the conference itself on the right. Both upper cylinders symbolize the database where information is kept. It can be seen, in the top panel of the figure, that the conferences are also kept in the database. In the lower panel accesses to the patients' clinical records (PCR) are represented, through information management modules and views, both capable of being specialized as revealed by the requirements engineering. It also shows the PCR data view from the conference to which you add a chat room.

In particular, and as a current example, the teledermatology application consists of specialized interfaces tailored to their needs and, speaking in terms of software architecture, a control layer that connects these interfaces to the kernel.

The development evolves from a prototype, as it seeks early interaction with users to achieve intuitive interfaces according to their needs. Furthermore, we use aspects of technologies under the Web 2.0, fundamentally AJAX, which allows a similar behavior to that of a desktop application using asynchronous communication between client and server. Figure 3 shows a diagram of the layers underlying the prototype of the Web application.

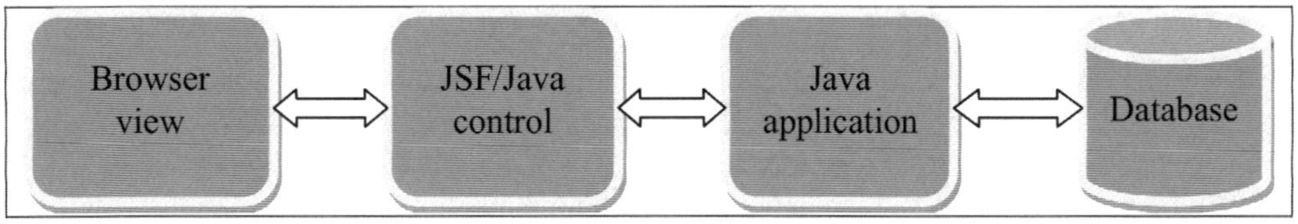

Figure 3. Diagram of layers

It should be noted that while in this paper we present a teledermatology application of the platform, the conditions which have led to the current characteristics, in terms of architecture and software design, have been derived from different medical specialties with particular visions of the best approach to the "remote " analysis of patients and their diseases. Thus, the design is based on a common core that manages patients' information, with interfaces tailored to different needs arising from the requirements set by each group of health workers.
The current state of part of the application interfaces is geared to all the information needed by specialists of the Department of Dermatology of the School of Medicine (Universidad Nacional de Tucuman). These are a set of forms ready for data entry, general history of the patient, current consultation cause, personal and family history of present illness, diagnosis (or diagnostic opinion), etc. Some of these are multimedia forms, that is to say, they allow the entry of text as well as pictures, videos, etc. To sum up, they make up a multimedia record of the patient clinical data, organized hierarchically.

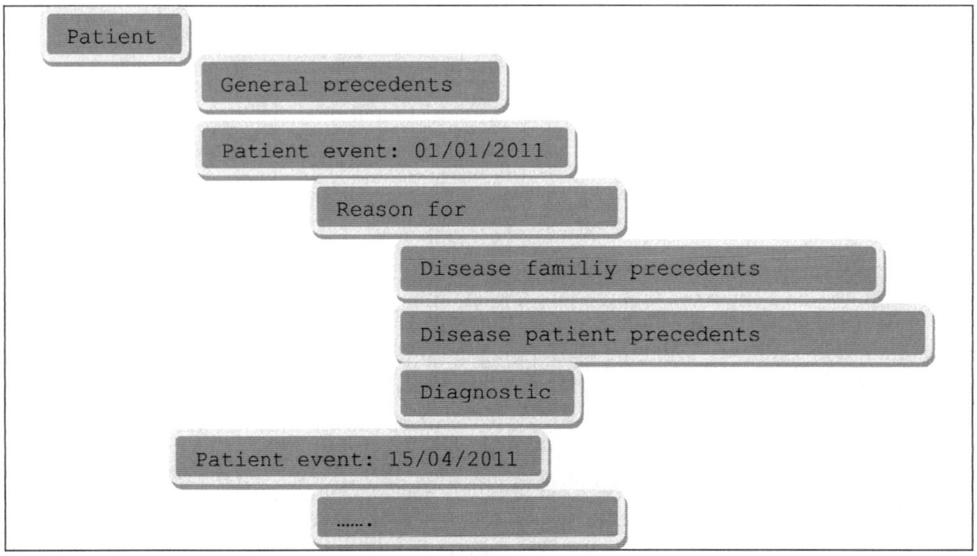

Figure 4. Tree structure

Scenario of rural teledermatology

Below is a case of typical system usage to show the different user interfaces that occur during the sequence of tasks.
Its aim is that a doctor at a central hospital (CHP-Central Hospital Doctor) assists in the diagnostic evaluation and proposes a treatment for the patient, based on information that the rural hospital doctor (RHP- Rural Hospital Doctor) entered in the PCR and what was discussed during the conference.
In particular it is a consultation about a skin disease, which originated in the town of Ranchillos, Tucumán, located 25 km from the capital. The specialists consulted are from the Department of Dermatology of the Medical School of the Universidad Nacional de Tucuman (UNT), which has a specialty service in Hospital Avellaneda in the capital. The consultation is done virtually, through the system.
We present the sample case using a sequence of steps.

Step 1:
The Rural Hospital physician enters data and photographs of injuries from the forms as shown in Figures 5 and 6 which show two different forms, corresponding to the operations of PCR.

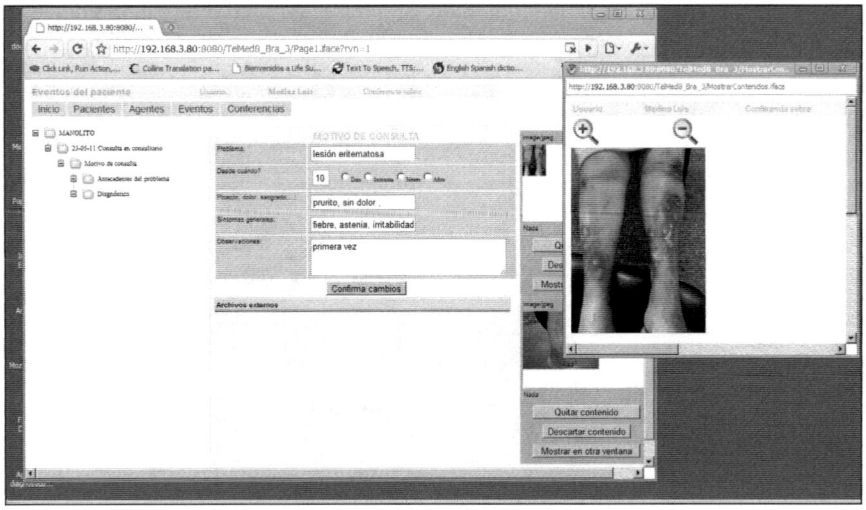

Figure 5. Patient's Clinical Record

In these forms, the RHP, in contact with the patient, introduces the consultation causes, images of the lesion and the patient's history derived from the interview in the PCR.

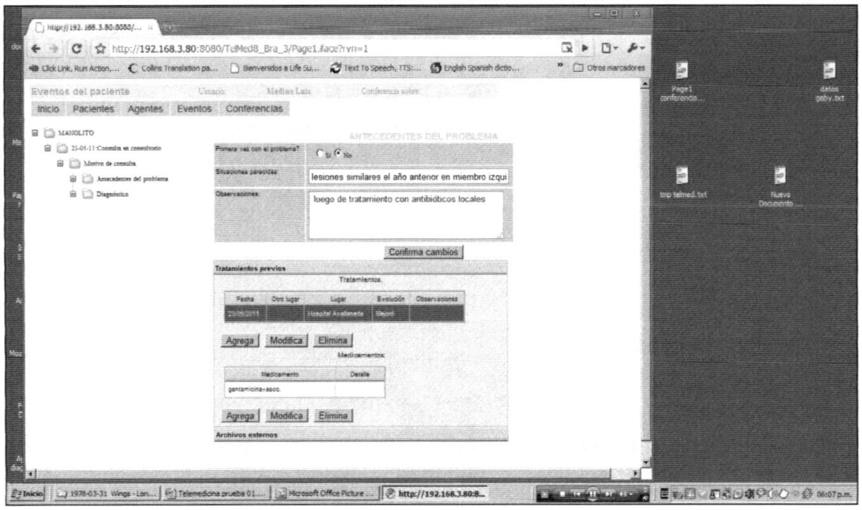

Figure 6. Patient's Clinical Registration Form

Step 2:
Later, he invites a specialist (or several) which he wants to lecture on the topic (Figure 7). He may invite a doctor from the Central Hospital (CHP) and a chair of the dermatology department. The invitation can be made at any time and there is no limit on the number of speakers. It is not necessary

for the guest speaker to be online at that moment; the invitation can be synchronous or asynchronous. But it is mandatory that the speakers have been admitted to the system previously.

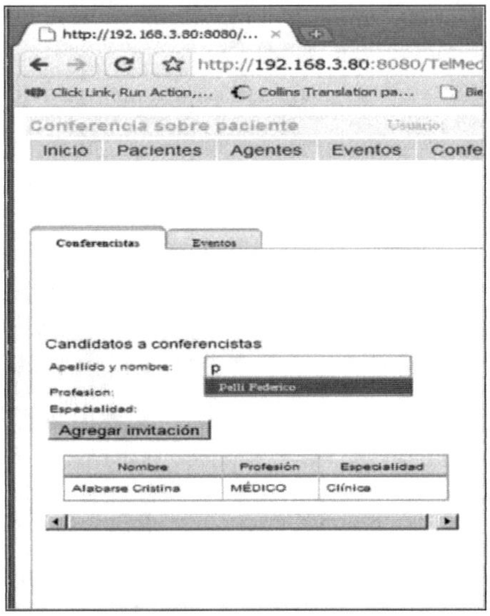

Figure 7. Call for Speakers

Step 3:
When a guest speaker accesses the system through the identification interface, the conferences to which he has been invited are presented as options (in the manner shown in the previous step), as well as those in which he is already involved by previous invitations or generated by the current speaker. Figure 8 is a composition of the screens of both speakers. The log is shown on the right of the system screen once the CHP has been identified. At the bottom is the conference to which he was invited by the RHP. On the left, the screen shows the RHP who is using the conference panel.

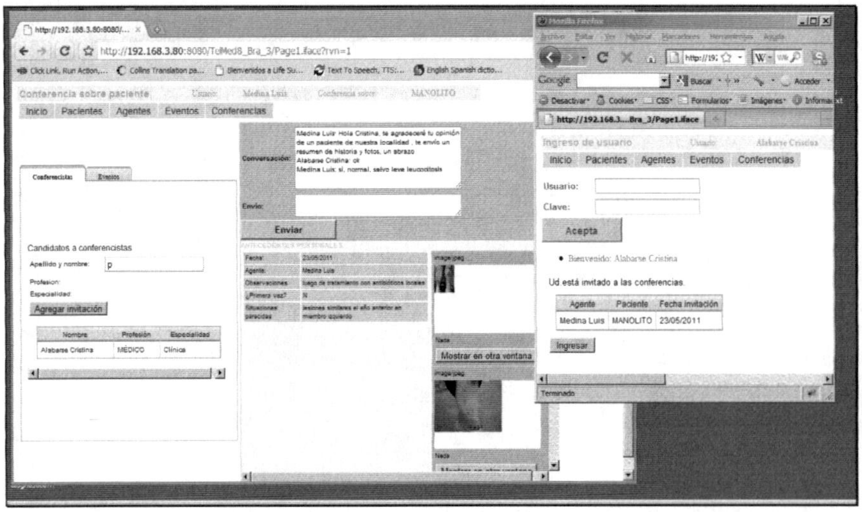

Figure 8. Initial Screen of guest lecturer

Step 4:
Figure 9 shows the screens of both the RHP and the CHP conferences. The right side of each screen shows the conference, which is the same in both cases and is updated each time a speaker sends patient's information from the PCR to the conference. Moreover, as the conference is stored, any guest speaker, entering at any time, accesses the total conference.

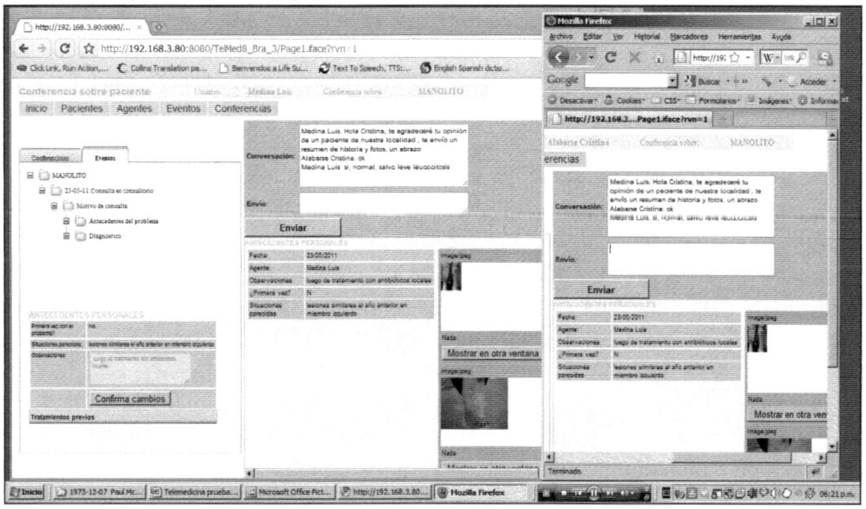

Figure 9. Conference Shared Screen

Discussion and Conclusion

In general in rural areas the proposals are intended to extend the uses of the platform to pathologies such as leishmaniasis, refractory epilepsy, some heart disease.
We are working to provide VoIP, as for online communication it is more convenient than chat. While there is quality technology to do it off the shelf, the goal is to manage users, when and from where communications are made.
Another goal is to make the specialist independent from a particular workplace, so we are developing aspects that make the application usable from smartphones.
Finally, the act of "going into production" of the system requires management to solve the security and confidentiality policies.
The short experience we have had using the system indicates that the development of a software architecture that allows the specialization of interfaces according to the user group is strategically vital, as the users seek specific information of their interest and all not going in that direction naturally removes the "virtual meeting" among health workers.
As a conclusion, it can be said that the teledermatological application has been well received by physicians and dermatologists of rural areas, especially because the interfaces could be adjusted to their requirements. Even though there is still a lot to be done in this field, since the requirements evolve as the applications are used, they consider the teledermatological application of the platform a useful tool to complete the process of referral and counter-referral.

Acknowledgments

This work was financially supported in part by grants from the UNT Research Council (CIUNT), 26/E422 Research Program, INSIBIO-CONICET and A/027435/09 AECID.

References

[1] "Primer estudio de Teledermatología en México. Una nueva herramienta de salud pública". (2004) *Gaceta Médica de México. Academia Nacional de Medicina de México*, A.C. Vol 14 Nro. 1.

[2] Pablo F. Solarz, Viviana I Rotger, Luis Medina Ruiz "Prototipo web para telemedicina" Proceedings *XVII Congreso Argentino de Bioingeniería y VI Jornadas de Ingeniería Clínica, SABI 2009*. Vol 1, Núm 1, pp 1-2, ISSN 1852-6292. 2009

[3] Recomendaciones Vol. 3 Telemedicina. *Serie Tecnologías en Salud. CENETEC*, Secretaría de Salud. Estado de México. 2007

[4] Medina Ruiz L; Pelli F.; Molina E.; Solarz P.; Rotger V.; Toll E.; Liato A.; Molina S; Nuñez L. "Red de Salud en Teledermatología" *APS y Redes. 4º Congreso Provincial de APS, 2º Encuentro Nacional de APS. Atención Primaria de la Salud* Mar del Plata. 2011

[5] Pablo F. Solarz, Viviana I Rotger, Luis Medina Ruiz "Prototipo web para telemedicina" Proceedings *XVII Congreso Argentino de Bioingeniería y VI Jornadas de Ingeniería Clínica, SABI 2009.* Vol 1, Núm 1, pp 1-2, ISSN 1852-6292. Rosario, Santa Fe, 14-16 de octubre de 2009

[6] Solarz P, Medina Ruiz L y Rotger V. "Software para Telemedicina". *XI Reunión de Investigación en Ciencias de la Salud. Facultad de Medicina. Tucumán. 1 de Diciembre de 2009 y A.P.S. y Equidad Social en el Bicentenario* 28 al 30 de Abril 2010 Mar del Plata.

AUTHOR INDEX

AUTHOR INDEX

Institute of Physics
Dirac House, Temple Back
Bristol BS1 6BE UK

CURRAN ASSOCIATES INC
proceedings.com